Human body dynamics: impact, occupational, and athletic aspects

EDITED BY

DHANJOO N. GHISTA

Professor of Medicine and Mechanical Engineering,
McMaster University, Hamilton, Ontario.

CLARENDON PRESS · OXFORD
1982

Oxford University Press, Walton Street, Oxford OX2 6DP
London Glasgow New York Toronto
Delhi Bombay Calcutta Madras Karachi
Kuala Lumpur Singapore Hong Kong Tokyo
Nairobi Dar es Salaam Cape Town
Melbourne Auckland
and associates in
Beirut Berlin Ibadan Mexico City Nicosia

Published in the United States by Oxford University Press, New York

British Library Cataloguing in Publication Data

Ghista, Dhanjoo N.
  Human body dynamics. – (Oxford medical
  engineering series)
  1. Bioenergetics   2. Human mechanics
  3. Exercise
  I. Title
  612′.76   QP301

  ISBN 0-19-857548-3

Photo Typeset by
Macmillan India Ltd., Bangalore.
Printed in Great Britain
at the University Press, Oxford
by Eric Buckley
Printer to the University

Oxford Medical Engineering Series

EDITORS: B. McA. Sayers
P. Cliffe

To P. R. Sarkar, the propounder of PROUT and harbinger of Renaissance Universal

and

My dear departed father and grandparents

Preface

Daily living entails a risk of injury from occupational tasks, from car crashes, and from taking part in sport. The study of human body dynamics reveals the mechanisms of injuries sustained in these situations and the measures that can be taken to avoid them. Moreover, in the area of athletics and sport, human body dynamics also plays an active part in the analysis of factors that contribute to the athlete's performance, and in the design of protective clothing and equipment. This book brings together the analyses of various types of human body dynamics involved in impact situations, occupational tasks, and athletics and sporting events, by invoking the disciplines of engineering mechanics, analytical mechanics, dynamics of structures, vibration, and impact mechanics.

The book is divided into three sections: impact; occupational tasks and environment; and athletics and sport. The first section deals with theoretical models of head, thorax, and spinal injury, to elucidate the mechanisms of injury and describe appropriate restraint and protective systems, as well as vehicle design for crashworthiness. The second section presents computerized human body models first to determine the relationships between applied forces and resulting displacements in simulated jumping, work motions in space, and response to impulsive forces; secondly to analyse slow moves in the handling of heavy materials in order to understand the mechanism of low-back injury and how it can best be prevented; and thirdly to determine maximal levels for load-lifting, pushing, and pulling. Experimental studies of the response of the body to vibrations from industrial machinery and vehicles are also described, together with suggestions for improving the design of, among other things, hand-held power tools, tractor suspension systems, and the interiors of ambulances. The last section deals with athletics and sport. Theoretical mechanics models are developed for the following: (i) to study impulse and momentum considerations in kicking and heading a soccer ball, and kinetic-to-potential energy conversion in maximizing pole-vaulting height; (ii) to analyse discus throw, javelin throw, and shot put, in order to demonstrate the influence on performance of disc spin–speed and moment of inertia, javelin inclination and throw angle and geometry, and the shot angle; (iii) to help in the design of hockey helmets, sticks, and jogging shoes; (iv) to describe the mechanics of the impact of golf-club head and ball, to maximize the velocity imparted to a golfball; (v) to calculate energy balance in assessing rope strength when arresting the fall of rock-climbers; and (vi) to estimate the energy of karate blows and kicks, for fracture of mediums.

The book is designed as a biomechanics course text as well as a reference source in body dynamics of impact, occupational manoeuvres, and athletics, for physicians, bioengineers, and physical educators. The text employs mechanics rigor at the requisite level for elucidating the mechanisms and applications, and also emphasizes the results of the analyses in setting out the causes of injury, design considerations for protective measures and devices, and factors for maximizing performance in sport and athletics.

I wish to thank the contributing authors for their painstaking efforts and Professor Collins for his editorial assistance.

September 1981   DNG
Hamilton, Ontario

Contents

List of contributors

S. H. ADVANI	Ohio State University, Columbus, Ohio, USA.
A. BHATTACHARYA	University of Cincinnati (Medical School), Cincinnati, Ohio, USA.
P. A. BOURASSA	University of Sherbrooke, Sherbrooke, Quebec, Canada.
D. B. CHAFFIN	G. G. Brown Laboratory, University of Michigan, Ann Arbor, Michigan, USA.
R. COLLINS	Université Paul Sabatier, Toulouse, France.
K. V. FROLOV	State Scientific Research Institute of Machinery, Moscow, USSR.
D. P. GARG	Duke University, Durham, North Carolina, USA.
D. N. GHISTA	McMaster University, Hamilton, Ontario, Canada.
T. C. HUANG	University of Wisconsin, Madison, Wisconsin, USA.
R. L. HUSTON	University of Cincinnati, Cincinnati, Ohio, USA.
W. JOHNSON	University of Cambridge, Cambridge, England.
D. KALLIERS	University of Heidelberg, Heidelberg, FRG.
Y. KING LIU	University of Iowa, Iowa City, Iowa, USA.
P. G. KIRMSER	Kansas State University, Manhattan, Kansas, USA.
Y. KIVITY	University of California, Los Angeles, California, USA.
A. G. MAMALIS	University of Cambridge, Cambridge, England.
R. MATHERN	University of Heidelberg, Heidelberg, FRG.
D. MOHAN	Highway Safety Research Institute, Ann Arbor, Michigan, USA.
A. K. OMMAYA	National Institutes of Health, Bethesda, Maryland, USA.
R. OUDENHOVEN	Neurological Surgery, 430 S. Webster, Green Bay, Wisconsin, USA.
C. E. PASSERELLO	Michigan Technological University, Houghton, Michigan, USA.
N. PERRONE	Office of Naval Research, Arlington, Virginia, USA.

B. A. POTEMKIN	State Scientific Research Institute of Machinery, Moscow, USSR.
G. RAY	Michigan Technological University, Houghton, Michigan, USA.
S. R. REID	University of Cambridge, Cambridge, England.
E. M. ROBERTS	University of Wisconsin, Madison, Wisconsin, USA.
E. ROHL	Institut fur Kraftfahrwesen, Aachen, FRG.
G. SCHMIDT	University of Heidelberg, Heidelberg, FRG.
T.-C. SOONG	Xerox Corporation, Rochester, New York, USA.
R. G. THERRIEN	University of Sherbrooke, Sherbrooke, Quebec, Canada.
H. S. WALKER	Kansas State University, Manhattan, Kansas, USA.
H. WOLFF	Institut fur Kraftfahrwesen, Aachen, FRG.
W.-J. YANG	University of Michigan, Ann Arbor, Michigan, USA.
Y. YOUM	Catholic University of America, Washington DC, USA.

Section I Impact

1. Head injury mechanisms— characterizations and clinical evaluation

Sunder H. Advani, Ayub K. Ommaya, and Wen-Jei Yang

1.1. Introduction

Head injury poses a grave health problem. The National Safety Council of the United States estimates that 6 per cent of all accidents involve the head but this figure is as high as 67 per cent in automobile accidents. In the United States such accidents cause about three million head injuries annually; 50 000 of these are fatal. The head injury literature is vast and varied since several disciplines are involved. These include neurophysiology, neurosurgery, neuropathology, kinesiology, engineering mechanics and computer modelling. With the objective of bringing these fields into an interdisciplinary focus, the *1966 Head Injury Conference Proceedings* [1] represent a milestone in terms of discipline cross-fertilization. This chapter essentially presents subsequent work which has emerged from this vantage point and is oriented towards the biomechanics of head injury stemming from a mechanical impact environment. Related work can also be found in a comprehensive review by Goldsmith [2] and studies by McElhaney, Stalnaker, and Roberts [3].

In order to understand the head injury mechanisms it is necessary that the mechanical input be related to the resultant pathophysiological responses. Dynamic loading to the head can generally be classified in two categories:

(i) Direct contact between the head and colliding object. In an automotive crash environment the head may impact the windscreen, dashboard, airbag, side door, head rest, etc.
(ii) Inertial impact such as loading transmitted to the head by torso motion via the head–neck junction.

Figure 1.1 presents an overview of the possible head injury loadings, mechanisms, and resulting clinical and pathological responses. A fundamental but controversial query still exists regarding the delineation of the translational and rotational components of the head centre of gravity acceleration vector and their relative contributions to head injury in various impacts. It is hoped that the following sections on anatomical, constitutive, modelling (analytical and experimental), and pathophysiological considerations will present the reader with a state-of-the-art appraisal. Selected literature is

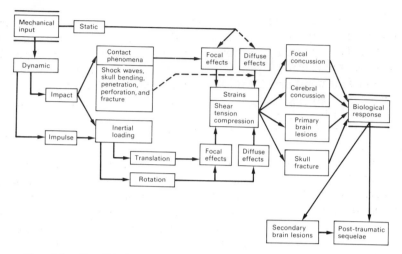

Fɪɢ. 1.1.   Possible head injury loadings, mechanisms, and responses.

directly cited wherever possible. Additionally, pertinent references are listed regularly by the National Institutes of Health, USA.

1.2. Anatomical considerations

A comprehensive anatomical description of the human head is not undertaken here since several standard textbooks on human anatomy such as Gray [4] and Woodburn [5] provide excellent qualitative descriptions and agree well. Useful anatomical information can also be found in various atlases [6] and specialized articles. Detailed measurements and landmark co-ordinates for a typical 50th percentile human skull can be found in the work by Hubbard and McLeod [7]. A brief anatomical summary pertinent to the biomechanics of head impact is presented here.

The human head is a complex structural system which on a simplified basis is composed of the scalp, the skull, the dura, the pia–arachnoid complex, the brain, the blood vessels, and the cerebrospinal fluid. Considering the general structure of the head with regard to the central nervous system (CNS) one is impressed with the protective packaging. In general the CNS can be considered as a viscoelastic mass hydraulically shock mounted in a stiff container with external shock-attenuating layers. The external protective layer, or scalp, is composed of integument, subcutaneous fat, superficial facia, galea aponeurotica, and pericranium. Compared with other body skin the scalp is relatively thick, especially in the occipital region. The scalp is also richly supplied with blood vessels which course over the entire cranial area. As a composite tissue the scalp, along with variable quantities of hair, presents the first line of defence from physical trauma.

The eight bones comprising the skull are knit together in a structurally sound design approximating a spherical shell. The composition of the cranial bones is characterized by two lamellae or inner and outer tables between which lies a bed of dipolic tissue. In certain portions of the bones, particularly at their edges and junctures, the tables join internally into soild bone devoid of diploë. The sutures joining the cranial bones of an adult are comparatively strong in compression loading, although, depending on the variables of force application effecting the trauma, they may shear prior to or along with bone fracture.

The meninges impose the next and last barrier to access to the brain. The outermost membrane, the dura mater, offers the most significant structural protection of the three meninges. The disposition of the dura mater varies over the entire surface of the brain (and spinal cord). Elements of this tissue elaborate to form the falx and tentorium partitions among others. On the cranial surface the dura mater is a tough elastic tissue tightly connected to the internal surfaces of the cranial bones. In contrast, the subdural surface is relatively smooth and unattached, being equipped with a layer of mesothelium lining. Communicating via the arachnoid and parallel with the dura mater is the pia mater. This membrane, which is much finer than the dura, conforms to the cortical folds and is invested with the vascular network supplying the nervous tissue. Its strength is secondary to the transport of blood over the surface of the brain. The last meningeal membrane is the arachnoid which lies between and traverses the space between the dura mater and the pia mater. It is a delicate avascular membrane of little structural value in itself. Between the arachnoid and the pia mater lies the subarachnoid space filled with cerebrospinal fluid.

The brain is a soft structure composed of nerve cells, the grey matter, and their axons, the white matter, both of which are supported by the glia, all of these being derived from the ectoderm. It consists of two cerebral hemispheres, the basal ganglia, the cerebellum, and the brain stem. The cerebral hemispheres are divided into the frontal, temporal, parietal, and occipital lobes. Inside the brain are cavities, called ventricles, containing cerebrospinal fluid. The ventricles are in continuity with each other, either directly or indirectly, and the fourth ventricle empties into the cisterns at the base of the brain. The basal cisterns are in continuity with the subarachnoid spaces within the spinal canal and over the surface of the brain. The cerebrospinal fluid is largely secreted within the ventricles and is largely absorbed over the surface of the cerebral hemispheres.

1.3. Constitutive considerations

The mechanical properties of the tissues of the head have been extensively investigated [8]. In particular, considerable effort has been devoted to the characterization of the constitutive properties of human skull and brain tissue because of their relative importance.

Average values of the skull elastic moduli E and the ultimate strength σ_{ult} adapted from ref. 8 are presented in Table 1.1. The scatter in the data, although not reported here, is wide in range and typical for such tests. The variability of the data compared with that in literature is evident. For example, Wood [10] has reported an average tensile strength of 10 000 lbf in^{-2} at a strain rate of 0.01 s^{-1}. It is noteworthy, however, that the values in Table 1.1 represent lumped material properties of the structure in view of the skull compact bone–diploë sandwich configuration. Typical average compression curves for skull specimens (biopsy, autopsy, frozen and embalmed) tested in the compressive and tangential modes are illustrated in Fig. 1.2. Flexure characteristics of layered cranial bone have been reported by Hubbard, Melvin, and Barodawala [11].

TABLE 1.1

Average static mechanical properties of skull bones

	SAMPLE SOURCE			
	Biopsy	Frozen biopsy	Autopsy	Embalmed
E (radial compression) (lbf in^{-2})	5.23×10^4	22.99×10^4	11.84×10^4	38.10×10^4
E (tangential compression (lbf in^{-2})	59.40×10^4	62.6×10^4	37.8×10^4	80.8×10^4
E (tangential tension) (lbf in^{-2})	–	–	–	127×10^4
σ (ultimate radial compression) (lbf in^{-2})	–	21.1×10^3	14.21×10^3	13.9×10^3
σ (ultimate tangential compression) (lbf in^{-2})	11.87×10^3	11.81×10^3	7.34×10^3	13.12×10^3
σ (ultimate tangential tension) (lbf in^{-2})	–	–	–	6.3×10^{3a}

[a] Value adapted from McElhaney *et al.*[9].

The rheological response of human brain tissue in pure shear has been reported by Shuck and Advani [12]. Mechanical shear strain failure levels of 0.035 rad have been identified at 10 Hz from dynamic torsion tests. Viscoelastic models characterizing brain tissue properties in the frequency range 5–350 Hz have also been formulated. A plot of the experimental storage G_1 and the loss modulus G_2 for human brain tissue in the yielded and unyielded states is illustrated in Fig. 1.3. Free-standing dynamic compression tests on brain tissue have been conducted by Estes and McElhaney [13]. The instantaneous dynamic elastic modulus at high strain rates is about 10 lbf in^{-2}. Creep and relaxation tests on brain tissue have been reported by Galford and McElhaney [14]. Compressibility and ultrasonic tests indicate that the bulk modulus of brain tissue is constant (305 000 lbf in^{-2}) over a wide frequency range [8].

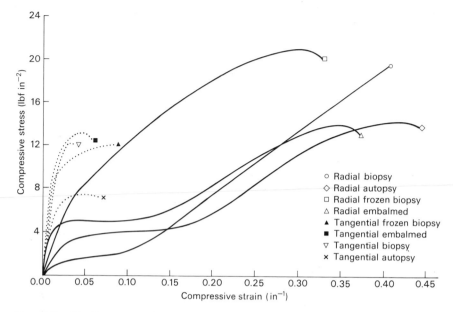

FIG. 1.2.   Skull compressive stress–strain curve; composite curve (radial and tangential).

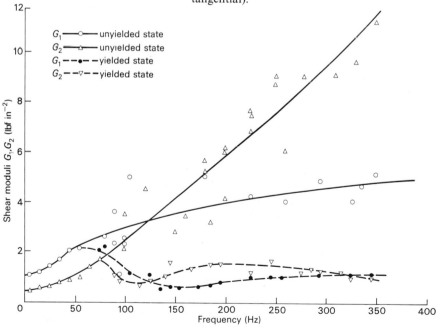

FIG. 1.3.   Human brain tissue dynamic shear moduli (yielded and unyielded state). The mean values of G_1 and G_2 for white and grey matter were obtained at 98.6°F.

Free vibration testing of human dura strips have revealed that the average elastic modulus is represented by $E_1 = 4570$ lbf in^{-2}, $E_2 = 500$ lbf in^{-2} at 22 Hz $(E = E_1 + iE_2)$ [8]. Uniaxial tensile tests on human dura at strain rates of 0.0666, 0.666, and 6.66 s^{-1} demonstrate little rate sensitivity effect. The initial average elastic modulus from these tests is 6000 lbf in^{-2} and the failure stress is 1100 lbf in^{-2}.

Free vibration tests on Rhesus monkey scalp specimens indicate that the average dynamic modulii are given by $E_1 = 210$ lbf in^{-2} and $E_2 = 74$ lbf in^{-2} at 20 Hz. Detailed results of relaxation and creep scalp tests have been reported by Galford and McElhaney [14]. Constant velocity free-standing compression tests at 65.0 s^{-1} indicate that the scalp thickness dynamic compressive modulus is of the order of 300 lbf in^{-2}. Limited tests of the pia indicate that the pia elastic modulus is about 20 lbf in^{-2} [8].

Extensive uniaxial tensile property investigations of cerebral blood vessels have been conducted by Chalupnik, Daly, and Merchant [15]. These tests reveal the non-linear stress–strain characteristics typical of most biological materials with the behaviour independent of strain rate from 0.001 to 50 s^{-1}. Sinusoidal testing also has indicated no frequency dependence of tensile properties up to 100 Hz. A typical relation between the Lagrangian stress (σ) and the extension ratio (λ) in the physiological stretch range is given by $\sigma = a\varepsilon^{b\lambda}$. No failure limits were identified in these studies since failure primarily occurred at the gripping ends of the vessels during these tests.

1.4. Head injury criteria

Two broad clinical classifications of cerebral trauma are generally recognized. The first type is open head injury which occurs when the skull fractures and the dura mater is penetrated. Tearing or shearing of brain tissue in contact with the projectile or skull bone fragments usually occurs. Concomitant large intracranial pressures resulting from the volume occupied by the intruding object are also created. The second type of injury, namely closed head injury, includes cerebral concussion, contusions, lacerations, and haematoma due to rupture of cerebral blood vessels. Closed head injury mechanisms, as reported in current biomechanical literature, are categorized in terms of the rotation hypothesis (angular acceleration), the pressure gradient–cavitation hypothesis (translational acceleration), and the bending–stretching hypothesis.

The rotation hypothesis advanced by Holbourn [16] attributes cerebral trauma to the damaging shearing strains produced by head angular accelerations and the relative rotational displacement between the brain and cranium. The rationale for this hypothesis is based on the fact that the bulk modulus of brain tissue (305 000 lbf in^{-2}, constant with frequency) is considerably larger than the shear modulus (1 lbf in^{-2} at 5 Hz to 12.8 lbf in^{-2} at 350 Hz).

According to the pressure gradient–cavitation hypothesis large pressure gradients are established in the brain owing to the propagation of steep frontal waves following impact. Brain damage can be produced as a result of linear acceleration involving both the absolute motion of the brain and its relative displacement with respect to the skull and concurrent large flows through the foramen magnum. The latter may induce shear stresses in the vicinity of the brain stem, collapse of the cavitation bubbles formed by the negative pressure at the contrecoup site, separation of the brain from the cranial wall, and neurovascular friction. One school of investigators adopts the view that the contrecoup type of injury is caused by the collapse of cavities due to the high attendant pressures created in the brain. The injury is formed at the contrecoup site (anti-pole) and consists of haemorrhages of varying size and shape.

The bending–stretching hypothesis describes only the cause leading to the damage of the cerebrospinal junction as being due to the flexion–extension and/or bending of the upper cervical cord during motion of the head–neck junction generated by an impulsive loading of the torso. Kinematic models of the head–neck–torso complex can be employed in assessing this injury.

The physical parameters used in the evaluation of head injury include the following:

(i) translational and/or rotational acceleration levels resulting from blunt impact or whiplash;
(ii) impact force, velocity and kinetic energy, impulse (or related average), and impact duration;
(iii) skull displacements and stresses; brain displacements, pressures, and strains/stresses; neck stretch/strain.

Items (i) and (ii) have been extensively used incorporating animal, human cadaver, and dummy experimental data to define gross tolerable and survival thresholds for head impact in translation and rotation. Item (iii) is usually related to analytical and experimental head model studies. A chronological overview of current severity indices, criteria, and associated threshold levels is now given.

The Wayne State tolerance curve [17], established from animal, cadaver, and human volunteer impact tests, has served as a benchmark for several head injury index formulations. In particular the Gadd severity index (GSI) [18], based on the concept of cumulative impulse averaging, can provide a semi-empirical basis for predicting skull fracture and possibly brain extrusion from foramen magnum. The GSI is defined by

$$\text{GSI} = \int_0^T a^{2.5}(t)\,dt \qquad (1.1)$$

where a is the head centre of gravity acceleration in g's and T is the impact pulse duration. A GSI value of 1000 is generally accepted as the head injury survival level. This level corresponds to a rectangular pulse of magnitude $100\,g$ and duration $10\,ms$.

Recently a modified criterion, namely the head injury criterion (HIC) has been proposed in the form [19]

$$\mathrm{HIC} = \left| \frac{\int_{t_1}^{t_2} a\,dt}{t_2 - t_1} \right|^{2.5} (t_2 - t_1) \qquad (1.2)$$

where t_1 is an arbitrary time in the pulse and t_2 is the time computed for a given t_1 which maximizes HIC. Analytical studies of HIC and its interrelation with GSI have been conducted by Chou and Nyquist [20].

Other head severity criteria include the maximum strain criterion (MSC) [21], the effective displacement index (EDI) [22], the Vienna Institute index (JIT) [23], and the revised brain model (RBM) [24]. These criteria essentially incorporate spring–mass–dashpot representations and have been discussed by Versace [25] and McElhaney et al. [3]. Such lumped-model representations are only valid in characterizing the gross inertial response of the system. Prediction of continuum phenomenon such as wave propagation characteristics, local displacements, strains, pressures, and displacements can only be accomplished by more sophisticated models. These considerations along with experimental validation are discussed next.

1.5. Head injury model formulation and simulation

Dynamic response modelling of the human head has received considerable attention in recent years. Head injury models representing various degrees of sophistication have ranged from spring–mass–dashpot systems [21–24] to elastic spherical shell–fluid/viscoelastic core representations [26–32] and finite-element characterizations [33, 34]. Structural responses such as frequencies, skull stresses, brain displacements, and pressure/stresses have been computed in these studies.

To characterize the impact of the head with an object, say an automobile dashboard, it is first necessary to define the acceleration load–time characteristics. This entails the delineation of the dynamic translational and rotational components of the total head centre of gravity accelerations based on kinematical considerations. Formulation and simulation of a head injury model generally involves (i) model selection based on experimental, geometric, and constitutive data and impact loading, (ii) solution of the governing equations of motion, and (iii) numerical representation of the cranio-cerebral injury response variables. Comparison of analytical responses with corresponding data on experimental models and animal/cadaver impact tests is essential for subsequent model validation. Two primary head injury model representations are studied here to determine the effects of pure translation and pure rotation.

The selected spherical shell–core translational model of the skull–brain system is illustrated in Fig. 1.4 [35] and the related analysis is summarized in Appendix 1.1. The radial displacement, radial stress, and shear stress are assumed to be continuous along the skull–brain interface. However, a similar model assumes a slip condition [32]. The spatial characteristics of the loading are indicated in the form of a cosine distribution. Since the scalp is not incorporated in the model, the impact load is assumed to act on a 60° cap of the shell surface. The scalp distributes the impact load and extends the pulse time along with a typical energy absorption of 10 per cent. The non-dimensionalized time history of the assumed loading is defined by $(t/t_0)\exp(-t/t_0)$. A comparison of the GSI curve for this pulse (of magnitude 1000) versus the experimentally obtained Wayne State curve is illustrated in Fig. 1.5. Table 1.2 gives the functional form of the GSI for five pulse shapes. The first pulse is a square wave, the second is triangular, and the fifth is a half sine wave. The third and fifth pulse shapes correspond to the form $(t/t_0)\exp(-t/t_0)$ selected here with durations $4t_0$ and $6t_0$ respectively. The pulse duration is not as obvious for this pulse shape because of the exponential decay, and a cut-off point expressed as a percentage of the peak value is chosen to define the duration. For the same GSI the maximum value of acceleration for the triangular pulse is 65 per cent greater than the maximum value of acceleration for the square wave. The effective acceleration is, however, about 20 per cent higher for the square wave than for the triangular wave for the same GSI.

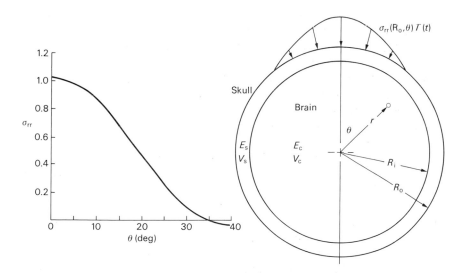

FIG. 1.4.   Head injury translational model.

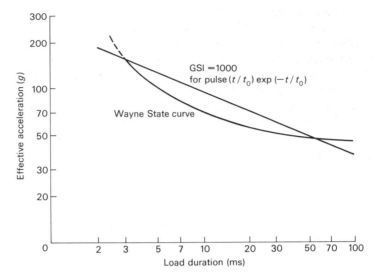

F$_{\text{IG}}$. 1.5.   Comparison of Wayne State curve and GSI curve.

T$_{\text{ABLE}}$ 1.2

Functional form of the GSI for five pulse shapes

Pulse shape	Maximum acceleration A_m	Effective acceleration A_e
A	$TA_m^{2.5}$	$TA_e^{2.5}$
	$T\left(\dfrac{A_m}{1.65}\right)^{2.5}$	$T(1.21A_e)^{2.5}$
	$T\left(\dfrac{A_m}{1.41}\right)^{2.5}$	$T(1.04A_e)^{2.5}$
	$T\left(\dfrac{A_m}{1.68}\right)^{2.5}$	$T(1.31A_e)^{2.5}$
	$T\left(\dfrac{A_m}{1.37}\right)^{2.5}$	$T(1.14A_e)^{2.5}$

The gross constitutive, geometric, and mass parameters for the simulated translational model are as follows:

Skull: structural elastic modulus $E_s = 2 \times 10^5 \text{ lbf in}^{-2}$; Poisson's ratio $v_s = 0.20$; weight $W_s = 2.35$ lb; $R_o = 3$ in; $R_i/R_o = 0.90$.
Brain tissue: bulk modulus $K = 3 \times 10^5 \text{ lbf in}^{-2}$; shear modulus $G(t)$ defined by Fig. 1.3; weight $W_B = 2.96$ lb.
Impact load: $T(t) = (t/t_0)\exp(-t/t_0)$. The magnitude of the acceleration and the duration of the pulse are variable. The duration of the load is defined by $t = 4t_0$. This time corresponds to the time required for the head to reach 90 per cent of the total velocity change.

For the selected configuration and pulse an effective acceleration A_e of 100 g corresponds to a peak stress of 250 lbf in^{-2} at the pole.

The idealized pure rotational acceleration model is illustrated in Fig. 1.6 [36]; the pertinent analysis is summarized in Appendix 1.2. The head is assumed to be subjected to a Heaviside exisymmetric torsional acceleration. The displacement at the skull–brain interface is prescribed to be zero. As an extension of this model, an asymmetric rotational model (Fig. 1.7) is also investigated [37]. Physically the applied loading can be interpreted in terms of a combination of a symmetric rotation as in Fig. 1.6 and an asymmetric translational acceleration.

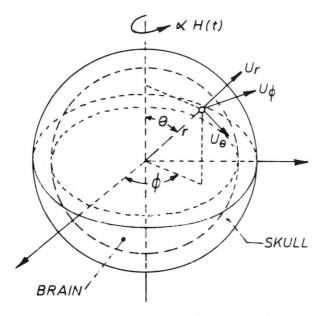

FIG. 1.6.   Axisymmetric head injury rotational model.

Response solutions for combined translational and rotational loadings can be superposed to yield results for generalized head injury models with arbitrary load–time characteristics. Such loadings include whiplash and selected contact impacts.

1.6. Cranio-cerebral injury parameter evaluations and comparisons

Kinematic considerations for the head–neck–torso complex are briefly presented. Numerical results for the selected head injury translational and rotational models are also given in this section. The responses are obtained by using the modal acceleration method [38] which entails solution of the

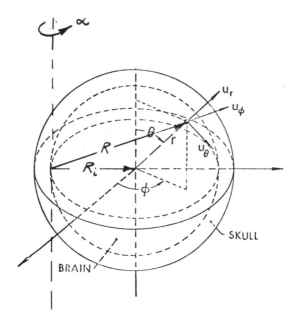

FIG. 1.7.   Asymmetric head injury rotational model.

equivalent static problem, solution of the free vibration problem, and the time function solution. Comparisons with experimental data and other models are also given.

Kinematical considerations

It is well recognized that, in general, a head impact produces both translational and rotational motion as well as deformation of the skull with resultant brain injury from both the absolute motion of the brain and its relative displacement with respect to the skull. The T-1 vertebra is generally considered to be a pivot point. Several discrete models consisting of masses, springs, and dashpots (linear and rotational) have been extensively used to examine the relative motion between the head, neck, and torso resulting from impact loading [39–41]. Roberts, Ward, and Nahum [42] have investigated an elastodynamic model with eleven degrees of freedom (describing the human skull, brain, spinal column, neck, and torso) under a triangular pulse applied to the mass centre of the head. Significant dependence of the response of the neck to the location and rise time of the applied load is revealed. McKenzie and Williams [43] have developed a two-dimensional model of the head, cervical spine, and torso to study their response to whiplash loading. An extensive evaluation of the head and neck response to inertial acceleration using human volunteers has been documented by Ewing and Thomas [44]. A three-dimensional vehicle occupant model developed by Bowman and Robbins [45] simulates the head–neck–torso region for frontal, side, and oblique impacts. The model has two ball-and-socket joints and the neck is allowed to stretch. Several dummy neck mechanical models permitting flexion and extension have also been fabricated [46, 47].

Translational acceleration response

The translational acceleration response variables include model structural stiffness, fundamental natural frequency, dynamic skull displacements and stresses, brain displacements, pressures, and stresses.

Head structural stiffness and fundamental frequency. The static structural stiffness is defined as the force acting on the head, on diametrically opposed sides, divided by the relative displacement. Reported experimental stiffness coefficients for cadavers, obtained by applying compressive loads at the anterior and posterior locations (A–P) of the skull, range from 8000 lbf in^{-1} to 20 000 lbf in^{-1} [3]. The corresponding stiffness coefficient for the selected model is 14 950 lbf in^{-1} [48]. Hardy and Marcal [49] also computed a comparable value (17 500 lbf in^{-1}) for front pressure loading of a skull *in vacuo* using the finite-element method. They determined a displacement of 0.20 in for a static load of 3500 lbf. Experimental stiffnesses of 4000–10 000 lbf in^{-1} are, however, obtained for side loading (left to right) [3]. This lower range can be attributed to local thicknesses and geometry.

Examination of the frequency spectrum for the selected skull–brain model reveals a fundamental axisymmetric frequency of 536 Hz [50]. A corresponding range of values from 400 to 700 Hz has been found experimentally by the driving point impedance method [51].

Skull structure response. The radial displacement–time response characteristics of the skull surface for $A_e = 100\ g$ (to convert the effective acceleration to the peak acceleration, for the assumed pulse, multiply by 1.48) and duration $t = 20$ ms are illustrated in Fig. 1.8. The computed peak displacement of 0.025 in at the impact pole, using the translational model, is consistent with the experimental value cited by Fan [24] and corresponds to a pole–antipole brain 'strain' of 0.0048 in in^{-1} using the MSC. This value also compares favourably with the lumped strain threshold value of 0.0061 in in^{-1} obtained by McElhaney *et al.* [3] using the driving point impedance method and the brain diameter as the gauge length. The skull response for short duration loadings of the order of 1 ms is considerably more oscillatory. However, rigid body and elastokinetic analysis reveals that the impact duration of the human head against a rigid flat surface is at least 3 ms [50].

The compressive and tensile skull stresses associated with an assumed peak contact stress of 1000 lbf in^{-2} at the impact pole are given in Table 1.3. These computed tangential stresses, for a skull thickness of 0.20 in, approach the experimentally determined ultimate tensile and compressive stress values

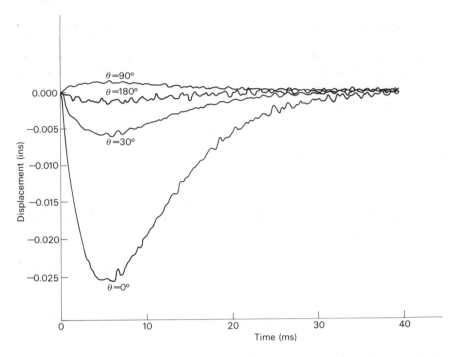

FIG. 1.8. Skull elastic radial displacement dynamic response at various locations: $A_e = 100\ g$; load duration, 20 ms.

TABLE 1.3

Peak tangential compressive and tensile skull stresses at the impact pole for a Peak Skull Contact Stress of 1000 lbf in^{-2}

Skull thickness (in)	Membrane stretch contribution (lbf in^{-2})	Bending contribution (outer and inner surface) (lbf in^{-2})	Total compressive stress (skull surface) (lbf in^{-2})	Total tensile stress (skull–brain interface) (lbf in^{-2})
0.30	−3495	± 5125	− 8620	1630
0.25	−4195	± 7380	−11 575	3185
0.20	−5240	±11 520	−16 760	6280

given in Table 1.1. This indicates that the skull fracture response is optimized in compression and tension simultaneously with the peak fracture load, for the assumed loading, corresponding to 2130 lbf. For more localized loading, lower fracture loads can be predicted by the model. The range of values is consistent with loads obtained by Hodgson *et al.* [52] and Melvin and Evans [53]. A more rigorous criterion of skull fracture thresholds would involve interrelationship between the fracture frequency and the fragility of the skull as discussed by Gurdjian and Webster [54] and consideration of the elasto-dynamic theory of cracks in brittle material [55]. Stress intensity factors and strain energy density computations are necessary for predicting accurate fracture thresholds [56].

Brain radial displacement and pressure response. The radial displacement time history in Fig. 1.8 also reveals the brain response at the skull–brain interface since the radial compression of the skull is negligible. The corresponding computed response for the 100 G 1 ms pulse can be shown to have wave propagation characteristics as opposed to the inertial response in Fig. 1.8 [50]. This contrast of response results from the fact that the wave transit time for a pulse through the head is approximately 0.10 ms.

To study the brain cavitation injury hypothesis it is necessary to understand the impact mechanics at the antipole. Cavitation (contrecoup injury) is generally attributed to the tensile pressure phenomenon at the antipole. It occurs if the local pressure is less than or equivalent to the vapour pressure of brain fluid. The average cavitation threshold for fresh water is 1 atm (frequencies up to 5 kHz) and the threshold increases with decreased pulse length. For head impact load durations typical of automotive collisions (i.e. greater than 5 ms) the maximum magnitude of the pressure ($p = \sigma_{ii}/3$) along the pole–antipole axis is predicted by Fig. 1.9. This variation is influenced by the inertia of the brain and its trend agrees with the linear pressure relation obtained experimentally across the skull pole–antipole axis [57]. The corresponding pressure–time profile with relatively small fluctuations follows the acceleration–time curve. The relation between the computed peak pressure (p in lbf in^{-2}) and the effective acceleration (A_e in g's) at the antipole is

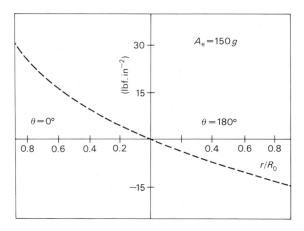

F$_{IG}$. 1.9.   Pole–antipole axis peak pressure distribution: effective acceleration, 150 g, load duration, $\geqslant$ 5 ms.

governed by [48].

$$p = -0.096A_e. \tag{1.3}$$

The cavitation threshold (defined at $-14.7\,\text{lbf in}^{-2}$) therefore corresponds to a limiting effective acceleration $A_e = 153\,g$. This value agrees well with the corresponding point on the Eiband curve (Fig. 1.10) [58] and a GSI value of 1000 (Fig. 1.5).

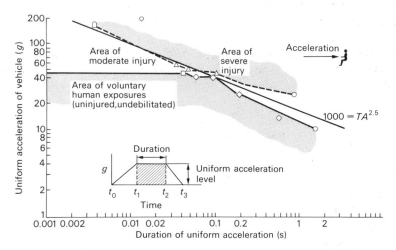

FIG. 1.10. Eiband curve: frontal deceleration versus duration threshold from experimental data.

For short pulse durations of the order of 1 ms the wave characteristics dominate in the computed pressure response at the antipole. The oscillations arise from the interaction of the flexural–extensional waves travelling circumferentially around the skull with the compressive wave transmitted through the brain. This phenomenon has been experimentally and/or analytically demonstrated by several investigators [26, 59, 60].

Recently Suh, Yang, and McElhaney [61] conducted an experimental study on the rarefaction of unheated liquids (water and dilute APO 30 polymer solutions having the same viscosity as brain) in a glass sphere due to a local radial ultrasonic disturbance of sinusoidal, square or impulsive impacts. By measuring pressure gradients at the onset of cavitation the possible vulnerable regions of the fluid have been determined. The test results for water agree with the results of ref. 26 but those for polymer solutions do not.

Brain shear strain response. The shear strain mode of brain deformation represents the most significant injury variable in view of the low brain shear moduli values. As previously indicated mechanical shear strain values of

0.035 rad have been identified at 10 Hz. This value presumably represents an upper bound for the pathophysiological failure shear strain.

The peak computed shear stresses and strains and their location for the 100 g 20 ms pulse are presented in Table 1.4. It should be noted that the computed shear strains are higher than the mechanical threshold at the interface and mid-brain region (the area controlling consciousness). Shear strains in the brain-stem area are also crucial in the evaluation of closed head injury. Gurdjian and Lissner [62] demonstrated the existence of shear strains attributable to intercranial flow at the simulated cerebrospinal junction when hammer blows were applied to a two-dimensional plastic skull model supplied with a tube to resemble the spinal cord. The shear strain magnitudes increase with decreasing pulse duration [30]. Cumulative (integral) stress–time thresholds are desirable for assessing damage from pressure and/or shear phenomena.

TABLE 1.4

Brain shear stress and strain magnitudes

Maximum brain shear stress location ($\theta = 30°$)	Shear stress magnitude (lbf in^{-2})	Shear strain magnitude $\|G\| = 1.57$ lbf in^{-2}
$r/R_0 = 0.9$, skull–brain interface	0.091	0.058
$r/R_0 = 0.70$	0.063	0.040
$r/R_0 = 0.50$ (mid-brain)	0.064	0.041

$A_e = 100\ g$; duration, 20 ms.

Rotational acceleration response

The formulated boundary-value problem for the model in Fig. 1.6 entails solution of the non-vanishing torsional displacement U_ϕ and the torsional shear stress $\sigma_{r\theta}$ [36] (Appendix 1.2). The computed transient displacement elastic response for the selected brain geometric and constitutive parameters with $|G| = 12$ lbf in^{-2} (since the Heaviside step angular acceleration is chosen) for an angular acceleration $\alpha_0 = 2000$ rad s^{-2} and $\theta = \pi/2$ is illustrated in Fig. 1.11. The peak displacement of 0.03 in occurs at $r/R_i = 0.4$ after an elapsed time of 5.5 ms. Figure 1.12 reveals the dynamic elastic shear stress–time history for $\theta = \pi/2$. The computed peak shear strain magnitude of 0.046 at the skull–brain interface is comparable to the value predicted for the translational impact model.

The rotational response for an arbitrary time pulse can be computed from a Duhamel integral representation of the step response. The corresponding viscoelastic response is obtainable by using the elastic–viscoelastic correspondence principle [63]. Using an identical model and comparisons with

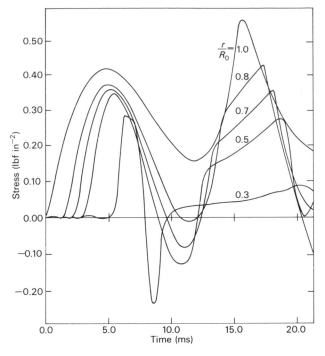

FIG. 1.11.   Elastic dynamic shear stress $\sigma_{r\phi}$ in the axisymmetric rotational model: $\alpha = 2000\,\text{rad s}^{-2}$; Heaviside loading.

experimental concussion data on monkeys Bycroft [31] deduced a shear strain value of 0.05 in the region of the upper reticular formation as that necessary to cause concussion. The model, scaled for humans, predicts that a step angular acceleration of at least 3500 rad s^{-2} is necessary for producing concussion.

The static shear stress response for the asymmetric model illustrated in Fig. 1.7 is optimized for the incompressible case (i.e. brain material with $v = 0.5$). The corresponding stresses are [37] (Appendix 1.3)

$$\sigma_{rr} = \sigma_{\theta\theta} = \sigma_{\phi\phi} = \rho\alpha R_i r \sin\theta \cos\phi$$
$$\sigma_{r\phi} = -\rho\alpha r^2 \sin\theta/5 \tag{1.4}$$
$$\sigma_{r\theta} = \sigma_{\theta\phi} = 0.$$

The pressure profile $p = \sigma_{ii}/3$ exhibits the familiar linear gradient across any diameter with peak values occurring at two sites $(a, \pi/2, 0)$ and $(a, \pi/2, \pi)$. The maximum shear stress $\sigma_{r\phi}$ occurs on the surface of the equatorial circle $(\theta = \pi/2)$. The corresponding static values for an angular acceleration of $\alpha = 3500\,\text{rad s}^{-2}$ are $p = 2.92\,\text{lbf in}^{-2}$ and $\sigma_{r\phi} = 0.58\,\text{lbf in}^{-2}$. The computed shear strain value is in the damaging threshold range of 0.035 rad [37].

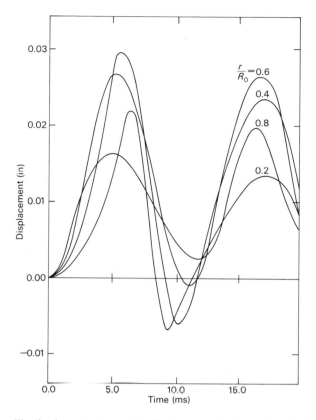

FIG. 1.12. Elastic dynamic torsional displacement in the axisymmetric rotational model: $\alpha = 2000$ rad s^{-2}; Heaviside loading.

Rotational acceleration experiments on squirrel monkeys by Unter-harnscheidt and Higgins [64] indicate that the concussive threshold is higher for 2–4 ms imputs than for 10 ms wave inputs. Ommaya and his coworkers [65–68], in a series of experiments on monkeys, have demonstrated the importance of rotational acceleration in producing cerebral concussion. This work includes threshold rotational accelerational–time plots with and without neck restraint, rotational velocity–impulse characteristics, and scaling of experimental data to man. Figure 1.13 illustrates a plot of the scaling relationships between subhuman primates and man for concussive levels of rotational acceleration. Pulse durations are not indicated in this figure since only inertial scaling laws are employed in the analysis [68].

Summary of head injury model responses

The preceding translational and rotational models provide an engineering link for quantifying critical head injury parameters. For the translational model

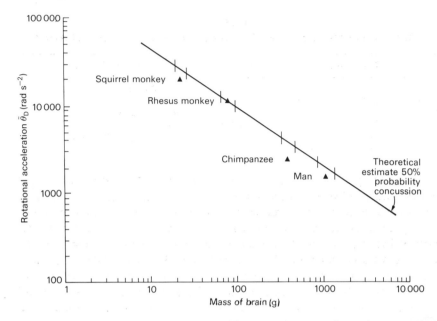

F$_{IG}$. 1.13.   Scaling relationship between subhuman primates and man for concussive
levels of rotational acceleration.

these parameters are (i) the skull tangential stresses, (ii) brain displacements
and contrecoup pressures, and (iii) brain shear strains at the surface and mid-
brain region. Peak values of these parameters with location, are summarized
for specified loadings in Table 1.5. This model primarily describes inertial
movement of the intracranial contents with combined variations in cerebral
blood volume, cerebrospinal fluid, brain volume, and associated pressures.
Contrecoup brain injury is interpreted by this model in terms of acceleration
time pulses. The rotational models single out the shear strain as a principal
mechanism of injury. Table 1.6 shows that the peak surface brain shear strains
are in the damage threshold range for angular accelerations around
2000–3000 rad s^{-2}. Distortion of the intracranial contents is characterized by
the model.

T$_{ABLE}$ 1.5
Critical parameters and values for closed head injury: translational model

	Peak value	Site	Loading
Skull–brain surface displacement	0.025 in	Impact pole	100 g, 20 ms
Brain pressure	−14.7 lbf in^{-2}	Contrecoup	153 g, > 15 ms
Brain shear strain	0.058 rad	Interface $\theta = 30°$	100 g, 20 ms
	0.041 rad	Mid-brain $\theta = 30°$	100 g, 20 ms

TABLE 1.6

Critical parameters and values for closed hand injury: rotational models

	Peak value	Site	Model and loading
Brain shear strain	0.046 rad	Interface	$\alpha_0 = 2000$ rad s^{-2} axisymmetric-step dynamic loading
	0.041 rad	Equatorial circle	Asymmetric static load $\alpha_0 = 3000$ rad s^{-2}

The models provide estimates of the deformable head motions such as skull depressions, selected cerebral blood vessel and ventricular system movements which can be experimentally verified by detailed high speed cinefluorographic investigations on animals and human cadavers subjected to controlled impacts. Subsequent neuropathological studies on the cerebral hemispheres, the cerebellum, the brain stem, and the cervical spinal chord can relate the sites and intensity of pathologic lesion patterns (e.g. haemorrhages with or without haemostasis, contusions, lacerations, etc) with the predicted displacements and stresses. The models indicate that the skull–brain interface is vulnerable in the shear mode.

1.8. Pathophysiological considerations—Neural trauma

Selected results from inertial experiments along with clinical and pathological observations are summarized here, with cerebral concussion as the thematic focus.

Results from recent experiments on animals by Ommaya and Gennarelli [69] and Gennarelli, Thibault, and Obbaya [70, 71] indicate that rotational accelerations of the head produce loss of consciousness (paralytic coma) as well as diffuse lesions in the brain. Equivalent levels of pure head inertial translational acceleration failed to produce such traumatic unconsciousness and caused fewer lesions, not in the brain stem, which were asymmetric and always focal. In animals subjected to rotational acceleration the lesions are diffuse, bilaterally symmetrical, and tend to occur at tissue interfaces. These results substantiate the rotational head injury hypothesis and also provide additional data on the use of the somato-sensory evoked response (SER) as a quantitative on-line indicator of traumatic unconsciousness and the development of secondary lesion effects [72, 73]. Evaluation of the SER before, during, and after independent rotational and translational acceleration experiments clearly indicates the coincidence of the period of SER abolition with traumatic unconsciousness in the rotational tests. In addition the return of the longer latency wave (P_2), which represents conduction in non-specific reticular sensory pathways, correlates well with the duration of traumatic unconsciousness (paralytic coma) in the animal. In the translated animals the P_2 wave is always preserved.

It appears from the above experiments that paralytic coma of cerebral concussion is associated with failure of activity in the mesencephalic reticular formation. Irrespective of such effects of trauma on the brain stem, severe and more long lasting effects were observed in the cortex of the concussed animals. Figure 1.14 illustrates the time taken for the P_2 wave to travel from one hemisphere to the other after head injury as a percentage of the time taken prior to head acceleration. Marked slowing of such interhemispheric cortico-cortical transfer was found only in the rotated animals and this persisted long after return of the P_2 wave indicated adequate conduction through the reticular formation. This observation suggested that telencephalic effects of concussive head injury are widespread, especially in the cortex, and are probably more severe than in the brain stem.

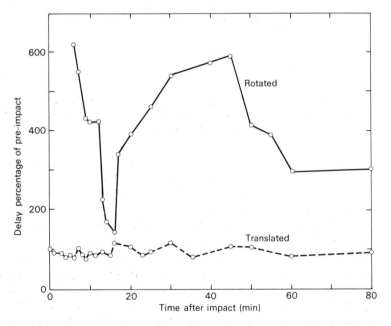

FIG. 1.14.   Percentage ratio of post-impact/pre-impact P_2 wave travel time.

Cerebral concussion

Normal consciousness can be defined as that state of awareness in the organism characterized by the maximum capacity to utilize sensory input and motor output potential while achieving accurate storage and retrieval of events in contemporary time and space. Cerebral concussion is then defined as a graded set of clinical syndromes following head injury wherein increasing severity of disturbance in the level and content of consciousness is caused by brain mechanical deformation in a centripetal sequence of disruptive effect on function and structure. The effects of this sequence appear to initiate at the brain surfaces in the mild cases and extend inward to the dien-cephalic–mesencephalic core at the most severe levels of trauma.

A possible classification of the grades of cerebral concussion thus produced is shown in Fig. 1.15.

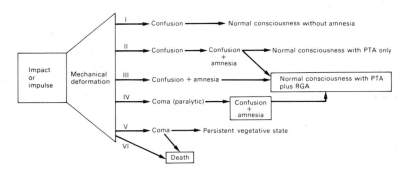

FIG. 1.15. Syndromes of cerebral concussion: I and II, cortical–subcortical disconnection; II and III, cortical–subcortical disconnection plus diencephalic disconnection; IV to VI, cortical–subcortical disconnection plus diencephalic discon-nection plus mesencephalic disconnection; PTA, post-traumatic amnesia; RGA, retrograde amnesia.

The extent of cortical and subcortical involvement in head injury is significant and the probability of peripheral damage increases proportionally when the amount of brain deformation is large enough to affect the rostral brain stem and produce the 'typical' case with the paralytic coma of traumatic unconsciousness (Grade IV). Extending on either side of this level of injury severity are cases with lesser and greater damage. On the one hand, less severe cases can be found where memory disturbance occurs without loss of motor control and only partially impaired awareness (Grades I to III); in such cases it is conjectured that significant deformation does not reach the reticular activating system. On the other hand, there are more severe cases with greater degrees of diffuse irreversible damage. When such diffuse damage reaches a critical amount, and not necessarily when the mesencephalon shows visible structural damage, the Grade V case is developed. This type of result is aptly described by the term 'persistent vegetative state' as suggested by Jennet and Plum [74].

Clinical observations

The complex pattern of clinical behaviour after head injuries is well recognized. It is possible, however, to discern a general pattern of recovery in many patients after severe but relatively uncomplicated head trauma which it is important to recapitulate. The patient falls unconscious and remains for a while in an unresponsive immobile coma. Emerging from this state of paralytic unconsciousness the patient successively passes through stages of stupor, confusion with or without delirium, and finally an almost lucid phase with automatism before becoming fully alert. Stated in another way, return to awareness to stimuli usually precedes motor and sensory recovery which in turn recover before restoration of memory and other cognitive functions. It is of great interest that this sequence of reintegration of normal consciousness is similar to a wide spectrum of head injuries and correlates with the severity of injury only in being much slower in cases with prolonged coma, such patients being of course more liable to have residual symptoms. The common element in such severe cases is that alertness always returns prior to full return of memory functions. Another feature of the 'recovered' case of severe head injury is the association of a labile affect with the difficulty in learning new material. These observations support the idea of a greater vulnerability of the cortex and particularly of the limbic and fronto-temporal cortices which occupy zones of great structural irregularity and variation in tissue density.

A large number of clinical observations at the less severe end of the spectrum of cerebral concussion syndromes can also be similarly discussed. The lesser grades of cerebral concussion (I to III) are quite common, particularly in contact sports such as American football and boxing. The majority of such concussions do not produce paralytic coma or traumatic unconsciousness; instead, confusion and amnesia are usually seen. It is a common experience for patients who have been briefly 'dazed' or confused to continue well co-ordinated sensory–motor activity after a sports accident without subsequent recall of the episode. Yarnell [75, 76] has recently reported a group of such mildly concussed patients examined immediately after impact who were confused and disoriented in time but possessed intact recall of events immediately prior to impact; retrograde amnesia did not develop until 5–10 min after the impact. He also described further cases similar to that reported by Fischer [77] in which the occurrence of severe amnesia after head impact was not associated with a loss of alertness. In the latter report the author writes: 'It must be concluded that a traumatic insult to the memory mechanisms can occur with complete sparing of the neural basis of alertness' and again, 'In moderate to severe concussion, although both systems are usually affected together, they are not equally vulnerable, for alertness is restored first as a rule, while impaired memory almost always persists for a longer period. It might be expected therefore that amnesia would occur in the presence of retained alertness' [77].

A reasonable explanation for the greater vulnerability of memory and the lesser vulnerability of alertness in head injury can also be offered. In a recent study on memory mechanisms in man [78] data have been presented which suggest that it is the hippocampal gyri rather than the hippocampus *per se* which form the brain structures critically involved in coding experiences for retrieval from storage via the associative neocortex. It would follow that if the hippocampal mesocortex and temporal neocortex bear the main brunt of the cortical damage, then the recovery of alertness should precede the return of telencephalic integrative functions controlling motor and sensory mechanisms with memory mechanisms being restored last of all. The observation by Torres and Shapiro [79] that electroencephelogram abnormalities occurring in a group of patients after non-impact inertial loading of the head (whiplash injury) were similar in nature and incidence to those seen after impact-produced head injury is inexplicable except by invoking the common role of inertial loading with maximal effect on the cortex in both conditions.

With regard to the relative contributions of primary and secondary effects to the ultimate outcome in patients surviving after head injury a recent paper by Overgaard *et al.* [80] is most illuminating. These authors show clinical data which suggest that the first neurological examination (done within 6 h in 70 per cent of the patients) was reliably predictive for the outcome of the case when only three clinical criteria were used: level of consciousness, motor pattern and neurophthalmologic signs. Their data support the hypothesis that the critical damage was primary and mainly in the cortical fields rather than in the reticular formation. A second observation made by this group is also noteworthy, i.e. that the presence of episodic systolic hypertension was significantly correlated with a poor prognosis. This would suggest that the additional volumes of cerebral edema, vascular congestion, and possibly also increased cerebrospinal fluid (CSF) due to decreased absorption constitute the main mechanism for secondary ischemic hypoxia which further damages neural tissues to add to the burden of irreversible primary insult. The lack of significant correlation between prognosis and intracranial pressure (ICP) or cerebral blood flow (CBF) measurements in man may well be due to the lack of correlated data on the critical rate of such extra volume expansion in the head.

Pathological observations

It is not possible to present a satisfactory description of the pathology of blunt head injury. In spite of the many detailed studies on various aspects of the problem carried out to date a completely adequate description of the relative three-dimensional distribution of all primary lesions throughout the brain at known intervals of time after head injury is not yet available either for man or experimental animals. Such a study is urgently required; data are needed in cases sustaining injury at three levels of injury severity, e.g. with reversible deficits, with irreversible deficits but survival, and with irreversible deficits plus death. However, a review of the available data does indicate certain trends. The observation that the prime location of contusions is in the temporal and frontal regions irrespective of the site of impact has not been satisfactorily explained by prior hypothesis. In a recent review of the pathology of brain damage in blunt head injuries Strich [81] has emphasized the diffuse nature of histological changes in axons and microglial stars around tissue tears as reported by investigators. In a recent attempt to provide more comprehensive data Grcevic and Jacob [82] have reported preliminary data on a serial section study in fatal human head injuries showing diffuse lesions in a rostral periventricular pattern. The consensus from such data would suggest that axons in white matter are more fragile than neurones or their synapses in mechanical trauma. However, in the absence of electron microscope or even adequate cell-count data from controlled experiments in animals it is difficult to decide on the relative vulnerability of cortical and subcortical structures to mechanical trauma, nor is it yet possible to envisage the three-dimensional distribution of the diffuse lesions at the light microscope level. An important study by Mitchell and Adams [83] on the incidence of primary brain stem lesions in patients dying after head injury supports the idea that such lesions probably do not occur in head injuries without severe damage elsewhere. It therefore seems that the brain stem may be the least vulnerable part of the brain for primary damage.

The difficulties in interpretation of pathological data from experimental studies have been pointed out on numerous occasions. Because of the smaller mass of brain in animals and the restriction implicit in impact techniques (skull failure being interposed) it has not been possible to duplicate all the clinical and pathological observations made in man. A critical review of the experiments claiming that primary brain stem damage and even primary rostral cervical cord damage are the substrate of acceleration concussion suggests that the data may equally well be interpreted as a consequence of the method used. It is important to stress that interpretations from experimental data must provide satisfactory explanations of the clinical observations on cerebral concussion. This point was also made by Denny-Brown [84] who, in reviewing the work of Friede, pointed out the lack of immediate disturbance of consciousness and of traumatic amnesias in contusions to the medulla or upper cervical cord in man.

1.9. Conclusions

In seeking a biomechanical framework for characterizing craniocerebral trauma mechanisms and their evaluation we have attempted to integrate results using balanced engineering and clinical methodologies. The classical approaches, which are also applicable to other human body components, entail (a) rheological and structural model formulation and response solution,

(b) experimental response determination using volunteers/isolated body components/animals/cadavers/dummies, and (c) clinical and/or pathophysiological documentation of accident victims and controlled experiments.

The simulated craniocerebral injury models provide a realistic continuum evaluation of skull response and brain displacement, pressure, and shear stress magnitudes. Since automotive crash injuries involve durations greater than 5 ms, the brain pressure and shear dynamic response is a consequence of the inertial response and can be quite accurately predicted by assuming that the skull is a rigid container. The models demonstrate that brain shear distortion is a crucial injury variable for both translational and rotational impacts. Cavitation–contrecoup injury is most likely for durations of less than 5 ms and for higher durations the inertial characteristics rather than the wave response dominate. This suggests that selected lumped-parameter representations can, in a gross pathological sense, serve as predictors of injury even though they do not actually model the associated injury phenomenon. Contact pulses for head impacts identified from the literature [3] correspond to 3–5 ms for rigid flat plate impacts, 12–18 ms for padded impacts, and 50 ms and higher for airbag tests. These contact durations have been predicted by analytical modelling [50].

A primary application for using biomechanics as a blueprint for craniocerebral trauma evaluation is in the optimization of automobile occupant head impact response. This includes design of conpartment intrusion, support and head-restraint systems (such as head rests and airbags), etc. Consideration of the viscoelasticity of brain tissue for multiple impacts (such as airbag–head-rest) merits particular consideration. Computer model simulations offer attractive economical possibilities over competing dummy and cadaver simulations. Secondary applications lie in the design of protective gear (helmets) for occupational and recreational activities.

An area of current concern is the biomechanics of head injury trauma in children resulting from falls, pedestrian accidents, and automotive occupant accidents. Age-related changes in mechanical properties of tissues and structures are of particular importance. Evaluation of several automotive child restraint systems is in progress.

Considerable progress in rigorously defining anatomical tissue, structure, and interface (such as brain–skull) properties needs to be made prior to the successful application of sophisticated finite-element models. Coupled with these micro-trauma experiments are essential in seeking to establish the physicochemical basis of reversible and irreversible damage to neural tissue at the cellular and molecular level. Because this approach necessarily can progress only as far as our current incomplete understanding of the structural correlates of normal function in the nervous system, it will readily be realized that head injury research in the broadest sense is also research at the frontiers of neurobiology [85].

Appendix 1.1. Translational impact head injury model

The linear elasticity equations governing the motion of the model (Fig. 1.4) are

$$\left(K_{\text{S,B}} + \frac{G_{\text{S,B}}}{3}\right)\nabla(\nabla \cdot \mathbf{u}) + G_{\text{S,B}}\nabla^2 \mathbf{u} + \mathbf{F} = \rho_{\text{S,B}}\ddot{\mathbf{u}} \qquad (\text{A1.1})$$

where $\mathbf{u}$ is the axisymmetric displacement vector with components U_r and U_θ in the r and θ directions respectively, ρ is the mass density, $\mathbf{F}$ is the translational acceleration body force, and ∇ is the gradient operator. K and G are the bulk and shear moduli respectively and subscripts S and B denote the skull and brain material respectively.

The associated boundary and interface conditions are

$$\sigma_{rr} = \sum_{n=0}^{N} \sigma_n P_n(\cos \theta) T(t) \qquad \sigma_{r\theta} = 0 \quad \text{at } r = R_0$$

and U_r, U_θ, σ_{rr} and $\sigma_{r\theta}$ are continuous across $r = R_{\text{i}}$. Quiescent conditions are assumed at $t = 0$.

Using the modal acceleration method [38] the solution to equations (A1.1) can be obtained by solving two different formulations: (i) the static problem and (ii) the free vibration and time function problem. Thus

$$\mathbf{u}(r, \theta, t) = \mathbf{U}_{\text{static}} + \sum_{m=1}^{\infty} \sum_{n=0}^{N} \mathbf{U}_{nm}(r, \theta) q_{nm}(t) \qquad (\text{A1.2})$$

where $\mathbf{U}_{nm}$ denotes the displacement modal function and $q_{nm}(t)$ denotes the time function. The static solution satisfies the imposed boundary conditions for the governing equations with $\rho_{\text{S,B}} = 0$. The dynamic solution satisfies the homogeneous boundary conditions and neglects the body force term. Details of the static free vibration and dynamic solutions can be found in ref. 35 and the computational results are presented in ref. 48. The viscoelastic response is obtainable by use of the correspondence principle [63].

Appendix 1.2. Axisymmetric rotational impact head injury model

The equation governing the uncoupled torsional dynamic displacement response U_ϕ for the axisymmetric rotation model (Fig. 1.6) is

$$G_{\text{B}}\left(\nabla^2 U_\phi - \frac{U_\phi}{r^2 \sin^2 \theta}\right) - \rho_{\text{B}}\alpha_0 r \sin \theta \, H(t) = \rho_{\text{B}}\ddot{U}_\phi \qquad (\text{A2.1})$$

with the boundary condition $U_\phi(R_{\text{i}}, \theta, t) = 0$. Elastic and viscoelastic solutions for the torsional displacement U_ϕ and the non-vanishing shear stress $\sigma_{r\phi}$ obtained by the modal acceleration method are presented in ref. 36.

Appendix 1.3. Asymmetric rotational static head injury model

The displacement equations governing the static asymmetric rotational model (Fig. 1.7) are

$$\left(K_B + \frac{G_B}{3}\right)\nabla(\nabla \cdot U) + G_B\nabla^2 u + \rho(\alpha \times R) = 0 \qquad (A3.1)$$

where u is the displacement vector with components U_r, U_θ, and U_ϕ. The assumed boundary conditions are

$$U_r(R_i,\theta,\phi) = U_\theta(R_i,\theta,\phi) = U_\phi(R_i,\theta,\phi) = 0. \qquad (A3.2)$$

The static displacement and stress solutions to equations (A3.1), using vector spherical harmonics, are presented in ref. 37.

Selected glossary

Amnesia: a condition in which memory is impaired

Axisymmetric: geometric and load symmetry about axis

Cavitation: the formation of a vacuum in a liquid

Cerebral concussion: clinical definition of impairment of neural function (e.g. vision, consciousness, balance, etc.)

Creep test: measurement of strain as a function of time for a constant stress

Consciousness: a state of wakefulness and responsiveness to environment

Constitutive equations: expressions describing a material property (usually stress versus strain and strain rate)

Contrecoup brain injury: injury occurring at skull–brain interface diametrically opposed to impact site

Contusions: alteration of brain structure integrity

Elastic modulus: the ratio of stress to strain usually in the linear range

Haemorrhage: discharge of blood

Lacerations (brain): gross tearing of neural tissue

Relaxation test: measurement of stress as a function of time for a constant strain

Severity index: a numerical quantity characterizing the level of injury

Storage, loss moduli: elastic and dissipative components of a complex modulus of elasticity expressed as functions of frequency

Viscoelastic: designation of time-dependent material property

Whiplash injury: injury resulting from excessive stretch and/or flection of the neck

References

1 CAVENESS, W. F., and WALKER, A. E. (eds.) (1966). *Head Injury Conf. Proc.*, Lippincott, Philadelphia, Pa.

2 GOLDSMITH, W. (1972). Biomechanics of head injury. In *Biomechanics—its foundations and objectives* (ed. Y. C. Fung, N. Perrone, and M. Anliker), pp. 585–634. Prentice Hall, Englewood Cliffs, NJ.

3 McELHANEY, J. H., STALNAKER, R. L., and ROBERTS, V. L. (1973). Biomechanical aspects of head injury. In *Human impact response measurement and simulation* (ed. W. F. King and H. J. Mertz) pp. 85–112. Plenum Press, New York.

4 GRAY, H. (1970). *Anatomy of the human body* (28th edn). Lea and Febinger, Philadelphia, Pa.

5 WOODBURN, R. (1965). *Essentials of human anatomy* (3rd edn). Oxford University Press, New York.

6 NETTER, F. H. (1968). *Nervous system, Vol. 1, The CIBA collection of medical illustrations.* CIBA Pharmaceutical Co., Summit, NJ.

7 HUBBARD, R. P., and McLEOD, D. G. (1973). A basis for crash dummy skull and head geometry. In *Human impact response measurement and simulation* (ed. W. F. King and H. J. Mertz) pp. 129–52. Plenum Press, New York.

8 Determination of the physical properties of the tissues of the head. *Final Rep., Contract No. PH43–67–1137*, West Virginia University, 1971.

9 McELHANEY, J., FOGLE, J., MELVIN, J., HAYNES, R., ROBERTS, V., and ALEM, N. (1970). Mechanical properties of cranial bone. *J. Biomech.* **3**, 495.

10 WOOD, J. (1971). Dynamic response of human cranial bones. *J. Biomech.* **4**, 1.

11 HUBBARD, R., MELVIN, J., and BARODAWALA, I. T. (1971). Flexure of cranial sutures. *J. Biomech.* **4**, 491.

12 SHUCK, L., and ADVANI, S. (1972). Rheological response of human brain tissue in shear. *J. basic Eng.* **94**, 905.

13 ESTES, M., and McELHANEY, J. (1970). Response of brain tissue to compressive loading. *ASME Pap.* 70–BHF–13.

14 GALFORD, J., and McELHANEY, J. (1970). A viscoelastic study of scalp, brain, and dura. *J. Biomech.* **3**, 211.

15 CHALUPNIK, J., DALY, C., and MERCHANT, H. (1971). Material properties of cerebral blood vessels, *Rep. No. ME 71–11*, University of Washington, Seattle, Wash.

16 HOLBOURN, A. H. S. (1943). Mechanics of head injuries. *Lancet* **ii**, 438.

17 GURDJIAN, E., LISSNER, H., and PATRICK, L. (1962). Protection of the head and neck in sports. *J. am. med. Ass.* **182**, 509.

18 GADD, C., (1966). Use of a weighted impulse criterion for estimating injury hazard, *Proc. 10th Stapp Car Crash Conf.*, pp. 164–174. Society of Automotive Engineers, New York.

19 Department of Transportation, NHTSA (49 CFR, Part 571), Docket No 69–7; Notice 17.

20 CHOU, C., and NYQUIST, G. (1974). Analytical studies of the head injury criterion. *Soc. automot. Eng. Pap.* 740082.
21 MCELHANEY, J., STALNAKER, R., ROBERTS, V., and SNYDER, R. (1971). Door crash worthiness criteria. *Proc. 15th Stapp Car Crash Conf.*, pp. 489–517. Society of Automotive Engineers, New York.
22 BRINN, J., and STAFFELD, S. (1971). The effective displacement index—an analysis technique for crash impact of anthroprometric dummies. *Proc. 15th Stapp Car Crash Conf.*, pp. 817–24. Society of Automotive Engineers, New York.
23 SLATTENSCHECK, A., and TAUFFKIRCHEN, W. (1970). Critical evaluation of assessment methods for head impact applied in appraisal of brain injury hazard in particular in head impact of windshields. *Int. Automobile Safety Conf. Compendium, SAE (Soc. automot. Eng.) Pap.* 700426.
24 FAN, W. (1971). Internal head injury assessment. *Proc. 15th Stapp Car Crash Conf.*, pp. 645–65. Society of Automotive Engineers, New York.
25 VERSACE, J. (1971). A review of the severity index. *Proc. 15th Stapp Car Crash Conf.*, pp. 771–96. Society of Automotive Engineers, New York.
26 ENGIN, A. (1969). The axisymmetric dynamic response of a fluid filled spherical shell to a local radial pulse—a model for head injury. *J. Biomech.* **2**, 324.
27 ADVANI, S., OWINGS, R., and SHUCK, L. Z. (1972). Dynamic response evaluation of translational and rotational head injury models. *Shock vibr. Dig.* **4**, 3.
28 KENNER, V., and GOLDSMITH, W. (1973). Impact on a simple physical model of the head. *J. Biomech.* **6**, 1.
29 CHAN, H. S. (1971). The asymmetric response of a fluid-filled spherical shell—a mathematical model simulation. Ph.D. Diss. Tulane University, New Orleans, La.
30 BENEDICT, J., HARRIS, E., and VON ROSENBERG, D. (1970). An analytical investigation of the cavitation hypothesis of brain damage. *J. Basic Eng.* **92**, 3.
31 BYCROFT, G. (1973). Mathematical model of a head subjected to an angular acceleration. *J. Biomech.* **6**, 489.
32 HICKLING, R., and WENNER, M. L. (1973). Mathematical model of a head subjected to an axisymmetric impact. *J. Biomech.* **6**, 115.
33 SHUGAR, T., and KATONA, M. (1973). Elastic axisymmetric finite element skull analysis. *Spec. Rep. No. SN 74–51–01.* Civil Engineering Laboratory, Naval Construction Battelian Center, Port Hueneme.
34 NICKELL, R., and MARCAL, P. (1973). *In vacuo* modal dynamic response of the human skull, *ASME Pap. No. 73–DET–112*; (Also in *J. Eng. Ind.* ASME Transactions **96**, 490 (1974)).
35 OWINGS, R. P. (1973). Axisymmetric dynamic response of a spherical shell

with an elastic core—a head injury model, Ph.D. Diss., West Virginia University, Morgantown, W. Va.

36 LEE, Y. C., and ADVANI, S. H. (1970). Transient response of a sphere to torsional loading—a head injury model. *Math. Biosci.* **6**, 473.

37' ADVANI, S. H., and McCLUNG, H. B. (1974). Asymmetric response of an elastic sphere—a rotational head injury model. *Adv. Bioeng. 73.*

38 REISMANN, H. (1967). On the forced motion of elastic solids. *Appl. sci. Res.* **18**, 156.

39 VON GIERKE, H. E. (1964). Biodynamic response of the human body. *Appl. mech. Rev.* **17**, 951.

40 MARTINEZ, J., WICKSTROM, J., and BARCELO, B. (1965). The whiplash injury—a study of the head–neck action and injuries in animals. *ASME Pap. 65–Wa/HUF 6.*

41 McHENRY, R., and NAAB, K. (1966). Computer simulation of the crash victim—a validation study. *Proc. 10th Stapp Car Crash Conf.* pp. 126–63. Society of Automotive Engineers, New York.

42 ROBERTS, S., WARD, C., and NAHUM, A. (1969). Head trauma—a parametric dynamic study. *J. Biomech.* **2**, 397.

43 McKENZIE, J., and WILLIAMS, J. (1971). The dynamic behavior of the head and cervical spine during whiplash (1971). *J. Biomech.*, **4**, 477.

44 EWING, C., and THOMAS, D. (1972). Human head and neck response to impact acceleration. *Army–Navy Joint Rep. NAMRL Monograph 21*, USAARL 73–1.

45 BOWMAN, B., and ROBBINS, D. (1972). Parameter study of biomechanical quantities in analytical neck models. *Proc. 16th Stapp Car Crash Conf.*, pp. 14–43. Society of Automotive Engineers, New York.

46 MELVIN, J., McELHANEY, J., and ROBERTS, V. (1973). Evaluation of dummy neck performance. In *Human impact response measurement and simulation* (ed. W. F. King and H. J. Mertz), pp. 247–61. Plenum Press, New York.

47 HAFFNER, M., and COHEN, G. (1973). Progress in the mechanical simulation of human head–neck response. In *Human impact response measurement and simulation* (ed. W. F. King and H. J. Mertz) pp. 289–320. Plenum Press, New York.

48 ADVANI, S., and OWINGS, R. (1975). Structural modelling of the human head. *J. Div. Eng. Mech.*, *ASCE* **101**, 257.

49 HARDY, C., and MARCAL, P. (1973). Elastic analysis of a skull, *ASME Pap.* 73–APMW–36.

50 ADVANI, S., and OWINGS, R. (1974). Evaluation of Head injury criteria. *Automotive Engineering Congr., SAE (Soc. automot. Eng.)* Pap. 740083. 1974.

51 STALNAKER, R., FOGLE, J., and McELHANEY, J. (1971). Driving point impedance characteristics of the head. *J. Biomech.* **4**, 127.

52 HODGSON, V. R., BRINN, J., GREENBERG, S., and THOMAS, L. M. (1970). Fracture behavior of the frontal bone against cylindrical surfaces. *Proc. 14th Stapp Car Crash Conf.*, p. 33. Society of Automotive Engineers, New York.

53 MELVIN, J., and EVANS, F. (1971). A strain energy approach to the mechanics of skull fracture. *Proc. 15th Stapp Car Crash Conf.*, pp. 666–85. Society of Automotive Engineers, New York.

54 GURDJIAN, E., and WEBSTER, J., *Head injuries*. Little, Brown, Boston.

55 SIH, G. C., and LIEBOWITZ, H., (1968). Mathematical theories of brittle fracture. In *Fracture, an advanced treatise—Mathematical fundamentals* (ed. H. Liebowitz), Vol. 2, pp. 67–190. Academic Press, New York.

56 SIH, G. C. (1973). *Handbook of stress intensity factors*. Institute of Fracture and Solid Mechanics, Lehigh University, Bethlehem, Pa.

57 ROBERTS, V., HODGSON, V., and THOMAS, L. (1966). Fluid pressure gradients caused by impact to the human skull. *ASME Pap.* 66–HUF–1.

58 EIBAND, M. (1959). Human tolerance to rapidly applied accelerations—a summary of literature. *NASA Memo.* 5–19–59E.

59 GROSS, A. (1958). A new theory on the dynamics of brain concussion and brain injury. *J. Neurosurg.* **15**, 548.

60 LINDGREN, S. O. (1960). Experimental studies of mechanical effects in head injury, *Acta chir. scand.* Suppl. 360, 1–100.

61 SUH, C., YANG, W., and McELHANEY, J. (1972). Rarefaction of liquids in a spherical shell due to local radial loads with application to brain damage. *J. Biomech.* **5**, 181.

62 GURDJIAN, E., and LISSNER, H. (1961). Photoelastic confirmation of the Presence of shear strain at the craniocerebral junction in closed head injury. *J. Neurosurg.* **18**, 58.

63 LEE, E. H. Viscoelasticity. In *Handbook of engineering mechanics* (ed. W. Flugge), pp. 53–1, 53–22. McGraw Hill, New York.

64 UNTERHARNSCHEIDT, F., and HIGGINS, L. (1969). Traumatic lesions of brain and spinal chord due to non-deforming angular acceleration of the head. *Univ. Texas Rep. Biol. Med.* **27**, (1).

65 OMMAYA, A. K., HIRSCH, A., and MARTINEZ, J. L. (1966). The role of whiplash in cerebral concussion. *Proc. 10th Stapp Car Crash Conf.*, pp. 314–24. Society of Automotive Engineers, New York.

66 OMMAYA, A. K., YARNELL, P., HIRSCH, A., and HARRIS, E. (1967). Scaling of experimental data on cerebral concussion in subhuman primates to concussion threshold for man. *Proc. 11th Stapp Car Crash Conf.*, pp. 47–52. Society of Automotive Engineers, New York.

67 OMMAYA, A. K., HIRSCH, A., FLAM, E., and MAHONE, R. (1966). Cerebral concussion in the monkey: an experimental model. *Science, N.Y.* **153**, 211.

68 OMMAYA, A. K., and HIRSCH, A. (1971). Tolerances for cerebral concussion from head impact and whiplash trauma. *J. Biomech.* **4**, 13.

69 OMMAYA, A. K., and GENNARELLI, T. (1974). Cerebral concussion and traumatic unconsciousness. *Brain* **97**, 633.

70 GENNARELLI, T., THIBAULT, L., and OMMAYA, A. K. (1971). Comparison of linear and rotational acceleration in experimental cerebral concussion. *Proc. 15th Stapp Car Crash Conf.*, pp. 797–803. Society of Automotive Engineers, New York.

71 GENNARELLI, T., THIBAULT, L., and OMMAYA, A. K. (1972). Pathophysiologic responses to rotational and translational accelerations of the head. *Proc. 16th Stapp Car Crash Conf.*, pp. 296–308. Society of Automotive Engineers, New York.

72 OMMAYA, A. K., CORRAO, P. G., and LETCHER, F. (1973). Head injury in the chimpanzee: Part I. Biodynamics of traumatic unconsciousness. *J. Neurosurg.* **39**, 152.

73 LETCHER, F., CORRAO, P. G., and OMMAYA, A. K. (1973). Head injury in the chimpanzee: Part II. Spontaneous and evoked epidermal potentials as indices of injury severity. *J. Neurosurg.* **39**, 167.

74 JENNET, B., and PLUM, F. (1972). Persistent vegetative state after brain damage: a syndrome in search of a name. *Lancet* **i**, 734.

75 YARNELL, P. R. (1970). Retrograde memory immediately after concussion. *Lancet* **i**, 863–4.

76 YARNELL, P. R., and LYNCH, S. (1973). The 'ding': amnestic states in football trauma. *Neurology*, **23**, 186–97.

77 FISCHER, C. M. (1966). Concussion amnesia. *Neurology* **16**, 826–30.

78 OMMAYA, A. K., and FEDIO, P. (1972). The contribution of cingulum and hippocampal structures of memory mechanisms in man. *Confin. Neurol.* **34**, 398–411.

79 TORRES, F., and SHAPIRO, S. K. (1961). Electroencephalographic abnormalities associated with whiplash injury. A comparison with the abnormalities present in closed head injuries. *Arch Neurol.* **28**.

80 OVERGAARD, J., CHRISTENSEN, S., HAASE, J., HEIN, O., HVID-HANSEN, O., LAND, A. M., PEDERSEN, K. K., and TWEED, W. A. (1973). Prognosis after head injury based on early clinical examination. *Lancet* **ii**, 631–5.

81 STRICH, S. J. (1969). The pathology of brain damage due to blunt head injuries. In *The late effects of head injury* (ed. A. E. Walker, W. F. Caveness, and MacD. Critchley), pp. 501–26. Charles C. Thomas, Springfield, Ill.

82 GRCEVIC, N., and JACOB, H. (1965). Some observations on the pathology and correlative neuroanatomy of sequels of cerebral trauma. *Proc. 18th Int. Congr. of Neurology, Vienna*, pp. 369–74.

83 MITCHELL, D. E., and HUME ADAMS, J. (1973). Primary focal impact

damage to the brain stem in blunt head injuries: does it exist? *Lancet* **ii**, 215–18.

84 DENNY-BROWN, D. (1961). Brain trauma and concussion. *Arch Neurol.* **5**, 1–2.

85 COHEN, L. (1973). Changes in neuron structure during action potential propagation and synaptic transmission, *Physiol. Rev.* **53**, 373.

2. Thoracic impact: injury mechanisms

R. Collins, Y. Kivity, D. Kalliers, R. Mathern, G. Schmidt, and D. Mohan

2.1. Introduction

In industrialized countries accidents are the third most common cause of death after cardio-vascular disease and cancer, and the second main cause of days lost from work. Over 15 million people are injured annually around the world and about 500 000 are killed in accidents. Of these, 6 million are injured and 200 000 killed in road accidents, the great majority due to thoracic injuries.

The great expansion of automobile traffic, new highways, more powerful engines and a growing population have all contributed to an increase in automobile fatalities in the United States which have now reached an annual level of approximately 53 000. Certain injuries, such as those to the head and neck, are evident although not well understood. Internal injuries, however, have gone largely undefined for many years. In order to reduce the incidence of injury and death due to these accidents it is very important to evaluate, analyse, and understand human response to trauma in mechanical, physiological, and even psychological terms.

2.2. Thorax anatomy

The thorax can be thought of as a shell which contains the following important organs: the heart, the lungs, the trachea, the oesophagus, the aorta and other great blood vessels, and the nerves. The size and shape of the thorax depend on the age and sex of the individual, but roughly it may be described as a truncated cone with its depth less than its base width.

The walls of the thorax consist of twelve pairs of ribs which articulate with the thoracic vertebrae at the back; the upper seven ribs articulate with the breastbone (sternum) in front. The eighth, ninth, and tenth ribs articulate with the costal cartilages of the ribs immediately superior, with the last two pairs floating with their cartilage embedded in the thoracic wall. The sternum itself is made of three parts (manubrium, body, and xiphoid process) joined by connective tissue. The upper opening of the thorax is bounded by the first pair of ribs, the first thoracic vertebra, and the jugular notch of the sternum. The lower opening is wider than the upper one and is bounded by the twelfth thoracic vertebra, the eleventh and twelfth pairs of ribs, and the xiphoid process of the sternum. Between the ribs are intercostal muscles, nerves, arteries, and veins; the chest wall is covered by skin on the outside and the parietal pleural membrane on its inner surface. The thorax is separated from the abdominal cavity by the diaphragm and into left and right sides by the membrane mediastinum. The heart is contained in a special membrane (pericardium) and the lungs are covered by a visceral pleural membrane. The chest cage is semi-rigid in structure and not only provides protection to the internal organs but also facilitates the mechanics of respiration.

2.3. Thorax trauma caused by impact—nature of the syndrome

The blunt thorax trauma in car accidents

Closed or blunt thoracic injuries are always the result of one or more force impacts against the thoracic region itself or other parts of the body which might transfer the trauma to the chest, e.g. chin tip, arms, or thighs. The severity in a thoracic trauma depends (among other factors) on impact velocity, deformation characteristics of the impinging parts of the vehicles and their resulting deceleration values, and the design of the cabin in relation to the place of impact with the occupant's body. The chance of survival also depends on the nature and severity of the accompanying injuries. More than three-quarters of all thoracic injuries have thorax trauma in combination with other injuries such as two-cavity injuries with ruptures of spleen, liver, kidneys or diaphragm or with severe skull–brain trauma or fractures of the skeleton, which produce special problems for first aid at the place of accident, transportability, and indication of operation and clinical therapy [1].

One-third of the patients with thorax trauma are unconscious when they reach hospital. The unconsciousness is caused, apart from accompanying injuries of the skull, by a severe bleeding shock due to intra-thoracical loss of blood or respiratory insufficiency as a result of hypoxia and CO_2 retention [2] and often combinations of both (Table 2.1). Even without injuries of the thoracic wall or internal chest organs, there remains the danger of lung oedema (fluid lung) in cases of severe blunt thoracic trauma. Also the blunt heart trauma without tissue damage may lead to a decrease of heart volume and consequently to death.

Injuries of the thoracic wall

Skin. Skin damage, half-formed or formed wounds, may give hints to identify the place of impact. Typical examples are the excoriations and contusions received by an impact against the steering wheel or steering wheel nave. Characteristic skin damage on the chest is caused by diagonal shoulder belts, three-point belts and four-point belts.

TABLE 2.1

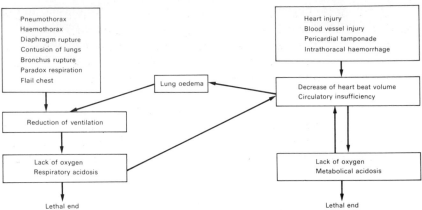

Pneumothorax	Heart injury
Haemothorax	Blood vessel injury
Diaphragm rupture	Pericardial tamponade
Contusion of lungs	Intrathoracal haemorrhage
Bronchus rupture	
Paradox respiration	
Flail chest	

Lung oedema

Decrease of heart beat volume
Circulatory insufficiency

Reduction of ventilation

Lack of oxygen
Respiratory acidosis

Lack of oxygen
Metabolical acidosis

Lethal end   Lethal end

Rib cage

Clinical aspects. Rib fractures with accompanying injuries of chest organs are especially complicated by the occurrence of a pneumothorax and/or a haemothorax. Pneumothorax means free air between the chest wall and the lung (pleural activity). The pulmonary lobe, which is normally kept unfolded by the negative pressure in the intrapleural space (and thus follows the respiratory movements of the chest wall) collapses on condition of pneumothorax owing to its elasticity and therefore is no longer available for oxygen exchange (Fig. 2.1). The result is a vehement disorder of respiration with undersaturation of oxygen and increase of CO_2 in the blood.

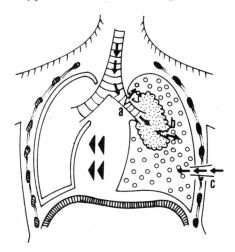

FIG. 2.1. Pneumothorax with mediastinal deviation: (a) rupture of bronchus; (b) laceration of lung; (c) open pleural laceration.

The air, which penetrates the intrapleural space in case of pneumothorax, comes either from a lung injury (caused for example by perforating rib fragments) or intrudes through an open injury of the chest wall (open pneumothorax). If the air is only entering the intrapleural space without any possibility to escape somewhere, this valve mechanism causes tension-pneumothorax with displacement of the mediastinum. In addition to the oxygen undersaturation, the venous bloodflow back to the heart is then also disturbed.

In the case of haemothorax, blood flows into the intrapleural space (Fig. 2.2). Like pneumothorax this leads to pulmonary collapse by means of displacement with the same consequences to respiration and circulation. Sources of haemorrhage may be injuries of lung and pleura caused by rib fractures and

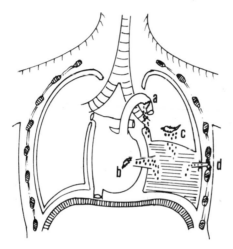

FIG. 2.2. Haemothorax: (a) rupture of aorta; (b) heart laceration; (c) lung laceration; (d) rupture of intercostal artery.

most of all injuries of intercostal arteries and veins. Often, however, the sources of haemorrhage and causes of distinct haemothorax are injuries of other chest organs such as the heart and the aorta or vena cava which will be discussed later. If several ribs are completely broken off at their connection, these fragments may be pulled inward during inspiration (due to intrathoracical decrease of pressure) and pushed outward during expiration (due to intrathoracical increase of pressure). This disorder of respiration is called 'paradoxical respiration'. In normal breathing inspiration leads to an expansion and expiration to contraction of the thorax (Fig. 2.3). Hence these paradoxical movements tend to disorder the ventilation of the lungs severely, leading to respiratory insufficiency.

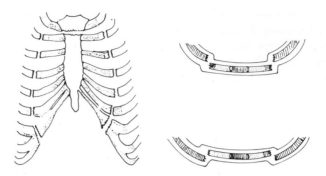

FIG. 2.3.   Paradoxical respiration with unstable thorax wall.

Biomechanical aspects of rib fractures. Rib fractures may be biomechanically classified as follows [3]: (1) direct bending fractures, if the impact and
fracture sites are identical (Fig. 2.4(a)); (2) indirect bending fractures if the site
of fracture does not coincide with the impact site (Figs. 2.4(b) and 2.4(c));
(3) torsion fractures.

Bending fractures occur at the site of maximal bending stress and generally
run at right angles to the longitudinal axis of the rib. Only the first rib is an
exception. The deformation of ribs is governed by their mode of fixation to the
vertebral column which allows small movements in the dorsal and ventral
direction of the paravertebral parts of the spine. Therefore in a frontal thorax
trauma either indirect bending fractures occur in the paravertebral region or
the processus transversus of the appertaining vertebrae break.

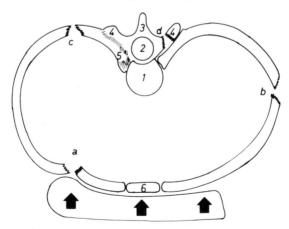

FIG. 2.4.   Biomechanical aspects of rib fracture: (1) vertebral column; (2) spinal cord;
(3) processus spinosus; (4) processus transversus; (5) costal–vertebral joint;
(6) sternum; (a) direct deflection fracture; (b) indirect deflection fracture; (c) indirect
deflection fracture; (d) indirect fracture of processus transversus.

Torsion fractures occur if an axial rotation of one rib segment takes place against another. In the ventral and lateral parts of the thorax they lead to long splintered fractures but they are seldom found in the upper four rib pairs.

Sternal fractures. Sternal fractures are found as direct or indirect bending fractures caused by ventro-dorsal impact. As a rule they are transverse fractures. In the place of impact or in close vicinity to it the broken ends can yawn widely so that there is danger of a rupture of the arteria thoracica interna which runs at the internal side of the frontal thorax wall in the peripheral region of the sternum. Furthermore, transfixing injuries of the heart and aorta may occur. In accidents where safety belts have been used we often find transverse sternal fractures corresponding to the direction and level of the bearing surface area of the belt on the thorax.

Thoracic vertebral column. Injuries of the vertebral column occur at the connection of the vertebral bodies to their osseous parts as well as to the vertebral discs, ligaments, and the spinal cord (Fig. 2.5). Compared with the cervical column, the thoracic vertebral column is relatively more protected against hyper-extension and hyper-flexion movements. Its movements are limited by the elastic rib system of the thorax which is disposed in front of the longitudinal axis of the vertebral column.

Processus transversus. Fractures of the processus transversus can be attributed to indirect force influences. The anatomical connection between

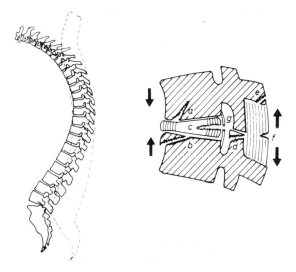

FIG. 2.5.   Deflection of vertebral column at frontal impact: left-hand side, survey; right-hand side, detail. (a) Infraction of cover plate; (b) edge fracture (tear drop fracture); (c) rupture of intervertebral disc; (d) fracture of joint processus; (e) fracture of processus spinosus; (f) rupture of interspinal ligament; (g) rupture of ligamentum longitudinale.

ribs and processus transversus allows a transmission of violence from the thorax wall through the elastic rib skeleton to the processus transversus in a ventro-dorsal trauma. As a rule such indirect bending fractures open at the ventral side (d in Fig. 2.4).

Vertebral bodies. If the protecting thoracic skeleton breaks, we find end-plate and compression fractures and even the complete shattering of vertebra bodies owing to segmental bending and axial compression. The vertebral bodies, which consist of only a small osseous compacta cover and mostly of spongiosa, are more prone to injuries than vertebral archs and joint processes which are considerably more resistant. Under conditions of hyper-extension (dorso-ventral impact) transverse laceration fractures of vertebral bodies occur owing to axial tractive forces.

Intervertebral discs and ligamentous system. Intervertebral discs consist of fibre cartilage and a gelatine nucleus. They connect an adjacent pair of vertebral bodies and by means of their elasticity are the basis of flexibility between two vertebral bodies. If the elasticity of these intervertebral discs is exceeded, one or more disc ruptures occur, the localization of which gives a clue to the direction of the flexion movement of the spine and thus to the impact direction.

Disc ruptures are found in the posterior part of the discus if the trauma is ventro-dorsal. The danger of disc injuries lies in a prolapse of the nucleus pulposus and formation of clasps in healing. Both may lead to compression symptoms of the spinal cord and nerve roots and thus cause paralysis.

After an appropriate impact the ligamentous system, which also confines movements of the spine, shows haemorrhages or even lacerations which may be connected with osseous and disc injuries and/or with luxations.

Spinal cord. There is no clear correlation concerning the severity of injuries of the vertebral column and of the spinal cord. However, in the case of axial bending or counter-displacement of segments haemorrhages, contusions, and even ruptures have to be expected. They may be part of the cause of irreversible paralysis.

Intra-thoracical injuries

Heart injuries

Clinical aspects. Heart injuries can range from temporary irregularities of the heart impulse to the most severe insufficiency of blood ejection with fatal heart failure without any morphological alterations of tissue. In blunt thorax trauma these conditions may occasionally lead to sudden death due to commotio cordis. Disorders of blood circulation occasionally develop in less serious cases [4, 5], leading to the morphological pattern of heart muscle necrosis and thus to myocardial infarction (traumatic myocardial infarction).

Heart injuries associated with blunt thorax trauma may involve damage to all parts of the heart, e.g. auricle, ventricle, auricle septum, ventricle septum, etc. (Fig. 2.6). In the case of complete rupture of one cavity, blood flows into the pericardium during the phase of heart contraction and thus causes heart tamponade provided the pericardium itself is not ruptured. The pericardium is thus filled with blood and under arterial blood pressure (in the case of left ventricular rupture) hinders the diastolic relaxation of the ventricles and thus the conveyance of venous blood to the heart. The consequences are storage of blood in front of the heart and in the lungs as well as reduced blood supply. In a very short time, these conditions cause fatal cardiovascular failure with symptoms of acute suffocation. If at the same time ruptures of the heart wall and pericardium exist, there will be a bleeding to death in the pleural cavity.

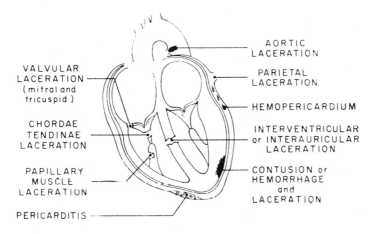

FIG. 2.6.   Types of heart injury [6].

Traumatic myocardial infarction may also be caused by coronary injuries [7, 8]. Its pathogenetic mechanism is generally myocardial infarction due to arterio-sclerosis and coronary sclerosis. This syndrome develops as a result of either complete rupture of a coronary artery or contusion and strain of a coronary artery with intima lesions. In the former case there is a decrease of pressure in that coronary region distal to the vessel rupture, leading to myocardial infarction. In the latter case the coronary circulation is not affected by the trauma, but a thrombosis in the damaged intima develops after some time leading to stenosis and coronary occlusion.

The laceration and rupture of the inferior vena cava at the site where it empties into the right atrium is still the most frequently considered heart injury [9, 10]. An injury at this site may cause thrombosis on the lacerated inner wall of the vein with the resulting danger of pulmonary embolism. In the case of complete rupture there could even be the danger of bleeding to death.

Biomechanical aspects. Dynamic conditions which lead to heart injuries are heart contusions due to sagittal–lateral and anterior–posterior thorax compressions [3] without associated fractures of the thoracic skeleton [11]. On the other hand, sternum and rib fractures are occasionally the cause of transfixing injuries of the heart owing to the intrusion of the fracture ends [12].

The hydrodynamic pressure effect of the blood volume in the contused heart can result in bursting ruptures, especially ruptures of the papillary muscle and valves. The cause of coronary rupture (as well as that of any transverse vessel rupture) is a longitudinal tension due to transverse deformation of the heart (see Fig. 2.7) which exceeds the breaking strength [8].

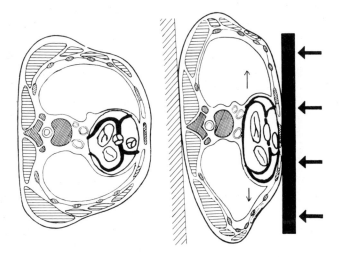

FIG. 2.7.   Biomechanics of coronary rupture.

Lesions of the large intra-thoracical vessels

All injuries to the aorta, the vena cava superior, the pulmonary arteries and veins, and the arteria and vena subclavia belong to this group.

Rupture of the aorta

(1) *Symptoms.* Ruptures (of circumferential type) are generally found as wall rupture of the ramus descendeus below the insertion of the ligamentum arteriosum botalli. The next most common rupture is the localization just above the aorta valves at the ramus ascendeus. The rupture may occur as complete rupture of the entire vessel wall or involve only parts of the circumference. It is also possible that the rupture involves only some parts of the wall layers so that the vessel does not leak and fatal haemorrhage into the mediastinum and pleural cavity does not occur immediately. Depending on the localization and extent of the rupture, the consequences are in general as follows: (i) profuse thoracic cavity haemorrhage with intra-thoracical loss of blood up to 3 l (haemothorax); (ii) pericardial

tamponade if the laceration of the ramus ascendeus is still within the pericardium; (iii) hemato mediastinum if the pleura mediastinalis resists; (iv) traumatic aortic aneurysm in the case of incomplete rupture of the wall develops a tearing haemorrhage in the muscular layer of the wall owing to arterial blood pressure. Later on this may lead to a complete rupture at another site and thus may be the cause of a sudden death. While the first two consequences lead to immediate death at the place of accident, hemato mediastinum and incomplete wall ruptures afford some time of survival.

(2) *Biomechanics of aorta rupture*. Aorta ruptures are attributed to the following causes: (i) increase of intra-vasal blood pressure caused by compression [13]; (ii) extension of the aortic arch with strain of fibrous tissue at the insertion of the ligamentum arteriosum botalli [14]; (iii) longitudinal tension of the vessel due to caudally directed motion of the abdominal organs; (iv) vibration of different amplitudes of the abdominal and chest organs; (v) ventrally directed motion of the aorta away from the vertebral column during frontal deceleration and hyper-flexion of the aortic arch [15]. Owing to compression of the chest with dislocation of the intra-thoracic soft parts (shovel effect) on to the aorta arch [3] (Fig. 2.8) the aortic arch is deflected so that at the point where it passes into the aorta descendeus the axial forces exceed the breaking strength of the aortic wall. Additional attributive factors could be the internal pressure of the aorta, shock waves in the blood column, and contusion due to impressed parts of the thorax. The tensile strength varies in different parts of the aorta. In the isthmus region, where the ruptures are often found, the tensile strength is about two-thirds to three-quarters of the average [16]. The critical tolerance limit for the thorax is $77g$ [15].

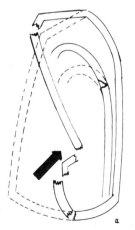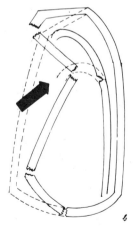

FIG. 2.8. Biomechanics of aortic rupture: (a) frontal impact at lower sternal part; (b) frontal impact at mid-sternal part.

Injuries of the vena cava inferior and superior

(1) *Symptoms.* In blunt thorax trauma the most frequent vessel injury is the lesion of the inferior vena cava. However, most of the injuries are not serious, as for example adventitial haemorrhages. Intima ruptures, which run at right angles to the longitudinal axis of the vessel, are more common than circumferential ruptures. Lesions of the superior vena cava are rare and mostly occur in combination with other serious injuries of the chest organs after a severe thorax trauma.

(2) *Biomechanics.* Injuries of the superior and inferior vena cava are caused by strain. Lesion to the inferior vena cava is caused by the deformation of the anterior chest wall with dislocation of the heart to the upper left (trauma to the lower half of the thorax). Dislocation of the heart to the lower left (trauma to the upper half of the thorax) results in lesions of the vena cava superior [3]. The vena cava superior is less susceptible to extensive damage than the superior vena cava because the cranial anchorage of the heart is much better protected and more resistant owing to short and less deformable ribs, little space for displacement of chest organs, and more resistant anchorage of the heart by the aorta.

Other intra-thoracical vessels. Injuries of the pulmonary arteries and veins are very seldom and only appear in most severe thorax traumas, generally in connection with a serious shattering of the osseous thorax wall and impression of the fractured ends.

Injuries of lungs and bronchi

Lungs. In blunt thorax trauma the most frequent lung injuries are caused by the perforating ends of fractured ribs resulting in lacerations of the pleura and pulmonary tissue. Subpleural haemorrhages occur below the fracture. Pulmonary parenchymal lacerations within the organ are a result of pulmonary contusions or compressions whereby one or both pulmonary lobes are contused between two approaching thoracic sections.

The effect of pushing aside the pulmonary lobes during deformation of the chest is enhanced by the sudden increase of pulmonary air pressure due to spasmodic closing of the epiglottis during the trauma [17]. Also the sudden increase of blood pressure in the pulmonary circulation of blood, caused by heart contusion, is supposed to be a reason for traumatic pulmonary haemorrhages.

The clinical consequences of the above-mentioned injuries are pulmonary parenchymal haemorrhages combined with reduction of the respiratory surface and increased danger of infection in the contused tissue mixed with blood as well as haemothorax and pneumothorax.

Bronchi. Bronchial injuries are due to shear of the trachea and bronchi between the impressed sternum and the spinal column, tension due to a lateral pushing aside of the lungs, and finally direct injuries caused by transfixation of sharp fracture ends of sternum and ribs [18]. Depending on their location and extent the consequences of bronchial and tracheal injuries, in case of complete ruptures, are heavy ventilation disorders in the supply district, pneumothorax, mediastinal emphysema, skin emphysema, and haemothorax (Fig. 2.2).

Injuries of the diaphragm. Diaphragm injuries are caused by intra-thoracical or intra-abdominal variations of pressure caused by compression of the mentioned body regions. Strain of the diaphragm is also caused by deformation of its attached margin owing to thorax deformation. Finally there are injuries of the diaphragm caused by penetration of the bone fragments in the case of thoracic skeleton fractures.

The clinical consequences are haemorrhages in thoracical and abdominal regions and the so-called traumatic diaphragmatical herniae, as well as displacements of abdominal viscera into the thoracic cavity caused by the ruptured diaphragm (Figs. 2.9 and 2.10) resulting in a reduced pulmonary ventilation with corresponding respiratory disorders. Rupture of the diaphragm is usually connected with other serious injuries.

Thorax injuries of restrained occupants

In restrained occupants blunt thorax trauma is caused by the contact area of the belt, resulting in the so-called 'seat-belt syndrome'. High deceleration values generally cause excoriations and subcutaneous haematomas on the skin parts under the bearing surface area of the belt. The position of rib fractures partly corresponds to the contact area of the belt. The belt which deforms the chest wall causes direct bending fractures which occur as fragmenting lesions

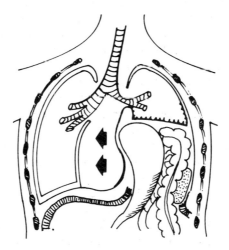

FIG. 2.9. Rupture of the diaphragm with mediastinal dislocation and displacement of the abdominal organs.

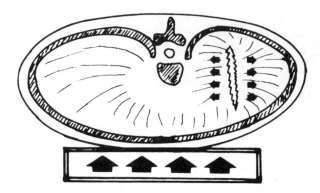

F IG. 2.10.   Biomechanics of diaphragm rupture at the frontal impact.

at high deceleration values, especially in cases involving elder individuals. Restrained occupants are also threatened by injuries of the thoracic viscera due to compression of the chest. In cadaver tests, an impact velocity of 50 km h^{-1} against a rigid wall caused laceration of the lung coat at the hilius as well as ruptures of the inferior vena cava pericardium, the outer skin of the heart, and the atrium. In one case a rupture of the aorta was found.

Injuries of the vertebral column correspond to those of hyperflexion, with ventral fractures of the vertebral sides, predominantly dorsal disc ruptures, and ruptures of the interspinal ligaments. Fractures of the processus transversus often occur in the upper part of the thoracic vertebral column.

2.4. Traumatic. rupture of the aorta

Intrathoracic injuries may result in damage to the heart, the lungs (causing compression or paradoxical respiration), the great vessels, the tracheo-bronchial tree, the oesophagus or the diaphragm (Table 2.1). While the mechanisms of associated injuries are easy to deduce, those resulting from non-penetrating thoracic injuries (which indeed occur very often) are not well understood and will be emphasized in this chapter.

Among the most serious of these is death caused by rupture of the aorta, generally followed by immediate exsanguination. The lesion is generally associated with accidents characterized by violent and sudden deceleration, and often results from impact of the driver's chest with the steering wheel (Fig. 2.11). Greendyke [19] reports on a statistical study of a sample of approximately 1253 deaths occurring in Monroe County, N. Y. over a four-year period. Of the cases examined, 420 were accidental, 267 or more occurring as a result of automobile accidents. It was demonstrated that traumatic rupture of the aorta had occurred in 16 per cent of these auto-accident victims. One can only infer that a comparable percentage of all automobile fatalities

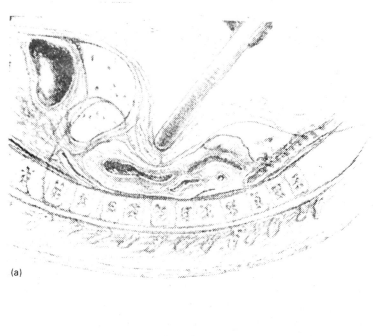

(a)

(b)

FIG. 2.11. (a) Before and (b) after impact of the driver's chest against steering wheel. (After Bright *et al.* [11])

occurring each year is due to aortic rupture. The accuracy of the extrapolation is admittedly uncertain on the grounds of this study alone, but much additional evidence is available in other studies to support the conclusion that aortic rupture occurs with remarkable frequency.

TABLE 2.2

Cause	Isolated aortic rupture	Combined with cardiac injury	Total
Automobile	114	42	156
Airplane	12	31	43
Vehicle versus pedestrian	9	7	16
Fall (long distance)	12	12	24
Motorcycle	3	1	4
Automobile versus train	0	2	2
Compression by heavy object	4	0	4
Buried by dirt	1	0	1
Direct blow by object	5	0	5
Unknown	11	9	20
Totals	171	104	275

TABLE 2.3

Nature of injury and site	Total injuries	Type of trauma		Blunt force injuries
		Penetrating injuries		
		Established	Probable	
Rupture of the ascending aorta within 2.5 cm. of aortic valve	65	0	9	56
Rupture of the remainder of the ascending aorta	16	0	2	14
Rupture through the arch of the aorta	10	0	2	8
Avulsion or transection of the vessels arising from the aortic arch	18	1	1	16
Rupture of the aorta distal to the origin of the left subclavian artery	51	1	5	45
Rupture of the descending aorta	37	2	6	29
Rupture of the aorta at the level of the diaphragm	5	0	1	4
Avulsion of intercostal or vertebral artery branches at origin	5	0	1	4
Multiple ruptures of the aorta	17	1	3	13
Fragmentation of the aorta	19	1	10	8
Site of rupture not clearly defined	17	3	8	6
Rupture of the pulmonary artery	21	2	5	14
Ruptures of the superior and inferior vena cava	30	3	3	24
Rupture of pulmonary veins	16	0	5	11
Totals	327	14	61	252

Patel *et al.* [20] have found that 'the lesion is far more common, particularly associated with automobile accidents, than is usually recognized . . . Frequently no evidence of chest injury is present, but deceleration stresses appear to be common in all cases.' (See Table 2.2.) In some instances, with early diagnosis by aortography partial or even complete rupture of the aorta (see Figs. 2.12–2.14) can be repaired surgically with a Dacron prosthetic graft. The most common site of aortic ruptures (see Tables 2.3 and 2.4), of which 61–83

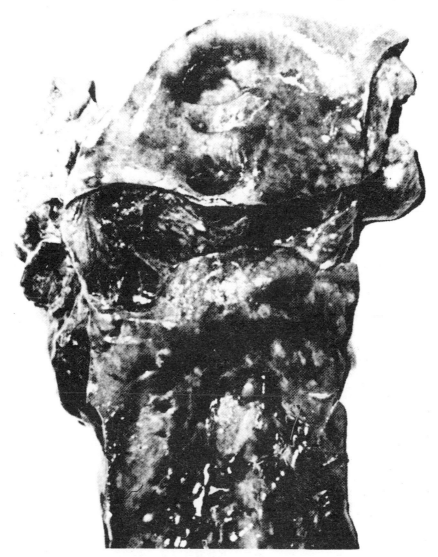

FIG. 2.12. Rupture of the aorta.

TABLE 2.4

Sites of rupture	DOA[a]	Periods of survival													Cured	Total survived
		>30 min <1 hr	1h <6 hr	6hr <24 hr	1 day <2 days	2 days <4 days	5 days <8 days	8 days <15 days	22 days	50 days	76 days	4 months	10 months	2–4 years		
Ascending aorta	60	1	1	—	1	—	—	—	—	1	—	—	—	—	—	4
Arch	18	—	1	1	—	—	1	—	—	—	—	—	—	—	—	3
Isthmus	101	2	1	3	1	3	3	3	—	—	1	1	1	3	2[b]	24
Thoracic aorta	31	—	1	—	—	—	1	2	—	—	—	—	—	—	—	4
Abdominal aorta	10	—	1	—	—	—	1	—	1	—	—	—	—	—	—	3
Multiple sites	17	—	—	—	—	—	—	—	—	—	—	—	—	—	—	0
Total	237[c]	3	5	4	2	3	6	5	1	1	1	1	1	3	2	38[c]

a Dead on arrival–immediate death or lived <30 min.
b Cured by surgery or living
c 273 necropsy cases; 2 patients living

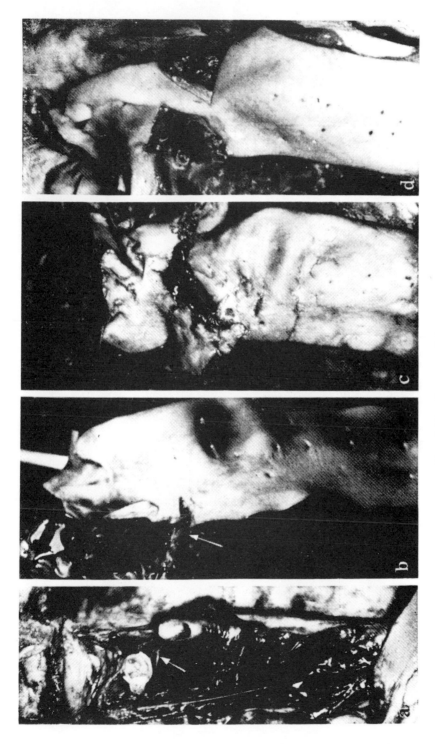

Fig. 2.13. Rupture of the aorta.

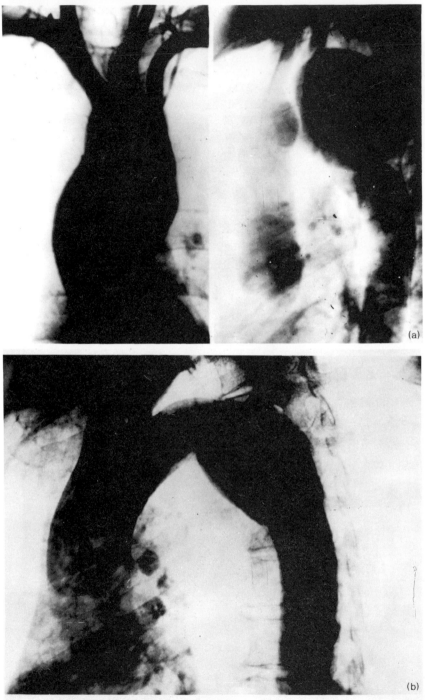

Fig. 2.14. (a) Aneurysm of the aorta; (b) aneurysm of the aorta.

per cent are estimated to be caused by automobile accidents (see refs. 21–23), is at or near the point of insertion of the ligamentum arteriosum. At this point (Fig. 2.15) a fibrous ligament connects the aortic isthmus to the pulmonary artery. Relative motion between the two arteries may cause tensile stresses in the connecting ligament which could cause weakening or perforation of the aortic wall. In most cases the rupture is propagated circumferentially.

Failure of the aorta wall may also be due to increased internal pressure due to wave propagation in the blood vessel. The forces developed in the walls due to structural deformations and in the fluid may combine to produce the observed failure.

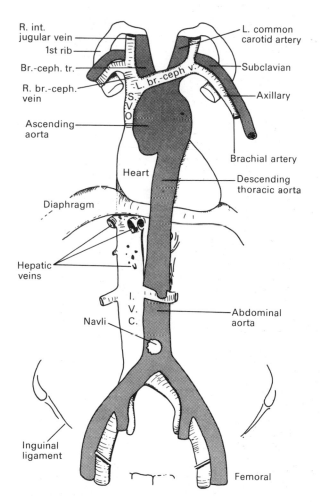

FIG. 2.15. Diagram of the great arteries and veins.

Forces applied to a passenger during collision

Full-scale automobile collisions using anthropometric dummies (Fig. 2.16) have been performed for a number of years by Severy and co-workers at the University of California at Los Angeles. In a series of head-on collisions Severy, Mathewson, and Siegel [24] measured the deceleration and belt load for restrained occupants.

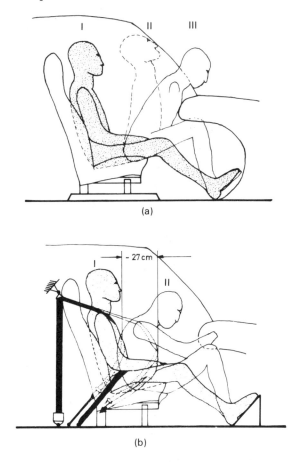

(a)

(b)

FIG. 2.16.   Kinematics of (a) unrestrained and (b) belted occupants in frontal barrier collision.

It is very possible (and generally true) that the deceleration of the chest, shoulders, and hips of the occupant exceeds that of the automobile owing to slackness in the restraint system. Curves of deceleration as a function of time were given in ref. 24 and are reproduced in Fig. 2.17. Belt loads of the order of 7500 lbf may be developed during a 20 mile h^{-1} impact. Vehicle decelerations

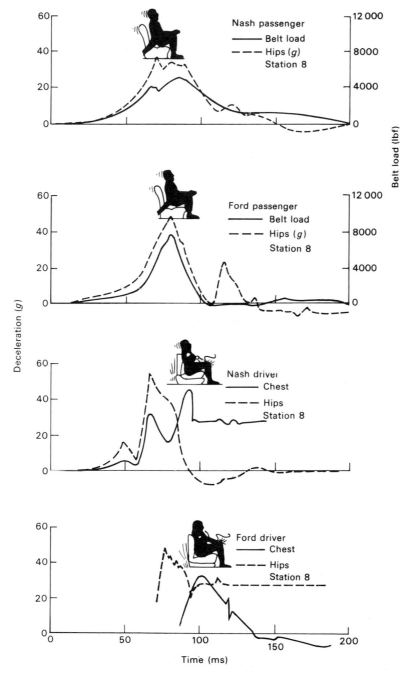

FIG. 2.17. Occupant deceleration and belt load recordings. (After Severy *et al.* [24].)

of 22 g may produce decelerations of the chest of 45 g and of the hips of 54 g in the case of unrestrained passengers. The major fluctuations in deceleration occur within a 50 ms period although the complete events last about three times longer. At impacts greater than 50 mile h^{-1} belt loads may increase to 12 000 lbf.

By the same reasoning deceleration of the internal organs of the cardiovascular system will be significantly higher than that of the chest since these organs will decelerate over a shorter distance. The forces tending to rupture blood vessels are then clearly significant.

Lundevall [25] has classified the mechanical forces acting on the aortic wall into three types: (a) longitudinal, radial, and torsional forces due to motion of the adjacent thoracic organs; (b) wave propagation in the aortic wall due to sudden stretching of the aorta; (c) intra-aortic pressure held. Lundevall has indicated that geometric distortion of the aorta in the sagittal plane during deceleration may cause local longitudinal stretching of the aortic wall particularly at two points of fixation, the base of the heart and the isthmus.

Traumatic rupture of the aorta is always observed to be transverse, normally indicating failure axially in tension. In addition it is expected that radial expansion of the aorta will be limited by the restoring forces of the surrounding tissue. Under normal physiological conditions the excursions in vessel radius are limited to 10 per cent, while the axial displacements of the wall are virtually negligible owing to the tethering effect of the many intercostal arteries branching from the aorta (see ref. 26).

Hypotheses concerning aortic rupture

Wilson and Roome (1933) (cited by McDonald and Campbell [27]) suggest that injury is most likely at the start of diastole when the aorta is fully distended with blood, whereas Warfield (1933) (quoted by McDonald and Campbell [27]) contends that the important factor is the condition of full inspiration when the heart is caught between the sternum and the fully inflated lungs. McDonald and Campbell support the theory of Rindfleisch concerning the importance of fixation of the aorta in localizing the rupture area. McKnight et al. (1964) (quoted by Patel et al. [20]) state that it is the aortic arch that is mobile, the descending aorta being attached to the left anterolateral border of the vertebral column. They agree with Rindfleisch only in that points of fixation in general determine the zone of concentrated stresses but differ on the ways in which fixation is produced.

Strassman [28] presented the findings of 72 cases of aortic rupture examined in New York City during the period 1936–42 from a total of approximately 7000 autopsies. The age distribution was from under 10 to over 80 years of age. In all cases of spontaneous rupture the tear started within the media. However, in all cases of traumatic rupture the aorta was completely severed, making it difficult to determine in which layer the tear had begun, although in some survivors the adventitia remained intact. It would appear that in some cases of traumatic rupture the tear begins in the intima. Strassman found that the majority of traumatic ruptures occurred at the isthmus where the aorta is narrower and relatively fixed by the ligamentum arteriosum. He concluded that the most likely explanation of aortic rupture in a few cases in which there was no evidence of bone fracture or external injury was the sudden increase in intra-arterial pressure caused by the blunt compression of the aorta against the vertebral column.

Gable and Townsend [29] found in a study of 459 cases of fatal injuries of the cardiovascular system resulting from accelerative forces that the aorta and its branches were the most commonly involved of all the major blood vessels. In fact the aorta is far more susceptible to injury than are the other major blood vessels. They also affirmed the importance of accelerative force in causing cardiovascular injury but were not able to choose definitively between that and the hypothesis of Rindfleisch (rupture due to intravascular pressure). However, they noted the remarkable concurrence of findings among many researchers that the highest incidence of injury was just distal to the left subclavian artery, i.e. in the region of the ductus, for cases of isolated aortic ruptures. When heart lesions were present the incidence of rupture above the aortic valve was double that of the ductus region.

Taylor [30] has demonstrated on pigs that during acceleration an emptying of the distal half of the thoracic aorta occurs with engorgement of the upper half and of the arch. This retrograde flow may increase the pressure sufficiently in the region of the arch to cause rupture here. Lundevall [22] has suggested that the geometric distortion of the aorta in the sagittal plane during deceleration will cause local longitudinal stretching of the aortic wall at the two points of fixation, i.e. at the base of the heart and at the isthmus.

In traffic accidents the contact of the lower portion of the steering wheel with the abdomen may push the abdominal viscera upwards. The left lung may press upward against the aortic arch causing increased bending or even kinking of the arch. A transverse rupture results near the isthmus. In addition the cervical vessels may stretch during head motion, exerting longitudinal forces on the aortic wall. Internal pressure in the aortic arch may rise suddenly owing to both compression of the heart and forward inertial motion of the blood already in the arch.

It is evident from this brief survey of the literature that plausible examples can be found for all three of the rupture mechanisms suggested by Lundevall. However, none of these has ever been analysed and quantified scientifically in a manner which permits one to progress beyond the realm of endless hypothesis. It is one of our specific intents in this chapter to explore in detail the consequences of wave propagation on the development of very high local stresses and strains in the aorta following sharp impact to the thorax. It will be demonstrated that rupture conditions may ensue under the loadings described earlier in this section and that these indeed depend critically upon the highly non-linear viscoelastic material properties of the arterial wall.

In the succeeding sections of this chapter we shall discuss methods of evaluating the physical properties of the greater blood vessels and subsequently present three mathematical analyses, of successively increasing complexity and generality, of blood flow in the aorta, culminating in the solution for impulsive flow in non-uniform viscoelastic vessels subjected to a wide range of initial and boundary conditions.

2.5. Physical and mechanical properties of the aorta

Model studies of impulsive flow in the aorta

Glass models scaled from X-ray records of the human aorta have been used to estimate the influence of vessel curvature. A syringe at the proximal end of the model aorta expelled fluid at time intervals of 40 ms by a falling weight. The pressure field in the fluid was recorded by semiconductor strain gauge pressure transducers with a frequency response in the range of hundreds of kilohertz. It is expected on the basis of a simplified one-dimensional unsteady flow analysis that high-pressure regions occur at the inlet to the model aorta as well as in the region distal to the arch. These are indeed the most prevalent sites of aortic rupture.

The use of a transparent model of the aorta makes possible the utilization of dye-injection techniques for direct observation of regions of turbulence and flow separation. Both these phenomena are known to contribute to an increase in fluid pressure. The velocity profile across a section of the aorta was measured by stationing the pressure transducers at various distances from the centre of the axis. Secondary flows developed in the curved tube causing a circulation of the fluid from the walls to the centre and back again. The result, confirmed by these experiments, is a flattening of the velocity profile from a parabolic change into a blunt slug-like flow profile.

Although only straight tapered vessels have been considered in the mathematical flow analyses given later in this chapter, the effects of vessel curvature described above may none the less be of particular importance in high impulsive flows and should eventually be incorporated into a more generalized, albeit complex, formulation.

Aortic stress–strain and strain rates under dynamic loading

During an automobile collision the driver's chest may impact upon the steering wheel. Forces are then transmitted through the thorax to the heart and aorta. Direct measurements of the incident and transmitted stresses permit the calculated behaviour of the aorta to be related to the intensity of the impact resulting in traumatic rupture.

   The mechanical constitutive relation of aortic wall material is one of the fundamental physical properties of the aorta which it is necessary to know for the understanding of the functioning of the cardiovascular system and for the prediction of its mechanical reactions to external impact. Specifically, this section concerns a determination of the stress–strain relationship of the human aorta through dynamic *in vitro* tensile tests at a number of different loading rates (or strain rates). In view of the nonlinearity and viscoelasticity of the aortic wall material, it is evident that the aorta will react quite differently under impact than under static loading. Following the usual classification of loading ranges, as summarized in Fig. 2.18, we have concentrated our experiments in the dynamic and impact loading ranges, especially around the strain rates of $10-100\,s^{-1}$. It is expected that under ordinary impact conditions, the aorta will be stretched at these loading rates.

   Most data presently available on the mechanical properties of aortae pertain only to static loading conditions or extremely low strain rates (see Tables 2.5 and 2.6). However, they are useful in determining the elasticity parameters

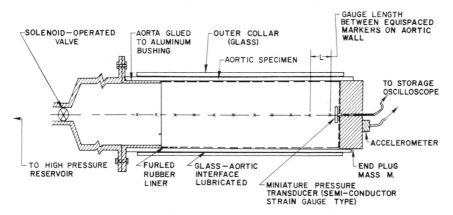

FIG. 2.18.   Installation of an aortic specimen.

TABLE 2.5

Mean values for static incremental modulus of elasticity*

$(dyn\,cm^{-2} \times 10^6 \pm SD\ of\ mean)$

Pressure (mm Hg)	Thoracic aorta	Abdominal aorta	Femoral artery	Carotid artery
40	1.2 ± 0.1(6)	1.6 ± 0.4(4)	1.2 ± 0.2(6)	1.0 ± 0.2(7)
100	4.3 ± 0.4(12)	8.9 ± 3.5(8)	6.9 ± 1.0(9)	6.4 ± 1.0(12)
160	9.9 ± 0.5(6)	12.4 ± 2.2(4)	12.1 ± 2.4(6)	12.2 ± 2.7(6)
220	18.1 ± 2.8(5)	18.0 ± 5.5(3)	20.4 ± 4.4(6)	12.2 ± 1.5(7)

* The number of measurements is shown in parentheses. Some additional specimens were studied at 100 mm Hg before making dynamic measurements, and these have been included.

TABLE 2.6
Tensile strength of aorta from 10 bodies

No.	Body age (years)	Sex	Horiz. length of arch (cm)	Int. diam. of aorta (cm) p. asc.	Isthm.	p. ascend. 1	2	3	4	5	isthmus 6	7	8	p. desc. 9	10	lower long.	p. desc. transv.
1	35	m	7.5	1.8	1.45	530	490 (mean)	730 538	400	490	420 (mean)	360 365	190	450 (mean)	320 385	410	895
2	36	m	8	2.05	1.55	375	735	825 598	455	505	215 (mean)	345 370 353	315				
3	39	f	6	1.4	1.1	845	690 (mean)	720 764	800		(mean)		335	400	320		
4	47	m	8	2.45	1.8	525	460	555 533	590	555	365 (mean)	430 404	265	430 (mean)	455 443 240		
5	61	m	11	2.8	2.25	565	440 (mean)	540 515			485 235 307 (mean)		200				
6	64	f	10	2.55	2.05	690	(mean)	545 575	490	355	250 (mean)	365 320	310	315 (mean)	365 340 650		
7	68	m	8	1.75	1.25	675	440 (mean)	490 493	565	535	430 (mean)	390 420	320				
8	79	m	10	2.45	1.9		495 (mean)	575 533	560		310 (mean)	380 325	285				
9	80	f	9	1.9	1.65	590	350 (mean)	335 475	625		640 (mean)	440 465	315	495 (mean)	380 438	285	595
10	84	m	11.5	2.8	2.45	375	300 (mean)	620 454	520	280	190 (mean)	125 204	220	400 (mean)	245 323		
Means	8.9			1.85	1.54			547.8				350.8			397.4		

appearing in the dynamic stress–strain–strain rate relation and are therefore presented here.

In order to determine the constitutive relations for aortic tissue under dynamic loading let us consider the simplest possible stress state, i.e. uniaxial tension. Lawton [31] and numerous other authors have indicated that a Poisson ratio of $\frac{1}{2}$ corresponds to most biological tissues. This implies deformation at constant volume.

Consider the longitudinal extension of a cylindrical segment of aorta of uniform cross-section with gauge length L, volume V_0, radius R, and wall thickness h where $h \ll R$. During an equivolumic extension of the aortic segment the wall thickness h will vary as

$$h = \frac{V_0}{2\pi\,RL} \tag{2.1}$$

For a cylindrical shell closed at one end and constrained radially by a well-lubricated outer collar as in Fig. 2.19, a force balance on the end plug yields

$$m\ddot{X}(t) = p(t)\pi R^2 - 2\pi Rh\sigma \tag{2.2}$$

where σ is the aortic wall stress at the end and $X(t)$ the axial co-ordinate measuring the absolute travel of the end mass. The wall stress σ is then obtained in terms of measurable quantities as

$$\sigma(t) = \frac{-\,m\ddot{X}(t) + \pi R^2(t)p(t)}{2\pi R(t)\,h(t)} \tag{2.3}$$

where the pressure $p(t)$ is measured from the oscilloscope tracing of the signal from a miniature pressure transducer, $\ddot{X}(t)$ is obtainable from an accelerometer attached rigidly to the end mass, $R(t)$ is estimated from high-speed cine-films, and the wall thickness h is calculated from equation (1) on the basis of a Poisson ratio of $\frac{1}{2}$.

A gauge length L is defined (Fig. 2.18) on the distal end of the aortic specimen by the segment between two wall markers spaced nominally 0.5 in apart. The strain ε is determined as a function of the extension of this gauge length L, at which locations the flow pressure and the acceleration of the end plug are measured:

$$\varepsilon(t) = \int_{L(t_0)}^{L(t)} \frac{\mathrm{d}L}{L} = \log\frac{L(t)}{L(t_0)}$$

and the strain rate $\varepsilon(t)$ can be calculated as the corresponding time derivative from cine-film data.

The aortic specimens, which were approximately 5 in long, were pared of their surrounding fat and bathed in a physiological saline solution to preserve their properties. The ends were dried and glued (Eastman 910 glue) to short segments of hollow tapered aluminum cylinders which were subsequently connected into the high-pressure surge line as shown in Fig. 2.18.

The distal end of the aortic segment was closed by an end plug and radial expansion of the aorta during impulsive injection of water was limited by a closely fitting outer glass collar. A furled rubber liner placed inside the aorta prevented leaks through the numerous fine holes in the aortic wall formed as a result of excision of the branching intercostal vessels. This liner always remained furled and therefore did not contribute to the strength of the aorta during dynamic loading. The extension of the aortic gauge length was filmed with a high-speed camera (400 frames s^{-1}).

A series of tests using fresh pig aortae was conducted in the strain-rate range $\dot{\varepsilon} = 1.0–3.5\,\mathrm{s}^{-1}$ by varying the chamber pressure. In addition a set of 'static' tests ($\dot{\varepsilon} = 0.05\,\mathrm{s}^{-1}$) was carried out to evaluate the creep characteristics of

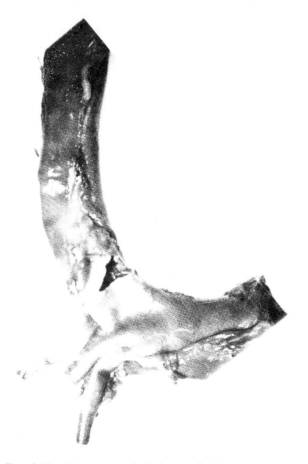

FIG. 2.19.   Rupture at the isthmus of a human aorta.

aortic tissue and to afford a direct comparison with the effects of dynamic loading.

In addition preliminary qualitative tests were performed on specimens of human aortae to assess the effects of fixation of the aorta in localizing the rupture area. Sudden motion of the aorta away from the point of insertion of the ligamentum arteriosum just distal to the aortic arch (Fig. 2.19) resulted in a localized rupture which rapidly propagated circumferentially in the presence of a moderate axial tension.

The stress–strain curves for dynamic loading for a limited range of strain rate are shown in Fig. 2.20. In the range $\dot{\varepsilon} = 1$–$3.5\,\text{s}^{-1}$ all the data appear to lie above the static stress–strain curve which indicates a dependence of the constitutive relationships on strain rate in this range. The slope of the stress–strain curve is generally steeper for dynamic than for static loading, indicating that the aortic tissue acts as a stiffer material when strained rapidly.

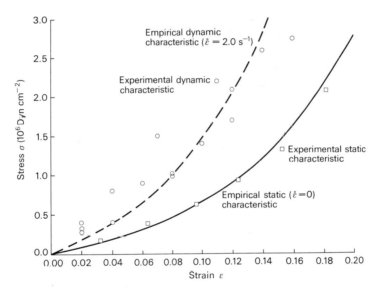

FIG. 2.20 Effect of the strain rate of stress–strain relations. Experimental results: □ 'static' test ($\dot{\varepsilon} = 0.005\,\text{s}^{-1}$); O dynamic test ($\dot{\varepsilon} = 1$–$3.5\,\text{s}^{-1}$). Empirical stress–strain law $\sigma = (0.28 + 0.18\,\dot{\varepsilon})(e^{12\varepsilon} - 1)$ dyn cm^{-2}: solid curve, 'static', $\dot{\varepsilon} = 0$; broken curve, dynamic, $\dot{\varepsilon} = 2.0\,\text{s}^{-1}$.

The stress–strain relation is represented here in the form

$$\sigma = 10^6 (0.28 + 0.18)(e^{12\varepsilon} - 1)$$

for $\dot{\varepsilon} < 3.5\,\text{s}^{-1}$. The aortic material clearly displays greater stiffness under dynamic loading, typical of other biological tissues such as muscles and visceral tissue. The results indicate a definite sensitivity to strain rate within the range $\dot{\varepsilon} = 3.5\,s^{-1}$. This fact may have particular relevance to automobile safety, where the proper design of passenger restraint systems must ensure limitation of stress levels in the aorta if traumatic rupture is to be avoided. The departure from a uniaxial stress state is expected to be small since radial excursions were limited to a maximum of 10 per cent; the non-linearity of the stress–strain relation renders it difficult to estimate this contribution more precisely.

The initial modulus of elasticity of pig aortae is estimated to be 4–10 $\times 10^6\,\mathrm{dyn\,cm^{-2}}$ at strain rates up to $3.5\,s^{-1}$. These values lie within the range of elastic moduli (1.2–$18.1 \times 10^6\,\mathrm{dyn\,cm^{-2}}$) estimated by Bergel [32] for dog aorta and compare favourably with the measurements of Peterson [26] of 3 $\times 10^6\,\mathrm{dyn\,cm^{-2}}$ for elastin and $30 \times 10^6\,\mathrm{dyn\,cm^{-2}}$ for collagen. The experiments indicate rupture stresses of the order of $30 \times 10^6\,\mathrm{dyn\,cm^{-2}}$ under static loading and 40–$45 \times 10^6\,\mathrm{dyn\,cm^{-2}}$ for dynamic loading at strain rates up to $3.5\,s^{-1}$.

As the strain rate increases rupture occurs at a higher stress and at lower strain. It has been noted that the aorta is 'tethered' to inhibit relative longitudinal motion. This would indicate that the longitudinal strain is limited and that tensile rupture may then occur only at higher rates of strain. Such information is of prime importance in the design of shock-absorbing material in the passenger compartment of vehicles. Passenger decelerations must be limited to the extent that the rupture stress will not be reached at the corresponding rate of strain. Restraint systems possessing some slackness may eventually produce a 'second impact' upon distension, with passenger deceleration exceeding that of the vehicle.

The present results, when supplemented with additional experimental data using human aortic specimens, will serve to establish working curves relating tensile stress to strain in the aortic wall at various rates of strain. These strain rates may be related to tolerance levels of deceleration, thus providing meaningful specifications for safe vehicle design.

In the §2.6 an alternative approach is described for assessing the dynamic material response of large blood vessels. It is based upon the establishment of a non-physiological steady state flow regime which is intended, through strong wave propagation, to achieve even higher levels of strain rate. Knowledge of the response of aortic tissue in the non-physiological impact range (Fig. 2.21) is crucial to the prediction of traumatic aortic rupture. It is indeed clear from Fig. 2.20 that the slope of the stress–strain curve will steepen markedly beyond the physiological range of $\dot{\varepsilon}$.

Strain-rate-dependent viscoelastic properties by shock experiments

A more general method [33] for determining the strain-rate dependent viscoelastic properties of the great vessels is based on *in vitro* measurements of the axial distributions of either cross-sectional area or intraluminal pressure

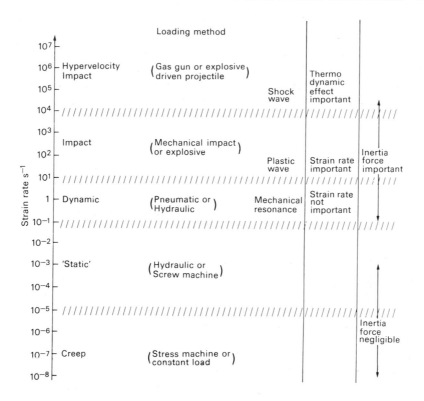

Fɪɢ. 2.21.   Dynamic loading regions.

upon passage of a shock wave into the vessel. The corresponding analysis uses a quasi-one-dimensional model of steady state wave propagation in a non-linear viscoelastic fluid-filled tube in which axial and bending stresses have been neglected. An approximate analytic expression is derived for the thickness of the shock transition in terms of the viscoelastic parameters, the shock velocity, and the pressure difference across the shock. It is predicted on theoretical grounds that the entry length and shock thickness will be impractically large for fluids such as blood or water having densities of order unity. In order to overcome this difficulty mercury is proposed as the filling fluid. Calculations for mercury show that in this unique case meaningful experiments are feasible using human aortic segments of lengths of the order of 15 cm. Material properties thus determined correspond to high rates of strain and may find direct application in studies of human tolerance in strong decelerative force fields.

It is known that the large vessels such as the aorta, although elastic at low rates of extension, behave as viscoelastic materials of ever-increasing stiffness as the rate of strain to which they are subjected increases [34, 35]. Meaningful calculations of the stresses to which these vessels are exposed depend most critically on these dynamic stress–strain–strain rate relationships. Results of acceleration experiments performed over a number of years by Stapp [36] on young military volunteers indicate the surprisingly wide range of forces and accelerations to which the subject may be exposed during falls, vehicular collisions, vibrations, impacts, and ejection from aircraft in addition to the less well-documented spectrum of forces encountered during space travel. For example rates of decelerations exceeding $1000\,g\,s^{-1}$ at levels higher than $25\,g$ become progressively more difficult to withstand, even in the backward-facing position, or with pelvic and shoulder girdle restraints eliciting transient musculo-skeletal or visceral pain changes in vision and perturbations of the cardiac and respiratory functions.

It is clear that theoretical and experimental investigations of accelerative phenomena will require accurate estimates of the dynamic material properties of biological tissues in these wide force regimes. At present such data are rare or non-existent. For the great vessels material properties have been determined from 'jerk-type' experiments in which the vessel is abruptly elongated from its position at rest. Large strains and strain rates occur only after significant stretching of the complete vessel segment, often to levels which are physiologically unrealistic.

In this section a more practical approach is suggested, based upon the passage of a shock front through a liquid-filled vessel. The specimen is strained significantly and rapidly only in the immediate vicinity of the propagating front. In the following a mathematical model is formulated for shock propagation in a fluid-filled distensible tube and resulting changes in intraluminal pressure and cross-sectional area are related directly, in closed analytical form, to the material properties of the tube wall. Shock experiments are then proposed which would yield the viscoelastic properties of arteries and veins on the basis of this analysis.

Shock waves in the greater vessels. The formation and propagation of shock waves in fluid-filled distensible tubes have recently been considered by Rudinger [37], Beam [38], Lambert [39] and Anliker, Rockwell, and Ogden [40]. The formation of a shock wave is a result of the steepening of a finite-amplitude continuous wave in an elastic tube having a non-linear pressure–area relation. When no mechanism is provided in the mathematical model to smooth rapid changes in the flow the wave steepens to the level at which a discontinuity eventually forms.

In the proposed experiments the tube segment must be sufficiently long for formation of a shock front. This length is related to the mechanical properties of the tube, the fluid density, and the initial rise time of the wave (see ref. 37, equation (12)). The dispersive effects of fluid viscosity and wall viscoelasticity in fact 'smear' the mathematical shock discontinuity over a finite thickness, the precise form of which then depends most critically upon the viscoelastic properties of the tube wall rather than of the fluid which is therefore taken as inviscid in this study. It will subsequently be shown, however, that the fluid density plays an important role in determining the entrance length and shock thickness and that these can be shortened significantly for heavy fluids.

In order to simplify the mathematical problem it is assumed that the shock wave propagates in a tube with uniform mechanical properties and cross-sectional area and that a steady state is achieved, i.e. the flow field becomes stationary for an observer moving with the shock. Under these assumptions the shock structure is governed by an ordinary differential equation which allows one to study the basic properties of the shock transition fairly easily.

Mathematical model. We adopt here a quasi-one-dimensional model for the flow of an inviscid incompressible fluid in a viscoelastic tube based upon the assumptions that (a) the wavelength is long compared with the tube diameter and (b) the tube is constrained from longitudinal motions. Under these assumptions the governing equations of motion can be expressed as

$$A_t + (Au)_x = 0 \qquad \text{(continuity)} \qquad (2.4)$$

$$u_t + uu_x + \frac{p_x}{\rho} = 0. \qquad \text{(conservation of momentum)} \qquad (2.5)$$

Here t denotes the time, x the distance along the axis of the tube measured in the flow direction, u the fluid velocity (averaged over the cross-section), A the cross-sectional area, p the pressure, and ρ the constant density of the fluid.

To these equations must be added a relation between the pressure and the cross-sectional area. For a viscoelastic tube the pressure depends on the cross-sectional area and its time rate of change $\eta = \partial A/\partial t$. The strain-rate-dependent pressure–area relation can be written as

$$p = f(A) + g(A, \eta) \qquad \eta = \frac{\partial A}{\partial t} \qquad (2.6)$$

where $f(A)$ corresponds to static loading and $g(A, \eta)$ accounts for the viscoelastic properties of the tube wall. g is a monotonically increasing function of η with

$$g(A, 0) = 0 \qquad (2.7)$$

For the analysis of the stationary shock wave it is convenient to write the differential equations in a co-ordinate system moving with the shock. Assuming that the shock wave is moving to the right with a constant velocity U we employ the transformation

$$Z = x - Ut$$
$$\tau = t \qquad (2.8)$$

The differential equations (2.4) and (2.5) become

$$A_\tau + (vA)_z = 0 \tag{2.9}$$

$$v_\tau + \left(\frac{v^2}{2} + \frac{p}{\rho}\right)_z = 0 \tag{2.10}$$

where

$$v = u - U \tag{2.11}$$

The functional relation (2.6) remains unchanged but now η is given by

$$\eta = \frac{\partial A}{\partial \tau} - U\frac{\partial A}{\partial z} \tag{2.12}$$

The steady state equations are obtained by putting $\partial/\partial\tau = 0$ in the differential equations (2.9) and (2.10). They then become ordinary differential equations which after one integration with respect to z give

$$vA = m \tag{2.13}$$

$$\frac{v^2}{2} + \frac{p}{\rho} = F \tag{2.14}$$

The pressure–area relation becomes

$$p = f(A) + g\left(A, -U\frac{dA}{dz}\right) \tag{2.15}$$

The shock jump conditions can now be obtained by considering the flow conditions far ahead of the shock and far behind the shock. Denoting these states by subscripts 1 and 2 respectively one has

$$v_1 A_1 = v_2 A_2 = m \tag{2.16}$$

$$\frac{v_1^2}{2} + \frac{p_1}{\rho} = \frac{v_2^2}{2} + \frac{p_2}{\rho} = F \tag{2.17}$$

where $p_{1,2} = f(A_{1,2})$ since dA/dz vanishes at both ends and equation (2.7) is noted. The relations for the shock velocity U and the change in the fluid velocity $(u_2 - u_1)$ follow from equations (2.16), (2.17), and (2.11):

$$\rho(U - u_1)^2 = \frac{f(A_2) - f(A_1)}{\frac{1}{2}\{1 - (A_1/A_2)^2\}} \tag{2.18}$$

$$\frac{u_2 - u_1}{U - u_1} = 1 - \frac{A_1}{A_2} \tag{2.19}$$

The set of equations (2.13)–(2.15) can be reduced to a single equation by eliminating p and v from equation (2.14) using equations (2.13) and (2.15). The resulting differential equation for A is

$$g\left(A, -U\frac{\mathrm{d}A}{\mathrm{d}z}\right) = \rho\left(F - \frac{1}{2}\frac{m^2}{A^2}\right) - f(A) \tag{2.20}$$

If the constants m and F from equations (2.13) and (2.14) are expressed in terms of the flow variables in states 1 and 2, equation (2.20) becomes

$$g\left(A, -U\frac{\mathrm{d}A}{\mathrm{d}z}\right) = f(A) - f(A_1) - \frac{f(A_2) - f(A_1)}{1 - A_1^2/A_2^2}(1 - A_1^2/A^2) \tag{2.21}$$

Once the functions g and f are known, equation (2.21) constitutes a first-order differential equation for $A(z)$ which may be integrated by simple quadrature. Alternatively, if $A(z)$ is measured equation (2.21) provides a relation between f and g. If in addition the function f is known, then the function g can be determined directly from equation (2.21). It is in fact this approach which forms the basis of the method proposed here.

The function $f(A)$ corresponding to the zero strain rate response can be measured accurately by static loading experiments. The function will have the following property. For physically realistic behaviour the cross-sectional area A increases with pressure p, i.e. $\mathrm{d}f/\mathrm{d}A > 0$. If in addition a continuous wave travelling in the tube is to steepen into a shock front one also requires that [38]

$$\frac{\mathrm{d}^2\bar{f}}{\mathrm{d}\alpha^2} > 0 \tag{2.22}$$

where

$$\alpha \equiv \tfrac{1}{2}(1 - A_0^2/A^2) \quad \text{and} \quad \bar{f}(\alpha) \equiv \bar{f}(A) \tag{2.23}$$

and A_0 is a reference cross-sectional area.

This condition can be shown to be equivalent to the fact that the velocity of sound c in a flexible tube generally increases with intraluminal pressure, or more rigorously $(1 + \rho c(\mathrm{d}c/\mathrm{d}p)) > 0$, as has been verified by the experiments of King [41] which were carried out over the physiological range of pressure. In fact, recent measurements [35] have corroborated the property equation (2.22) at higher intraluminal pressures.

Particular solutions. The general relation (2.21) can be solved for specific functional forms of g and f. From the experiments of Collins and Hu [35] for the aorta, the resulting stress–strain relation would indicate that the function $g(A, \eta)$ is linear in η so that it can be written as

$$g(A, \eta) = G(A)\eta \tag{2.24}$$

where $G(A) > 0$ since the stiffness increases with increasing strain rate. The differential equation for the shock structure now becomes

$$dz = U(A_1, A_2)\frac{G(A)\,dA}{H(A)} \tag{2.25}$$

where $H(A)$ is given by the right-hand side of equation (2.21), i.e.

$$H(A) = f(A) - f(A_1) - \frac{f(A_2) - f(A_1)}{1 - A_1^2/A_2^2}\left(1 - \frac{A_1^2}{A^2}\right) \tag{2.26}$$

and $f(A)$ is known from static measurements.

For the discussion of the solution of (2.25) it is advantageous to replace A by the Beam parameter α (equation (2.23)). Equation (2.25) becomes

$$dz = U(\alpha_1, \alpha_2)\,\overline{G}(\alpha)\frac{d\alpha}{\xi(\alpha)} \tag{2.27}$$

where

$$\xi(\alpha) = \overline{f}(\alpha) - \overline{f}(\alpha_1) - \frac{\overline{f}(\alpha_2) - \overline{f}(\alpha_1)}{\alpha_2 - \alpha_1}(\alpha - \alpha_1) \tag{2.28}$$

$$\overline{G}(\alpha) = G\{A(\alpha)\}\frac{dA(\alpha)}{d\alpha} \tag{2.29}$$

and U is given from equation (2.18) as

$$\rho\{U(\alpha_1, \alpha_2) - u_1\}^2 = \left(\frac{A_0}{A_1}\right)^2 \frac{\overline{f}(\alpha_2) - \overline{f}(\alpha_1)}{\alpha_2 - \alpha_1}. \tag{2.30}$$

It is clear that equation (2.27) possesses singularities at the ends of the interval $\alpha_1 \le \alpha \le \alpha_2$ and that a special treatment is required there before numerical integration can be attempted.

Since ξ vanishes at $\alpha = \alpha_{1,2}$ and also $\xi''(\alpha) > 0$ in $\alpha_1 \le \alpha \le \alpha_2$ it follows that

$$\xi(\alpha) \ne 0 \quad \text{for } \alpha_1 < \alpha < \alpha_2.$$

Near the ends ($\alpha = \alpha_{1,2}$) $\xi(\alpha)$ approaches zero as $B_{1,2}(\alpha - \alpha_{1,2})$ where from equation (2.28)

$$B_{1,2} = \frac{d\overline{f}}{d\alpha}\bigg|_{(\alpha_{1,2})} - \frac{\overline{f}(\alpha_2) - \overline{f}(\alpha_1)}{\alpha_2 - \alpha_1}. \tag{2.31}$$

The solution of (2.27) near α_1 and α_2 is therefore

$$z + \text{const.} = \frac{U\overline{G}(\alpha_{1,2})}{B_{1,2}}\ln|\alpha - \alpha_{1,2}|. \tag{2.32}$$

for $\alpha \to \alpha_{1,2}$.

Since the function $\xi < 0$ for all α and behaves as $B_{1,2}(\alpha - \alpha_{1,2})$ near

THORACIC IMPACT: INJURY MECHANISMS 75

$\alpha = \alpha_{1,2}$, it follows that $B_1 < 0$ and $B_2 > 0$. Then $z \to \infty$ when $\alpha \to \alpha_1$ and $z \to -\infty$ when $\alpha \to \alpha_2$ which implies that the mathematical shock transition extends from $-\infty$ to $+\infty$. However, it will be shown that the principal variations in the flow field occur over a finite extent and this may be considered for all practical purposes as the shock 'thickness'.

The solution of (2.27) for general functions $\overline{G}(\alpha)$ and $\overline{f}(\alpha)$ can be obtained by quadrature with the help of the limiting forms (2.31). In a similar way limiting forms can be derived for the solution of the general case embodied in equation (2.21).

Solutions of equation (2.27) can now be derived for polynomial representations of $\overline{G}$ and $\overline{f}$. The simplest closed form solution is obtained for

$$\overline{f}(\alpha) = B(\alpha + \Gamma\alpha^2) \tag{2.33}$$

$$\overline{G}(\alpha) = \text{const.} = \mu. \tag{2.34}$$

The parameters B and Γ in the function $\overline{f}$ can be selected so that $\overline{f}(\alpha)$ closely follows the experimentally measured curves. However, the experimental results are usually expressed in the form of a relation between the speed of sound c and the pressure p. From the Moens–Korteweg relation for the speed of sound in an elastic tube

$$c^2 = \frac{A}{\rho}\frac{\mathrm{d}f}{\mathrm{d}A} \tag{2.35}$$

which for the form of $\overline{f}(\alpha)$ given by equation (2.33) yields

$$c_0 \equiv c|_{p=p_0} = \sqrt{\left(\frac{B}{\rho}\right)} \tag{2.36}$$

and

$$\rho c_0 \frac{\mathrm{d}c}{\mathrm{d}p}\bigg|_{p=p_0} = \Gamma - 1 \tag{2.37}$$

Relations (2.36) and (2.37) serve to determine B and Γ in the function $\overline{f}(\alpha)$ of relation (2.33) in terms of the measured c_0 and $\mathrm{d}c/\mathrm{d}p$ at $p = p_0$.

The differential equation (2.27) becomes

$$\mathrm{d}z = \frac{\mu U}{\rho c_0^2} \frac{\mathrm{d}\alpha}{\Gamma(\alpha - \alpha_1)(\alpha - \alpha_2)}$$

whose solution is

$$\ln\left(\frac{\alpha_2 - \alpha}{\alpha - \alpha_1}\right) = (\alpha_2 - \alpha_1)\frac{\Gamma z}{\nu U} \tag{2.38}$$

where ν is the ratio of the viscoelastic to the elastic moduli at the reference area

A_0 and is given by

$$v = \mu/\rho c_0^2. \tag{2.39}$$

An alternate form of equation (2.38) which is analogous to the classical relation for the structure of weak gas dynamical shocks (see ref. 42) is

$$\alpha - \alpha_1 = \alpha^* \left(1 - \tanh \frac{\alpha^* \Gamma}{vU} z \right) \tag{2.40}$$

where

$$\alpha^* \equiv \frac{\alpha_2 - \alpha_1}{2}.$$

With this solution for the axial distribution of cross-sectional area through the shock transition one can now proceed to an estimate of the length of the shock transition, i.e. the shock thickness over which most of the pressure jump occurs.

A general relation between p and A for the stationary flow is given by equations (2.20) and (2.6):

$$p - p_1 = \rho \left(F - \frac{1}{2} \frac{m^2}{A^2} \right).$$

On inserting the expressions for F and m from equations (2.16) and (2.17) this becomes

$$p - p_1 = \rho (U - u_1)^2 \frac{1}{2} \left(1 - \frac{A_1^2}{A^2} \right)$$

or

$$p - p_1 = \rho (U - u_1)^2 \left(\frac{A_1}{A_0} \right)^2 (\alpha - \alpha_1) \tag{2.41}$$

which shows that the pressure drop is proportional to $\alpha - \alpha_1$. This fact is used to define the shock-wave 'thickness' δ as

$$\delta \equiv \frac{p_2 - p_1}{|dp/dz|_{max}} = \frac{\alpha_2 - \alpha_1}{|d\alpha/dz|_{max}} \tag{2.42}$$

Using the solution (2.40) one obtains the shock thickness

$$\delta = \frac{4vU}{(\alpha_2 - \alpha_1)\Gamma} \tag{2.43}$$

The thickness δ corresponds to the distance over which three-quarters of the total pressure jump takes place.

Proposed experimental determination of viscoelastic properties.

The function $g(A, \eta)$ characterizing the viscoelastic material behaviour in the pressure–area relation (2.15) can be determined by a series of careful experiments in which either (a) the wall profile is measured giving $A = A(x)$ or (b) the axial distribution of pressure $p = p(x)$ is recorded for a given position of the shock front.

A choice between (a) and (b) might be made purely on the basis of experimental accuracy and facility. For alternative (a) one uses the measured distributions $A = A(z)$ over a range of shock strengths. Numerical differentiation carried out as accurately as possible then allows one to evaluate both arguments of $g(A, -U\,\mathrm{d}A/\mathrm{d}z)$ in eqn (2.21) for a range of values of $\mathrm{d}A/\mathrm{d}z$ so that the surface g can be constructed. The right-hand side of eqn (2.21) is evaluated using the function f determined from the usual static tensile tests. If desired, relation (2.15) can then be easily converted to an expression for stress in terms of strain and strain rate.

Method (b) requires differentiation of the measured pressure distribution $p(z)$. The pressure and its derivative are calculated in terms of cross-sectional area through relations (2.41) and (2.23), and the procedure for method (a) outlined above is followed directly.

The tube diameter, and hence the cross-sectional area, can be measured during the passage of the uniform shock front by recording by X-ray cinematography the motion of two fine metallic wires disposed axially and diametrically opposite one another along the outer wall surfaces of the tube. If the fluid which fills the tube is radio-opaque, X-ray cinematography will record the instantaneous diameter directly. The pressure can be measured by miniature strain-gauge-type semiconductor transducers mounted in a catheter and oriented along the axis of the lumen.

Wave-propagation experiments in water-filled rubber and polyvinyl chloride tubes 10 m long have been performed by Lorentz and Zeller [43] for weak shocks which did not produce the high rates of strain of interest here.

Errors in determination of the dynamic material properties in this manner can emanate only from inaccuracies in the measurement of diameter (or pressure) and in numerical differentiation of these quantities which should be carried out carefully. It is noted that azimuthal isotropy is implicit in this quasi one-dimensional formulation.

The tube must be sufficiently long for formation of a shock front and for its development into an equilibrium profile. The distances that the wave must travel in order to fulfil this requirement can be estimated as follows:

(i) The distance S_1 for shock formation (ref. 37, equation 12) is

$$S_1 = \frac{\rho c_0^3}{(\mathrm{d}p/\mathrm{d}t)|_{x=x_0}(1 + \rho c_0(\mathrm{d}c/\mathrm{d}p))} \tag{2.44}$$

where

$$1 + \rho c_0 \frac{dc}{dp} = \Gamma = 3$$

from the measurements of Collins and Hu [35].

(ii) The distance S_2 for the formed shock to reach an equilibrium state has been calculated using the general numerical solution of Kivity and Collins [44] for a blood-filled viscoelastic tube of uniform cross-section with $B = 0.08$ s, $\rho = 1.05$ g cm^{-3} and $c_0 = 300$ cm s^{-1}.

The boundary condition at the proximal end was chosen to simulate the flow pattern when the tube is connected to a high pressure reservoir, i.e. the total pressure $\rho_0 = p + \frac{1}{2}\rho u^2$ is prescribed as a function of time. The tube is assumed to be sufficiently long to ensure that no signals are transmitted back from the distal end. This calculation predicts a shock thickness of approximately 22 cm and an entrance length of 45–55 cm (depending on the rise time of the proximal pressure P which varied in the calculations between 2 and 20 ms) for a reservoir pressure of 0.66 atm.

For these values of the parameters eqn (2.44) predicts a distance S_1 for shock formation of less than 1 cm which could be disregarded in estimating the required minimum lengths of test specimens. However, the distance S_2 (entrance length and shock thickness) is clearly too long for testing aortic segments 10–15 cm long, corresponding for instance to the human aortic arch, along which the material properties do not vary significantly. Fortunately it is possible to reduce the entrance length and shock thickness to acceptable levels by increasing the density of the fluid. This follows from an exact scaling law, which is directly derivable from the equations of motion (2.4)–(2.6) and relation (2.35), which shows that distances vary inversely as the square root of the fluid density. In fact mercury with a density of 13.6 g cm^{-3} is ideal for this purpose. The calculated entrance length and shock thickness then become 12–15 cm and 6 cm respectively. A shock thickness of 6 cm is quite compatible with a test segment of length 15 cm. Shorter shock thicknesses would begin to violate the basic assumption of a large ratio of wavelength to diameter. A furled plastic liner introduced into the aorta (cf. Collins and Hu [35]) will prevent direct contact between the mercury and the intimal layer of the vessel wall.

2.6. Non-linear wave propagation in the aorta

With this preliminary knowledge of the dynamic material properties of the aorta, we can now embark upon an analysis of the stress and strain fields developed in terms of the basic rupture mechanism considered here, i.e. wave propagation following impact to the thorax.

Tethered isotropic wall

We adopt a quasi-one-dimensional model [45, 37, 39] for the flow of an incompressible fluid in a distensible tube based on the assumptions that (a) the wavelength is long compared with the tube diameter and (b) the tube is constrained from longitudinal motion. The wall material is assumed to be viscoelastic but fluid viscoelasticity is neglected since its effect on the flow in the large arteries is insignificant. Under these conditions the governing equations of motion are

$$\frac{\partial A}{\partial t} + \frac{\partial}{\partial x}(Au) = 0 \text{ (continuity)} \tag{2.45}$$

$$\frac{\partial u}{\partial t} + u\frac{\partial u}{\partial x} + \frac{1}{\rho}\frac{\partial p}{\partial x} = G \text{ (momentum).} \tag{2.46}$$

Here t denotes the time and x the distance along the axis of the tube (see Fig. 2.22), u the fluid velocity (averaged over the cross-section), A the cross-sectional area, p the pressure, ρ the constant density of the fluid and G the body force resulting from acceleration or a gravitational field.

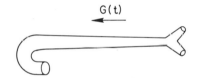

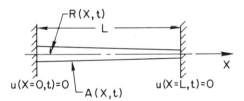

FIG. 2.22. Geometric idealization of the aorta in an accelerative force field.

To these equations one must add a relation between pressure and cross-sectional area. If the pressure p is assumed to be a function of A alone, the resulting differential equations are hyperbolic, admitting discontinuous solutions (shock waves). In the present study for a viscoelastic tube the pressure depends also on the time rate of change of the cross-sectional area:

$\eta = \partial A/\partial t$. The pressure–area relation is expressed as

$$p = f(A) + g(A, \eta) \qquad \eta = \frac{\partial A}{\partial t} \tag{2.47}$$

where the function $f(A)$ corresponds to static loading and the term $g(A, \eta)$ accounts for the viscoelastic properties of the wall where g is a monotonically increasing function of η with $g(A, 0) = 0$.

With the relation (2.47), the differential equations take on a parabolic character and mathematical discontinuities are not admissible. However, shock-like transitions may develop in the form of very steep (but continuous) wavefronts.

To complete the mathematical description one must specify initial and boundary conditions. For the problem considered in this study the initial conditions are

$$u(x, 0) = 0 \tag{2.48}$$

$$A(X, 0) = A_0(x) \tag{2.49}$$

$$\eta(x, 0) = \frac{\partial A}{\partial t}(x, 0) = 0 \tag{2.50}$$

The pressure relation employed here (see equation (2.69)) implies for these initial conditions that $p(x, 0) = 0$, referred to some base pressure. The boundary conditions are

$$u(0, t) = 0$$
$$u(L, t) = 0 \tag{2.51}$$

The physical reasons underlying the choice of these conditions are described later.

Once $A(x, t)$ is determined in the above problem, the axial strain and strain rate follow directly. The inside tube radius $R(x, t) = \{A(x, t)/\pi\}^{1/2}$ whence

$$\varepsilon_a = \left\{ 1 + \left(\frac{dR}{dx} \right)^2 \right\}^{1/2} - 1 \tag{2.52}$$

and

$$\dot{\varepsilon}_a = \partial \varepsilon_a / \partial t$$

The initial value of the axial strain has been neglected in the above expression since it is negligibly small. In fact the initial radius of the tapered tube is given as $R = R_i \exp(-0.0225 x)$ from which one can estimate $\varepsilon_a(t = 0) = 2.5 \times 10^{-4}$, a value too small to appear in the results.

Numerical solution

Equation (2.46) can be written as

$$\frac{\partial u}{\partial t} + \frac{\partial}{\partial x} \left(\frac{u^2}{2} + \frac{p}{\rho} + \Phi \right) = 0 \tag{2.53}$$

where

$$\frac{\partial \Phi}{\partial x} = -G. \tag{2.54}$$

The set of equations (2.45) and (2.53) is in conservation form and therefore the numerical solution can be based on the Lax–Wendroff difference scheme [46].

Without performing a rigorous stability analysis of the set of equations (2.45), (2.47), and (2.53) it appears sufficient to impose two stability criteria relating to both the hyperbolic and parabolic aspects of the equations. The time interval for integration is determined at each step as the more restrictive of the two and has been found satisfactory in all cases computed.

The hyperbolic stability criterion can be expressed as

$$\Delta t = \frac{\lambda \Delta x}{(|u| + c)_{max}} \tag{2.55}$$

where c is the speed of sound based on the pressure–area relation for static loading

$$c^2 = \frac{A}{\rho} \frac{\mathrm{d} f(A)}{\mathrm{d} A} \tag{2.56}$$

and λ is a constant less than or equal to 1. In our calculations λ was taken to be equal to $\frac{1}{2}$. This condition implies that the time step should not be larger than one-half the transit time for a wave propagating through a particular calculational cell.

The parabolic stability criterion is based upon the following derivation. First we define a new variable

$$F = A u \tag{2.57}$$

and write the differential equation in terms of A and F. The continuity equation becomes

$$A_t + F_x = 0. \tag{2.58}$$

The momentum equation is multiplied by A and is added to u times the continuity equation (2.58). This gives

$$F_t + \left(\frac{F^2}{A}\right)_x + c^2(A)A_x + \frac{A}{\rho}\left(\frac{\partial g}{\partial A} A_x + \frac{\partial g}{\partial \eta} \eta_x\right) = 0 \tag{2.59}$$

where the constitutive relation (2.47) has been used to eliminate p. The quantity η_x, can be expressed in terms of F using equation (2.58) as

$$\eta_x = \frac{\partial^2 A}{\partial x \, \partial t} = -\frac{\partial^2 F}{\partial x^2}$$

so that equation (2.59) becomes

$$F_t - \left(\frac{A}{\rho}\frac{\partial g}{\partial \eta}\right)F_{xx} + \left(\frac{F^2}{A}\right)_x + \left\{c^2(A) + \frac{A}{\rho}\frac{\partial g}{\partial A}\right\}A_x = 0. \qquad (2.60)$$

The form of this equation is obviously similar to that of the heat equation if one disregards the last two terms. The parabolic stability criterion is given by (ref. 46, p. 195) pp. 195, 200.

$$\frac{\sigma \Delta t}{\Delta x^2} < \frac{1}{2} \qquad (2.61)$$

where

$$\sigma = \frac{A}{\rho}\frac{\partial g(A, \eta)}{\partial \eta}. \qquad (2.62)$$

Alternatively it is possible to formulate the flow problem in terms of a single function ψ defined from equation (2.58) by

$$\psi_x = A, \qquad \psi_t = -F. \qquad (2.63)$$

The resulting third-order differential equation for ψ follows from equations (2.60) and (2.63). Although this form was not used here it may be of interest in devising other computational schemes.

The governing differential equations (2.45) and (2.53) are written in difference form according to the two-step Lax–Wendroff scheme which is known to possess useful properties of stability.

Step I:

$$A_{j+1/2}^{n+1/2} = \tfrac{1}{2}(A_{j+1}^n + A_j^n) - \tfrac{1}{2}\frac{\Delta t^n}{\Delta x}\{(Au)_{j+1}^n - (Au)_j^n\} \qquad (2.64)$$

$$u_{j+1/2}^{n+1/2} = \tfrac{1}{2}(u_{j+1}^n + u_j^n) - \tfrac{1}{2}\frac{\Delta t^n}{\Delta x}(q_{j+1}^n - q_j^n)$$

where

$$q = \left(\frac{u^2}{2}\right) + \frac{p}{\rho} + \Phi$$

and q_j^n is defined by

$$q_j^n = \left(\frac{u^2}{2}\right)_j^n + \Phi(x_j, t^n) + \frac{1}{\rho}\{f(A_j^n) + g(A_j^n, \eta_j^n)\}$$

and

$$\eta_j^n = \frac{A_j^n - A_j^{n-1}}{\Delta t}.$$

Step II:

$$A_j^{n+1} = A_j^n - \frac{\Delta t^n}{\Delta x}\{(Au)_{j+1/2}^{n+1/2} - (Au)_{j-1/2}^{n+1/2}\} \tag{2.65}$$

$$u_j^{n+1} = u_j^n - \frac{\Delta t^n}{\Delta x}(q_{j+1/2}^{n+1/2} - q_{j-1/2}^{n+1/2})$$

where

$$q_{j+1/2}^{n+1/2} = \left(\frac{u^2}{2}\right)_{j+1/2}^{n+1/2} + \Phi\left(\frac{x_j + x_{j+1}}{2}A^n + \frac{1}{2}\Delta t^n\right)$$

$$+ \frac{1}{\rho}\{f(A_{j+1/2}^{n+1/2}) + g(A_{j+1/2}^{n+1/2}, \eta_{j+1/2}^{n+1/2})\}$$

and

$$\eta_{j+1/2}^{n+1/2} = \frac{A_{j+1/2}^{n+1/2} - \frac{1}{2}(A_{j+1}^n + A_j^n)}{(\frac{1}{2}\Delta t^n)}.$$

The j subscripts indicate intervals in distance x along the tube while the n superscripts denote the time interval.

The initial and boundary conditions (2.48)–(2.51) are simply stated as

$$u_j^0 = 0$$
$$A_j^0 = A_0(x_j); \ x_j = (j-1)x$$
$$\eta_j^0 = 0 \tag{2.66}$$
$$u_1^n = 0$$
$$u_j^n = 0$$
$$u_{jm}^n = 0$$

where the subscripts $j = 1$ and $j = jm$ denote the ends $x = 0$ and $x = L$ respectively. At the end points Step II must be modified as follows:

$$u_1^{n+1} = 0$$
$$A_1^{n+1} = A_1^n - \frac{\Delta t^n}{\frac{1}{2}\Delta x}(Au)_{11/2}^{n+1/2} \tag{2.67}$$
$$u_{jm}^{n+1} = 0$$
$$A_{jm}^{n+1} = A_{jm}^n + \frac{\Delta t^n}{\frac{1}{2}\Delta x}(Au)_{jm-1/2}^{n+1/2}$$

The system is integrated in a stepwise marching procedure in time. The initial values of equations (2.66) are used to evaluate the right-hand sides of equation (2.64) from which one obtains the new values of A and u at the next half-time step for each position x. These in turn are used to evaluate the right-hand sides of (2.65) which then yield the values of A and u advanced to the full-time step. Relations (2.67) are used to determine u and A at the proximal and distal ends of the aorta at each time step. One then returns to the first step and continues in time.

Constitutive model for the aorta under dynamic loading

Generally soft tissue and muscles behave as non-linear viscous materials whose stiffness increases with increasing strain rate. For the dynamic loading rates associated with thoracic impact it appears that the influence of strain rate on the resulting tissue stresses is of paramount importance. Measurements by Collins and Hu [35] for fresh aortic tissue have resulted in a dynamic stress–strain relation, valid for strain rates up to 3.5 s^{-1} in the form

$$\sigma = 0.28 \times 10^6 (1 + 0.644 \dot{\varepsilon}) (e^{12\varepsilon} - 1) \text{dyn cm}^{-2} \qquad (2.68)$$

where the strain rate $\dot{\varepsilon}$ is measured in inverse seconds.

For an isotropic material one can deduce from the above expression a relation between the transmural pressure $p - p_0$ and the intraluminal cross-sectional area A.

The true strain is defined by

$$\varepsilon \equiv \ln \frac{L}{L_0} = \tfrac{1}{2} \ln \frac{A}{A_0}$$

where L is the extended length of an elemental segment and L_0 its original length. From the force balance on a thin cylindrical element of thickness h the transmural pressure is given by

$$p - p_0 = \frac{\sigma h}{R}$$

where R is the radius of curvature of the element. Using a Poisson ratio of $\tfrac{1}{2}$ (isovolumetric deformation) which is typical of soft biological tissue

$$hR = h_0 R_0$$

whence

$$p - p_0 = \Delta p_2 = f(A) + g(A, \dot{A}) \qquad (2.69)$$

where

$$f(A) = \frac{pc_0}{n} \left\{ \left(\frac{A}{A_0} \right)^{n-1} - \frac{A_0}{A} \right\} \qquad n = 6$$

and

$$g(A, \dot{A}) = B \left| f(A) \right| \frac{\dot{A}}{A}; \qquad B = 0.322 \text{ s}.$$

In the above expression for $f(A)$ the multiplying factor has been determined from the definition for the velocity of sound in a distensible tube:

$$c_0^2 = \left(\frac{A}{\rho} \frac{df}{dA} \right)_{A = A_0}$$

where ρ is the constant fluid density.

The absolute modulus appearing in the function g has been added to deal with the range of wall deformation in which $A/A_0 < 1$ so that the condition of increasing stiffness at increasing strain rate $(g > 0$ for $\dot{A} > 0)$ can be maintained even if the aorta should constrict below its reference cross-sectional area A. This 'reflection' of the stress–strain curve for $A/A_0 < 1$ appears reasonable in the absence of other experimental data. The variation of sound speed c_0 and reference cross-sectional area A_0 with distance x along the axis of the vessel can be given in the form

$$c_0(x) = c_i(1 + \beta_1 x) \tag{2.70}$$

$$A_0(x) = A_i \exp(-\beta_2 x) \tag{2.71}$$

where the parameters have been estimated by Anliker, Rockwell, and Ogden [40] as $c_i = 300 \text{ cm s}^{-1}$, $\beta_1 = 0.02 \text{ cm}$, $\beta_2 = 0.045 \text{ cm}^{-1}$ for dog aortae, and $A_i = \pi \text{ cm}^2$ in experiments of Hanson [47].

However, for large decelerative fields the strain rate may well exceed the maximum value of 3.5 s^{-1} obtained in the experiments of Collins and Hu [35] by 1.5 decades $(10^{1.5})$. To the knowledge of the authors no data are available in that range of strain rate which could serve to guide the extrapolation of the test results. One must then turn to other sources of data. It is known that an equivalence can be established between tensile tests at high strain rates and those at low temperatures.

When a viscoelastic material is stretched at a particular rate two competing processes act to determine the stress–time response: (a) the progressive deformation of internal bonds which tends to increase the stress and (b) a continuous relaxation which alleviates the stress. If the material is strained slowly (and at high temperatures) the relaxation process is very rapid relative to the time scale of the deformation and a continuous equilibrium is maintained. However, the departure from equilibrium becomes progressively greater as the test temperature is decreased or the extension rate is increased. At high strain rates or at low temperatures the material cannot accommodate rapidly enough to reach a continuous state of equilibrium. The effective stiffness of the material increases as one departs from the equilibrium state by either of these means. The availability of stress–strain data at low temperatures and low strain rates (e.g. ref. 48) tempts one to deduce from them the corresponding stress–strain response at high strain rates and normal ambient temperatures.

A method for doing this has been described by Macgregor and Fisher [49] and summarized in the book by McClintock and Argon [50]. The result may be stated simply: the stress corresponding to a test at an arbitrary strain rate $\dot{\varepsilon}$ and an arbitrary temperature T will be the same as for a test at the strain rate $\dot{\varepsilon}_0$ and temperature $T(1 - k \log \dot{\varepsilon} \dot{\varepsilon}_0)$, where k is a constant. That is, if the stress corresponding to strain rate $\dot{\varepsilon}$ and temperature T is plotted against the modified temperature abscissa $T(1 - k \log \dot{\varepsilon} \dot{\varepsilon}_0)$, the resultant curve will coincide with the stress versus temperature curve for strain rate $\dot{\varepsilon}_0$. Thus stress–strain data at low temperatures can be interpreted in the same way as data at high strain rates. However, even here, the published experimental data presently available are not grouped suitably to carry out such a conversion.

A more promising approach was found in the work of Smith [51] who measured stress-relaxation data at large strains for certain elastomers. He showed that the time and strain dependence are separable, so that a constant strain rate modulus $E(t)$ can be introduced in the form

$$E(t) = \frac{H(\varepsilon)\,\sigma(\varepsilon,\,t)}{\varepsilon}$$

and the time t required to reach the selected strain at each strain rate is obtained as $\varepsilon/\dot{\varepsilon}$. The experimentally determined relation of Collins and Hu [35] is quite fortuitously of the same functional form for large ε with

$$E(t) = \frac{1}{t} = \frac{\dot{\varepsilon}}{\varepsilon}$$

$$H(\varepsilon) = \frac{1}{0.18\,(e^{12\varepsilon} - 1)} \tag{2.72}$$

Now we wish to extrapolate the measured relation through 1.5 decades ($10^{1.5}$) of $\dot{\varepsilon}$ up to about $100\,\mathrm{s}^{-1}$. Then according to Smith, the function $H(\varepsilon)$ is independent of time and hence strain rate divided by strain provided one remains within the time span represented by the experimental data, i.e. the function $H(\varepsilon)$ should not change its form even at higher strain rates. The relative invariance of $H(\varepsilon)$ with time has been observed for a number of amorphous elastomers (e.g. styrene butadiene rubber vulcanizate and NBS polyisobutylene) where $H(\varepsilon)$ depends only on strain over about three decades of time for temperatures between -42.8 and $93.3\,^{\circ}\mathrm{C}$.

Since the measured stress–strain relation of Collins and Hu [35], which was based purely on experimental results, just happens to possess the same functional form as that of Smith, it is quite justifiable to broaden its validity directly to higher strain rates by virtue of the properties of Smith's function $H(\varepsilon)$ described above. The experimentally determined relation is therefore used in its original form in the numerical solution of the general equations of motion.

Results and discussion

A number of examples were computed with a view to illustrating the dependence of strain and strain rate (and hence wall stress) on: (a) the magnitude of the viscoelasticity of the wall, (b) the amplitude and duration of the acceleration history, and (c) the length and taper of the tube.

The geometry of the aorta and its idealization into a straight tapered segment are shown in Fig. 2.22. The aorta is subjected to an accelerative field directed from the thoracic aorta towards the aortic arch which corresponds to the type of deceleration measured by Hanson [47] in which a series of anaesthetized beagle dogs were exposed to a head-on impact $(-G_z)$ over a range of 5–60 g. Such accelerations occur in humans during, for example, a head-on vehicular collision in which the passenger is pitched forward over the steering wheel, his back becoming horizontal, and parallel to the direction of motion of the vehicle. The abrupt deceleration causes forward motion of the blood in the descending aorta towards the aortic arch.

A typical acceleration–time history is shown later (see Fig. 2.29), the shape of which can be very closely approximated by two sine curves of equal amplitude at time t_m but differing wavelength with total duration $t = T$.

One can express equation (2.54) as

$$\frac{\partial \Phi}{\partial x} = -G(t)$$

$$= -G_m \begin{cases} \sin \dfrac{\pi}{2} \dfrac{t}{t_m} & \text{for } t < t_m \\[2em] \cos \dfrac{\pi}{2} \dfrac{t - t_m}{T - t_m} & \text{for } t_m < t \leqslant T \end{cases}$$

according to the measurements in Hanson's experiment 4-F. The parameters T, t_m and G_m are approximated by

$$T = 49.5 \text{ ms} \qquad t_m = 17 \text{ ms}$$
$$G_m = 50 \text{ g} = 49.050 \text{ cm s}^{-2}.$$

Hanson gives measurements of the acceleration, the intra-aortic pressure, and the intrapleural pressure as a function of time. From the latter two measurements one can calculate the transmural pressure across the aortic wall. Measurements of intrapleural pressure are particularly susceptible to error as a result of stress concentrations which can develop around the embedded pressure transducer (see ref. 52). However, knowledge of the transmural pressures is essential for comparison with the analysis of the stresses which develop in the aortic wall. The results of the impact experiments of Kroell, Schneider, and Nahum [53] are therefore of very limited use in this respect since intrapleural pressures were not recorded in their work.

The physical behaviour of a fluid-filled viscoelastic tube is shown in Figs. 2.23–2.28 in which Hanson's experiment with dog No. 4 and acceleration pattern F (see his Fig. 6) was computed using his initial data as a means of checking the present analysis. For this purpose it was found that $B = 0.08$ s was more appropriate for dog aortae, as will be described shortly, than $B = 0.322$ s (equation (2.69)) as implied in the experiments of Collins and Hu [35] using pig and human aortae. This initial taper of the aorta is plotted in Fig. 2.23 for time $t = 0$. For later times (with $B = 0.08$ s) a rapid change in cross-section indicates the formation of a shock wave reflecting from the fixed end $x = 0$ and travelling to the right against the direction of blood flow. This steepening of the slope of the aortic wall leads to high axial stresses and strains which may eventually culminate in transverse rupture of the aorta. The corresponding fluid velocities (directed to the left) are shown in Fig. 2.24. Their position of steepening corresponds to that of the wall steepening of Fig. 2.23 and appears to confirm the existence of a 'shock' front in the flow. Behind the front the flow velocity reaches a plateau at early times and is gradually altered as the effect of the boundary condition at $x = L$ transmits its influence towards the aortic root.

The experiments of Hanson do not reveal the correct boundary condition at $x = L$; however, it appears plausible to take $u(L, t) = 0$ since the vascular bed cannot respond rapidly to a 50 ms event. Furthermore, such a boundary condition corresponds closely to the classic steering wheel impact [54] in

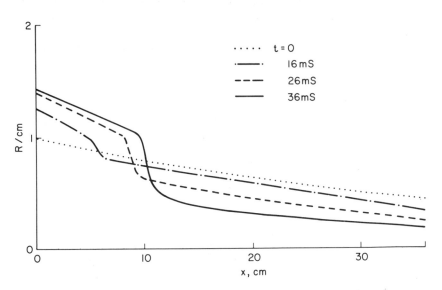

FIG. 2.23. Calculated variation of the inside radius of the aorta for Hanson's experiment 4-F with $B = 0.08$ s.

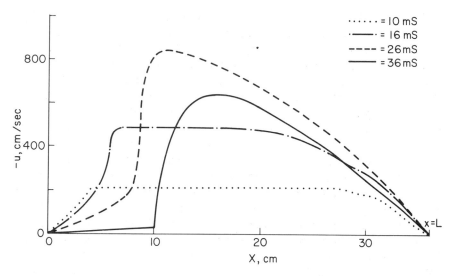

FIG. 2.24.   Calculated variation of blood velocity for Hanson's experiment 4-F with $B = 0.08$ s.

which the steering wheel of the colliding vehicle penetrates into the lower abdomen of the driver as he is subjected to a head-on deceleration. It can be seen in Fig. 2.24 that the peak velocity decreases after 26 ms. The distributions of axial wall strain and strain rate along the length x of the tube are plotted in Figs. 2.25 and 2.26 revealing peaks which again correspond to the passage of the shock. Axial strain rates are seen to diminish after 26 ms, as did the peak fluid velocity. Figures 2.27 and 2.28 show the time variation of the maximum

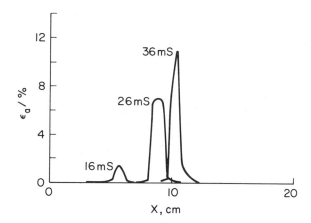

FIG. 2.25.   Calculated variation of axial wall strain for Hanson's experiment 4-F with $B = 0.08$ s.

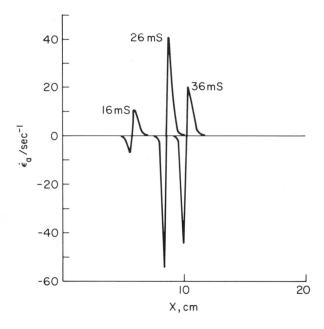

F<small>IG</small>. 2.26.   Calculated variation of axial strain rate for Hanson's experiment 4-F with $B = 0.08$ s.

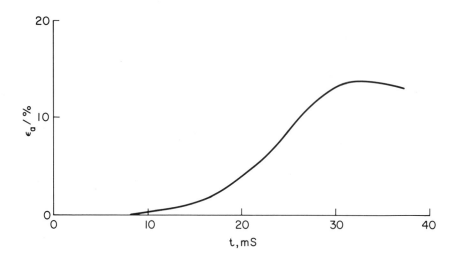

F<small>IG</small>. 2.27.   Variation of maximum wall strain with time for Hanson's experiment 4-F with $B = 0.08$ s.

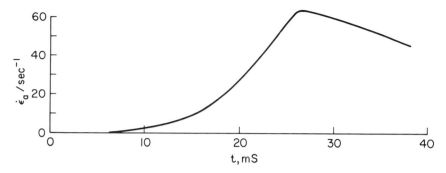

FIG. 2.28.   Variation of maximum strain rate for Hanson's experiment 4-F with
$B = 0.08$ s.

values of strain and strain rate over the tube length. It should be noted that the curve of Fig. 2.28 is not simply the slope of the curve of Fig. 2.27 since the maxima of ε_a and $\dot{\varepsilon}_a$ do not necessarily occur at the same x positions at a given time t. The maximum value of $\dot{\varepsilon}$ occurs at about 27 ms, preceding the maximum strain by about 5 ms.

Calculations of the transmural pressure at $x = 0$ as a function of time (Fig. 2.29) show a definite sensitivity to the value of the viscoelasticity B. For $B = 0.08$ the amplitude of the pressure variation and its phase lag relative to the acceleration curve closely match the experimental results of Hanson for dog No. 4-F once the origins have been made to coincide. Computations with this value of B also agree reasonably well with the results of Hanson's experiment 4-C and appear to lend adequate credence to the predictive ability of the present analysis and numerical scheme. Again, the inherent inaccuracies associated with the difficulty of measuring intra-pleural pressures should be

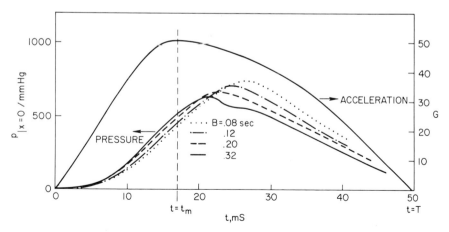

FIG. 2.29.   Accelerations and intra-aortic pressures.

noted [52]. The transmural pressure is shown (Fig. 2.30) to be quite sensitive to even very slight tapering of the aorta; it is clear that the aortic taper must always be reported if associated experimental results are to be interpreted meaningfully.

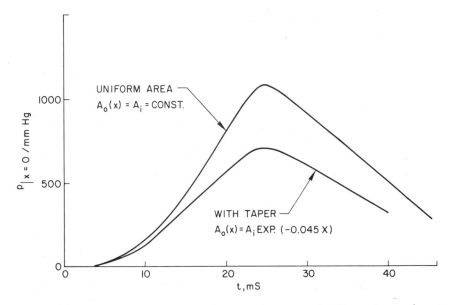

FIG. 2.30.   Influence of wall taper on intra-aortic pressure for Hanson's experiment 4-F with $B = 0.12$ s.

The maximum values of the strain and strain rate over both x and t are shown in the semi-logarithmic plots of Figs. 2.31 and 2.32 to be much more sensitive to variations in the viscoelasticity B than is the transmural pressure. In fact it would appear very worthwhile to formulate an experimental procedure for determining the unknown viscoelastic properties of materials on

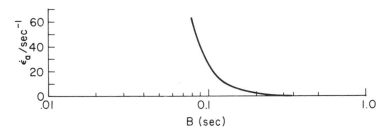

FIG. 2.31.   Variation of axial strain rate with viscoelasticity for Hanson's experiment 4-F.

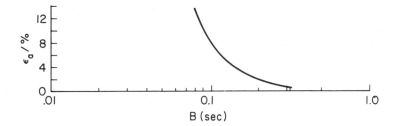

FIG. 2.32.   Variation of axial strain with viscoelasticity for Hanson's experiment 4-F.

the basis of shock experiments in which axial strains and strain rates are recorded. As the viscoelasticity decreases the shock steepness increases with concomitant increases in the local strain and strain rate.

The remaining calculations were carried out with a value of $B = 0.12$ s instead of $B = 0.08$ s for the sake of economy as the computational time, which increases markedly for decreasing B, doubles between these two values. The results for $B = 0.12$ s are equally indicative of the sensitivity of ε_a and $\dot\varepsilon_a$ to changes in the acceleration.

The maximum strain and strain rate depend very critically upon the level and duration of the acceleration. For a constant duration of 49.5 ms, which is quite typical of automobile accidents in which aortic rupture occurs, Figs. 2.33 and 2.34 depict a smooth monotonic increase of ε_a and $\dot\varepsilon_a$ with acceleration within the range 25–100 g. It is clear that the total kinetic energy expended in the impact also increases with G if the duration is held constant. In Fig. 2.35 the variation of ε_a and $\dot\varepsilon_a$ with acceleration has been calculated for constant initial impact velocity $u_1 = \int_0^T G\,dt \approx 15.7$ m s^{-1} and appears to be almost linear. In this case higher levels of acceleration are applied over shorter periods so that the kinetic energy of impact remains constant. These results are useful in the design of vehicles and restraint systems to protect passengers in a collision from a specified initial velocity. Representative data on aortic rupture

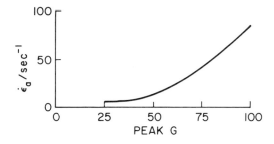

FIG. 2.33.   Variation of axial strain rate with peak acceleration for constant duration: $T = 49.5$ ms and $B = 0.12$.

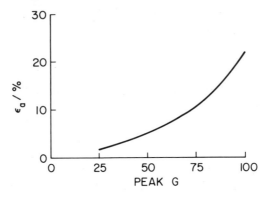

F$_{IG}$. 2.34.   Variation of axial strain with peak acceleration for constant duration: $T = 49.5$ ms and $B = 0.12$ s.

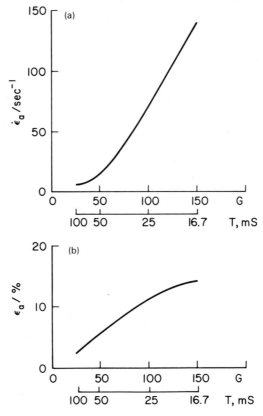

F$_{IG}$. 2.35.   (a) Variation of axial strain rate with peak acceleration for constant $\int G \mathrm{d}t = 15.7 \, \mathrm{m \, s}^{-1}$ and $B = 0.12$ s. (b) Variation of axial strain with peak acceleration for constant $\int G \mathrm{d}t = 15.7 \, \mathrm{m \, s}^{-1}$ and $B = 0.12$ s.

strengths, when available, will permit definite tolerance levels on impact velocities and acceleration levels to be set.

2.7. Impulsive flow in untethered non-orthotropic vessels

The impulsive injection of blood from the heart into the aorta, and subsequently the connecting great vessels, engenders wave propagation in the vessel walls, the motion of which in turn modifies the blood flow. This interaction between fluid and vessel wall motions will depend strongly upon the material properties of the system, such as fluid density and viscosity and the dynamic viscoelastic properties of the wall, in addition to the mechanical constraints imposed on the vessel by the surrounding connective tissue and intercostal arteries. Pathologically induced changes in these properties will be shown by modified pressure and flow variations in the distal portions of the arterial system. Alternatively, such variations may serve as a basis for estimating such changes in the material properties.

Although radial excursions of 8–10 per cent of the aortic lumen have been generally observed in *in situ* measurements under normal physiological conditions, little if any longitudinal displacements of the vessel wall have been detected. In the previous section a numerical method was described for determining instantaneous variations in vessel cross-section, pressure, and fluid velocity under conditions of complete longitudinal tethering. However, in laboratory testing of excised biological specimens, large longitudinal motions of the wall are known to occur.

This section provides a generalized treatment of viscous fluid motion in a non-linear orthotropic viscoelastic thin-walled tube. A numerical solution based upon a two-step Lax–Wendroff finite difference scheme is described and is shown to be very stable and rapid. The following analysis, in conjunction with experiments, provides a means of assessing the role of longitudinal displacements of the aortic wall in modifying the stresses developed during impact.

Mathematical formulation

In this present generalized treatment of the axisymmetric motion of an incompressible fluid in a non-linear viscoelastic tube, the fluid viscosity is accounted for and hence the action of wall friction on longitudinal displacements of the vessel wall. The actual ratio of wall thickness to diameter for the human aorta is about one-tenth. In the thin-wall approximation this ratio is considered to be insignificantly small, leading mathematically to the neglect of the shear stresses and bending moments. Patel and Fry [55] showed that the aorta and carotid arteries developed shearing strains that were very much smaller than the corresponding longitudinal and circumferential strains when the vessel was inflated to physiological pressures. Thus the vessel is orthotropic and shear strains can be neglected. In this case the three orthogonal strains, circumferential, longitudinal, and radial, are the only important ones. For an incompressible wall (Poisson ratio of $\frac{1}{2}$) they are interrelated.

A theory of large deflections for thin shells of revolution has been given by Reissner [56] and a more generalized treatment is described in the book by Krauss [57]. Under the foregoing assumptions the equilibrium of an element of the shell is expressed by equation (28) of ref. 56 as

$$\frac{\partial}{\partial \xi}(rN_\xi \mathbf{j}_\xi) + \frac{\partial}{\partial \theta}(\alpha N_\theta \mathbf{j}_\theta) + r\alpha \mathbf{p} = 0 \tag{2.73}$$

where ξ is a Lagrangian coordinate along the generator of the middle surface of the tube, θ is the polar angle of a point on the middle surface (see Figs. 2.36 and 2.37), and x, y, and z are the corresponding Eulerian co-ordinates; $r = (x^2 + y^2)^{1/2}$ is the Eulerian co-ordinate of a point on the generator of the surface and must be regarded as functions of (ξ, t) since the problem is time dependent. α is defined by

$$\alpha^2 = \left(\frac{\partial r}{\partial \xi}\right)^2 + \left(\frac{\partial z}{\partial \xi}\right)^2 \tag{2.74}$$

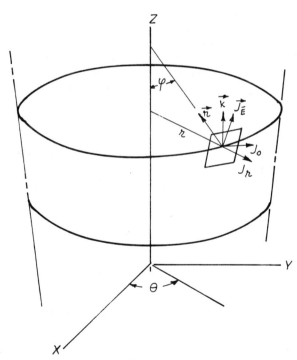

Fig. 2.36.   Middle surface of the shell showing co-ordinates on the middle surface and unit vectors associated with the middle surface.

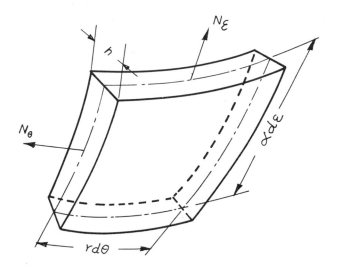

FIG. 2.37.   Element of a shell showing stress resultants.

N_ξ and N_θ are the stress resultants in the ξ and θ directions, and **p** is the load intensity vector (external load per unit area).

In the context of the theory of shells, the load **p** is usually specified. In the present problem, however, the load **p** results from the interaction between the fluid and the tube. The wall motion is therefore coupled to that of the fluid through the load vector **p** per unit area. A number of external forces contribute to the load intensity vector **p**. The normal component $\mathbf{p}_n$ arises from the fluid hydrodynamic pressure acting normal to the inner layer of the wall in addition to possible normal forces imposed on the outer layer by visceral resistance to radial motions. The tangential component p_ξ derives from the shear stress τ due to viscous drag at the inner wall as well as from longitudinal restraints on the outer wall associated with the 'tethering' effects of connective tissue and intercostal arteries [55].

The quasi one-dimensional viscous flow equations for an incompressible fluid in a distensible axisymmetric duct can be expressed as

$$\frac{\partial r^2}{\partial t} + \frac{\partial}{\partial z}(r^2 v) = 0 \quad \text{continuity} \tag{2.75}$$

$$\frac{\partial v}{\partial t} + \frac{\partial}{\partial z}\left(\frac{v^2}{2} + \frac{p_{\text{hydro}}}{\rho} + \Phi\right) = -F \quad \text{momentum} \tag{2.76}$$

where v is the fluid velocity (averaged over the cross-section), ρ is the constant density of the fluid, p_{hydro} is the hydrostatic pressure which is equal but of opposite sign to the normal component of the external load p_n and Φ the gravitational force potential defined by

$$\frac{\partial \Phi}{\partial z} = -G. \qquad \text{(acceleration)}$$

F represents the friction per unit mass of fluid due to viscous fluid drag at the wall. For circular tubes

$$F = \frac{2}{\rho r} \tau \qquad (2.77)$$

where τ is the shear stress at the wall.

For steady flow in a pipe one obtains from the corresponding formulae of Poiseuille and Blasius (ref. 58, pp. 79, 561) for laminar and turbulent flow the shear stress

$$\tau = C_f \tfrac{1}{2} \rho v^2 \, \text{Re}^{-m} \qquad (2.78)$$

where the Reynolds number $\text{Re} = \rho r |v| / \mu$, C_f is the dimensionless skin friction coefficient, and μ is the fluid viscosity. For laminar flow $m = 1$ and $C_f = 8$, while for turbulent flow $m = \tfrac{1}{4}$ and $C_f \approx 0.07$.

In the present calculations the laminar form of the shear stress has been used corresponding to the generally accepted principle that turbulent regions are rare in most normal physiological blood flows. For example, it was pointed out by Ling, Atabek, and Carmody [59] that according to hot-film anemometric measurements carried out in the arteries of living animals the flow is normally non-steady and laminar, i.e. flow profiles are developed locally and transiently during the passage of each pulsatile wave. The rise time of the pressure-gradient wavefront is of the order of 0.02 s (under normal physiological conditions) and the apparent propagating velocity of the wavefront is approximately $600\,\text{cm s}^{-1}$, implying a physical width of the travelling wave of about 12 cm. In general, flow profiles were found to be blunt and axially symmetric. Both turbulence and secondary flows were found to be inhibited and localized by the nature of the pulsatile flow, and although weak turbulence was observed in the aortic arch of small dogs it was quickly dissipated within each heart cycle. None the less, in the regions near the shock front where conditions are clearly not physiological this may no longer be the case. Although these expressions are strictly valid only for steady flow, they will be applied here for the unsteady flow conditions to be described subsequently.

To complete the system of equations (2.73), (2.75) and (2.76) one must add a constitutive relation linking wall stresses to strains and rates of strain within the vessel wall. For this purpose a mathematical expression relating the stress resultants N_ξ and N_θ to the material co-ordinates r and z and their time derivatives would be desirable. However, even for small deformations such a highly developed constitutive relation is not available at present without mentioning the still greater lack of knowledge for the large deformation high strain rate range corresponding to shock experiments.

A simplified material model is proposed in which the stress is related to strain and strain rate in the circumferential and longitudinal directions separately. This effectively decoupled representation is not inconsistent with the assumption of orthotropy implicit in the thin-walled approximation described above.

Physically, this representation of the state of stress corresponds to a model of the tube wall composed of two distinct layers: a layer of 'fibres' which carries the stress in the tangential direction and a layer of 'rings' which carries the hoop stress. Similar ideas have been proposed by Apter, Rabinowitz, and Cummings [60].

The stress–strain rate relation in each direction is written in the form

$$\sigma(\lambda, \dot{\lambda}) = f(\lambda) + g(\lambda, \dot{\lambda}) \tag{2.79}$$

with

$$g(\lambda, 0) = 0 \tag{2.80}$$

so that $f(\lambda)$ represents the static loading relation. As a measure of strain the extension ratio is employed here with

$$\lambda = L/L_0. \tag{2.81}$$

where L is the length of a strained element and L_0 is its length at some reference state. For the measurements of Collins and Hu [23], the explicit expression (2.79) becomes

$$\sigma(\lambda, \dot{\lambda}) = E(1 + B\dot{\lambda}/\lambda)(\lambda^n - 1) \tag{2.82}$$

with $n = 12$, $B = 0.64\,\mathrm{s}$ and $E = 0.28 \times 10^6$ dynes/cm^2 along the axial direction.

Relation (2.82) may be converted to an expression for the stress resultants in terms of the coordinates r, z. The circumferential extension is simply

$$\lambda_0 = \frac{r(\xi, t)}{r(\xi, 0)} = \frac{r}{r_0} \tag{2.83}$$

The axial extension may be expressed in terms of the quantity α of eqn (2.2) by

$$\lambda_\xi = \frac{\alpha(\xi, t)}{\alpha(, 0)} = \frac{\alpha}{\alpha_0} \tag{2.84}$$

the stress resultants are then related to the stresses by

$$N = h\sigma \tag{2.85}$$

where h is the wall thickness. The variations of the wall thickness h with the tube displacements may be found in a simple manner if one uses the Poisson ratio $v = 1/2$ typical of biological tissue. In this case, the material is incompressible, so that

$$\frac{\partial}{\partial t}(r\alpha h) = 0$$

or

$$r(\xi, t)\alpha(\xi, t)h(\xi, t) = r_0(\xi)\alpha_0(\xi)h_0(\xi) \qquad (2.86)$$

where

$$f_0(\xi) = f(\xi, t = 0)$$

Equations (2.73), (2.75), (2.76), and (2.82) and their associated relations constitute a complete system governing the coupled fluid–wall motions. In the next section these will be reduced to a simpler set whose numerical solution, with illustrative values of the physical parameters, will be described in the remainder of the work.

Reduction of the governing equations

It is convenient to decompose the vector equation (2.73) into its components in the ξ and n directions. These directions refer to the generator of the middle surface (see Fig. 2.37). The hydrostatic pressure in the fluid contributes to the normal component p_n of $\mathbf{p}$ and the wall friction (due to fluid viscosity) to the tangential component p_ξ of $\mathbf{p}$. Carrying out the differentiations in (2.73) gives

$$\frac{\partial}{\partial \xi}(rN_\xi)\mathbf{j}_\xi + \alpha N_\theta \frac{\partial}{\partial \theta}\mathbf{j}_\theta + r\alpha p_\xi \mathbf{j}_\xi$$

$$+ rN_\xi \frac{\partial}{\partial \xi}\mathbf{j}_\xi + r\alpha p_n \mathbf{j}_n = 0.$$

Let $\mathbf{i}$, $\mathbf{j}$, and $\mathbf{k}$, be unit vectors in the x, y, and z directions respectively (see Fig. 2.37). The radial and circumferential unit vectors $\mathbf{j}_r$ and $\mathbf{j}_\theta$ are then related to the former by [56]

$$\mathbf{j}_r = \mathbf{i}\cos\theta + \mathbf{j}\sin\theta; \quad \mathbf{j}_\theta = -\mathbf{i}\sin\theta + \mathbf{j}\cos\theta \qquad (2.87)$$

and the tangential and normal unit vectors by

$$\mathbf{j}_\xi = \mathbf{j}_r\cos\phi + \mathbf{k}\sin\phi$$
$$\mathbf{n} = -\mathbf{j}_r\sin\phi + \mathbf{k}\cos\phi \qquad (2.88)$$

Then from equation (2.88) one finds

$$\frac{\partial}{\partial \xi}\mathbf{j}_\xi = \mathbf{j}_n\frac{\partial\phi}{\partial \xi} \qquad (2.89)$$

and from equation (2.87)

$$\frac{\partial}{\partial \theta} \mathbf{j}_\theta = -\mathbf{j}_r = -\mathbf{j}_\xi \cos \phi + \mathbf{j}_n \sin \phi \qquad (2.90)$$

where $\mathbf{j}_q$ denotes a unit vector in the q direction and ϕ is the angle between the tangent to the meridian curve and the z-axis with

$$\tan \phi = \frac{\partial z / \partial \xi}{\partial r / \partial \xi}. \qquad (2.91)$$

The scalar components of equation (2.73) in the ξ and n directions are now readily found to be

$$\frac{\partial}{\partial \xi}(rN_\xi) - \alpha N_\theta \cos \phi + r\alpha p_\xi = 0 \qquad (2.92)$$

$$rN_\xi \frac{\partial \phi}{\partial \xi} + \alpha N_\theta \sin \phi + r\alpha p_n = 0 \qquad (2.93)$$

For convenience we replace the variables N_ξ and N_θ by

$$T_\xi \equiv rN_\xi = \frac{r_0 h_0 \alpha_0}{\alpha} \sigma_\xi = \frac{r_0 h_0 \sigma_\xi}{\lambda_\xi} \qquad (2.94)$$

$$T_\theta \equiv \alpha N_\theta = \frac{r_0 h_0 \alpha_0}{r} \sigma_\theta = \frac{\alpha_0 h_0 \sigma_\theta}{\lambda_\theta}. \qquad (2.95)$$

The scalar wall equations (2.92) and (2.93) then become

$$\frac{\partial}{\partial \xi} T_\xi - T_\theta \cos \phi + r\alpha p_\xi = 0 \qquad (2.96)$$

$$T_\xi \frac{\partial \phi}{\partial \xi} + T_\theta \sin \phi + r\alpha p_n = 0. \qquad (2.97)$$

One can eliminate $p_n (= -p_{\text{hydro}})$ between equations (2.76) and (2.97) yielding

$$v_t + \left\{ \frac{v^2}{2} + \frac{1}{\rho r\alpha} \left(T_\xi \frac{\partial \phi}{\partial \xi} + T_\theta \sin \phi \right) + \phi \right\}_z + \frac{2}{\rho r} \tau = 0. \qquad (2.98)$$

The governing system then reduces essentially to the three simultaneous equations (2.75), (2.96), and (2.98) in conjunction with the auxiliary relations (2.74), (2.77), (2.78), (2.82)–(2.86), (2.91), (2.93), and (2.95). p_ξ is the sum of all tangential loads acting on the wall; in the absence of tethering $p_\xi = \tau$ where τ is the shear stress due to fluid viscosity.

Initial boundary value problem

The system of equations (2.75), (2.96), and (2.98) is solved numerically with the following illustrative boundary and initial conditions corresponding to an

arterial segment open at one end ($\xi = 0$) with an applied stagnation pressure at
the other end ($\xi = L$), i.e.

at $\xi = 0$ $\qquad\qquad$ $p = 0$ $\qquad$ (open end)

$\qquad\qquad\qquad\qquad$ $T_\xi = 0$ $\qquad$ (tube unrestrained longitudinally) $\qquad\qquad$ (2.99)

at $\xi = L$ $\qquad$ $\dfrac{p}{\rho} + \tfrac{1}{2}v^2 = H \begin{cases} \sin^2\left(\dfrac{\pi}{2}t\right) & \text{for } t < 20 \text{ ms} \\ 1 & \text{for } t \geq 20 \text{ ms} \end{cases}$ $\qquad$ (2.100)

$$Z = \text{const.} = Z_0(L)$$

where ρH is constant and equal to the stagnation pressure in the reservoir.
At $t = 0$

$$v(\xi, 0) = 0$$
$$r(\xi, 0) = r_0(\xi) \qquad\qquad\qquad (2.101)$$
$$z(\xi, 0) = z_0(\xi).$$

Stability conditions

The numerical stability of the set of governing equations can be assessed in
approximate form in a manner similar to that of Kivity and Collins [44], i.e.
the system can be cast in both hyperbolic and parabolic forms, and the
appropriate stability criteria developed for each. However, it has been
confirmed in subsequent calculations that it is invariably the stability criterion
associated with the parabolic form which is more restrictive and hence only the
derivation of that form is outlined here.

One considers the system of equations (2.75), (2.96) and (2.98) with the
assumption for simplicity that $\phi = 0$ and $\tau = 0$. Then equation (2.96) becomes

$$\frac{\partial T_\xi}{\partial \xi} = T_0 \cos \phi$$
$$T_\xi = \int T_0 \cos \phi \, d\xi + \text{const.} \qquad\qquad (2.102)$$

Multiplying equation (2.75) by v and equation (2.100) by $A = r^2$ and adding
gives, in terms of a new variable defined by $Q = Av$,

$$Q_t + \left(\frac{Q^2}{A}\right)_z + A\left\{\frac{1}{\rho r \alpha}\left(T_\xi \frac{\partial \phi}{\partial \xi} + T_\theta \sin \phi\right)\right\}_z = 0. \qquad (2.103)$$

For small displacements ϕ is very close to $\pi/2$, so that $\sin \phi \approx 1$. Also from
(2.102) the change in T_ξ is small compared with T_θ so that T_ξ is approximately
constant. The quantity $(1/\alpha)/(\partial\phi/\partial\xi)$ appearing in equation (2.103) equals the
curvature $1/R_\xi$ and is also very small under the above assumption so that the

term $T_\xi(\partial\phi/\partial\xi)$ can be neglected with respect to T_θ. Equation (2.103) then becomes

$$Q_t + \left(\frac{Q^2}{A}\right)_z + A\left(\frac{T_\theta}{\rho r \alpha}\right)_z = 0.$$

T_θ can be expressed in terms of A and $\dot{A} = -Q_z$ using the stress–strain relation (2.82) employed in this study. The resulting differential equation takes on the form

$$Q_t - S Q_{\xi\xi} + W = 0 \qquad (2.104)$$

where W is a function of lower order ξ-derivatives which do not enter into this first-order estimate of a numerical stability criterion and

$$S = \frac{f(\lambda_\theta)}{\lambda_\theta^2} \frac{h_0}{r_0} \frac{B_\theta/2}{\rho\lambda_\xi} \left/ \left(\frac{\partial z}{\partial\xi}\right)^2 \right. \qquad (2.105)$$

$$\frac{\partial z}{\partial\xi} \approx \lambda_\xi. \qquad (2.106)$$

The stability criterion for the parabolic equation (2.104) is

$$\frac{S\Delta t}{(\Delta\xi)^2} \le 1/2$$

which implies a time step

$$\Delta t < \frac{\rho\lambda_\xi^3\lambda_\theta^2}{B_\theta f(\lambda_\theta) h_0/r_0} (\Delta\xi)^2. \qquad (2.107)$$

Numerical solution

For the solution of the system (2.75), (2.96), and (2.98), one employs a numerical algorithm based on the two-step Lax–Wendroff difference scheme. In each step the radius r and the particle velocity v are explicitly advanced in time using the continuity equation (2.75) and the momentum equation (2.98) respectively. The new value of r serves to determine a new value of T_θ from relation (2.95). T_ξ is then determined by integration of (2.96). Finally z is computed implicitly using a simple Newton method.

This procedure has proved to be reliable provided that one observes the stability criterion (2.107). In the actual calculation Δt was taken as 0.8 times the right-hand side of (2.107). The initial estimate for z in the Newton method was obtained from the value of z at the previous step.

The Lagrangian coordinate ξ was chosen to be the initial co-ordinate of a point, i.e. $z(\xi, 0) = \xi$. To simplify the presentation of the difference scheme the following abbreviations are used.

$$\delta^n_{j+1/2} = \frac{1}{2} \frac{\Delta t^n}{z^n_{j+1} - z^n_{j-1}}$$

$$\delta^n_j = \frac{\Delta t^n}{z^n_{j+1/2} - z^n_{j-1/2}}$$

$$f^n_{j+1/2} = \tfrac{1}{2}(f^n_j + f^n_{j+1}) \quad \text{unless a different}$$
definition is explicitly specified

$$\pi = \tfrac{1}{2}v^2 + \frac{1}{\rho r \alpha}\left(T_\xi \frac{\partial \phi}{\partial \xi} + T_\theta \sin \phi\right)$$

$$\pi^n_j = \tfrac{1}{2}(v^n_j)^2 + \frac{1}{\rho r^n_j \alpha^n_j}\left\{(T_\xi)^n_{j+1/2}\frac{\phi^n_{j+1/2} - \phi^n_{j-1/2}}{\Delta \xi} + (T_\theta)^n_j \sin \phi^n_j\right\}$$

$$\pi^{n+1/2}_{j+1/2} = \tfrac{1}{2}(v^{n+1/2}_{j+1/2})^2 + \frac{1}{\rho r^{n+1/2}_{j+1/2} \alpha^n_{j+1/2}}[\tfrac{1}{2}\{(T_\xi)^{n+1/2}_j$$

$$+ (T_\xi)^{n+1/2}_{j+1}\}\frac{\phi^n_{j+1} - \phi^n_j}{\Delta \xi} + (T_\theta)^{n+1/2}_{j+1/2}\sin \phi^n_{j+1/2}]$$

$$(\alpha^n_{j+1/2})^2 = \frac{(z^n_{j+1} - z^n_j)^2 + (r^n_{j+1} - r^n_j)^2}{(\Delta \xi)^2}$$

$$(\alpha^n_j)^2 = \tfrac{1}{2}(\alpha^n_{j+1/2} + \alpha^n_{j-1})$$

$$\phi^n_{j+1/2} = \text{arc cos}\left(\frac{r^n_{j+1} - r^n_j}{\alpha^n_{j+1/2}\Delta \xi}\right)$$

$$\phi^n_j = \tfrac{1}{2}(\phi^n_{j+1/2} + \phi^n_{j-1/2})$$

$$(\lambda_\theta)^n_j = \frac{r^n_j}{(r_0)_j}$$

$$(\lambda_\xi)^n_{j+1/2} = \frac{\alpha^n_{j+1/2}}{(\alpha_0)_{j+1/2}}.$$

The difference scheme consists of the following steps.

Step 1

$$(r^2)^{n+1/2}_{j+1/2} = (r^2)^n_{j+1/2} - \delta^n_{j+1/2}\{(r^2 v)^n_{j+1} - (r^2 v)^n_j\}$$

$$v^{n+1/2}_{j+1/2} = v^n_{j+1/2} - \delta^n_{j+1/2}(\pi^n_{j+1} - \pi^n_j) - \frac{\Delta t^n}{\rho r^n_{j+1/2}}\tau(v^n_{j+1/2}, r^n_{j+1/2})$$

$$(T_\theta)_{j+1/2}^{n+1/2} = (h_0 \alpha_0)_{j+1/2} \sigma_\theta \left\{ (\lambda_\theta)_{j+1/2}^{n+1/2}, \frac{(\lambda_\theta)_{j+1/2}^{n+1/2} - (\lambda_\theta)_{j+1/2}^n}{\frac{1}{2}\Delta t^n} \right\} \bigg/ (\lambda_\theta)_{j+1/2}^{n+1/2}$$

$$(T_\xi)_{j+1}^{n+1/2} = (T_\xi)_j^{n+1/2} + \Delta \xi \left\{ (T_\theta)_{j+1/2}^{n+1/2} \cos \phi_{j+1/2}^n - r_j^n \alpha_j^n \tau (v_j^n, r_j^n) \right\}.$$

To determine $z_{j+1/2}^{n+1/2}$ it is necessary to solve the following implicit equation for $(\lambda_\xi)_j^{n+1/2}$:

$$\sigma_\xi \left\{ (\lambda_\xi)_j^{n+1/2}, \frac{(\lambda_\xi)_j^{n+1/2} - (\lambda_\xi)_j^n}{\frac{1}{2}\Delta t^n} \right\} \bigg/ (\lambda_\xi)_j^{n+1/2} = (T_\xi)_j^{n+1/2} \bigg/ (r_0 h_0)_j$$

where the function $\sigma_\xi(\lambda, \dot{\lambda})$ gives the stress in terms of the extension λ_ξ and its time rate of change. Since $(\lambda_\xi)_j^{n+1/2}$ is known $z_{j+1/2}^{n+1/2}$ is found from the relation

$$(z_{j+1/2}^{n+1/2} - z_{j-1/2}^{n+1/2})^2 + (r_{j+1/2}^{n+1/2} - r_{j-1/2}^{n+1/2})^2 = \{(\lambda_\xi)_j^{n+1/2} (\alpha_0)_j\}^2.$$

Step 2

$$(r^2)_j^{n+1} = (r^2)_j^n - \delta_j^{n+1/2} \left\{ (r^2 v)_{j+1/2}^{n+1/2} - (r^2 v)_{j-1/2}^{n+1/2} \right\}$$

$$v_j^{n+1} = v_j^n - \delta_j^{n+1/2} \left\{ \pi_{j+1/2}^{n+1/2} - \pi_{j-1/2}^{n+1/2} \right\}$$

$$- 2 \frac{\Delta t^n}{\rho r_{av}} \tau \left(\frac{v_{j+1/2}^{n+1/2} + v_{j-1/2}^{n+1/2}}{2}, r_{av} \right)$$

where

$$r_{av} = \tfrac{1}{2}(r_{j+1/2}^{n+1/2} + r_{j-1/2}^{n+1/2}).$$

$$(T_\theta)_j^{n+1} = (h_0 \alpha_0)_j \sigma_\theta \left\{ (\lambda_\theta)_j^{n+1}, \frac{(\lambda_\theta)_j^{n+1} - (\lambda_\theta)_j^n}{\Delta t^n} \right\} \bigg/ (\lambda_\theta)_j^{n+1}$$

$$(T_\xi)_{j+1/2}^{n+1} = (T_\xi)_{j-1/2}^{n+1} + \Delta \xi \left\{ (T_\theta)_j^{n+1} \frac{r_{j+1/2}^{n+1/2} - r_{j-1/2}^{n+1/2}}{\alpha_j^n \Delta \xi} \right.$$

$$\left. - r_{av} \alpha_j^n \tau \left(\frac{v_{j+1/2}^{n+1/2} + v_{j-1/2}^{n+1/2}}{2}, r_{av} \right) \right\}.$$

To determine z_j^{n+1} one first solves for $(\lambda_\xi)_{j+1/2}^{n+1}$ from the equation

$$\sigma_\xi \left\{ (\lambda_\xi)_{j+1/2}^{n+1}, \frac{(\lambda_\xi)_{j+1/2}^{n+1} - (\lambda_\xi)_{j+1/2}^n}{\Delta t^n} \right\} \bigg/ (\lambda_\xi)_{j+1/2}^{n+1} = (T_\xi)_{j+1/2}^{n+1} \bigg/ (r_0 h_0)_{j+1/2}$$

and then z_j^{n+1} is determined from

$$(z_{j+1}^{n+1} - z_j^{n+1})^2 + (r_{j+1}^{n+1} - r_j^{n+1})^2 = \{\lambda_{j+1/2}^{n+1} (\alpha_0)_{j+1/2}\}^2.$$

The difference analogues of the boundary conditions (2.99) and (2.100) are as follows. For the open-end case

$$r_1^{n+1} = \text{const.} = r_0(0)$$

$$v_1^{n+1} = v_1^n - 2\delta_{3/2}^n(\pi_2^n - \pi_1^n) - 2\frac{\Delta t^n}{\rho r_{3/2}^{n+1/2}}\tau(v_{3/2}^{n+1/2}, r_{3/2}^{n+1/2}).$$

z_1^{n+1} is computed by the general procedure.

For the fixed-end case (with applied stagnation pressure)

$$z_{im}^{n+1} = L \qquad \text{(vessel length)}$$

$$v_{im}^{n+1} = v_{im}^n - \delta_{im-1/2}^n(\pi_{im}^{n+1/2} - \pi_{im-1/2}^{n+1/2})$$

$$-\frac{2\Delta t^n}{\rho r_{im-1/2}^{n+1/2}}\tau(v_{im-1/2}^{n+1/2}, r_{im-1/2}^{n+1/2})$$

where $\pi_{im}^{n+1/2}$ is computed from the given applied stagnation pressure as a function of time $t^n + \frac{1}{2}\Delta t^n$. Then r_{im}^{n+1} is calculated by first solving the implicit equation for $(\lambda_\theta)_{im}^{n+1}$

$$\sigma_\theta\left\{(\lambda_\theta)_{im}^{n+1}, \frac{(\lambda_\theta)_{im}^{n+1} - (\lambda_\theta)_{im}^n}{\Delta t^n}\right\} \bigg/ \{(\lambda_\theta)_{im}^{n+1}\}^2$$

$$= \frac{\{\pi_{im}^{n+1} - \frac{1}{2}(v_{im}^{n+1})^2\}\rho(r_0)_{im}(h_0)_{im}\alpha_{im-1/2}^{n+1/2}/(\alpha_0)_{im-1/2}}{\sin\phi_{im-1/2}^{n+1/2}}$$

and then

$$r_{im}^{n+1} = (r_0)_{im}(\lambda_\theta)_{im}^{n+1}$$

Note. In calculating the variables r and v one progresses from $\xi = 0$ to $\xi = L$ ($i = 1$ to $i = im$), whereas in the calculation of λ_ξ and z one starts at the fixed end $\xi = L$ since z is prescribed there.

Results and discussion

Illustrative computations have been carried out for an aorta considered as a non-linear viscoelastic material with a dynamic constitutive stress–strain–strain-rate relation of the functional form given in equation (2.82). The physical parameters selected are given in Table 2.7.

TABLE 2.7

$E_\xi = E_\theta = 0.28 \times 10^6$ dyn cm^{-2}	
$B_\xi = B_\theta = 0.16$ s	$r_0(\xi) = 1$ cm.
$n_\xi = n_\theta = 12$	
$\mu = 0$ P	$H = 0.63 \times 10^6$ dyn cm^{-2}
$\rho = 1.05$ g cm^{-3}	
$C_0 = 300$ cm s^{-1}. (sound speed in the aorta at zero transmural pressure)	

Figure 2.38 shows the evolution of shape changes at 10 ms intervals, with the shock wave progressing along the tube from right to left, as a function of the Lagrangian axial coordinate ξ. The end $\xi = 100\,\text{cm}$ is held fixed while the remainder of the tube is free to move longitudinally as shown in Fig. 2.39. Such longitudinal motion, although inhibited by tethering for vessels *in situ*,

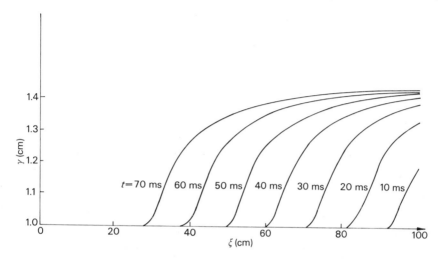

FIG. 2.38. Radius of the tube at selected times.

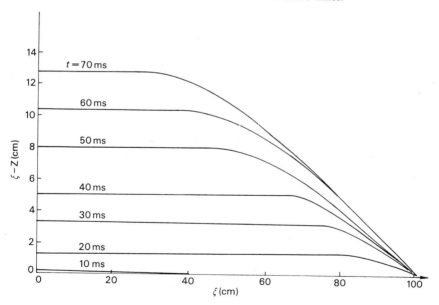

FIG. 2.39. Axial wall displacement at selected times.

must be considered in *in vitro* laboratory experiments using excised tissue. The radial and longitudinal displacements vary smoothly but the demarcation between 'shocked' and undisturbed regions of the vessel wall is quite evident. The axial distributions of fluid velocity shown in Fig. 2.40 resemble the shape changes of Fig. 2.38. The pressure distribution in the tube, although not shown here, can be calculated directly from the variation of the extension ratio α (Fig. 2.41) and from the distributions of longitudinal and circumferential wall stresses in Figs. 2.42 and 2.43.

The computational method possesses wide generality, well beyond that implicit in the above example calculation. The present analysis, in addition to admitting large longitudinal motions of a tapered tube, also included the effects of fluid viscosity and orthotropicity of the vessel wall. The influence of variations of these latter parameters is summarized in Table 2.8 for the steady state solution of flow in a tube of uniform cross-section obtained numerically from the unsteady solution as the asymptotic limit for long times. The velocity dz_e/dt of the distal end of the tube $\xi = 0$ is calculated from a simple finite difference at times of 70 and 75 ms.

The steady state shock speed U can also be estimated from the velocity of the distal end of the tube in terms of the extension ratio λ. In a small time interval Δt the tube elongates in the region over which the shock has passed by an amount equal to the strain $\lambda - 1$, multiplied by the speed U of the wave front, i.e.

$$\Delta z_e = (\lambda_\xi - 1) U \Delta t$$

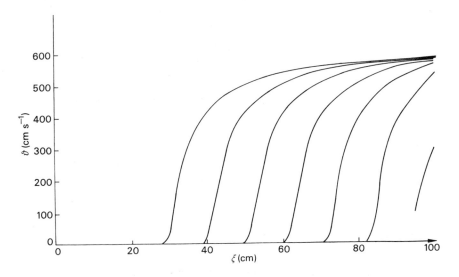

FIG. 2.40.   Fluid velocity at selected times.

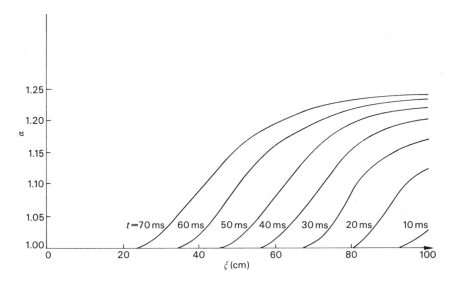

FIG. 2.41. Longitudinal extension at selected times.

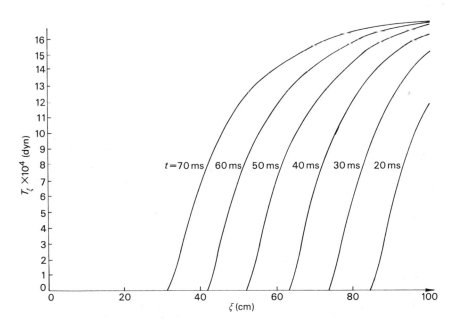

FIG. 2.42. Longitudinal stress resultant at selected times.

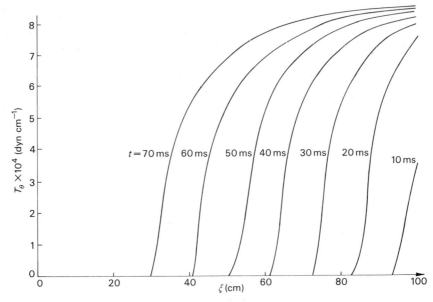

Fɪɢ. 2.43.   Circumferential stress resultant at selected times.

TABLE 2.8

Case no.	$-dz_e/dt$ (cm s^{-1})	λ_ξ	$-U$ (cm s^{-1})	Notes
0	250	1.242	1040	Parameters as in Table 2.1.
1	185	1.175	1055	Parameters as in Table 2.1 except $E_\xi = 2E_\theta$.
2	316	1.317	1000	Parameters as in Table 2.1 except $E_\xi = \frac{1}{2}E_\theta$.
3	250	1.245	1020	Parameters as in Table 2.1 except $\mu = 0.05$ P. (laminar friction law)
4	183	1.154	1190	Parameters as in Table 2.1 except $B_\xi = 10\,B_\theta$.

or

$$U = \frac{1}{\lambda_\xi - 1}\frac{dz_e}{dt}.$$

This estimate of U, which is given in Table 2.7 agrees favourably with that computed from the full solution.

Upon comparing the extension ratio λ_ξ for cases 0 and 3, it would appear that the fluid viscosity and consequent drag on the tube wall are not significant, although this conclusion remains tentative until further examples are examined. This would indicate that the wall forces are much more important than viscous drag in determining longitudinal displacements. However, Table 2.7 brings out the considerable influence of the orthotropicity on the extension ratios. In fact both changes in the axial wall elasticity E_ξ (cases 0, 1 and 2) and in the axial viscoelastic modulus B_ξ (cases 0 and 4) effect important changes in the extension ratio λ_ξ. None the less the steady state shock velocity appears relatively insensitive to such changes with the exception of the viscoelastic parameter B_ξ. It may well be possible to exploit the sensitivity of the wall elongation as an indirect measure of the circumferential or axial wall moduli of biological vessels subjected to rapidly changing loads.

2.8. Injuries to the rib cage

Introduction

Biomechanically, the rib cage serves as a load-carrying structure and the main injuries experienced are rib and sternal fractures (Fig. 2.4). These are rather common and may occur due to compression of the chest at low strain rates as in external cardiac massage or due to high speed impacts as in automobile accidents or falls from heights. Rib fractures result in loss of the structural integrity of the chest which causes it to lose its elastic recoil thus making respiration very difficult. In addition, other complications like pulmonary embolisms and haemorrhage may also result from secondary injuries caused by fractured ribs. If we know the injury tolerance limits for the human thorax, we can then design safer car interiors, more effective safety belts, better external cardiac massage techniques, etc.

For a given force the extent of chest wall depression is dependent on the amount of air filling the lungs, on whether the glottis is open or closed, on the impact velocity, and on the tenseness of muscles. Chest impact studies have been conducted by a number of researchers [61–65]; however, the exact influences of inflated lungs, status of the glottis, and muscle tone are still not clear. Although impact velocity clearly influences the thoracic response, the exact reasons for these differences in load–penetration characteristics are again not known. This is because injury tolerance work can only be done on animals and human cadavers. Results from animal studies cannot be very easily extrapolated to humans, and cadavers do not have the necessary muscle tone and other lifelike properties. Human volunteers can only be used for static chest load–deflection tests for low deflections. The relationship between the response of human cadavers and live humans can be determined by performing static load–deflection tests on cadavers and human volunteers. The cadavers themselves can be made more 'lifelike' by inflating their lungs and pressurizing their vascular system. Over and above all these factors, there is yet another factor even more difficult to quantify and that is the effect of variation between people. This variation can exist owing to age, sex, and pathological condition.

Several parameters have been suggested for evaluating injuries to the thoracic cage, in particular acceleration, force, displacement, or some combination of these. The experimental data indicate that while maximum forces on chest impact vary with impact velocity and the condition of the subject, the rib fractures usually occur only when chest deflections exceed 2 in in front or side impact. These fractures occur owing to bending of ribs in front or due to torsion nearer the spinal column as the rib cage moves downwards anteriorly on impact and causes the ribs to rotate about the spinal column. As the deflection limit does not change appreciably from quasi-static deflection rates to dynamic impact velocities of $30 \, \text{ft s}^{-1}$, it appears that rib fractures are primarily dependent on the extent of chest deflection and not on impact forces.

There are some problems associated with the use of the cadaver chest for obtaining tolerance information and careful consideration must be used when interpreting chest impact data. Recent studies indicate that the effects of muscle tension and air-filled lungs can contribute significantly to the load-carrying ability of the thorax. The effects of tensing of thoracic muscles are indicated in Fig. 2.44. The data bands cover the range for all the data gathered

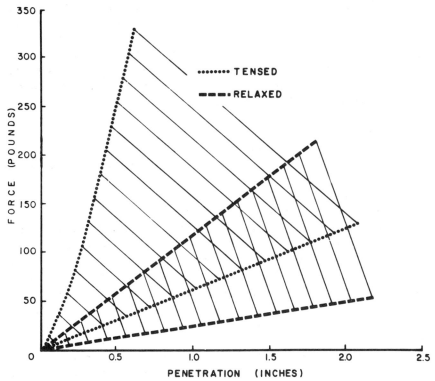

FIG. 2.44.   The effects of tensing of thoracic muscles on the load-carrying ability of the thorax.

by various investigators, and therefore the wide range is due to both anatomical differences among volunteers and different testing procedures. It is worth noting that the maximum stiffness of the tensed volunteers is about twice that of the maximum stiffness of the relaxed volunteers and almost eight times their minimum stiffness. When these curves are compared with those obtained for embalmed and unembalmed cadavers the stiffness of the chests of unembalmed cadavers is found to lie in the lower range of relaxed volunteers' chest stiffnesses. This is probably due to the lack of muscle tone.

Specification of thoracic impact tolerance is complicated further by the lack of biomechanical fidelity of current dummy chest structures. A deflection criterion for thoracic impact can be used directly only if the dummy chest responds to load in a manner similar to the living human thorax. Under certain well-controlled loading environments an equivalent deflection response of a dummy chest might be usable, but it would not be useful in general applications with a variety of loading conditions. For frontal chest impact tolerance a deflection of 1.75 in (4.4 cm) may be used if rib fracture is to be avoided. A tolerance value based on the AMA Abbreviated Injury Scale (AIA) level of 3 (severe, but not life-threatening) would be in the 2.5–3.0 in (6.4–7.6 cm) range for the average male. This corresponds to an approximately 30–35 per cent reduction in the chest depth. A similar value for percentage reduction in chest width was found for side impact studies by Stalnaker and Mohan [65] using experimental lower primates. The corresponding deflection levels for an average male would be 2.65 in (6.7 cm) for a non-fracture level and 3.72 in (9.45 cm) for an AIS level 3 injury in side impact.

Tolerance to acceleration

The earliest and most reliable experiments to measure human response to acceleration are reported in the classic studies of Stapp [66, 67]. These studies are based on the Holloman rocket sled tests and Daisy track tests. In the Holloman runs, human volunteers were seated in various configurations with different restraint systems. These volunteers were able to withstand accelerations varying from 35 to 50 g with pulse durations exceeding 100 ms. Volunteers in the Daisy track tests were subjected to decelerations ranging from 19.6 g to a maximum of 39.3 g for a pulse duration of 65 ms in the forward facing direction. In the most severe case the volunteer suffered compression fractures of the vertebral bodies probably caused by interaction with the restraint system. In the rearward-facing direction the decelerations varied from 37.3 to 82.6 g. In the most severe case the volunteer suffered severe chest pains and clearly the safe tolerance level had been exceeded.

Stapp did not quantify any differences between the A–P (anterior–posterior) and p–a (posterior–anterior) thoracic tolerance limits. Another series of tests was conducted by Mertz and Gadd [68] in which a professional high diver dived on to a 3 ft deep mattress from heights ranging from 27 to

57 ft. Decelerations were measured at the sternum and the P–A component ranged from 28.5 to 46 g. No unusual discomfort was experienced in any of these tests. The maximum onset of the P–A component was recorded at 5900 $g s^{-1}$. Judging from this information and other papers cited in the bibliography, it should be safe to assume that a deceleration of 50 g for an impulse duration of 100 ms may be considered as a safe tolerance limit for normal healthy males.

The aviation industry and the space programme have been rather interested in human tolerance limits to Gz acceleration. Initially the interest in this area stemmed from problems encountered in aircraft ejection-seat systems. Most of the studies do not specifically relate human thorax tolerance to Gz acceleration. Swearingen et al. [69] recorded cases of severe chest and stomach pains in subjects when shoulder accelerometer recorded 10 g with a rate of onset of slightly over 600 $g s^{-1}$. In standing impacts with knees locked, limits of tolerance were found to be identical to sitting except that input loads were lower. This difference may be accounted for by the rigidity of the skeletal system.

All of the above measured thoracic tolerance levels in cases of whole-body acceleration and actual forces and deflections at the chest were not really measured. It was not until the 1960s that investigators started looking at dynamic load deflection characteristics of the human thorax and even now these numbers are not available.

Analytical models

At present there are no analytical models of the thorax which can be used for predicting injuries to the rib-cage or the thoracic organs. We still do not have enough information on injury tolerance for all ages and sizes of people and this makes it very difficult to construct any general model.

However, based on the current state of the art a few models have been developed which try to approximate limited aspects of chest dynamics. These models have been developed by Coerman and co-workers [70, 71], Morris, Lucas, and Bresler [72], Fletcher [73], Olsen [74] Roberts and Chen [75], and Lobdell et al. [75]. Of these, we shall briefly discuss the Roberts and Chen model, since they are more detailed attempts at description of the human chest in mechanical terms.

Chen and Roberts [77] used a thin-walled ellipse to approximate the cross-section of ribs in modelling the human thorax. The finite-element model they have developed (Thorax II) is limited to the mid-sagitally symmetric case. The system is elastic, linear, and conservative and is based on small deformation theory. The component-mode synthesis technique is used in which the structure is modelled as an assemblage of connected structural components or substructures. Since the problem is mid-sagittally symmetric, only half of the thorax need be modelled. The model includes the head, neck, thorax, lumbar-sacral region, and pelvis. All these have been idealized as a system consisting of 27 connected components with 1104 unconstrained degrees of freedom. Each component is modelled as a three-dimensional finite-element structure consisting of from 1 to 13 beam-type elements successively connected at nodal points where the lumped masses are attached. Each element has uniform cross-sectional properties along its longitudinal axis.

For a given force input on the thorax, this model is capable of predicting displacement time histories of the ribs and sternum, the state of strain and stress at any given time in the structural members, and fractures of the ribs and sternum. The response of the model is compared with the data of Kroell, Schneider, and Nahum [63], Kroell and of Nahum *et al.* [78] in Fig. 2.45. The shortcomings of this model are that it does not incorporate the viscoelastic characteristics of thoracic soft tissue nor the effects of change in inflation of the lungs and the muscle tone in intercostal muscles. It is mid-sagitally symmetric and hence cannot respond to any eccentric loading conditions. A serious drawback is that endothoracic organs are not included in the modelling and so we obtain no information about internal injuries. This is a serious problem since rib injuries alone are considered moderate and it is the internal injuries which can be life threatening. It is hoped that relevant additions can be made to this model to make it more useful in predicting visceral injuries.

2.9. Experimental investigations of restrained and unrestrained occupants

Frontal impact: kinematics and test devices

In practice most frontal impacts take place centrally and contribute to 90 per cent of head injuries. The influence of the safety belt is most predominant in this type of accident. In the frontal collisions the kinematics of unrestrained and restrained occupants differ considerably as shown in Fig. 2.16.

The most representative results can undoubtedly be obtained by accident simulation with fabricated cars. However, for economical reasons, deceleration and acceleration sleds have been developed which allow repeated utilization [79–83]. For the same reasons, most sled runs are carried through without body parts [84]. It is possible on those sleds to install the seat in such a way to ensure various crash directions (frontal, lateral, and rear).

In order to determine the tolerance limits of the human body at least the volunteer's threshold of pain has to be exceeded. Patrick [85] discovered considerable similarity between volunteers and unembalmed cadavers when comparing high-speed films. Contusions and abrasions of the cadaver's soft tissues are similar to those seen in human volunteers and in real accident evaluations. However, owing to the low blood pressure, contusions do not spread as much as they do in the living human.

Simulation of blood pressure on the vascular system increases the test

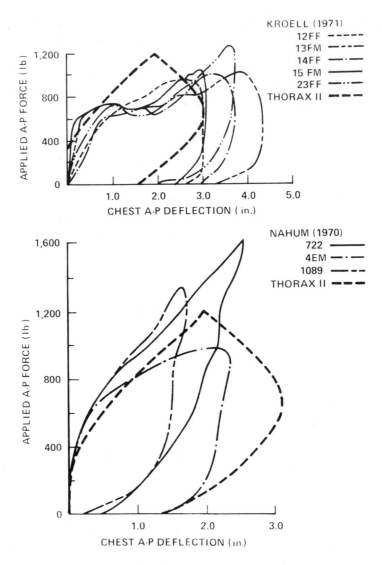

FIG. 2.45. The force *versus* displacement response of the Thorax II model of Chen and Roberts [77] compared with that of Kroell *et al.* [63] and Nahum *et al.* [78].

reality. Furthermore, by filling the lungs and other organs with gas or liquid it is possible to simulate realistic conditions [85].

Test results on restrained occupants

The following results have been obtained with the aid of the deceleration sled construction of the Institute of Legal Medicine at Heidelberg, using unembalmed human cadavers of both sexes and ages. Frontal impacts in front-seat passenger position were simulated for an impact velocity of 50 km h^{-1}. The deceleration–time characteristics and the belt forces are shown in Fig. 2.46. The restraint systems employed were three-point belts with retractor and diagonal shoulder belts with retractor combined with a kneebar (belt width 50 mm, 17 per cent elongation at 1135 kp) (Fig. 2.47).

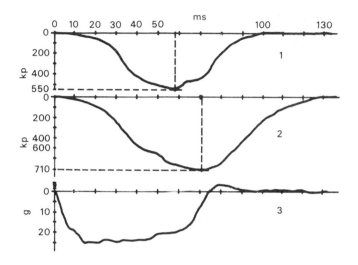

FIG. 2.46.   Deceleration–time (3), lap belt force–time (1), and shoulder belt force–time (2) characteristics in a simulated frontal impact (impact velocity 50 km h^{-1}).

The arrangement of rib fractures was governed by the belt direction on the thorax during the strain. Direct bending fractures at the site of the bearing surface as well as indirect bending fractures beyond the contact area were produced predominantly in the right lower and left upper thorax parts and in the paravertebral region. Fractures of the sternum mostly occurred in combination with rib fractures. However, isolated sternum fractures were also found predominantly as transverse fractures parallel to the belt direction.

Injuries of the intra-thoracic organs of restrained cadavers in front-seat passenger position at impact velocity of 50 km h^{-1} generally occurred as pleura transfixings and pleura injuries in connection with rib fractures (Table 2.9). Pulmonary ruptures also occurred owing to shear and tractive forces caused by a lateral pushing away of the lungs during the compression of the thorax. Osseous injuries of the thoracic vertebral column occurred in the transition regions of the cervical spine and lumbar vertebral column. On corresponding areas injuries of the intervertebral discs as well as of the muscle and ligamentous systems have been found.

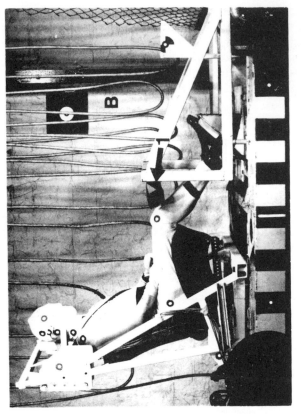

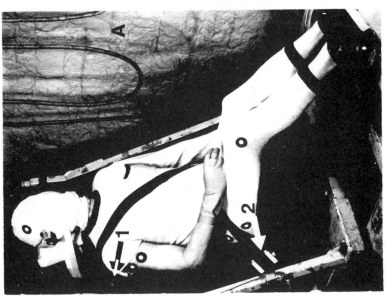

FIG. 2.47.   (a) Three-point belt system; (b) diagonal shoulder belt–knee bar system.

TABLE 2.9
Frequency of injuries of intra-thoracic tissues

Transfixing of the pleura parietalis	14
Transfixing of the pleura visceralis	4
Rupture of the lung root	4
Rupture of the hollow vein	2
Contusion of the thymus	2
Intima laceration of the artery	2
Mediastinal haemorrhage	1
Rupture of the aorta	1
Rupture of the atrium	1

55 cadaver tests; impact velocity 50 km h^{-1}.

Analytical studies of impact kinetics

During simulated crash accidents, motions of vehicle occupants are pictured by high-speed cameras and with high vision frequencies (500 pictures s^{-1}, 1000 pictures s^{-1}). Slow-motion pictures furnish detailed information on motion phases of the body parts and on maximal displacements and injury mechanisms. With the help of film analysers, linear and angular velocities and accelerations of the body parts (with the help of markers on the occupant's body) are determined which (in combination with the weights of the body parts) enable determination of forces on the body parts.

2.10. Recommendations for restraint systems

The purpose of restraint systems

Three advantages of restraint systems in automobiles during an impact have been recognized. Restraint systems prevent the ejection of occupants and transmit the deceleration of the automobile to the occupants during the first milliseconds during which the energy-absorbing plastic deformation of the vehicle takes place; they prevent to a great extent secondary impact of the occupants inside the automobile.

Wearing of safety belts by automobile occupants was suggested by Straith [86] at the end of the 1930s. De Haven [87] and Stapp [88] have given conclusive proof that restraint systems are most valuable in increasing the deceleration tolerance of the human body during sudden braking. Systematic injury studies at the end of the 1950s comparing non-safety-belt wearers and safety-belt wearers indicated a 60 per cent reduction of injuries [89, 90] and a 35 per cent reduction of most serious or fatal injuries [91] with the use of a restraint system.

According to the impact velocity and the type of injury, safety belts do indeed prevent serious injuries and fatalities but on the other hand they can also constitute sources of injuries for the person using them [91].

Belt types installed in automobiles

The three belt systems used up to now, lap belt, two-point belt, or diagonal shoulder belt and three-point belt, can cause their own characteristic injury patterns.

Even a small belt slack in a belt system can reduce its efficiency during a crash. The automatic belt reduces the belt slack to a minimum owing to the spring force of the retractor. When using static belts the occupant should press his upper torso against the belt before the collision in order to eliminate the belt slack; thereby, the deceleration distance will be reduced. Another way to eliminate belt slack is to preload the belt by means of gas cartridges which inflame sensors during collision in order to connect the occupants as quickly as possible with the vehicle.

The evaluation of real accidents has shown that a combination of diagonal and lap belts, the so-called three-point belt systems, are the most effective restraint systems currently available for injury protection.

The limitation of the belt force

During simulated accidents with fresh cadavers as well as during real accidents, highly dangerous or fatal injuries may occur according to the crash velocity, the kind of crash, and the age of the occupants. In order to prevent an increase of the belt load beyond the human tolerance level force limiters are installed in belt retractors or in the belt ribbon. Such force limiters (e.g. torsion bars) are installed in the automatic retractor and are also provided as folded and sewed up belt parts. However, force limitation is provided at the expense of a larger displacement and therefore it is possible that parts of the body may crash against the interior equipment of the vehicle. Hence, an optimal design synthesis of the restraint system and the interior equipment is called for.

Experimentally studied restraint systems installed in test vehicles

Four-point belt. Up to now four-point belts have only been installed experimentally with cadavers [93] or with volunteers at low collision velocity at 24 kmh^{-1} [94]. A reduction of injuries has been noticed with the use of this belt system. The four-point belt has not been installed on a large scale in standard vehicles since its application is somewhat complicated for everyday use. The deceleration forces of the four-point belt are transmitted to the shoulders. Therefore compared with the three-point belt lesser injuries result for the same crash situation.

Air bag: principle of operation. Investigations with the air bag have been predominantly carried out in the U.S.A. up to 1972 [95–97]. It was shown by test vehicles that the application of the air bag, combined with a lap belt, offered satisfactory accident protection.

The folded plastic wrapper is placed in a hollow space of the steering wheel or the dashboard. The air bag is connected to a compressed air cylinder by means of a valve. At the moment of the accident the valve opens electronically and the wrapper inflates in front of the occupants in about 30 ms. The occupants, thrown forward because of the impact, are stopped by the air bag, which thus prevents their impacting on the steering wheel, the dashboard, or the windscreen. The wrapper of the air bag is porous or has fissures through which the gas can escape if a body impacts against it and hence the body is stopped as gently as possible.

The air bag is considered only for a single use during a crash and is not suitable for successive automobile collisions.

Recommended restraint systems

By consideration of the experimental results of the three-point belt with a force limiter on the one hand and the air bag on the other hand, it is to be expected that a combination of both systems can offer optimal protection during frontal crashes and lateral crashes of up to 30°. This combination would also be effective during successive collisions. However, this restraint system is not sufficient for lateral crashes up to 90°. For those a shell seat with a lateral support would be necessary.

References

1 GOGLER, E. (1968). *Chirurgie und Verkehrsmedizin* (ed. K. Wagner, U. H. J. Wagner). Springer, Berlin.
2 SAEGESSER, M. (1972). *Spezielle chirurgische Therapie*, Part 9. Verlaghaus Huber, Bern.
3 VOIGT, G. E. (1968). Die Biomechanik stumpfer Brustverletzungen, besonders von Thorax, Aorta und Herz. *Hefte zur Unfallheilkunde,* Heft 96. Springer, Berlin.
4 WEIGEL, E. (1965). Die Herzlasion als Folge stumpfer Brustkorbtraumen. *Zentraebl. Chir.* **90**, 2509.
5 ADEBAHR, G. (1966). Histologische Befunde am Herzmuskel bei Wider-belebungsversuchen. *Dtsch. Z. ges. Gericht. Med.* **57**, 205–11.
6 LASKY, I. I., NAHUM, A. M., and SIEGEL, A. W. (1969). Cardiac injuries incurred by drivers in automobile accidents. *J. forens. Sci.* **14**, 13.
7 ROSENKRANZ, K. A. and FRITZE, E. (1967). Herzinfarkt und Brustkor-btraume, *H. Unfallheilk.* **75**, 29–34.
8 MATTERN, R. (1974). *Zum traumatischen Herzinfarkt.* Vortrag ober-rheinische Rechsmediziner, Mainz.
9 KLEIN, H. (1973). *Verkehrsmedizin im Handwörterbuch der Rechtsmedizin.* Ferdinand Enke, Stuttgart.
10 VOIGT, J. (1967). *Laesio traumatica cordis. Eine Frage von locus electus,*

Vol. 46. Thagung der Deutschen Gesellschaft für ger. und soziale Medizin, Kiel.

11 BRIGHT, E. F. and BECK, C. S. (1934). Non-penetrating wounds of the heart. A clinical and experimental study. *Am. Heart J.* **10**, 293.

12 BECK, C. S. (1935). Contusion of the heart. *J. am. med. Ass.* **104**, 109.

13 OPPENHEIM, F. (1918). Gibt es eine Spontanruptur der gesunden Aorta und wie kommt sie zustande? *Münch. med. Wschr.* **65**, 1234.

14 RINDFLEISCH, E. (1884). Über klammerartige Verbindungen zwischen Aorta und Pulmonalarterie (Vincula aortae). *Virchows Arch. path. Anat. Physiol.* **96**, 302.

15 ZEHNDER, M. A. (1960). Unfallmechanismus und Unfallmechanik der Aortenruptur im geschlossenen Thoraxraum. *Thoraxchirurgie* **8**, 47.

16 MATTERN, R. (1972). Beitrag zur Zugfestigkeit der Aorta. Inaug. Diss. Heidelberg.

17 KRAUSS, H. (1966). Brustkorb, Lunge, Zwerchfell. In *Handbuch der gesamten Unfallheilkunde* (ed. H. Burkle de la Camp and M. Schwaiger) Vol. 2, Part 3. Ferdinant Enke, Stuttgart.

18 LLOYD, J., HEYDINGER, K. K., KLASSEN, K. P., and ROTTIG, L. C. (1958). Rupture of the main bronchi in closed chest injury. *Archs Surg., Chicago* **77**, 597.

19 GREENDYKE, R. M. (1966). Traumatic rupture of aorta. *J. am. med. Ass.* **195**(7).

20 PATEL, J. W. *et al.* (1968). Traumatic rupture of the thoracic aorta. *J. am. med. Ass.* **203** (12), 118–20.

21 STRASSMANN, G. (1947). Traumatic rupture of the aorta. *Am. Heart J.* **33**, 508–15.

22 LUNDEVALL, J. (1964). The mechanism of traumatic rupture of the aorta. *Acta pathol. microbiol. scand.* **62**, 34–46.

23 JENSEN, O. M. (1964). Traumatisk aortarupture. *Nord. Med.* **71**, 337–41.

24 SEVERY, D. M., MATHEWSON, J. H. and SIEGEL, A. W. (1962). Automobile side-impact collisions, Series II, *Natl. Auto Week*, No. 491A.

25 LUNDEVALL, J. (1963). The mechanism of traumatic rupture of the aorta. *Norsk Laegeforen. Tidsskr. March,* 440–4.

26 PETERSON, L. (1960). Mechanical Properties of Arteries *in vivo. Cir. Res.* **8**, 622–639.

27 MCDONALD, J. B. and CAMPBELL, W. A. (1945). Traumatic rupture of the normal aorta in young adults. *Am. Heart J.* **30**(4), 321–4.

28 STRASSMAN, G. (1947). Traumatic rupture of the aorta. *Am. Heart. J.* **33**, 508–15.

29 GABLE, W. D. and TOWNSEND, F. M. (1963). An analysis of cardiovascular injuries resulting from accelerative force. *Aerosp. Med.* 929–34.

30 TAYLOR, E. R. (1962). Thrombocytopenia following abrupt deceleration.

Rep. ARL-TDR-62-30, 6571st ARL, AMD, AFSC, Holloman Air Force
Base, New Mexico.

31 LAWTON, R. W. (1955). Measurements on the elasticity and damping of
isolated aortic strips of the dog. *Circ. Res.* **3**, 403–8.

32 BERGEL, D. (1961). The static elastic properties of the arterial wall. *J.
Physiol., Lond.* **156**, 445–57.

33 KIVITY, Y., and COLLINS, R. (1974). Steady state fluid flow in viscoelastic
tubes: applications to blood flow in human arteries. *Arch. Mech.* **26**, 921–
31.

34 FUNG, Y. C. (1970). Mathematical representation of the mechanical
properties of the heart muscle. *J. Biomech.* **3**, 381–404.

35 COLLINS, R. and HU, W. C. L. (1972). Dynamic deformation experiments
on aortic tissue. *J. Biomech.* **5**, 333–7.

36 STAPP, J. P. (1970). Voluntary human tolerance levels. In *Impact injury and
crash protection,* (ed. E. S. Gurdijian *et al.*), Chap. 15, pp. 308–49. C. C.
Thomas, Springfield, Ill.

37 RUDINGER, G. (1970). Shock waves in mathematical models of the aorta.
J. appl. mech. **37**, 34.

38 BEAM, R. M. (1968). Finite amplitude waves in fluid-filled elastic tubes:
wave distortion, shock waves and Korotkoff sounds. *NASA Tech. Note TN
D-4803.*

39 LAMBERT, J. W. (1958). On the nonlinearities of fluid flow in non-rigid
tubes, *J. Franklin Inst.* **266**, 83.

40 ANLIKER, M., ROCKWELL, R. L., and OGDEN, E. (1971). Nonlinear analysis
of flow pulses and shock waves in arteries. *J. appl. Math. Phys. (Z. angew.
Math. Phys.)* **22**, 563–81.

41 KING, A. L. (1947). Waves in elastic tubes: velocity of the pulse wave in
large arteries. *J. appl. Phys.* **18**, 255.

42 HAYES, W. D. (1958). The basic theory of gas dynamic discontinuities. In
Fundamentals of gas dynamics (ed. H. W. Emmons), Sect. D, p. 461.
Princeton University Press, NJ.

43 LORENTZ, J., and ZELLER, H. (1972). An analogous treatment of wave
propagation in liquid-filled elastic tubes and gas-filled rigid tubes. *Proc.
Int. Conf. on Pressure Surges, Canterbury, Sept. 1972,* pp. B 5-45–59.

44 KIVITY, Y., and COLLINS, R. (1974). Nonlinear wave propagation in
viscoelastic tubes: application to aortic rupture. *J. Biomech.* **7**, 67–76.

45 OLSEN, J. H. and SHAPIRO, A. H. (1967). Large amplitude unsteady flow in
liquid-filled elastic tubes. *J. Fluid Mech.* **29**, 513–38.

46 RICHTMYER, R. D. and MORTON, J. W. (1967). *Difference methods for
initial value problems,* Sect. 12.7, 13.4. Wiley Interscience, New York.

47 HANSON, P. G. (1970). Pressure dynamics in thoracic aorta during linear
deceleration. *J. appl. Physiol.* **28**(1), 23–7.

48 POLMANTEER, K. E., SERVAIS, P. C. and KONKLE, G. M. (1952). Low temperature behavior of silicone and organic rubbers. *Ind. Eng. Chem.* **44**, 1576–81.

49 MACGREGOR, C. W., and FISHER, J. C. (1946). A velocity modified temperature for the plastic flow of metals. *J. Appl. Mech. A*, 11–16.

50 MCCLINTOCK, F. A., and ARGON, A. S. (1966). *Mechanical behavior of materials*, Sect. 5.10. Addison-Wesley, Reading, Mass.

51 SMITH, T. L. (1962). Nonlinear viscoelastic response of amorphous elastomers to constant strain rates. *Trans. Soc. Rheol.* **6**, 61–80.

52 COLLINS, R., LEE, K. J., LILLY, G. P., and WESTMANN, R. A. (1972). Mechanics of pressure cells. *Exp. Mech.* **12**(11), 514–19.

53 KROELL, C. K., SCHNEIDER, D. C., and NAHUM, A. M. (1971). Impact tolerance and response of the human thorax. *Proc. 15th Stapp Conf.* Soc. Automotive Engineers, New York.

54 BECK, C. S. (1935). Contusions of the heart. *J. Am. med. Assoc.* **104**(2), 109–14.

55 PATEL, D. J., and FRY, D. L. (1969). The elastic symmetry of arterial segments in dogs. *Circ. Res.* **24**, 1–8.

56 REISSNER, E. (1949). On the theory of thin elastic shells. In *Reissner Anniversary Volume* (ed. Polytechnic Inst. of Brooklyn). J. W. Edwards, Ann Arbor, Mich.

57 KRAUSS, H. (1967). *Thin elastic shells*, Chap. 2. Wiley, New York.

58 SCHLICHTING, H. (1968). *Boundary-layer theory* (6th edn) (trans. J. Kestin). McGraw-Hill, New York.

59 LING, S. C., ATABEK, H. B., and CARMODY, J. J. (1969). Pulsatile flow in arteries. In *Applied mechanics, Proc. 12th Int. Congr. of Applied Mechanics, Stanford University, Aug. 1968*, pp. 227–91. Springer, Berlin.

60 APTER, J. T., RABINOWITZ, M., and CUMMINGS, D. H. (1966). Correlation of viscoelastic properties of large arteries with microscopic structure. *Circ. Res.* **19**, 104–21.

61 PATRICK, L. M. and MERTZ, H. J. (1967). Cadaver knee, chest and head impact loads. *Proc. 11th Stapp Car Crash Conf. 1967.* Soc. Automotive Engineers, Warrendale, Pa.

62 PATRICK, L. M., and ANDERSSON, A. (1974). Three point harness accident and laboratory data comparison. *Proc. 18th Stapp Car Crash Conf. 1974.* Soc. Automotive Engineers, Warrendale, Pa.

63 KROELL, C. K., SCHNEIDER, D. C., and NAHUM, A. M. (1971). Impact tolerance and responses of the human thorax. *Proc. 15th Stapp Car Crash Conf.*, Paper 710851, p. 84. Soc. Automotive Engineers, New York.

64 STALNAKER, R. L., ROBERTS, V. L., and MCELHANEY, J. H. (1973). Door crashworthiness criteria. *Rep. No. DOT HS-031-2-382,* National Technical Information Service, Virginia.

65 STALNAKER, R. L., and MOHAN, D. (1974). Human chest impact protec-

tion criteria *SAE 3rd Int. Conf. on Occupant Protection, Troy, Michigan,* Paper No. 740589. Soc. Automotive Engineers, New York.

66 STAPP, J. P. (1949). Human response to linear deceleration: Part I. Preliminary survey of aft-facing seated position. *AF Tech. Rep. No. 5915.*

67 STAPP, J. P. (1951). Human exposures to linear deceleration: Part II. The forward facing position and the development of a crash harness. *AF Tech. Rep. No. 5915, Part 2.*

68 MERTZ, H. J., and GADD, C. W. (1971). Thoracic tolerance to whole body deceleration. *Proc. 15th Stapp Car Crash Conf., 1971,* p. 39. Soc. of Automotive Engineers, Warrendale, Pa.

69 SWEARINGEN, J. J. *et al.* (1960). Human tolerance to vertical impact *Aerosp. Med.* **31**, 989–98.

70 COERMANN, R. R., ZIEGENRUECKER, G. H., WITTWER, A. I., and VON GIERKE, H. E. (1966). The passive dynamic mechanical properties of the human thorax–abdomen system and of the whole body system. *Aerosp. Med.* **31**, 443–55.

71 COERMANN, R. R. (1962). The mechanical impedance of the human body in sitting and standing position at low frequencies. *Hum. Factors,* **4**.

72 MORRIS, J. M., LUCAS, D. B., and BRESLER, B. (1961). Role of the truck in stability of the spine. *J. bone jt Surg. Am.* Vol. **43** (3), 327–51.

73 FLETCHER, E. R. (1971). A model to simulate thoracic responses to air blast and to impact. *Proc. Symp. on Biodynamical Models and their Applications. Rep. AMRL-TR 71–29.* Wright Patterson Air Force Base, Ohio.

74 OLSEN, G. A. (1970). The interaction of the components of the human back. *Proc. Symp. on Biodynamic Models and their Application. Rep. AMRL-TR-71–29.* Wright–Patterson Air Force Base, Ohio.

75 ROBERTS, S. B., and CHEN, P. H. (1970). Elastostatic analysis of the human thoracic skeleton. *J. Biomech.* **3**, 527–45.

76 LOBDELL, T. E., *et al.* (1973). Impact response of the human thorax. In *Human impact response.* Plenum Press, New York.

77 CHEN, P. P. H., and ROBERTS, S. B. (1974). Dynamic response of the human thoracic skeleton to impact. *UCLA Paper Eng.-0274.* University of California, Los Angeles.

78 NAHUM, A. M., GADD, C. W., SCHNEIDER, D. C., and KROELL, C. K. (1971). The biomechanical basis for chest impact protection: I. Force–deflection characteristics of the thorax. *J. Trauma* **11**, 84.

79 SWEENEY, H. M. (1951). Human decelerator. *J. aviat. Med.* **22**, 39.

80 KROELL, C. K., and PATRICK, L. M. (1964). A new crash simulator and biomechanics research program. *Proc. the 8th Stapp Car Crash Conf.* Soc. Automotive Engineers, Warrendale, Pa. Wayne State University Press, Detroit.

81 LANGE, W. (1970). Simulation schwerer Auffahrunfalle mit einer elektro-hydraulischen Katapultanlage. *Automobiltech. Z.* **72** (5), 162–7.

82 HONTSCHIK, H., and SCHMID, J. (1972). Untersuchung von Einrichtungen zur Sicherung von Kindern in Kraftfahrzeugen. *Dtsch. Kraftfahrtforsch. Strassenverkehrstech.* **226**, 162–7.

83 KALLIERIS, D. (1974). Eine Fallgewichtsbeschleunigungsanlage zur Simulation von Aufprallunfallen: Prinzip und Arbeitsweise. *Z. Rechtsmed.* **74**, 25–30.

84 KALLIERIS, D., and SCHMIDT, GG. (1974). Belastbarkeit gurtgeschützter menschlicher Körper bei simulierten Frontalaufprallen. *Z. Rechtsmed.* **74**, 31–42.

85 PATRICK, L. M. (1974). Unembalmed cadaver trauma study. Presented at the IRCOBI Meeting, Goteborg, Sweden.

86 STRAITH, C. L. (1937). Automobile injuries *J. Am. med. Ass.* 348.

87 DeHAVEN, H. (1942). Mechanical analysis of survival in falls from heights of 50 to 150 feet. *War Med.* **2**, 586.

88 STAPP, J. P. (1951). Human tolerance to deceleration. *J. Aviat. Med.* **22**, 42.

89 BRAUNSTEIN, P. W. (1957). Medical aspects of automotive crash injury research. *J. Am. med. Ass.* **163**, 249.

90 LINDGREN, S., and WARG, E. (1962). Seat belts and accident prevention. *Practitioner* **188**, 467–73.

91 WILLIAMS, J. S. (1970). The nature of seatbelt injuries. *Proc. 14th Stapp Car Crash Conf.*

92 TOURIN, B., and GARRETT, J. W. (1960). Safety belt effectiveness in rural California automobile accidents. *Automotive Crash Injury Research of Cornell University, New York.* Cornell Aeronautical Laboratory Report.

93 HINZ, P. (1970). Die Verletzung der Halswirbelsäule durch Schleuderung und durch Abknickung. In *Die Wirbelsäule in Forschung und Praxis.* Hippokrates, Stuttgart.

94 PATRICK, L. M., and TROSIEN, K. R. (1971). Volunteer, anthropometric dummy and cadaver responses with three and four point restraints. *SAE Trans.* **80**, Pap. 710079.

95 CLARK, C., BLECHSCHMIDT, C., and GORDON, F. (1964). Impact protection with the airstop restraint system. *Proc. 8th Stapp Car Crash Conf. 1964.* Soc. Automotive Engineers, Warrendale, Pa. Wayne State University Press, Detroit.

96 CLARKE, T. D., SPROUFFSKE, J. F., TROUT, E. M., GRAGG, C. D., MUZZY, W. H., and KLOPFENSTEIN, H. S. (1970). Baboon tolerance to linear deceleration: airbag restraint. *Proc. 14th Stapp Car Crash Conf.* Soc. Automotive Engineers, Inc. New York.

97 PATRICK, L. M., NYQUIST, G. W., and TROSIEN, K. R. (1972). Safety performance of shaped steering assembly airbag. *Proc. 16th Stapp Car Crash Conf.* Soc. Automotive Engineers, Inc., New York.

3. Dynamic spinal injury: mechanisms, modelling, and systems for minimizing trauma

Y. KING LIU

3.1. Introduction

The spine, in its role as the main structural and kinematic member of the human torso, has intrigued the physician since time immemorial. Recently biomedical engineers and scientist have shared with the physician the concern over the structural integrity of this part of the musculoskeletal system. Much remains to be known about the quantitative behaviour of the spine in health, disease, and injury. Schultz [1] surveyed the present knowledge of the spine under *quasi-static* normal and pathological loading conditions. His review and expository paper is commended to all students and research workers on the biomechanics of the spine. The present article, however, is concerned mainly with the quantitative *dynamic* mechanical behaviour of the spine. Stated another way, while quasi-static problems emphasize loads which are of the order of magnitude of the body weight, dynamic problems give rise to inertial loads on the spine which are orders of magnitude higher than the body weight. Thus parameters which are of primary importance in the quasi-static situation may be of secondary importance under dynamic conditions.

Some topical examples of dynamic spinal loading are pilot ejection from disabled aircraft, automotive accidents of all sorts, manned launchings and recoveries on water or land, etc. To orient the reader on the nature of the problem, the case involving the inertial loading on the spine which has been most studied is discussed below.

The advent of high-speed aircraft during the early 1940s necessitated the development of powered pilot extraction systems. The ejection systems are capable of separating a man from an aircraft without his striking the vertical stabilizer and attaining sufficient height to allow parachute deployment during low-level ejection. In order to accomplish this during the stroke length of the ejection device the man–seat unit is subjected to a relatively high-amplitude short-duration acceleration pulse, a typical one of which is shown in Fig. 3.1. Unfortunately, this procedure generates large forces and deformations in the spinal column which may sometimes exceed safe levels. The large number of vertebral fractures which have been reported following ejection bear testimony to the dangers of this mode of escape.

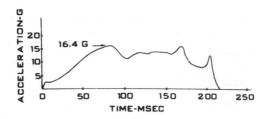

F IG. 3.1.   The acceleration–time profile of the typical Martin–Baker ejection seat used by pilots to escape from disabled high-performance aircraft.

Moffatt and Howard [2] made an extensive compilation (over 1000 incidents) of US Air Force and Navy aircraft ejections during the period 1959–67. They found, as shown in Fig. 3.2, that 17 per cent of the ejections resulted in vertebral fractures with most (70 per cent) occurring in the lower thoracic region (T7–T12). Shannon [3] studied 561 combatant and non-combatant ejections which occurred during 1967 and 1968. In this study 8.5 per cent of the ejectees demonstrated vertebral fractures which were caused entirely by the ejection system. Again the lower thoracic and upper lumbar regions were found to be the primary injury sites.

At a recent (1973) Ejection Seat Committee meeting of AGARD [4] the continuing seriousness of the ejection problem was reported by Professor R. P. Delahaye of France on behalf of all the participating NATO countries. These included the US, British, Hellenic, Italian, German, and French Air Forces. The US was represented by statistics from the Army and Air Force but not the

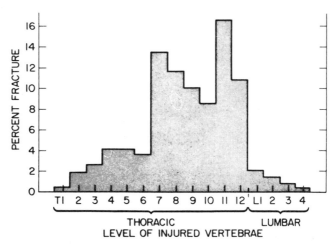

F IG. 3.2.   Spinal injury statistics from over 1000 cases of U.S. Navy and Air Force aircraft ejections (1959–1967) as compiled by Moffatt and Howard [2].

Navy. Of 678 ejection episodes in 1974, there were 114 deaths; 88 pilots sustained spinal injuries which resulted in 152 vertebral fractures. The distribution of 152 fractures in 88 pilots is shown by the solid lines in Fig. 3.3. The distribution of fractures in pilots with multiple spinal fractures is denoted by dotted lines. Figure 3.3 shows the greatest incidence of injury to be at T7–T8 and T12–L1, the thoracolumbar transition vertebra. Professor Delahaye further stated that the frequency of multiple fractures was increasing.

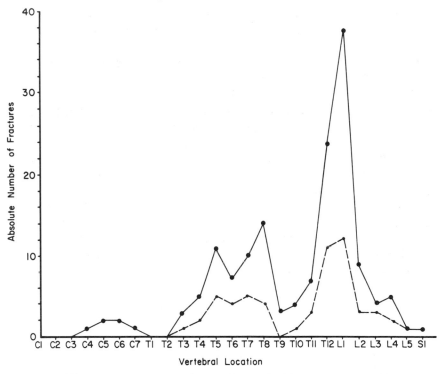

FIG. 3.3. Distribution of single and multiple vertebral fractures sustained by NATO pilots in one calendar year (1972–1973) according to the compilations of Delahaye [4]: solid lines, distribution of 152 fractures in 88 pilots; broken lines, distribution of 64 multiple fractures.

In addition to the medical aspects, there is a substantial economic cost associated with unnecessary fractures. Ewing [5] calculated an average yearly cost to the US Navy of $6 797 718 for aviators who had sustained vertebral fractures.

3.2. Anatomic overview

The supporting structure of the human body is a joint framework of bone called the skeleton which assist in body movement, supports the surrounding tissue, protects the vital organs and other soft tissues, manufactures blood cells, and provides a storage area for mineral salts to supply the body needs. The vertebral column is the main load-carrying part of the skeleton when sitting erect. In any other posture it also carries a substantial portion of the load imposed on the human body.

A detailed study of the vertebral column reveals it to be an extremely complicated structure consisting of fairly rigid bone segments, the vertebrae, connected together by means of ligaments and intervertebral discs. As shown in Fig. 3.4 the vertebral column consists of 33 vertebrae superimposed one on another in a series which provides a strong but flexible supporting column for the trunk and head. The vertebrae are separated by fibrocartilaginous intervertebral discs and are united by articular facet capsules and by ligaments. Of the fundamental 33 vertebrae, five are fused into the sacrum and four combine in forming the coccyx in the adult. The remaining vertebrae are classed according to location from the head downward: seven cervical, twelve thoracic, and five lumbar. Various views of the typical thoracic and lumbar vertebrae are shown in Fig. 3.5.

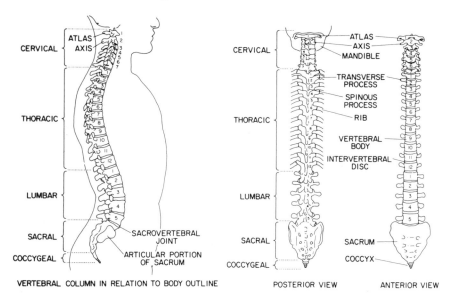

FIG. 3.4.   Lateral, posterior, and anterior views of the vertebral column. (Modified from ref. 6.)

In accordance with their weight-bearing function, the vertebrae become larger and more massively constructed towards the caudal end of the column but are all built on the same fundamental plan. They are composed of two main parts: the anteriorly placed body and the posteriorly placed arch which

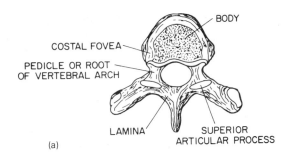

COSTAL FOVEA

PEDICLE OR ROOT
OF VERTEBRAL ARCH

BODY

LAMINA   SUPERIOR
ARTICULAR PROCESS

(a)

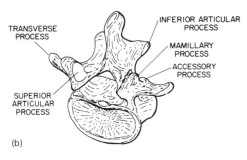

TRANSVERSE
PROCESS

INFERIOR ARTICULAR
PROCESS

MAMILLARY
PROCESS

ACCESSORY
PROCESS

SUPERIOR
ARTICULAR
PROCESS

(b)

FIG. 3.5.   (a) A typical thoracic vertebra viewed from above. (b) A representative
lumbar vertebra viewed from above and behind. (Modified from ref. 6.)

encloses the vertebral foramen through which the spinal cord passes. The body
is the main load-carrying part of the vertebra and has the form of a short
cylinder. The more or less planar end sections of the vertebral body are called
the end-plates. The bodies are separated by and also bound together with the
fibrocartilaginous intervertebral disc and are connected to one another by the
long anterior longitudinal and posterior longitudinal ligaments. These
ligaments are attached to their anterior and posterior surfaces of the body
throughout the entire series of movable vertebrae.

The superior and inferior articular processes of the vertebrae project
upward and downward from the vertebral arch. The overlapping of the
inferior process of one vertebra with the superior process of the vertebra below
limits the range and kind of motion between adjacent vertebrae. The
intervertebral discs are composed of fibrocartilage and vary in thickness and
size along the length of the spine and together constitute about one-quarter of
its length. Each disc is composed of an outer fibrous ring, called the annulus
fibrosus, which surrounds the nucleus pulposus. The fibrous ring consists of
concentrically arranged fibre bundles and of fibrocartilage. The fibres pass
diagonally from the vertebral body above to that below and alternate in their

diagonal direction in successive lamina. The more central lamina tend to be incomplete and less distinct and degrade over into the more cartilaginous nucleus pulposus. The nucleus pulposus is a soft pulpy highly compressible substance and has a high water content. With increasing age it exhibits an increase in fibrocartilage and decrease in its fluid content, i.e. degeneration.

We note that the spine is far from straight, with the curvature in the thoracic region being opposite to the curvature in the cervical and lumbar regions. In the seated position the human frame is supported at the ischial tuberosity, i.e. the lowest point of the pelvis. In this position the pelvis is subjected to some rotation since the ischial tuberosity in general does not lie on the plumb line through the lumbosacral joint. At the other end of the spine the head is supported by the first cervical vertebra, the atlas, which is essentially a hinge. The eccentric mass of the head is supported by the tension in the neck muscles.

3.3. Biomechanical data

In the mathematical modelling of the human system one should have access to the relevant biomechanical data before the modelling process can be meaningful. Frequently, however, such data are simply not available, or at least not in a form which can be directly used by those interested in developing the mathematical model.

A very extensive literature survey of the mechanical properties of vertebrae and intervertebral discs was given by Higgins, Enfield, and Marshall [7]. Since the subject is discussed elsewhere in this book, only a brief review of some *typical* results performed on parts of spinal column are given here. Unless otherwise stated, these results are based on quasi-static rates of loading. The values of the breaking strength of individual vertebrae of a large number of individuals has been compiled and reported by Ruff [8]. The units of failure load are given in kilograms force. In Ruff's results age was not a discernible factor in the strength of the vertebrae and the data reflect surprisingly little spread in the failure load for individual vertebrae. Lissner and Evans [9] obtained axial compression data on the intact lumbar spine which showed force–deformation curves similar to a non-linear hardening spring. A typical result is shown in Fig. 3.6. Brown, Hanson, and Yorra [10] conducted an experiment on the lumbosacral spine with particular emphasis on the intervertebral disc. They found that the ultimate static compressive load for the lumbar discs ranges from 1000 to 1300 lbf. Based on the axial compressive tests conducted on five specimens taken fresh from human cadavers, they found that the stiffness value for the discs range from about 500–800 lbf in^{-1} initially to 12 000–20 000 lbf in^{-1} after the applied load has reached the neighbourhood of 200–400 lbf. Failures under axial load invariably took place in the vertebral end-plates even when well-developed ruptures of the annulus

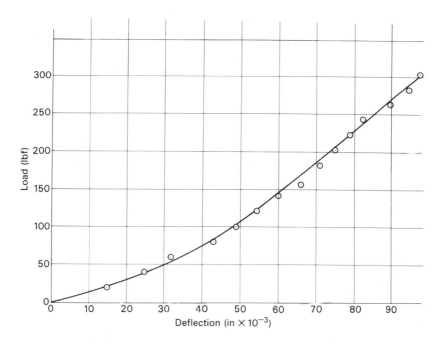

F IG. 3.6.   Typical load–deflection curve of an intact lumbar vertebral column.
(From ref. 9.)

fibrosus were present prior to testing. Failure modes ranged from impercep-
tible cracks to more or less complete collapse of the end-plate depending on the
conditions of the bone and the magnitude of the applied load. We note that the
curve is linear for most of the loading range. The initial non-linearity is
referred to a stress-free state of the test specimen. In reality static preload is
always present although it varies with posture and initial configuration. Thus
the major slope associated with the linear portion of the curve shown in
Fig. 3.6 is more indicative of the stiffness of the intervertebral joint. Crocker
and Higgins [11] performed static and dynamic tests for isolated vertebrae and
have found that the dynamic stiffness increased with increasing strain rate as
illustrated in Fig. 3.7. Their results (Fig. 3.8) also showed that the vertebral
bodies are much stiffer in axial compression than are the intervertebral discs by
a ratio of as much as 5:1.

Most of the mechanical properties of intervertebral discs were obtained
under fairly slow rates of strain and thus represent essentially the static
properties of the discs. However, Hirsch [12] has tested discs under constant
load and measured the resulting deformations, i.e. creep response, as a
function of time. A typical creep curve for a disc subjected to a 100 kgf load is
shown in Fig. 3.9. Hirsch made the following interesting observations:

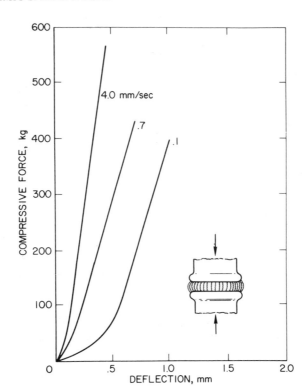

FIG. 3.7.   The effect of strain rate on the load–deflection curve of an intervertebral joint in axial compression. (From ref. 11.)

(a) the curve are essentially level after 5–10 min load application;
(b) most of the deformation occurs in the first 30 s;
(c) the nature of the curve is independent of the level of load (implying linearity with load);
(d) the discs recovered *completely* after the load was removed.

Unfortunately, the detailed behaviour at early times (such as the presence of an instantaneous elastic response) was not given. The inverse of the creep experiment, the stress relaxation test, was performed on a specimen consisting of an intervertebral disc bounded by segments of the adjoining vertebrae (see Fig. 3.10). Under a constant strain (maintained in a materials test machine) the stress in the specimen 'relaxed' to some non-zero value. A good first approximation to the time-dependent material behaviour observed is that of a three-parameter viscoelastic solid (see Flügge [14]).

Evans and Lissner [15] also tested intact fresh human lumbar spines. The

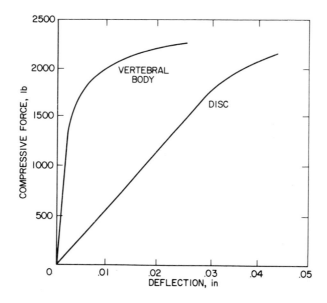

F IG. 3.8.   Load–deflection curves comparing the relative stiffness of the vertebral body and the intervertebral disc. (From ref. 11.)

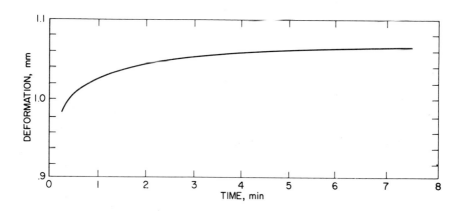

F IG. 3.9.   Creep response of an intervertebral disc subjected to a constant 100 kg compressive axial load. (From ref. 12.)

lumbar spine was simply supported as a beam and a concentrated load was applied to the anterior aspect at the midpoint of its length. A typical load–deflection curve is reproduced in Fig. 3.11.

Liu, Ray, and Hirsch [16] performed the direct shear test on the intervertebral joints of fresh human lumbar spine. The differential deflection

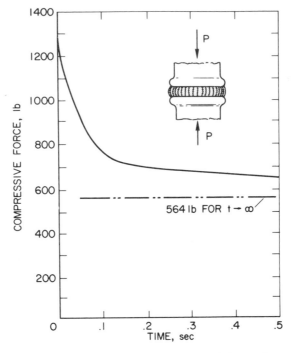

F<small>IG.</small> 3.10. Stress relaxation test for an intervertebral joint under constant strain. (From ref. 13.)

was directly measured by a transducer placed at a slight distance on both sides of the intervertebral disc. Typical load–deflection curves are shown in Fig. 3.12.

Under dynamic loading the most common failures of the intervertebral joint, according to Kazarian, Von Gierke, and Mohr [17], are either in anterior lip (or anterior wedge) or compression mode as illustrated in Fig. 3.13. These clinical and experimental observations strongly indicate the importance of eccentric anterior compression.

The load–deformation characteristics of the vertebral column given above are meant to be typical rather than exhaustive. Readers interested in more detailed information are urged to consult the references given in refs. 1 and 18.

3.4. Problem formulation

No real problem in nature can or even should be treated in all its complexity. Usually the most that can be hoped for is the identification of the more important dependent and independent variables of the problem and the possible functional relation of one to the other.

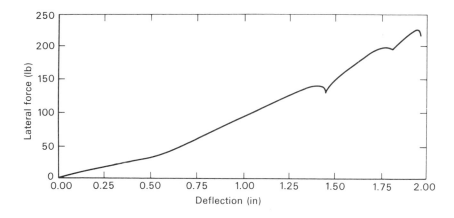

FIG. 3.11.   A typical load–deflection curve in anterior bending of the lumbar spine. The ligamentous lumbar vertebral column was simply supported and loaded at the mid-point of the anterior aspect of vertebral body. (From ref. 15.)

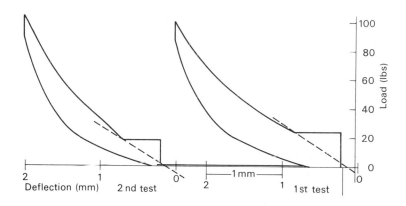

FIG. 3.12.   Typical load–deflection curves in direct shear test. Note that the hysteresis loop is smaller in the second test compared with the first. (From ref. 16.)

In common with any analysis of the dynamics of deformable bodies we focus upon three fundamentals: (1) kinematics of compatible deformations, (2) dynamic equilibrium, and (3) stress–strain constitutive relations. In the interest of clarity, only the planar version of these topics will be presented. The following formulation resembles closely the version given by Cramer, Liu, and von Rosenberg [19].

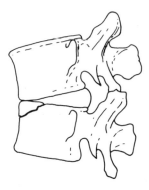

ANTERIOR LIP FRACTURE

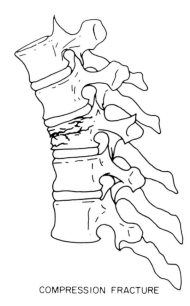

COMPRESSION FRACTURE

FIG. 3.13. The two most common modes of vertebral fracture as a result of pilot ejection: (a) anterior lip (or wedge) and (b) compression fracture. The importance of anterior bending is obvious. (Modified from ref. 17.)

Kinematics

The spinal element to be analysed is considered as an initially curved beam-column. There appears to be a general consensus among research workers that geometric non-linearities are significant. Referring to Figs. 3.14(a) and 3.14(b) we define the following quantities: z, co-ordinate along the vertical reference axis; S, initial arc length along the spine; A, initial angle between the tangent to

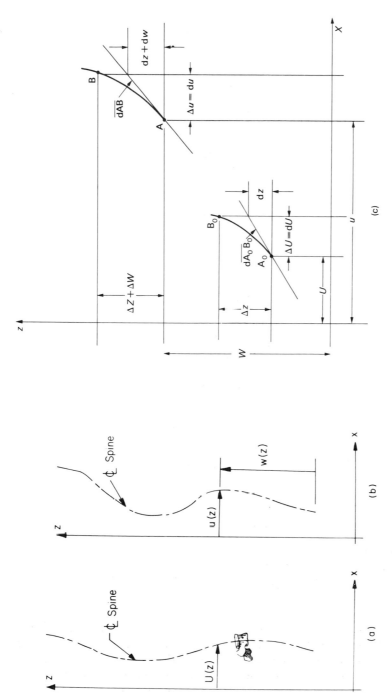

FIG. 3.14. Kinematics of vertebral column deformation: (a) initial configuration of the spinal centroidal axis; (b) the instantaneous deformed reference configuration of the spinal centroidal axis; (c) representation of the strain resulting from the deformation of the spinal centroidal axis. (From ref. 19.)

the curve and the vertical axis; U, initial horizontal displacement of the spine from the vertical reference axis z. It is assumed that z and U are continuous and single-valued functions of S, i.e.

$$z = z(S) \qquad U = U(S).$$

Hence there exists a function U such that

$$U = U(z).$$

The arc length S is related to $U(z)$ by

$$S(z) = \int_0^z \{1 + (dU/d\xi)^2\}^{1/2} d\xi. \tag{3.1}$$

Also

$$\tan A = dU/dz \tag{3.2}$$

and from (3.1)

$$dS/dz = \{1 + (dU/dz)^2\}^{1/2} = \sec A. \tag{3.3}$$

From the definition of curvature K we obtain

$$K = |dA/dS| = |d^2U/dz^2| \{1 + (dU/dz)^2\}^{-3/2}.$$

In the instantaneous deformed configuration the following analogous quantities are defined: s, arc length along the spine; α, angle between the tangent to the curve and the vertical; u, horizontal displacement from the vertical reference axis z; w, vertical displacement of the spine. A set of formulae similar to the initial configuration results:

$$u = u(z)$$

$$s = \int_0^z \{1 + (du/d\xi)^2\}^{1/2} d\xi$$

$$\tan \alpha = du/dz \tag{3.4}$$

$$ds/dz = \{1 + (du/dz)^2\}^{1/2} = \sec \alpha \tag{3.5}$$

$$k = |d\alpha/ds| = |d^2u/dz^2|^{1/2} \{1 + (du/dz)^2\}^{-3/2}$$

Representation of strain

The axial strain can be defined in engineering terms as the unit extension (or compression) of an arc segment. The axial strain, defined as the unit compression, is

$$\varepsilon_a = \frac{dS - ds}{dS} = 1 - \frac{ds}{dS}. \tag{3.6a}$$

Since it is more convenient to deal with the squares of the differential arc segments (in the sense that it is not necessary to make absolute value distinctions) the above is rewritten as

$$ds/dS = 1 - \varepsilon_a$$

or

$$(ds/dS)^2 = (1 - \varepsilon_a)^2 = 1 - 2\varepsilon_a + \varepsilon_a^2. \tag{3.6b}$$

For *small strains* $\varepsilon_a^2 \ll 2\varepsilon_a$ and we obtain

$$\varepsilon_a = 1 - \frac{(ds/dS)^2}{2}. \tag{3.6c}$$

When an initial arc segment $\Delta z \to 0$ (Fig. 3.14(c)):

$$(dS)^2 = (dU)^2 + (dz)^2$$
$$(ds)^2 = (du)^2 + (dz + dw)^2.$$

Substituting the above into (3.6c) and rearranging gives

$$\varepsilon_a = -\left\{ \frac{dw/dz + 0.5(du/dz)^2 - 0.5(dU/dz)^2 + 0.5(dw/dz)^2}{1 + (dU/dz)^2} \right\}. \tag{3.6d}$$

If one assumes that the first derivative of the w component of the displacement is small and of order ε, i.e. $dw/dz = O(\varepsilon)$, then $(dw/dz)^2 = O(\varepsilon^2)$. If, in addition, the initial curvature is assumed to be small, i.e. $(dU/dz) \ll 1$, then both the above assumptions give rise to the following strain measure:

$$\varepsilon_a^* = -\{dw/dz + 0.5(du/dz)^2 - 0.5(dU/dz)^2\} \tag{3.6e}$$

This is the strain term used in classical curved beam-column theory (see for example ref. 20). The quadratic terms in equation (3.6e) express the contribution to the axial strain due to bending of the beam-column, i.e. the coupling between the axial and bending deformations.

In defining the shear strain we follow the assumptions of Timoshenko beam theory [21]:

$$\varepsilon_s = \alpha - A - \phi \tag{3.7a}$$

where ϕ is the angular rotation of a plane cross-section of the beam from its initial configuration to the reference configuration. The strain due to bending of an initially curved beam-column has been studied extensively. The Winkler–Bach formulae have been summarized by many authors and the version given by Seely and Smith [22] is typical:

$$\varepsilon_b = \varepsilon_a + (\omega - \varepsilon_a) \frac{y}{R + y} \tag{3.7b}$$

where ε_b is the strain induced by pure bending at a distance of y from the centroidal axis, ε_a is the axial strain along the centroidal axis as a result of the displacement of the neutral axis towards the centre of curvature, ω is the angular strain, and R is the distance from the centre of curvature to the centroidal axis.

Dynamics

An axial force F, a shear force V, and a bending moment M are assumed to act along any segment of the spine. A differential element of the spine is assumed to be inertially loaded by an associated differential rigid segment of the trunk having its centre of mass anterior to the spine. A free-body diagram is shown in Fig. 3.15. The axial force F acts tangent to the spinal axis and the shear force V acts perpendicular to it. It is further assumed that the distance e from the generic point O on the centroidal axis of the spine to the centre of mass G is not a function of time. This assumption is based on the premise that the location of the centre of mass is unchanged during deformation, i.e. the associated segments of the trunk are *rigid*.

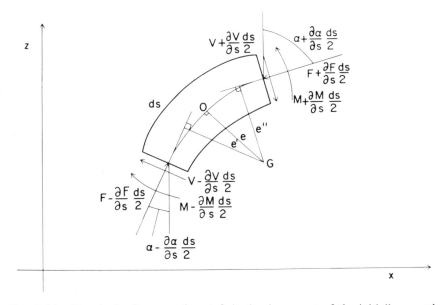

F_{IG}. 3.15.   Free-body diagram of an infinitesimal segment of the initially curved centroidal axis of the vertebral column. (From ref. 19.)

Referring to Fig. 3.15 the sum ΣM of the moments about the point G gives the following equation:

$$\Sigma M = I^*(s)\,ds\,\ddot{\theta} = \left(M - \frac{M_s\,ds}{2}\right) - \left(M + \frac{M_s\,ds}{2}\right)$$

$$+\left(F-\frac{F_s\,ds}{2}\right)e'-\left(F+\frac{F_s\,ds}{2}\right)e''+\left(V-\frac{V_s\,ds}{2}\right)\frac{ds}{2}$$

$$+\left(V+\frac{V_s\,ds}{2}\right)\frac{ds}{2}$$

where the subscript notation is used to denote partial differentiation, I^* is the mass moment of inertia per unit arc length and $\ddot{\theta}$ is the angular acceleration. In the limit, as $\Delta s \to ds$,

$$e' \to e \qquad e'' \to e$$

and we obtain

$$I^*(s)\ddot{\theta} = M_s - e\,F_s + V. \tag{3.8a}$$

The sum $[\Sigma F]^x$ gives the following equation:

$$[\Sigma F]^x = m(s)\,ds\,\dot{v}_2 = \left(F-\frac{F_s\,ds}{2}\right)\sin\left(\alpha-\frac{\alpha_s\,ds}{2}\right)$$

$$-\left(V-\frac{V_s\,ds}{2}\right)\cos\left(\alpha-\frac{\alpha_s\,ds}{2}\right)$$

$$-\left(F+\frac{F_s\,ds}{2}\right)\sin\left(\alpha+\frac{\alpha_s\,ds}{2}\right)$$

$$\left.\left(V+\frac{V_s\,ds}{2}\right)\cos\left(\alpha+\frac{\alpha_s\,ds}{2}\right)\right.$$

where $m(s)$ is the mass per unit length and $\dot{v}_2$ is the lateral acceleration component of point G. Neglecting second-order infinitesimals, i.e.

$$\sin\left(\alpha\pm\frac{\alpha_s\,ds}{2}\right)\approx \sin\alpha\pm\cos\alpha\,\frac{\alpha_s\,ds}{2}$$

$$\cos\left(\alpha\pm\frac{\alpha_s\,ds}{2}\right)\approx \cos\alpha\mp\sin\alpha\,\frac{\alpha_s\,ds}{2}$$

we obtain

$$m(s)\dot{v}_2 = -F\cos\alpha\,\alpha_s - F_s\sin\alpha - V\sin\alpha\,\alpha_s + V_s\cos\alpha. \tag{3.8b}$$

The sum $[\Sigma F]^z$ of the forces in the z direction gives the following equation:

$$[\Sigma F]^z = m(s)\,ds\,\dot{v}_1$$

$$=\left(F-\frac{F_s\,ds}{2}\right)\cos\left(\alpha-\frac{\alpha_s\,ds}{2}\right)$$

$$+\left(V-\frac{V_s\,ds}{2}\right)\sin\left(\alpha-\frac{\alpha_s\,ds}{2}\right)$$

$$-\left(F+\frac{F_s\,ds}{2}\right)\cos\left(\alpha+\frac{\alpha_s\,ds}{2}\right)$$

$$-\left(V+\frac{V_s\,ds}{2}\right)\sin\left(\alpha+\frac{\alpha_s\,ds}{2}\right)-m(s)\,ds\,g$$

where $\dot{v}_1$ is the vertical acceleration component of point G and g is the acceleration component due to gravity. Proceeding as above, we obtain

$$m(s)\,\dot{v}_1 = F\sin\alpha\,\alpha_s - F_s\cos\alpha\;V\cos\alpha\,\alpha_s - V_s\sin\alpha - m(s)\,g. \qquad (3.8c)$$

To change from the intrinsic co-ordinate to the Cartesian co-ordinate, we use the relation

$$\lim_{\Delta s\to 0}(\Delta z/\Delta s) = dz/ds = \cos\alpha.$$

Also from the principle of mass invariance we have

$$m(s)\,ds = m(z)\,dz.$$

Hence

$$m(s) = m(z)\cos\alpha$$

and similarly

$$I^*(s) = I^*(z)\cos\alpha.$$

The final governing differential equations of motion then become

$$I^*(z)\ddot{\theta} = -M_z - eF_z + V/\cos\alpha \qquad (3.9a)$$

$$m(z)\,\dot{v}_2 = -F\cos\alpha\,\alpha_z - F_z\sin\alpha - V\sin\alpha\,\alpha_z$$
$$+ V_z\cos\alpha \qquad (3.9b)$$

$$m(z)\,(\dot{v}_1 + g) = -V\cos\alpha\,\alpha_z + F\sin\alpha\,\alpha_z$$
$$- V_z\sin\alpha - F_z\cos\alpha. \qquad (3.9c)$$

The axial and lateral acceleration components of the centre of mass G can be found in terms of the displacements of the generic point O on the centroidal axis of the spine. The assumption of rigid-body rotation is made to yield the following acceleration components:

$$\dot{v}_1 = w_{tt} - e\cos(A+\phi)\phi_{tt} + e\sin(A+\phi)\phi_t^2 \qquad (3.10a)$$

$$\dot{v}_2 = u_{tt} - e\sin(A+\phi)\phi_{tt} - e\cos(A+\phi)\phi_t^2 \qquad (3.10b)$$

$$\ddot{\theta} = \phi_{tt}. \qquad (3.10c)$$

Constitutive equations

The constitutive equations are formulated from the experimental biomechanical data which showed the materials to be rate dependent. Mathematically, one can express the various material properties described by the constitutive equations of viscoelasticity. As a first approximation it is natural to limit the following consideration to *linear* viscoelasticity. Following Flügge [14] one obtains the general expressions

$$\overline{P}(\sigma) = \overline{Q}(\varepsilon_a) \qquad (3.11a)$$

$$\overline{R}(\tau) = \overline{S}(\varepsilon_s) \qquad (3.11b)$$

$$\overline{P}(M) = \overline{Q}(\Delta k)\,I \qquad (3.11c)$$

where $\overline{P}, \overline{Q}, \overline{R}$, and $\overline{S}$ are differential operators with respect to time of the form

$$\overline{P} = \sum_{k=0}^{m} P_k\,\mathrm{d}^k/\mathrm{d}t^k, \ \overline{Q} = \sum_{k=0}^{m} q_k\,\mathrm{d}^k/\mathrm{d}t^k, \dots \qquad (3.11d)$$

where p_k and q_k are material coefficients and σ, τ, ε_a, ε_s, M, Δk, and I are the normal stress, shear stress, normal strain, shear strain, moment, change of curvature, and area moment of inertia respectively. Restricting ourselves to the three-parameter viscoelastic linear solid in order to be more specific and brief, we can write the constitutive relations in terms of the forces and moments as

$$F/A + (p_1/A)F_t = q_0\varepsilon_a + q_1(\varepsilon_a)_t \qquad (3.12a)$$

$$V/A + (r_1/A)V_t = p_0\varepsilon_s + p_1(\varepsilon_s)_t \qquad (3.12b)$$

$$M + p_1 M_t = q_0\,\Delta k\,I + q_1 I(\Delta k)_t \qquad (3.12c)$$

In equations (3.11c) and (3.12c) we have implicitly assumed that the moment–curvature relation is the same as in an initially straight beam since the distance between the neutral and centroidal axes is negligible for the shallow curve which characterizes the normal spine.

The means whereby the kinematics, dynamic equilibrium, and constitutive equations are grouped, modified, and used characterizes the nature of the dynamic model of the torso-loaded vertebral column.

3.5. Survey and exposition of recent dynamic models of the spine

Discrete-parameter models

Recent dynamic models of the spine can be classified according to the extent to which the above fundamentals are either included or modified in the formulation. In this survey we have excluded all gross motion models with one or two degrees of freedom which cannot account for the effective behaviour at each intervertebral joint, i.e. the shortest wavelength of interest in the model should be at least the thickness of the intervertebral disc.

The paper of Orne and Liu [23] delineated a two-dimensional discrete-parameter vertebral column which considered many of the above-mentioned elements.

(a) axial, bending, and shear deformations of the massless discs were considered with the vertebrae assumed to be part of a rigid body.

(b) The inertial loading on the spine consists of idealizing the torso as rigid segments eccentrically attached to the vertebrae. This is a conservative assumption in that the calculated loads based on this representation are higher than those actually observed. The inertial data used have been given by Liu and Wickstrom [24] who determined the inertial property distribution of the torso by sectioning cadavers and correlated the results found with radiographs of volunteers.

(c) A three-parameter viscoelastic solid model of the discs was used, i.e. equation 3.12 was assumed to be valid.

(d) The variable size of the vertebrae and discs was taken into consideration as well as the natural curvature of the spine. The initial curvature is generated from straight-line segments approximating centroidal axis of the vertebrae.

(e) A trapezoidal acceleration pulse with finite rise and decay times was assumed.

(f) Any direction of planar acceleration loading was allowed.

Orne and Liu [23] used their model mainly to exercise the pilot-ejection problem, i.e. the spinal response to $+G_z$ accelerational ejection. McKenzie and Williams [25] took advantage of the flexibility of the model in terms of the direction of loading and adapted the Orne–Liu approach to study the whiplash problem, i.e. $+G_x$ acceleration. To account for the tethering effects of the passive and active neuromusculature of the neck they *arbitrarily* increased the axial, bending, and shear stiffnesses of the cervical column in order to match some human volunteer motion data. The loads sustained by the cervical intervertebral joints in this model are not very realistic because of the artificial increase in the material stiffness.

Liu, Pontius, and Hosey [26] have studied the effects of constraints on the Orne–Liu model by adding the paraphernalia of pilot ejection, i.e. face curtain, shoulder harness, and seat-back restraints, in the form of linear springs appropriately placed on the head, shoulder, and back. The results indicated that a state of nearly uniform axial stress exists in the vertebral column during ejection and thus the location of the maximum bending stress dictates the location of the most frequent fracture sites. Initial spinal alignment, in terms of the curvature of the column, is a major determinant of the location and magnitude of the maximum normal stresses encountered by the vertebral column during inertial loading.

As a result of some rather sophisticated experimental work involving instrumented cadavers Prasad, King, and Ewing [27] have found that the posterior elements, especially the articular facets, played rather important roles in load support and transmission under caudocephalad $(+G_z)$ acceleration. Based on this evidence they modified the Orne–Liu model to simulate the encapsulated facet joints. Two load paths exist in their model: one along the serially connected vertebra–disc–vertebra chain in the fashion of Orne and Liu and the other through the facets and laminae idealized as springs connected to the vertebral bodies by massless rigid rods. Furthermore, the interposition of the facets one on another at different contact angles along the spine plus the roles of the various ligaments lead to interference and ultimately locking, and thus limit extension. If one were to prehyperextend the vertebral column prior to ejection, then the number of g's which the pilot could sustain without injury to his vertebral column were shown to increase.

The prehyperextension idea has been in existence ever since the beginning of the ejection seat problem. Latham [28] reported work done by the German Air Force during the early 1940s which ascribed anterior fractures of the vertebrae to excessive flexion of the trunk. He further stressed the need for a lumbar pad to attain correct spinal alignment prior to ejection. Bosee and Payne [29] studied the Martin–Baker ejection system and postulated that the common anterior lip fracture was due to bending stresses which resulted from poor spinal alignment. The action of the so-called 'Improved McDonnell Lumbar Pad' on spinal alignment was qualitatively discussed by Sanford and Kellet [30]. Even though Latham [28], Bosee and Payne [29], and many others recognized the importance of bending, they did not incorporate this feature into any of their models or experiments. It remained for Vulcan, King, and Nakamura [31] to measure the amount of bending moment present quantitatively and Prasad *et al.* [27] to quantify the advantage of prehyperextension of the posterior elements in $+ G_z$ acceleration. The work of the Wayne State University's Biomechanics Research Center exemplifies the sort of systematic approach to prevention of spinal injury.

Bowman [32] proposed an analytical model for a vehicle occupant for use in three-dimensional crash simulation work. Included in the model was an n-mass torso joined by $n - 1$ ball and socket joints along the 'spine'. The joint model allows for resistance to bending in two planes and twisting, i.e. resisting moments are provided for the angular co-ordinates but *no* restoring forces were given for the corresponding translational co-ordinates. This simplification was perhaps justified by the conjecture that the magnitude of the axial and shear forces are small for the predominant G_x acceleration problem considered. However, Begeman, King, and Prasad [33] have recently shown, both experimentally and theoretically, the existence of a substantial *axial* force in the spine due to a $- G_x$ acceleration of the torso. Although no measurements were made, one can infer the presence of substantial shear forces as well. Furthermore, the spinal curvature was an important factor in the generation of this axial load. Panjabi [34] has presented a general matrix method for linking a large number of rigid bodies in three-dimensional motion through any combination of serial and parallel spring and dashpot connections between them as a possible model of spinal dynamics. At this point in time the Panjabi proposal should be viewed as an interesting abstract formalism for the *small* displacement analysis of linked rigid bodies since no *specific* model of the spine was presented in his paper.

A few three-dimensional *static* lumped-parameter force–deflection analyses of the vertebral column have appeared in the recent literature. Schultz and Galante [35] have described purely geometrical motions of the *in situ* normal spines. Panjabi and White [36] have used a three-dimensional mathematical analysis to delineate their experimentally obtained motion segment data. Force–deflection analyses of the isolated ligamentous spine have been given by Belytschko *et al.* [37] and Schultz *et al.* [38]. However, *none* of the above spinal models consider the relations between the externally acting accelerations and the forces arising in the tissues surrounding the spine, i.e. the eccentric inertial loading aspect is neglected. This simplification is justified since these models are used only for the kinematic and/or equilibrium analysis of the vertebral column where the loads anticipated are of the same order of magnitude as the body weight. Some preliminary results obtained through the adaption of the above static models to include the dynamic aspect were reported by Schwer, Belytschko, and Schultz [39].

Continuous-parameter models

The obvious extension of the discrete-parameter treatment described above is to formulate continuum models of the spine. One could begin with a tapered, curved, viscoelastic, serially layered beam-column. Unfortunately such an elaborate formulation would defy analytical solution. One retreats to the simplest mathematical formulation consistent with explaining the injury phenomena. By making the *homogeneous* assumption for beam-column the derivations given in the previous section for the kinematics, dynamic equilibrium, and constitutive relation combine to form the governing partial differential equations of motion.

The advantages of a distributed-parameter formulation are many:

(a) such an idealization has an infinite number of degrees of freedom and is thus capable of characterizing the propagation of compressive and shear waves upon inertial loading;

(b) direct determination of the dynamic stress history;

(c) the potential of determining the where, when, and how of the injury mechanism.

The shallow, elastic, sinusoidally curved beam-column under dynamic loading has been considered as a problem in applied mechanics by a number of investigators. These studies were motivated by the behaviour of an imperfect beam-column loaded in a universal material testing machine (see e.g. ref. 40). The first adaptation of the formulation for the spine was made by Li, Advani, and Lee [41]. These authors have attempted the solution of the dynamic curved-beam problem by the assumed-mode (first-mode) method. Moffatt, Advani, and Lin [42] experimentally determined the effective static material properties of the vertebral column as a beam and used them to exercise the model of Li et al. [41]. However, Liu and von Rosenberg [43] showed, using a finite-difference method, that while the assumed first-mode technique might be suitable for the slow loading rate in a testing machine of an initially curved beam-column, it was not appropriate for the rapid loading associated with aircraft ejection. Cramer et al. [19] have formulated the deformation of the spine as a small-strain large-deflection rod problem in plane motion. The assumptions and derivations detailed in § 3.4 summarize the contents of the above paper and thus will not be reiterated here.

Finite-element models

The success of the finite-difference solution obtained depends very much on the ability to extend the governing differential equations derived for an infinitesimal element to the entire continuum, i.e. in this case the initially curved beam-column. While it is theoretically possible to obtain a finite-difference solution to the impact of a continuous but inhomogeneous beam column, in practice it is difficult to obtain solutions where the geometry is complicated, e.g. the layered media which represent the vertebrae and discs of the spine. However, there exists an alternative numerical method known as the finite-element method with features which seem ideally suited to the complications at hand.

The finite-element method of numerical analysis has proved successful in the solution of a number of structural problems, particularly those with complex geometrical shape, material properties, and boundary conditions (e.g. ref. 44).

The basic advantage of the finite-element method lies in the fact that it bypasses the classical procedure of formulating the differential equation for the entire continuum with complicated geometry, constitutive relations, and boundary conditions. Unless the problem is highly idealized, the classic differential equations turn out to be intractable or quite difficult to handle. Usually, either approximations are made to drop a few terms from such equations in order to effect an analytical solution or a finite-difference solution is obtained. In contrast, in finite-element analysis the continuum is divided into basic simple elements of quite small but finite size. The elements are assumed to be joined only at the corners or nodes and all the material and inertial parameters are distributed to these nodes. For example, the inertial force, which normally should act at the centre of mass of the element, is proportioned by a suitable scheme and distributed to the nodes. The scheme for apportionment should be guided by experimental work on the inertial property distribution of the body segments.

It is to these basic elements that the physical laws, such as the equations of motion, energy, and momentum, are *directly* applied together with the assumed or known constitutive relations. Any limiting assumptions or approximations, such as the absence of viscosity, are immediately incorporated into the elemental property, i.e. any special assumptions or limitations are introduced at the element level. Once the element properties have been derived no further approximations are allowed. The resulting equations can then be solved simultaneously together with the associated boundary and initial conditions.

Using the finite-element technique Liu and Ray [45] constructed layered-medium models of the vertebral column. The following important geometrical and material properties of the human spine were accounted for:

(a) actual initial curvature of the spine;
(b) the layered construction of the spine with alternate discs and vertebrae;
(c) variable area and thickness of the discs and vertebrae along the length of the spine;
(d) different material properties (E, ρ, and v) for each disc and vertebra;
(e) simultaneous propagation of the axial, shear, and bending deformations including the contribution of rotary inertia and shear deformation.
(f) two distinctly different wave speeds, a faster one $c_1 = (E/\rho)^{1/2}$ for axial and bending waves and a slower one $c_2 = (kG/\rho)^{1/2}$, were considered for the shear wave;
(g) the inertial properties of the head were lumped into the nodes of an arbitrary 'last' element, but any arbitrary constraint on the deflection and rotation boundary conditions on this so-called ('last') head element could be imposed.

The soft tissue surrounding the spine was previously idealized as rigid segments in the models discussed thus far. This assumption overestimates the stresses on the vertebral column as a result of inertial loading. Using an extension of the finite-element model Liu and Ray [46] removed the rigidity assumption. Instead the segments are now assigned soft-tissue material properties. As well as geometric similarity to the *in vivo* human spine, dynamic similarity for each segment was insured by demanding that the following three equations be satisfied:

$$M_s = \sum_i m_i \tag{3.13}$$

$$e_s = \sum_i m_i x_i / M_s \tag{3.14}$$

$$I_y = \sum_i m_i x_i^2 \tag{3.15}$$

where M_s is the mass of the soft-tissue segment surrounding the intervertebral joint, m_i is the mass at the ith node of an element representing the segment, e_s is the eccentricity of the centre of mass with respect to the geometric centre of the vertebral column, and x_i is the location of the nodes with respect to the geometric axis of the spine.

The segmental values of M_s, e_s, and I_y were given by Liu and Wickstrom [24]. As well as satisfying conditions (3.13)–(3.15) certain density assumptions were made with respect to both the vertebrae and the surrounding soft tissue. Because of the difference in wave speeds between the vertebral column and the soft tissue, a distinct propagating bulge in the soft tissue was displayed in the time-dependent configuration of the soft-tissue-loaded spine. Such a bulge motion was observed by Kazarian, Hahn, and von Gierke [47] in primate acceleration experiments.

The role of neuromusculature

In the dynamic models of the human body considered so far the role of the neuromusculature has been considered to be of secondary importance. Because of the high applied acceleration the inertial loads induced on the spine are so predominant as to make all the other sources of load contribution secondary. The above physical argument appears to be partially substantiated by the few papers available on the interaction of the skeletal muscle with finite displacements of articular members of the body. The second reason for the lack of accounting for the neuromuscular forces is the difficulties in its analysis. Because of the large number of muscular origins and insertions the problem is highly indeterminate, i.e. there are far more unknowns than equations of motion available to solve them.

Soechting and Paslay [48] proposed a model for the human spine under $-G_x$ acceleration in the form of a straight rod in which the effect of the musculature is simulated as lateral surface tractions. They showed that in the particular situation investigated the flexural rigidity of the column can be neglected in a first approximation. Soechting [49] obtained a singular perturbation-type solution to account for the fact that the flexural rigidity of the column is important at the boundaries, i.e. the pelvic and head end.

Seireg and Arviker [50] have reported on the use of linear programming techniques to handle the large indeterminate problem associated with the neuromuscular forces acting on the vertebral column during quasi-static loading. The criterion function used for the optimization is not related to functional anatomy and/or physiology but in spite of this their results appear to be in qualitative agreement with the electromyogram data of Inman [51] and the 1953 research report of the Biomechanics Group at the University of California, Berkeley [52] as well as the intravital discometric measurements of Nachemson and Morris [53].

The thoracic spine is joined to the rib cage through encapsulated articulations. The box-like construction of the rib cage provides considerable flexural rigidity to the thoracic spine. The lumbar spine is anchored in the pelvic mass in addition to being relatively stiff with respect to bending, shear, and axial forces. In contrast, the cervical spine has intervertebral discs which have very little flexural rigidity. In addition, the articular facets have large moment arms about the mid-sagittal plane. The head and neck remain upright only because of the presence of the neuromusculature spanning the head and neck junction.

Stapp [54], in his pioneering experiments on linear deceleration, noted the importance of 'flexing the head and neck forward as near to horizontal as possible while holding the hand-holds tightly and bracing the knees together' in raising the threshold of tolerance to crash velocities without significant pain or injury. Hendler *et al.* [55] repeated Stapp's experiments with better restraints and instrumentation on volunteers. Two of their restrained and braced subjects were exposed to crash velocities of 30 mile h^{-1} without significant discomfort. Severy, Mathewson, and Bechtol [56] compared the response of anthropomorphic dummies and human volunteers in simulated rear-end automobile collisions. In the absence of neuromusculature the dummy head and neck response was considerably higher. Patrick and Trosien [57] and Mertz and Patrick [58] reported on similar tests with volunteer, cadaver, and dummy subjects. The role of cervical musculature in the head and neck response appears to be critical.

Because of the above considerations, Pontius, Liu, and Van Buskirk [59] and Pontius and Liu [60] took into account the active and passive musculature of the cervical spine in a dynamic model of the so-called whiplash injury. The model is an adaption of that of Orne and Liu [23] with the muscles added. The neuromusculature consists of *equivalent* flexor and extensor muscles together with the sternocleidomastoid, which crosses the cervical vertebral column. The model, shown in Fig. 3.16, succeeded in describing human volunteer data and indicated that the primary action of the neuromusculature of the neck is to increase the stability of the cervical spine during low-level $+ G_x$ acceleration at the expense of greatly increased axial compressive forces. Heuristically one would expect that as the acceleration level increases the inertial forces will become dominant so as to make the neuromuscular forces secondary. However, even with $10g$ applied to the $+ G_x$ direction the tethering action of the active neuromusculature was quite manifest. A parametric study was also conducted using the model. Decreased neuromuscular strength is associated with increased bending stresses in locations of high stress concentration, i.e. the lower and upper cervical column. An increase in the (stretch reflex) neural delay time increases bending stresses in the lower cervical spine.

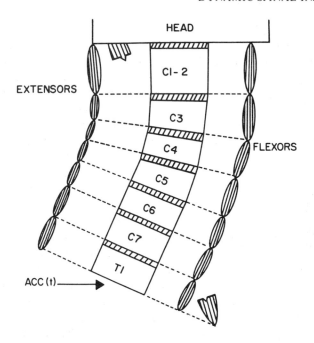

F IG. 3.16. Equivalent muscle system for the cervical spine. SCM, sternocleidomastoid muscle. The model was used by Pontius and Liu [60] to study the effect of neuromusculature on the biodynamic response of the cervical column. ACC (t) — timevarying acceleration.

3.6. Recommendations for future work

A great deal of work, both experimental and theoretical, on the three-dimensional inertial loading of the spine can be done. The basic two-dimensional problem formulation given in the previous sections can be extended to three dimensions through the general non-linear theory of rods, (see e.g. ref. 61).

In extending the discrete-parameter model of Orne and Liu [23] the 'disc' will incorporate the three-dimensional kinematics and constitutive relation of short viscoelastic rods. The inclusion of the articular facet forces in a spatial model, i.e. a refinement of the model of Prasad *et al.* [27], should reflect the geometrical peculiarities corresponding to each region of the vertebral column. The following are typical of these peculiarities.

(a) Veleanu [62] showed that the somato-articular area ratio, i.e. the ratio of the vertebral body to the articular facet area, increases as one examines the vertebral bodies cephalocaudally. The functional implication of this anatomic fact appears to be that the cervical articular facets normally take on an appreciable part of the load imposed on the cervical spine.

(b) The articular surfaces of the neural arch joints lie in different planes, i.e. in the cervical region the articular facets are interposed in a nearly coronal plane while in the lumbar region they face in a nearly sagittal plane. In the thoracic region, however, the facets are positioned on an arc of a circle whose centre is situated in the body of the vertebrae. This arrangement is ideally suited for rotation. In fact it has been asserted that significant rotation in the torso is confined to the thoracic region.

While the three-dimensional discrete-parameter model proposed above allows for the distribution of bending and torsional moments and of axial and shear forces, it cannot compute the dynamic stresses *directly*. To do so requires either a continuum or finite-element model. To extend the two-dimensional continuous-parameter model of Cramer *et al.* [19] to three dimensions one would probably still have to retain the homogeneous rod assumption. The dynamic stress histories thus obtained are 'effective' principal stresses because of the homogeneity assumption.

In the extension of both the discrete-parameter model of Prasad *et al.* [27] and the distributed-parameter model of Cramer *et al.* [19], the inertial loading can be distributed along the column such that *dynamic similarity* with respect to the inertial data of Liu and Wickstrom [24] will be maintained. In the discrete-parameter model the rigid segments are assumed to be attached to the vertebra, while in the distributed-parameter model one can idealize the initially curved continuous beam-column to have 24 point masses set at a fixed distance anterior to the centroidal axis of the column.

Similar remarks apply to the extension of the finite-element models of Liu and Ray [45, 46]. In a three-dimensional finite-element model the skin and viscera should be modelled as having typical soft-tissue properties, i.e. very soft behaviour initially and then an exponential rise (see ref. 63). The viscera can be attached to the vertebral column which in turn will be modelled as an alternating series of wedges representing the vertebrae and discs. The mass distribution of the soft tissue will, as in the two-dimensional model, be required to satisfy the *necessary* condition of dynamic similarity with the inertial data distribution, i.e.

$$M_s = \Sigma\, m_i \qquad\qquad (3.16)$$

$$e_x = \Sigma\, m_i x_i / M_s \qquad\qquad (3.17)$$

$$e_y = \Sigma\, m_i y_i / M_s \qquad\qquad (3.18)$$

$$I_x = \Sigma\, m_i x_i^2 \qquad\qquad (3.19)$$

$$I_y = \Sigma\, m_i y_i^2 \qquad\qquad (3.20)$$

$$I_z = \Sigma\, m_i (x_i^2 + y_i^2) \qquad\qquad (3.21)$$

where m_i is the mass of the ith node representing the segment, M_s is the mass of the soft-tissue segment, e_x and e_y are the x and y eccentricities of the centre of mass with respect to the centroidal axis of the vertebral column, x_i and y_i are the locations of the nodes used to designate the torso segment, and I_x, I_y, and I_z are the mass moments of inertia about the x, y, and z axes. The segmental values of M_s, e_x, e_y, I_x, I_y, and I_z have been obtained by Liu and Wickstrom [24]. Since there are many nodes in a segment, the necessary conditions for dynamic similarity given in equations (3.16)–(3.21) are not sufficient to determine the density distribution. Additional assumptions needed depend on the number of nodes used per segment. The experience gained in the two-dimensional finite-element model of Liu and Ray [46] has shown that typically one will need to include the following:

(a) the specific gravity of the vertebral bone material is estimated from Clauser, McConville, and Young [64] to be roughly 1.38, i.e. an average between cancellous and cortical bone;

(b) the soft-tissue density between the line drawn through the transverse processes of the posterior elements and the dorsal skin surface remains constant in any particular segment;

(c) in the region between the line drawn through the transverse processes and the ventral skin surface the density in any element is the average of four adjacent elements;

(d) where the cross-sectional anatomy indicates a predominance of one type of soft tissue, e.g. muscle, fat, etc., the specific gravity values given by Clauser et al. [64] are assigned.

The above list is by no means exclusive. Additional factors are generated consistent with the known specific gravity values of different connective tissues.

The role of the neuromusculature in biodynamics of the spine has received minimal attention. As pointed out earlier, because of the high acceleration the inertial loads induced in the spine were so predominant as to make the neuromusculature secondary. Even if the assertion were true, there exist gradations of importance by region, by muscle, and by function.

The extensors of the vertebral column are placed posterior to the transverse process and are dominated by the erector spinae. As its name implies this large bulk of muscle plays a major role in the stability of the spine. Its fibres run from vertebra to vertebra and it extends from the sacrum to the skull. Normal flexion or bending over is controlled by the gradual relaxation of the erector spinae against the force of gravity. The flexors are virtually confined to the cervical and lumbar regions, reflecting the fact that these are the two mobile regions of the vertebral column. In the cervical region flexion is mainly produced by the sternocleidomastoid and to a lesser extent by the scalene muscle, while the most effective flexors of the lumbar spine are the rectus abdominis assisted by the external and internal oblique muscles.

For neuromuscular modelling with mid-sagittal plane symmetry the accounting of the above muscles will suffice to account for the primary effects. However, very few practical situations exist for which mid-sagittal plane symmetry can be maintained. To include neuromusculature in a three-dimensional model of the vertebral column is to increase the computation time by an order of magnitude. The mechanical principle of importance in such complex situations is that one cannot bend a column already bent in one plane into another without concomitant rotation. Similarly, one cannot rotate or bend an initially curved beam-column without a concurrent axial deformation.

A few of the important muscles involved in non-mid-sagittal plane motion will be mentioned in accordance with the movement they produce. Lateral flexion in the cervical column is produced by the flexor muscles of one side and in the lumbar region primarily by the quadratus lumborum. Rotation is practically non-existent in the lumbar column but is confined to the thoracic and cervical spine. The external and internal obliques, which run obliquely in opposite directions between the pelvis and rib cage, are the most powerful rotators of the thoracic region. Rotation is achieved by the contractions of the external oblique on one side in unison with the internal oblique on the opposite side. Seventy per cent of the possible rotation in the cervical region occurs at the atlanto-axial and atlanto-occipital joints (C1, C2) with the remaining 30 per cent between the remaining vertebrae. The sternocleidomastoid is the most powerful single rotator followed by parts of the erector spinae running between the transverse process of one vertebra and the spinous process of another.

One practical example from traumatology suffices to illustrate what a three-dimensional neuromuscular model of the spine should be able to accomplish. It has long been known to clinicians and other experimentalists in traumatologic research that the aetiology of hyperextension (or whiplash) injuries lies in the mechanics of the head and neck region. The facet joints in the cervical column are normally in sliding contact. In forward flexion the head practically stops moving when the chin contacts the sternum, i.e. the chest becomes a head restraint. Similarly, in lateral flexion the head is restrained when the ear contacts the shoulder. Both these motions can be accomplished by any healthy neck with little strain. In extension of the head and neck, however, no restraint is present until the occiput hits the posterior chest wall. This feat is impossible to accomplish normally, i.e. beyond the physiological limit. Furthermore, according to McNab [65] a neck rotation of 45° in the transverse plane cuts the permissible homeostatic extension to about half of its normal sagittal-plane displacement. An inertial load under such conditions can easily result in either subclinical or clinical injuries of the central nervous system. Thus far, the modelling and experimental efforts have concentrated on the vertebrae and discs of the spinal column with little attention paid to the load transmitted to the spinal cord. However, it is spinal-cord injuries, especially those which result in paraplegia and quadriplegia, which constitute an economic problem of staggering magnitude. Biomedical scientists studying traumatic injury to the spinal cord have generally used an *open* animal preparation in which varying weights are dropped from different heights on to the exposed cord. Hung *et al.* [66] have used high-speed cinematography to determine the sequence of deformations of the meningeal invested intact spinal cord while simultaneously measuring the pressure wave propagation in the subarachnoid cerebrospinal fluid. This class of experiments is necessary because it is easily reproducible and controlled. However, such experiments merely constitute stepping stones towards the understanding of *closed* traumatic spinal cord injuries. Questions such as the amount of protection provided for the intact cord by the soft tissue, the vertebral column, and the epidural fat and vasculature can be answered by both experimental and modelling studies in tandem. It is such basic understanding which might prove helpful in the design of protective devices.

3.7. Conclusions

In spite of the fact that the biodynamics of the spine has been the object of research for many years, a comprehensive quantitative understanding of the dynamics of the spine is still lacking. The quasi-static work (see ref. 1) has begun to provide insights into a whole range of clinical problems in orthopaedics. Similarly, biodynamic investigations are beginning to yield understanding into the mechanics of spinal trauma and methods for its prevention. These two parallel and simultaneous studies have taken place relatively independently of each other because within the spectrum of spinal problems, covering health, disease, and injury, the primary factors in one are the secondary factors in the other. As the severe idealizations of modelling begin to relax, quantitative understandings from the two ends of the spectrum are spreading towards the centre. The decade 1965–75 has seen a surge of biomechanical engineering interest in the spine and thus has laid the difficult foundations. It is hoped that the next decade will provide the outline of the edifice for the structural biomechanics of the spine. This chapter has given a brief overview of the foundations of spinal dynamics without becoming involved in the many important details that concern every investigator in the different aspects of the spine. Of necessity, it also reflects the personal bias of the author. To some, many of the above statements will ring a bell while others may ignite fires of controversy. The central motivation, however, has been to expose the primary factors in the biodynamics of the spine from a long list of possible factors. It is hoped that the above exposition will attract experimentalists and theoreticians in structural mechanics to this practical and yet significant problem in biomechanics.

Review of contents from the physics/engineering science viewpoint.

The structural dynamics of an initially curved and inertially loaded beam-column is used to simulate the behaviour of the spinal column under impact. The geometric non-linearity of the posterior elements of the vertebral column substantially affects the load-bearing capability of the spine.

Review of contents from the biomedical viewpoint.

The morbidity of anterior lip and compression fracture of the vertebrae as a result of high acceleration impact is studied as a biomechanics problem. It was found that the initial configuration of the articular facets is the major determinant in reducing the incidence of these dynamically induced fractures.

Acknowledgements

The author was a NIH Research Career Development Awardee (Grant No. GM-40723-04) during the preparation of this paper. The additional support of NIH Grant No. GM-19107-03 from the National Institute of General Medical Sciences is gratefully acknowledged.

Glossary/Terminology

Inertial loading	The load induced in any material as a result of the acceleration imposed on it. In spinal dynamics the torso mass, when accelerated, induces inertial loads on the vertebral column.
Transition vertebra	The first and last vertebra of the cervical, thoracic and lumbar regions, i.e. C1, C7, T1, T12, L1, and L5.
Material non-linearity	The generalized load–deflection curve of a material is not a straight line.
Geometric non-linearity	The non-linear load–deflection curve is a result of certain geometric constraints.
Neutral axis	The curve connecting the loci of zero strain in a beam-column.
Centroidal axis	The curve connecting the cross-sectional centres of gravity of a beam-column.
Prehyperextension	Hyperextension of the vertebral column prior to the application of the impact force.
Inertial property	The mass, centre of mass, and the mass moments of inertia of a rigid body.

Dynamic similarity   The inertial property values are equivalent even though they are distributed differently.

Traumatology   The study of wounds.

References

1 SCHULTZ, A. (1974). Mechanics of the human spine. *Appl. Mech. Rev.* **27**, 1487–97.

2 MOFFATT, C. A. and HOWARD, R. H. (1968). The investigation of vertebral injury sustained during aircrew ejection. *Quarterly Rep. No. 2, NASA Contract NAS 2-5062*. Technology Inc., San Antonio, Texas.

3 SHANNON, R. H., (1970). Analysis of injuries incurred during emergency ejection—extraction combat and noncombat. *Aerosp. Med.* **41**, 798–802.

4 DELAHAYE, R. P., (1973). Personal Communication at Ejection Seat Committee Meeting of AGARD, Pensacola, Fla.

5 EWING, C. L., (1971). Non-fatal ejection vertebral fracture, U.S. Navy, fiscal years 1959 through 1965: costs. *Aerosp. Med.* **42**(11), 1226–8.

6 *Grant's Atlas of anatomy.* Williams and Wilkins, Baltimore, Mol.

7 HIGGINS, L. S., ENFIELD, S. A., and MARSHALL, R. J. (1965). Studies on vertebral injuries sustained during aircrew ejection. *Final Rep. to Office of Naval Research, Contract No. NONR-4675(00)*. Technology Inc., San Antonio, Texas.

8 Ruff, S. (1950). Brief accelerations: less than one second, *German Aviation Medicine, World War II*, Vol. 1, Chap. VI-C, pp. 584–597. US Government Printing Office, Washington, DC.

9 LISSNER, H. R. and EVANS, F. G. (1963). Effects of acceleration on the human skeleton. *Prog. Rep. to Public Health Service for Grant No. AC-00054-06.* University of Michigan.

10 BROWN, T., HANSON, R. J., and YORRA, S. M., (1957). Some mechanical tests on the Lumbosacral Spine with particular reference to the intervertebral discs. *J. bone jt Surg. am.* Vol. **39**(5), 1135–64.

11 CROCKER, J. F. and HIGGINS, L. S. (1966). *Phase IV: Investigation of strength of isolated vertebrae. Final Rep.* Technology Inc., San Antonio, Texas.

12 HIRSCH, C. (1965). The reaction of intervertebral discs to compression forces. *J. bone and jt Surg. am.* Vol. **37**(6), 1188–96.

13 MOFFATT, C. A., (1968). Private communication.

14 FLÜGGE, W. (1967). *Viscoelasticity.* Blaisdell, Waltham, Mass.

15 EVANS, F. G. and LISSNER, H. R. (1959). Biomechanical studies on the lumbar spine and pelvis, *J. bone jt Surg. am.* Vol. **41**(2), 278–90.

16 LIU, Y. K., RAY, G., and HIRSCH, C. (1975). The resistance of the lumbar spine to direct shear. *Orthop. Clins N. Am.* **6**(1), 33–48.

17 KAZARIAN, L. E., VON GIERKE, H. E., and MOHR, G. C. (1968). Mechanics of

vertebral injury as a result of G_z impact. Presented at 39th Annual Meeting of Aerospace Medicine Association, Oal Harbour, Flo. Aerospace Medical Association, Washington DC.

18 LIN, H. S. (1976). A system identification analysis of the intervertebral joint subjected to experimental complex loading. Ph.D. Diss. Tulane University.

19 CRAMER, H. J., LIU, Y. K., and VON ROSENBERG, D. U. (1976). A distributed parameter model of the inertially loaded human spine. *J. Biomech.* **9**, 115–30.

20 HOFF, N. J. (1951). The dynamics of the buckling of elastic columns. *J. Appl. Mech.* **18**, 68.

21 TIMOSHENKO, S. P. (1921). On the correction for shear of the differential equation for transverse vibrations of prismatic bars. *Phil. Mag.* **41**, 744–6.

22 SEELY, F. B. and SMITH, J. O. (1952). *Advanced mechanics of materials* (2nd edn). Wiley, New York.

23 ORNE, D. and LIU, Y. K. (1971). A mathematical model of spinal response to impact. In *J. Biomech.* **4**(1), 49–71.

24 LIU, Y. K. and WICKSTROM, J. K. (1972). Estimation of the inertial property distribution of the human torso from segmented cadaveric data. In *Perspectives in biomedical engineering* (ed. R. M. Kenedi) pp. 203–13. University Park Press, Baltimore, Md.

25 MCKENZIE, J. A. and WILLIAMS, J. G. (1971). The dynamic behavior of the head and cervical spine during whiplash. *J. Biomech.* **4**(6), 477–90.

26 LIU, Y. K., PONTIUS, U., and HOSEY, R. (1973). The effects of initial spinal configuration on pilot ejection. *Final Rep. on Res. Contract No. DABC01-73-C-1068*. US Army Aeromedical Research Lab., Fort Rucker, Alabama.

27 PRASAD, P., KING, A. I., and EWING, C. L. (1974). The role of articular facets during $+G_z$ acceleration. *J. appl. Mech.* **41**(E2), 321–6.

28 LATHAM, F. (1957). A study of body ballistics: seat ejections. *Proc. R. Soc. B* **147**, 121–39.

29 BOSEE, R. A., and PAYNE, C. F., Jr. (1961). The mechanism and cause of vertebral injuries sustained on ejection from U.S. naval aircraft. *Rep. NAML-ACEL-467*. Air Equipment Laboratory, Naval Air Material Center, Philadelphia, Pa.

30 SANFORD, R. L., and KELLET, G. L. (1966). A technical review and evaluation of a proposed lumbar pad for U.S. Navy configuration of MK-H5/H7F-1 ejection seats: Phase I-MK-H5. *Rep. NAEC-ACEL-538*. U.S. Naval Air Engineering Center.

31 VULCAN, A. P., KING, A. I., and NAKAMURA, G. S. (1970). Effects of bending on the vertebral column during $+G_z$ acceleration. *Aerosp. Med.* **41**, 294–300.

32 BOWMAN, B. M. (1971). *An analytical model of a vehicle occupant for use in*

crash simulation. Ph.D. Diss. Dept. Eng. Mech., University of Michigan.

33 BEGEMAN, P. C., KING, A. I., and PRASAD, P. (1973). Spinal loads resulting from $-G_x$ acceleration. *Proc. 17th Stapp Car Crash Conf.*, pp. 343–60. Soc. Automotive Engineers, New York.

34 PANJABI, M. M. (1973). Three-dimensional mathematical model of the human spine structure. *J. Biomech.* **6**(6), 671–80.

35 SCHULTZ, A. B., and GALANTE, J. O. (1970). A mathematical model for the study of the mechanics of the human vertebral column. *J. Biomech.* **3**(4), 405–17.

36 PANJABI, M. M., and WHITE, A. A. (1971). A mathematical approach for three-dimensional analysis of the mechanics of the spine. *J. Biomech.* **4**(5), 203–12.

37 BELYTSCHKO, T. B., ANDRIACCHI, T. P., SCHULTZ, A. B., and GALANTE, J. O. (1973). Analog studies of forces in the human spine: computational techniques. *J. Biomech.* **6**(4), 361–72.

38 SCHULTZ, A. B., BELYTSCHKO, T., ANDRIACCHI, T. P., and GALANTE, J. O. (1973). Analog studies of forces in the human spine: mechanical properties and motion segment behavior. *J. Biomech.* **6**(4), 373–84.

39 SCHWER, L., BELYTSCHKO, T., and SCHULTZ, A. (1975). A three dimensional dynamic model of the spine. *Proc. Biomechanics Symp., Troy, N.Y.*, pp. 139–41. ASME, New York.

40 SEVIN, E. (1960). On the elastic bending of columns due to dynamic axial forces including effects of axial inertia. *J. appl. Mech.* **27**, 125–31.

41 LI, T. F., ADVANI, S. H., and LEE, Y. C. (1971). The effect of initial curvature on the dynamic response of the spine to axial acceleration. *Proc. Symp. Biodynamic Modeling and Its Applications, Dayton, Ohio. Tech. Rep. Aeromedical Res.* Wright-Patterson, AFB, Ohio.

42 MOFFATT, C. A., ADVANI, S. H., and LIN, C. J. (1971). Analytical and experimental investigations of human spine flexure. *Am. Soc. mech. Eng. Pap.* 71-WA/BHF-7.

43 LIU, Y. K., and VON ROSENBERG, D. U. (1974). The effects of caudoce-phalad (G_z) acceleration on the initially curved spine. *Comp. Biol. Med.*, **4**(1), 85–106.

44 ZIENKIEWICZ, O. C. (1971). *The finite-element method* (2nd edn). McGraw-Hill, New York.

45 LIU, Y. K., and RAY, G. (1973). A finite element analysis of wave propagation in the human spine. Aerospace Medical Res. Lab., Wright–Patterson *Rep. AMRL-TR-73-40.* Air Force Base, Dayton, Ohio.

46 LIU, Y. K., and RAY, G. (1975). Dynamic response of human spine in $(+G_z)$ acceleration—a two dimensional model. *Proc. Biomechanics Symp., Troy, N.Y.*, pp. 135–8. AMD, Vol. 10, ASME, New York.

47 KAZARIAN, L. E., HAHN, J. W., and VON GIERKE, H. E. (1970). Biomechanics of the vertebral column and internal organ response to

seated spinal impact in the rhesus monkey. *Proc. 14th Stapp Car Crash Conf., Ann. Arb., Mich* p. 121. Soc. Automotive Engineers, Warrendale, Pa.

48 SOECHTING, J. F., and PASLAY, P. R. (1973). A model for the human spine during impact including musculature influence. *J. Biomech.* **6**(2), 195–203.

49 SOECHTING, J. F. (1973). Response of the human spinal column to lateral deceleration. *Am. Soc. Mech. Eng. Pap.* 73-APM-3. *J. appl. Mech. Trans. ASME* **30**, Ser. E., 643–9.

50 SEIREG, A., and ARVIKAR, R. J. (1975). The prediction of muscular load sharing and joint forces in the lower extremities during walking. *J. Biomech.* **8**, 89–102.

51 INMAN, V. T. (1947). Functional aspects of the abductor muscles of the hip. *J. Bone jt Surg.* **29**(3), 607–19.

52 University of California (1953). The pattern of muscular activity in the lower extremity during walking. *Prosthetics Devices Res. Rep. (Series II, Issue 25)*. University of California, Berkeley, California.

53 NACHEMSON, A., and MORRIS, J. M. (1964). *In Vivo* measurement of intradiscal pressure, *J. bone and jt Surg. am.* Vol. **46**(5), 1077–92.

54 STAPP, J. P. (1951). Human exposures to linear deceleration. Part II. The forward-facing position and the development of a crash harness. *USAF Tech. Rep. 5915, Part II*. U.S.A.F., Wright Air Development Center, Wright–Patterson Air Force Base, Dayton, Ohio.

55 HENDLER, E., O'ROURKE, J., SCHULMAN, M., KATZEFF, M., DOMZALSKI, L., and RODGERS, S. (1974). Effect of head and body position and muscular tensing on response to impact. *Proc. 18th Stapp Car Crash Conf.* Soc. Automotive Engineers, Warrendale, Pa.

56 SEVERY, D. M., MATHEWSON, J. H., and BECHTOL, C. O. (1955). Controlled automobile rear-end collisions: An investigation of related engineering and medical phenomena. *Can. Serv. med. J.* **11**(10), 727–59.

57 PATRICK, L., and TROSIEN, K. (1971). Volunteer, anthropometric dummy and cadaver responses with three and four point restraints. *SAE Trans.* **80**, paper no. 710079.

58 MERTZ, H. J., and PATRICK, L. M. (1971). Strength and response of the human neck. *Proc. 15th Stapp Car Crash Conf.* p. 39. Soc. Automotive Engineers, New York.

59 PONTIUS, U. R., LIU, Y. K., and VAN BUSKIRK, W. C. (1972). The effect of the cervical neuromusculature on the dynamics of whiplash. *Proc. 25th Annual Conf. on Engineering in Medicine and Biology, Bal Harbour*. Alliance for Engineering in Medicine and Biology, Vol. 17, Washington DC.

60 PONTIUS, U., and LIU, Y. K. (1975). Neuromuscular effects in a dynamic model of the cervical spine. *Proc. NSF Workshop on Voluntary Human*

Effort, Gainsville, April 14–15. pp. 147–74. Report INVHM-Proc 75. National Science Foundation, Washington DC.

61 HUANG, N. C. (1973). Theories of elastic slender curved rods. *Z. angew. Math. Phys.* **24**, 1–19.

62 VELENAU, C. (1971). Vertebral structural peculiarities with a role in the cervical spine mechanics. *Folia Morphol.* **4**, 388–93.

63 FUNG, Y. C. B. (1968). *Biomechanics. Appl. Mech. Rev.* **21**(1), 1–20.

64 CLAUSER, C. E., McCONVILLE, J. T., and YOUNG, J. W. (1969). *Weight, volume, and center of mass of segments of the human body. Rep. AMRL-TR-69-70.* Aerospace Medical Res. Lab., Dayton, Ohio.

65 McNAB, I. (1969). Acceleration–extension injuries of the cervical spine. *AAOS Symp. on the Spine*, pp. 11–17. Mosby, St Louis.

66 HUNG, T. K., ALBIN, M. S., BROWN, T. D., BUNEGIN, L., ALBIN, R., and JANNETTA, P. J. (1975). Biomechanical responses of open experimental spinal cord injury. *Surg. Neurol.* **4**(2), 271–6.

4. Aspects of car design and human injury

W. JOHNSON, A. G. MAMALIS, and S. R. REID

4.1. Introduction

Public concern about safety, and in particular public demand for better protection against injury from vehicles undergoing collision, has become stronger with the passage of years. In every country people are becoming better educated, and thus more able to express their concern when they are involved in accidents. People are also becoming better informed technically.

Agitation for better passive safety, which started with motor cars, has now grown to include aircraft and, to a lesser extent, other vehicles. Damage and loss of life in aircraft and rail disasters are nearly always sensational and on a relatively large scale, and are therefore newsworthy and attention-demanding. Reducing collision damage and redressing it is now a matter of great interest from the legal, commercial, and insurance points of view.

This chapter brings together some widely scattered knowledge and information about aspects of collision mainly in relation to the consequences for human passengers in cars. Those interested in aspects of collision relevant to a wider range of vehicles might refer to the short monograph by Johnson and Mamalis (1978).

The literature on this topic is vast and is growing very rapidly, so that only a small number of aspects can be dealt with here. There are annual conferences on automobile crash situations and design mechanics [1]. It is difficult to discuss motor car collision design in full, since this has only drawn a lot of attention during the last ten years or so. The volume of such activity is now very large, but the varieties differ from one manufacturer to another. However, the number of components to which considerable crashworthy design attention is *not* given is large and such matters only become evident when unforeseen accidents occur.†

While some motor car manufacturers have long endeavoured to engineer

---

† Recall the notorious case of the Ford Pinto car in which three people were killed when a rear end impact caused the expulsion of gasoline, with consequent fire, which was claimed to be due to a dangerous and ill-located fuel tank.

safety against impact into their models, attention to vehicle safety design seems first to have been highlighted, and great attention drawn to it, by Ralph Nader in his book *Unsafe at any speed* [2]; the updated 1973 edition is well worth reading.

4.2. Structural response to impact

This topic has important and widespread application; the reader is referred to an excellent review which contains a section specifically devoted to motor cars [3]. Also, the review by Johnson and Reid [4] gives information about general features of dynamic structural response and about specific devices which have been designed to absorb energy in impact situations.

Head-on collision with barriers or between cars moving with the same speed, in the range $40-72$ km h^{-1} ($25-45$ mile h^{-1}) for many British cars (which are generally somewhat smaller than those in the US), shows a time for retardation of about 0.1 s with an approximately linear rise to, and fall from, the greatest retardation of about $40\,g$, which occurs after about 0.05 s; this involves a crushing distance of about 0.6 m (2 ft) for each car. The mean crushing strength of a British car is probably 206 kN m^{-2} (30 lbf in^{-2}) of its minimum cross-sectional area, say 1.5 m^2 (16 ft^2). From results of Grime and Jones [5], it may be deduced that the crushing distance for these cars in frontal collision with a rigid barrier is approximately given by,

$$S \text{ (in inches)} = 1.2\,u - 10 \tag{4.1}$$

where u is the speed at impact in mile h^{-1}.

Plastic failure of tubes and struts in compression

There is a vast amount of literature on the buckling of solid struts and thin-walled tubes of circular or non-circular section when subjected to static or dynamic axial compression, and there is a need for relevant structural analysis in many fields of engineering besides that of vehicle impact—see, for example, the work summarized by Johnson and Reid, Johnson and Mamalis, and Johnson [4, 6, 7].

Crumple zones

Many manufacturers of cars now design into their models, regions at the front (but behind the bumper (fender)) and rear of the car, which crumple during collision in a pre-determined mode at a pre-selected rate, thereby aiming to minimize the retardation force (but preferably maintain it at a constant level) which the passenger may have to bear (see Fig. 4.1). For information concerning the dissipative characteristics of the crumple zones, see Saczalski and Angus in refs. [8, 9]. Finite element analysis and computer-aided design

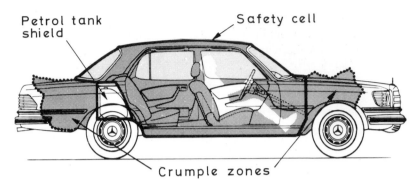

Fɪɢ. 4.1.   Rigid passenger cell with impact absorbing crumple zones.

are now commonly used for calculating the extent and degree of impact deformation, see for example ref. [10].

Bumpers and a crash-cushion design

Bumper designs are manifold—these typically having curved overriders encased in rubber. Bumper mountings also include strong ribbed steel cones which concertina in impact situations and hydraulic mountings are occasionally used. Bumpers are now often covered with resilient plastic layers so that they can also reduce injuries to pedestrians and cyclists. Information about bumper design can be found in an article in the *Guardian* newspaper for 14 November 1977.

   An interesting vehicle impact attenuation system is the *modular crash cushion* which consists of a regular arrangement of 55 gallon tight-head drums, see Fig. 4.2, so positioned as to protect motorists from driving hard into nominally rigid obstacles [11]; the aim is to arrest cars in 3.6 m (12 ft) to 5.5 m (18 ft). Full-scale tests showed the proposed arrangements to be effective.

Driver and passenger compartment

The 'survival space' or a minimum residual volume after impact, which envelops occupants taking account of occupant size, driving posture, and seats is of almost constant magnitude regardless of car model. Franchini [12] stated that designed maintenance of survival space for collision speeds of at least up to 48 km h^{-1} (30 mile h^{-1}) is generally possible.

   Strong steel compartments—cages or safety cells—are now frequently designed into cars (see Fig. 4.1); the passengers are protected from above and below, between the front and rear bumpers and from the sides. The sound functioning of the cell depends on good bumper and crumple zone design [13].

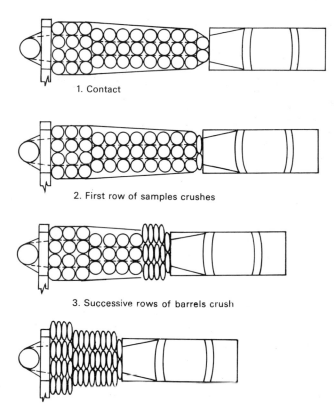

1. Contact

2. First row of samples crushes

3. Successive rows of barrels crush

FIG. 4.2.   Successive crushing of modular crash cushion.

4.3. Passive safety systems

Accident avoidance or *active safety*, includes improvement of vision limitations, braking, tyres, handling and aspects of driver control; the economic benefit of such measures is unknown. *Passive safety* refers to reducing the frequency and severity of a particular injury to a road user when once the collision event has occurred. Passive safety embraces the following features [6]:

  (i) *Seat belts*
     Investigations into the use of diagonal and lap belts have shown them to be beneficial in 30–55 per cent of cases; this applies much more for frontal than for side impacts. The effectiveness of three-point belts is well established.

  (ii) *Anti-burst door latches*
     A major source of fatal and serious injury in a collision is the ejection through a door space. On British motorways 56 per cent of fatalities

(1964), were by ejection, but the environment in which they occur also matters, (only 4 per cent in urban collisions).

(iii) *Energy absorbing steering column assemblies*
A number of designs of collapsible steering columns are discussed by Perrone [14], and the use of 'invertubes' or frangible tubes has also been proposed [4]. Typical column collapse loads are 6 to 11 kN with rearward penetration of 127 mm.

(iv) *Head restraints and seats*
Non-adjustable head restraints are said to be worth installing principally for protection against rear collision, and so that the frequency of cervical spine injuries is reduced.

Seat mounting failures are frequent and the present average $20\,g$ standard does not prevent them.

(v) *Windscreen glass*
Nader [2] summarizes the requirements of a windscreen as (a) not being so hard that the impinging head snaps back to cause fracture or concussion and (b) not allowing windscreen disintegration and consequent laceration. A transparent, tough polyester film (of tensile strength $8.4\,\mathrm{kN\,m^{-1}}$ in sheets 0.05 mm thick and which stretches by nearly 100 per cent) when fixed to glass panes with adhesive becomes shatter-proof. Damage remains local when it is hit by a stone, and under blast, up to at least $76\,\mathrm{kN\,m^{-2}}$ for 1 ms. The glass fragments are held together by the film.

(vi) *Instrument panels*
Perrone [14] commented that with the advent of the collapsible steering column, impact with the instrument panel would become one of the major sources of injury in automobile accidents. He proposed the use of tube bundles for shock protection within this particular internal structure as well as suggesting that tubes might also be used to mitigate the effects of striking other rigid structural elements.

(vii) *Vehicle structure* See Section 4.2.

(viii) *Air bags*
The chief advantage of air bags, see Fig. 4.3, is that the co-operation of the passenger is not required, i.e. they are a passive restraint system. Their disadvantages include an initial cost which is said to be high, a firing system which may be too loud in application and unreliable in operation, costly when requiring specialized testing and repair, and not very protective in roll-over or side-impact situations.

4.4. Bus coaches

In the USA, the fatality rate per 100 000 000 passenger miles (1970–2), was 0.09 for buses, 0.10 for airlines, 0.28 for railways, and 2.0 for motor cars. The

FIG. 4.3.   An airbag performing in a simulated crash.

excellence of these figures for buses and trains is *not*, however, due to the provision of passenger restraint systems or to designed-in crashworthiness. Broadly, this is due to a size effect. The severely crushed zone length in road vehicles undergoing frontal impact is a constant fraction of the vehicle length for a given impact speed, so that retardation magnitudes tend to decrease inversely with length [7]. Thus, human body damage tends to be the less the larger the transporting body, for a given speed.

The safety features advocated for incorporation into bus coaches with collisions in mind are [6]:

(i) The improvement of overturn crashworthiness and seat belts for inter-city buses. (Seat belts would prevent ejection and impact with hard interior surfaces but are disadvantageous in roll-over situations where passengers could be anchored and suspended in circumstances difficult to escape from.)

(ii) Improvements to seat anchorages, design against window ejection and high-backed padded seats have been suggested specifically.

(iii) It is proposed that for some uses, seats be re-designed to compartmentalize the occupant in a protective cocoon. This requires energy absorbing surfaces but is limited in value to frontal crashing.

The result of a head-on collision of a bus with a concrete overpass support column at 108 km h (67 miles h^{-1}), is shown in Fig. 4.4, the column penetrat-

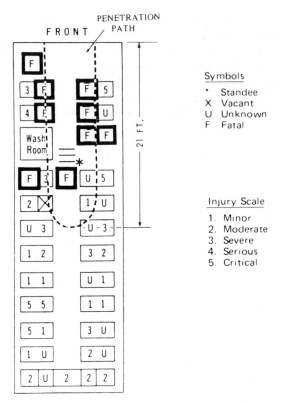

Fig. 4.4. Diagram of injuries by seating arrangement. (From Highway Accident Report NTSB-HAR-74-5, National Transportation Safety Board.)

ing 6.4 m (21 ft), into the bus. Of 45 passengers, 13 were fatally injured. The uniform retardation rate was approximately $(67 \times 22/15)^2/2 \times 21/ = 230 \, \text{ft s}^{-2} \simeq 7 \, g$. In relation to the localization of the impact zone, note the injury severity distribution and in particular that it is small towards the rear of the bus, see Fig. 4.4.

Forces experienced by vehicles in roll-over situations are not well known but a load of twice its own weight is often assumed. In some accident situations, a coach is likely to fall on to a top corner before there is complete roll-over, so that corner loading could be the most serious design problem [6]. For design against bus roll-over situations using computer techniques, see [15].

4.5. Human injury

Head injury

Head injury may be sustained during collisions or emergency situations in all vehicles and is an extremely important consideration for anyone concerned in crashworthiness studies. (For multi-collisional situations for transported bodies, see Appendix 1.1).

The skull–brain system is roughly that of a hard, more-or-less elastic shell containing a fairly uniformly dense substance of low modulus of rigidity. The skull is 10 to 13 mm ($\frac{3}{8}$ in to $\frac{1}{2}$ in) thick and the brain weighs about 1.5 kg and is approximately 165 mm long and 140 mm wide.

Some reasons suggested for the creation of brain damage due to a blow, are [6]:

   (i) Large compressive pressure gradients may be produced at the site of the blow, as a result of giving linear acceleration to the skull. Due to the inertia of the brain material, it 'piles-up' at the site of the blow and causes contusion and injury [16]. Also, because the skull is relatively rigid, the brain material and skull tend to separate on the opposite side of the brain giving rise to a tensile effect which leads to cavitation and/or tearing.

  (ii) *Compressive* waves delivered at a pole, or induced at the site of the blow,—a 'coup'—may be reflected as tensile or rarefaction waves at an anti-pole and cause a counter blow or 'contre-coup', see Fig. 4.5(a). Hence there may be 'proximal' damage at the initiating location due directly to the blow or there may be 'distal' damage due to the counter-blow [16, 17]. Thus, because of (i) and (ii), a non-penetrative direct blow on one side of the head may give rise to unexpected injury on the opposite side.

 (iii) Other causes of brain damage can be:
       (a) skull deformation due to local depression, indentation, penetration or vibration of parts of the skull with or without fracture; and
       (b) rotational acceleration of the head due to a tangential blow or as a result of whiplash, see Fig. 4.5(b),—similar to the effects of a 'hook' or 'upper-cut'—leading to relative angular motion between the skull and the fluid-like brain. This would introduce significant shear strain between them with the tearing of any connecting structures, which can also result in severe injuries to the neck [18].

The contre-coup damage of (i) and (ii) is usually relatively localized while that due to (iii) (b) tends to be widely distributed, (provided the angular acceleration α is sufficiently large) over the large area between the brain and skull. Thus the whiplash (or hook) type of injury tends to be much more damaging than the frontal blow. Also, the larger is α the more severe is the damage inflicted because the closer does it reach to the cerebellum.

Cavitation damage due to a blow

Several investigators have endeavoured to consider head impact in terms of the elementary engineering theory of longitudinal elastic stress waves as applied to

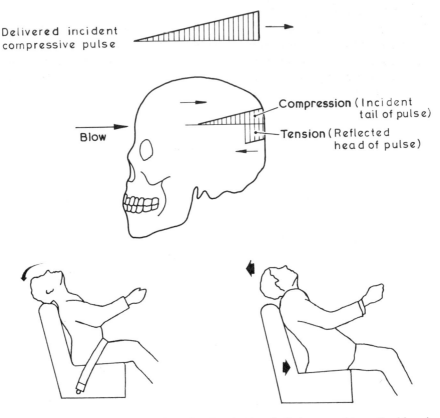

FIG. 4.5. (a) Incident and reflected pulses in the skull due to a blow; the blow is assumed to reflect from the far side of the brain as it was a free surface. (b) Rotational acceleration of the head due to a tangential blow: whiplash effect.

bars. If the theory associated with the notion of spalling† is used (refer Appendix 1.2), the blow being conceived of as compressive, triangular in shape and of length l (and with the bar being identified as the brain), then after reflection at a stress free end, the tensile stress generated is first greatest at $l/2$ from this distal end and its magnitude is $\sigma_0 = \rho c v$; the corresponding speed of particles is $\sigma_0/\rho c$, where ρ denotes the density of brain substance. Assuming $c = 150 \, \mathrm{m \, s^{-1}}$ as a typical speed for pressure waves in the brain mass, letting $\sigma_0 = P_c$ the cavitation pressure‡ (this is the pressure at which gases dissolved in fluid come out of solution) then with $P_c = 12 \, \mathrm{N \, cm^{-2}}$ the speed of the blow at which cavitation occurs is $\sim 80 \, \mathrm{cm \, s^{-1}}$.

† For an explanation of this, see Appendix 1.2.
‡ This is also an important element in injury caused by bullets penetrating human material.

Gadd Severity Index

The Wayne State Tolerance Curve widely used in automobile safety research, see Fig. 4.6, defines the level at which acceleration or retardation of the head causes concussion and skull fracture. It is based on an average acceleration of the skull (a square pulse) at the occipital bone for impacts of the forehead against a plane, unyielding surface.

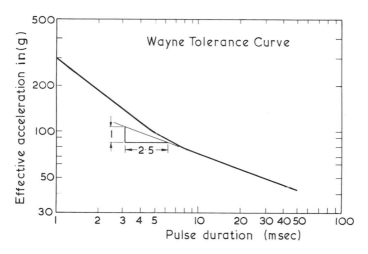

Fig. 4.6. Wayne State Tolerance Curve for head impact.

The Wayne curve is the basis of several indices of injury severity, the most popular index being that due to Gadd. Numbers are identified to represent tolerance for various regions of the body. Its main recommendation is for use by safety design engineers who have to work to certain numerical specifications.

For the head, the Gadd Severity Index is

$$\int_{0}^{T} a^{2.5} \, dt,$$

where a is acceleration expressed in terms of the number of g, t is time, and T the pulse length, in milliseconds.

A value of 1000 is taken as the threshold for serious internal head injury for a concentrated frontal impact and 1500 for distributed or non-concentrated impact. This index purports to identify the level that could be tolerated without permanent brain damage or skull fracture being incurred by a normal healthy adult.

The index 2.5 is primarily a straight line approximation to the Wayne State Tolerance Curve, between 2.5 and 50 milliseconds. Recent work on skull fracture (1970) tends to validate (as far as one number can) the use of this index and a level of 1000, and it was accepted by the Society of Automobile Engineers about 1966. The value of 1500 is a more recent introduction.

The concentrated impact criterion for avoiding fatality is then:

$$\int_0^T a^{2.5}\, dt < 1000;\qquad(4.2)$$

for its application the pulse time T must be such that $0.25\,\text{ms} < T < 50\,\text{ms}$.

If a rigidly seated driver, who is restrained by a seat belt, was brought to rest from speed v with uniform acceleration in a distance s, the Gadd Severity Index would be given by,

$$\text{GSI} = \frac{v^4/g}{(2sg)^{1.5}}.\qquad(4.3)$$

If the frontal crushing distance of a car is 0.6 m (2 ft) and s is equated to this, the speed required to attain the tolerance limit would be 92 km h (57 mile h^{-1}). Note, however, the following remarks by Mackay et al. [19] that 'no serious head or neck injury has been seen without direct contact . . . For the (belt) restrained driver, face contact with the wheel is the rule . . .' A 'car occupant wearing a three-point lap and diagonal safety belt . . . suffers concussion . . . [as a] result of impact of the head with the inside of the car' [20].

Thoracic and abdominal injury criteria

Results gained from work on volunteers and cadavers, and through accident studies, lead to the belief that rib fracture occurs for distributed loads of between 5.3 kN (1200 lbf) and 8.9 kN (2000 lbf). Rupture of the great vessels in the thoracic cage sometimes occurs but volunteer tests suggest that 60 g for up to 100 ms can be tolerated.

Load distribution over the chest is important. For many cars it is held that axial steering column collapse systems are, in truth, ineffective in limiting chest loads.

Lower limbs

The knee, femur, and pelvis skeletal complex is injured in about 40 per cent of seriously injured front seat car occupants. It has been suggested that 6.2 kN (1400 lbf) is the appropriate tolerance level for the femur. It is, however, also thought to be too high and, furthermore, that traumatic fracture is not the appropriate criterion but rather post-traumatic arthritis of the knee and pelvic joints; this will only occur, however, long after an accident.

It is noteworthy that in the UK, 68 per cent of front seat passengers are female and 83 per cent are male. Thus different injury criteria might be appropriate for the front seat positions. An alternative criterion which has been specified for the chest and pelvis is that they should not have to endure 60 g for more than 3 ms.

Aside from other references [6] the treatment of impact insensitivity for structures, for animals, and for men, by Kornhauser [21] is important and useful.

Crash helmets

A maximum peak force of 19.6 kN (4400 lbf) acting on the skull is used as a design criterion for crash helmets; it represents the force required to fracture the average cadaver skull when acting through the scalp on an area of about 13 cm^2 (2 in^2) as Gurdjian *et al.* determined from experiments by dropping cadaver skulls on to hard flat surfaces [22]. However, live skulls are believed to perform better than those of cadavers. A good measure of local protection is afforded by a stiff helmet with suitable load-spreading and energy absorbent liners when the impact energy reaches 168 to 204 Nm (120 to 150 lbf ft).

Helmets should be made penetration-resistant, smooth, and spherical enough to deflect a high proportion of blows to the head. The whole assembly should, however, have a low coefficient of restitution ($e < 0.3$) so that change of head speed is minimized. See also the section on helmets in the previous chapter, and Ref. [6]. The recent review by Goldsmith [23] surveys head and neck injury and protection in great detail.

Fire

Though the relative frequency of motorists killed or injured due to vehicle burn is not great, it calls for mention in the context of this chapter.

Crash-induced fires usually start within 5 s of impact due to the ignition of fuel-pressurized vapour: explosion is rarely encountered. The best preventive approach is to try to so design as to avoid or reduce the possibility of a fuel spill at impact, or to try not to have it occur where ignition may be encountered, (e.g. by energized light filaments and a hot exhaust system). The separation of tank connections (of filler vent and fuel lines), of tank puncturing and its crushing, are causes of spillage.

It is thought that the safest location for fuel tanks is between strong frame channels. Rapid egress from an automobile compartment is essential for survival against fire risk; it should also have good isolation from flame, heat, and toxic gases.

In fires the following four factors threaten life:

(i) hyperthermia: the effects of the build-up of heat/temperature;
(ii) toxification: the ingestion of poisonous gases;
(iii) anoxia: inadequate supply of oxygen;
(iv) protein de-naturization: direct burns from heated or flaming gases.

This subject has been discussed at great length by Severy, Blaisdell, and Kerkhoff [24].

Appendix 1.1

The multi-collisional situation for transported bodies [6]

Any transporting body may usefully be thought of as consisting of a bounding envelope of fixed or well-defined outer shape, and the contents which are to be carried. To reduce damage during transit, minimum relative velocity between the envelope and its contents is required throughout the whole journey. In a collision it is the envelope which first suffers the impact and undergoes damage locally in the impact region, or remotely as with spalling (see Fig. 4.7 (ix)). The contents only later suffer the effects of the impact; they continue to move as the envelope is arrested and, depending on their nature, distribution (and attachments) within the envelope, are later involved in secondary collisions with the inner surface of the envelope or other portions of the contents. The encounter of the envelope with an external body is the first collision, and subsequent collisions undergone by the contents within the envelope are collectively referred to as the second collision.

When a driven car undergoes collision it is the car surface envelope or exterior which is first arrested by an external object; this is the first collision. The driver (and other persons or objects) may later impinge with fixtures to the inside of the envelope (e.g. steering column and dashboard, etc.) and probably injury follows; this is a second collision.

The driver himself may, however, be considered as an 'envelope and contents' in respect of his head. Damage to the skull may be regarded as damage to an envelope and, due to the semi-liquid nature of the brain and its ability to transmit stress waves or undergo sloshing inside its bony container, the brain contents may also undergo damage; this would be a third collision. The set of phenomena may be referred to as a three-collision situation. The outer surface of any moving body undergoes the primary collision and consecutively the contents within sustain the secondary and subsequent collisions.

Appendix 1.2

Propagation of stress-waves in a long bar (see Fig. 4.7)

If the end A of the rod AB in (i) is pushed to the right with constant speed v, then after time t the length of the bar compressed, see (ii), will be ct where c is the speed with which pulses are transmitted along the bar. The compression of length ct will be vt, so that the compressive strain, $e = vt/ct$. Hence the applied

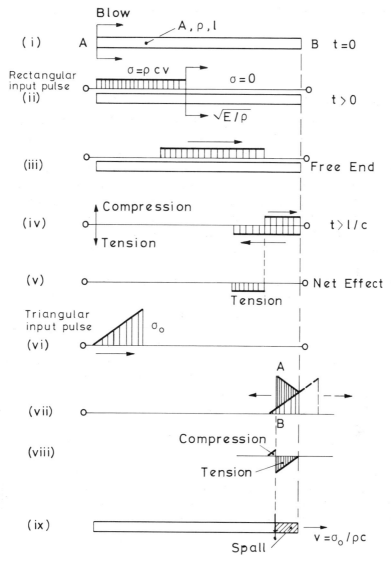

F IG. 4.7.   Incident and reflected pulses in a bar. If the net tension in a section e.g. AB exceeds the tensile fracture stress of the bar, the end portion flies off at speed V as a spall.

compressive stress, σ is

$$\sigma = Ee = Ev/c. \tag{4.4}$$

E is Young's modulus. If the area of the bar is A and ρ is its density, then the impulse applied at A is $(A\sigma . t)$ and the momentum acquired is $(A\rho ct . v)$. Since these are equal,

$$\sigma = \rho c v. \tag{4.5}$$

Eliminating v from eqns (4.4) and (4.5) gives,

$$c = \sqrt{(E/\rho)}.$$

Typical values of c for metals are about 3000 m s^{-1} or $10\,000 \text{ ft s}^{-1}$.

In (ii) and (iii) a compressive wave or pulse of uniform intensity σ and length l has been introduced into the bar. This wave, in process of reflection at the free end, is shown in (iv); note that the reflected wave is tensile and must be so in order that the *total* stress on the end of the bar shall be zero—which is the implication of free-endedness. Note in (v) how tension has been developed and that it first arises when one half of the length of the wave has been reflected; tension also first arises at one half the pulse length into the bar from the free end.

Many compressive blows are ramp-shaped as in (vi). Note in (vii) and (viii) how the higher intensity front portion of the blow, during reflection at the free end (as a tensile pulse), develops an increasingly larger tension. If the net tensile stress at a section becomes equal to the tensile strength of the material of the bar, fracture occurs. The end portion of the bar that then flies off, is known as a spall (see (ix)); it has trapped in it a fraction of the momentum of the original blow. Note (see (vi)) that a *compressive* blow delivered at the proximal end of the bar, is first developed into a *tensile* (or disintegrative) blow at the opposite or distal end.

References

1 AMERICAN SOCIETY OF AUTOMOBILE ENGINEERS (1976). *Proc. 20th Stapp Car Crash Conf.* Society of Automobile Engineers, New York.

2 NADER, R. (1973). *Unsafe at any speed.* Bantam, London.

3 RAWLINGS, B. (1974). Response of structures to dynamic loads. *Conf. Mechanical Properties at High Rates of Strain*, p. 279. Institute of Physics, Bristol.

4 JOHNSON, W., and REID, S. R. (1978). Metallic energy dissipating systems. *Appl. Mech. Rev.* **31**, 277.

5 GRIME, G., and JONES, I. S. (1970). *Car collisions: the movement of cars and their occupants in accidents.* Institution of Mechanical Engineers, London. Automobile Division, AD P5/70.

6 JOHNSON, W. and MAMALIS, A. G. (1978). *Crashworthiness of vehicles.* Mechanical Engineering Publications, London.

7 JOHNSON, W. (1972). *Impact strength of materials.* Arnold, London.

8 SACZALSKI, K. J., and PILKEY, W. D. (ed.) (1976). Measurement and

prediction of structural and biodynamic crash-impact response (ed. K. J. Saczalski and W. D. Pilkey). American Society of Mechanical Engineers Conference, New York.

9 DANNER, M., ANSELM, D., and BECHTER, F. (1979). Das Crashverhalten als Maszsta für PKW-Reparaturen. *Der Maschinenschaden* **52**, 140.

10 WARDILL, G. A. (1979). The use of large deflection finite element programme in cab impact calculations. 121 Euromech Colloquium on Dynamics and Crushing Analysis of Plastic Structures, Warsaw, Poland, August, 1979.

11 HIRSCH, T. J., and IVEY, D. L. (1969). *Vehicle impact attenuation by modular crash cushion*. Research Report, Texas Transportation Institute, pp. 146–7.

12 FRANCHINI, E. (1969). *The crash survival space*. American Society of Automotive Engineers, New York.

13 BEHR, P. (1979). Passenger compartment related to safety. 121 Euromech Colloquium on Dynamics and Crushing Analysis of Plastic Structures, Warsaw, Poland, August, 1979.

14 PERRONE, N. Crashworthiness and biomechanics of vehicle impact, dynamic response of biomechanical systems. Proc. Winter Annual Meeting, American Society of Mechanical Engineers.

15 VOITH, A. (1979). Bus roll-over simulation using computer techniques on the basis of measurement results. 121 Euromech Colloquium on Dynamics and Crushing Analysis of Plastic Structures, Warsaw, Poland, August 1979.

16 UNTERHARNSCHEIDT, F. T. (1975). Injuries due to boxing and other sports. *Handbook of clinical neurology* Ch. 26, Vol. 23. North Holland, Amsterdam.

17 JOHNSON, W., and MAMALIS, A. G. (1977). Aspects of mechanics in some sports and games. *Fortschritt-Berichte der VDI-Z* Reihe 17, No. 4, Düsseldorf.

18 Society of Automotive Engineers (1979). *The human neck—anatomy, injury mechanisms and biomechanics*. SP-438, Society of Automotive Engineers, New York.

19 MACKAY, G. M., GLYONS, P. F., HAYES, H. R. M., GRIFFITHS, D. K., and RATTENBURG, S. J. (1975). Serious trauma to car occupants wearing seat belts. *2nd. Int. Conf. Biomechanics of Serious Trauma*, Sept. 1975.

20 GRIME, G. (1975). Head and neck injuries to car occupants wearing seat belts in frontal collisions. *2nd. Int. Conf. Biomechanics of Serious Trauma*, p. 30.

21 KORNHAUSER, M. (1964). *Structural effects of impact*. Cleaver-Hume Press, London.

22 RAYNE, J. M. (1969). Dynamic behaviour of crash helmets. Royal Aircraft Establishment *Tech. Rep.* 69/60—ARC 31726.

23 GOLDSMITH, W. (1979). Some aspects of head and neck injury and protection. In *Progress in biomechanics* (ed. N. Akkas) pp. 337–77. Sijthoff and Noordhoff, Leiden.

24 SEVERY, D. M., BLAISDELL, D. M., and KERKOFF, J. F. (1974). Automotive collision fires. *Proc. Stapp Car Crash Conf.* pp. 113–99. Society of Automotive Engineers, New York.

5. Biomechanical and structural aspects of design for vehicle impact

NICHOLAS PERRONE

5.1. Introduction

The greatest advances in an increasingly technological society often bring with them associated risks. The generation of power by nuclear reactors has enormous potential for making us independent of fossil fuels, but it also carries the danger of radiation leakage, thermal pollution, and so on. Horse-drawn vehicles were an improvement on walking, but they also increased the occurrence of people being injured during their journeys. Surface motor vehicles such as automobiles, trains, and trucks, offer higher speeds and greater convenience, but also possess a greater potential for human injury when they crash. Indeed, the same is true of aircraft, which achieve much greater speeds, and where, of course, an accidental impact is potentially catastrophic for large numbers of people.

At the present time, with the many highly industrialized societies in the world about one quarter of a million people are killed a year as a result of motor-vehicle-related accidents and many more millions suffer serious, permanently debilitating, and costly injuries. In view of the social and economic loss associated with this huge toll in damaged lives, it behoves the engineering community to exert increased efforts in developing safer and yet economical motor vehicles.

This challenge is complicated by the rapid changes which are taking place in vehicle design. In the United States, a definitive trend is occurring towards lighter weight and smaller vehicles, presenting, at least for a number of years, a situation where there is a great mix of vehicles of various sizes and structural configuration among which impacts may occur. In the immediate future there are good prospects of using more new materials such as lightweight composites. In the United States, the introduction of a national speed limit of 55 m.p.h. in the early 1970s had a dramatic effect on reducing the number of impact-related fatalities. New vehicle safety standards which require the use of air bags or passive belt systems may have a very beneficial influence on vehicle safety, but also may cause some accidents if inadvertently fired.

The complex subject of vehicle design for impact protection is truly interdisciplinary, encompassing a variety of fields including biomechanics, dynamic plastic structural response, occupant motion simulation, impact-related injury criteria, numerical methods, finite element methods of structural analysis, etc. In effecting improved designs, two extremes could be contemplated: a totally analytical approach or a completely empirical trial and error technique. Neither extreme is desirable; some combination of rational analysis in concert with thoughtful testing offers the greatest potential benefit and is probably most cost effective. If one had to characterize the current state of the art, it probably leans more towards the empirical than the analytical extreme.

In this chapter an attempt will be made to provide a broad perspective of biodynamical and structural considerations pertaining to vehicle design for impact. In the next section, attention is focused on determining the response of structures where loading is dynamically applied, and plastic as opposed to elastic material response dominates. One might differentiate between kinetic energy absorbed over a large area associated with a vehicle impacting a fixed object or another vehicle, and energy absorbed in a small area where a body or a portion of a body component impacts some part of a vehicle interior. Both of these possibilities, which are concerned with external crashworthiness and internal vehicle design respectively, are dealt with in the next two sections. Biodynamical response problems pertaining to vehicle impact and the related important question of injury criteria are also discussed.

5.2. Dynamic plastic structural response

We might partition the vehicle impact problem into two broad categories: structural and biodynamical. Let us first consider the structural aspect and initially we will review the problems attendant with complex structural response from a general viewpoint. In the next section, we will link these concepts more directly to the actual vehicle response problem.

A typical engineering material when subjected to a uniaxial or tension test will demonstrate a stress–strain curve as shown in Fig. 5.1. For very small extensions or strain, up to the order of 0.1 per cent, the material behaves in a linear and elastic manner. In other terms, when the load is removed the material responds like a spring and returns to its original position. When loading is applied at stresses above the elastic limit stress, Point A on the curve in Fig. 5.1, then permanent or plastic deformation results. The material can normally sustain much higher elongation, of the order of 35 per cent of its original length, before fracture of the material occurs. For most structural designs including building type, vehicles, machinery, etc., it is assumed that the structure will operate within the elastic range. It is normally only accidental or unanticipated extreme conditions that will cause the structure to be loaded into the plastic range.

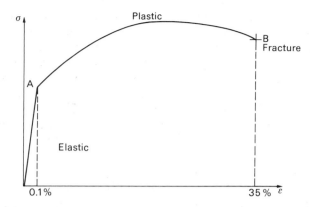

F<small>IG.</small> 5.1.   Stress–strain diagram for typical engineering material for static load.

It is useful to introduce idealized or mathematical models of material behaviour which maintain the essence of material behaviour for engineering utility and at the same time are mathematically expedient for rational analysis and design purposes. Figure 5.2 displays a number of such idealized material behaviour including elastic perfectly plastic, OAP, elastic linear hardening, OAH, and rigid perfectly plastic, OCP. The angle HAP is representative of the amount of strain hardening that the material manifests. Should elastic considerations be relatively insignificant when compared to plastic action such as is the case in vehicle impact response, then the rigid perfectly plastic material (or even a rigid linear hardening material, not shown) can be most useful. For that situation the important parameter is σ_0, the yield stress. It is important to realize that the actual material behaviour in Fig. 5.1 and the idealized representations in Fig. 5.2 all apply for a static or slow load applications. In dynamical problems where plastic response is the more important parameter, rate sensitivity effects can be readily manifested. Specifically, the yield stress σ_0

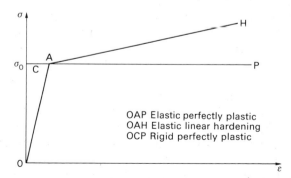

F<small>IG.</small> 5.2.   Useful idealizations of engineering materials for static applications.

of Fig. 5.2 could vary with strain rate. For example, dynamic tests on structural steel [1] were fitted by an exponential expression by Cooper and Symonds [2], eqn (5.1), Fig. 5.3.

$$\sigma/\sigma_0 = 1 + (\dot{\varepsilon}/40.4)^{1/5} \qquad (5.1)$$

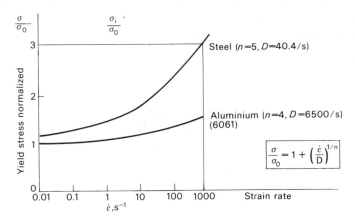

FIG. 5.3.   Yield stress variation with strain rate for steel (and aluminium).

The representation of the above equation for steel for strain rates up to $1000\,\text{s}^{-1}$ demonstrate that the yield stress for steel changes vary significantly with dynamic effects, indeed, increases by of the order of 200 per cent. When one considers that for vehicle impacts, strain rates of the order of 10 to $100\,\text{s}^{-1}$ occur, it becomes obvious that these effects must be incorporated even from a first order viewpoint. It should be further observed that a primary material component of most automobile, truck, train, and ship structures is steel; hence dynamic material effects should be included wherever possible in analytical and design models.

An examination of eqn (5.1) reveals that the variation of yield stress with strain rate is highly nonlinear; indeed it can only be displayed on a logarithmic plot (Fig. 5.3). When this material representation is included in any analytical models, the resulting equations are necessarily highly nonlinear and do not readily submit to routine numerical analysis. Let us turn our attention next to simplified approaches which can address this aspect of structural response.

In an actual structure, say a vehicle impacting a rigid barrier at a high speed, the strain rates throughout the structure are changing from point to point as well as with time during the impacting events. Such temporal and spatial variations also occur with a problem as 'simple' as a beam in view of the fact that the strain rate will vary even through the depth of the beam. To obtain a better physical feel of rate sensitivity effects we consider the problem of the impulsively loaded circular ring [3] (Fig. 5.4). This ring is assumed to suddenly have a uniform radial velocity u_0 applied to it. As a result, the ring will have an initial strain rate of u_0/r_0. It is further assumed that the ring is made of steel, a rate sensitive material as characterized by eqn (5.1) or in Fig. 5.3. At any given instant of time, the hoop stress in the ring is constant because of radial symmetry. A simple explanation of the physics of the event is as follows: the initial kinetic energy imparted to the ring as a result of the impulsive velocity is converted into plastic work associated with the radial expansion of the ring. When all of that kinetic energy is expended into plastic work the ring will be at its final resting position.

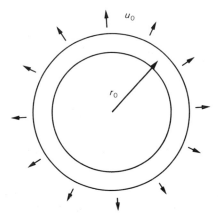

Fig. 5.4.   Impulsively loaded circular ring.

A rigorous calculation of the final deformation of the ring can be readily made by setting up the equation of motion of the ring and integrating in many small time steps using eqn (5.1) to describe the yield stress variation with the strain rate for the rigid perfectly plastic material being utilized. Let us call such an accurate numerical solution the exact one for purposes of discussion. A second approximate set of solutions is obtained by making what appears to be a very drastic assumption: that the yield stress in the ring is assumed to be constant associated with the initial strain rate. The associated equations of motion are then trivially integrated and the solution listed as an approximate one. Both sets of values are shown in Table 5.1 for purposes of comparison. Perhaps, surprisingly, the differences between both results are quite small being of the order of only 3 per cent for a range of initial stress and strain rate values for rings.

The clue to the success of this approximation is that the yield stress in Fig. 5.3 varies logarithmically with strain rate. If one considers that the ring strain energy is proportional to the velocity squared (or strain rate squared)

TABLE 5.1

Comparison of exact and approximate rate sensitive ring solutions

Initial conditions		Final ring strain		
Stress σ/σ_0	Strain rate (s^{-1})	Exact	Approximate	Difference %
2.0	40.4	0.00513	0.00501	2.3
2.38	202	0.01749	0.01694	3.1
3.0	1293	0.01385	0.01342	3.1

then by the time the stress–strain rate point in Fig. 5.3 moves but a small distance from its original position, most of the kinetic energy in the ring has been converted into plastic work; hence, the initial yield stress assumed to be a constant associated with the initial strain rate is a powerful approximation technique which is, in fact, capable of ready extension.

Let us consider next the problem of a massless beam made of a rate sensitive material with an attached tip mass, M, which is impulsively loaded (Fig. 5.5). We wish to determine the final damage angle θ_f when the beam ultimately comes to rest. Most of the bending action in the beam takes place at the fixed end. A modification of the classic plastic hinge action is occurring [4]. Assuming a rectangular beam cross section, the stress distribution for a so-called plastic hinge is shown in Fig. 5.6(b) with the related yield moment M_0. The yield stress at this time is interpreted as being the static value. Using the simplified approach as applied to the impulsively loaded ring, we would like to devise a procedure for this problem to obtain the dynamic value of the yield moment accounting for rate sensitivity effects.

Unlike the classic plastic hinge which occurs at zero length of the beam, for a dynamic situation accounting for rate sensitivity it is necessary to make some

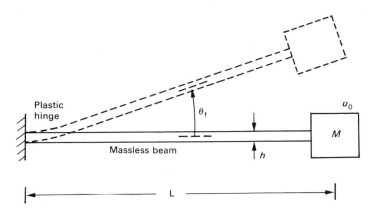

FIG. 5.5.   Massless beam with impulsively loaded tip mass.

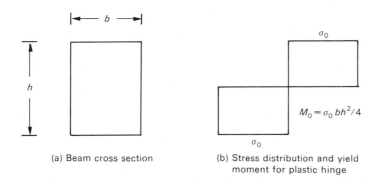

(a) Beam cross section

(b) Stress distribution and yield
moment for plastic hinge

$M_0 = \sigma_0\, bh^2/4$

FIG. 5.6.   Beam of rectangular cross-section with plastic hinge stress distribution.

crude estimate of the hinge length (otherwise the strain rate will turn out to be infinite). Let us assume for current considerations that the effective length of the plastic hinge is ξh (note that h is the depth of the beam whose width is b, Fig. 5.6). If R is the radius of curvature for the hinge region, then the maximum fibre strain is as shown in eqn (5.2a), Figure 5.7. α is the central angle associated with the plastic hinge region; it is obviously apparent from simple geometry that equation (5.2b) is valid.

$$\varepsilon = h/(2R) \qquad (5.2a)$$
$$R\alpha = \xi h \qquad (5.2b)$$
$$\varepsilon = \alpha/(2\xi). \qquad (5.2c)$$

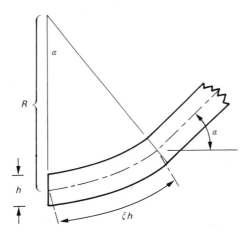

FIG. 5.7.   Effective length plastic hinge of length ξ h.

Eliminating R between eqns (5.2a) and (5.2b) we find the strain is as given by equation (5.2c); but really what we are striving for is to determine and estimate the initial strain rate. We can determine the strain rate by simply differentiating equation (5.2c) once with respect to time, eqn (5.3a). $\dot{\alpha}$ is essentially the angular velocity of the beam which is initially u_0/L, eqn (5.3b). A reasonable estimate of ξ is 4 [5] resulting finally in eqn (5.3c).

$$\dot{\varepsilon} = \dot{\alpha}/(2\xi) \tag{5.3a}$$

$$\dot{\varepsilon} = u_0/(2\xi L) \tag{5.3b}$$

$$\dot{\varepsilon} = u_0/(8L). \tag{5.3c}$$

Let us recall the simplified approach from the ring problem. We can assume that the yield stress is a constant associated with the initial strain rate of the material for a dynamically loaded problem. In the case of the impulsively loaded beam of Fig. 5.5 it behoves us to estimate the dynamic yield moment at the plastic hinge region. With eqn (5.3c) we can determine the initial strain rate; from eqn (5.1) we can calculate the dynamic initial value of the yield stress which when substituted into eqn (5.4a) gives the dynamic value of the plastic hinge moment. It is reasonable to assume this moment to be a constant during the entire response of the beam for the same argument used previously with the ring, namely that most of the kinetic energy is dissipated before the yield stress–strain rate point departs appreciably from its original position

$$M_d = \sigma_d bh^2/2 \tag{5.4a}$$

$$\theta_f = Mu_0^2/(2M_d). \tag{5.4b}$$

Finally, a simple energy balance equating the initial kinetic energy in the beam system to the plastic energy dissipated in the dynamic hinge moment results in eqn (5.4b) which provides the final damage angle. The problem in Fig. 5.5 is instructive from two viewpoints: firstly, as to how to use a generalized procedure for dynamic structural response with bending and rate sensitivity, and secondly, because the problem of a deforming seatback in a rear-end impacted vehicle is much like the mathematical model of the contilever beam.

In a sense, the solution obtained for the problem of Fig. 5.5 might also be termed a modal approximation approach because the form of the beam damage was assumed to be a straight line, i.e. one with all of the plastic deformation occurring at the end region. A modal approximation technique simply suggests that the response of a structure is the product of a function of space variables by a function of time. It has been used with some success on complicated structural problems [6, 7].

The simplification to account for rate sensitivity discussed initially for the ring and subsequently for the beam has also been used successfully in pulse loaded structures, and plates undergoing small deformations by essentially assuming the yield stress is a constant associated with the initial strain rate which varies from point to point [8, 9].

Our discussion to this point has been confined to structures wherein small deformations occur, that is where nonlinear membrane effects which would significantly influence the response of the structure do not arise. For example, in a plate constrained at its boundaries and loaded transversely undergoing deformations of the order of a few thicknesses or more, membrane effects dominate the response and must be considered from a first order viewpoint. Moreover, with the simplifying approximation discussed to this point in which we assume the yield stress is a constant associated with the initial strain rate, an obvious inadequacy exists for this large deflection problem because the initial strain rate is clearly zero.

A recent modified approach has been introduced to account for plastic rate sensitivity with large deformations [10]. Based on a comparison of an exact and appropriate solution of a string-supported central mass which has an impulsive velocity applied to it, this modified analysis demonstrates that the nonlinear forces arising in the string are reasonably constant during the bulk of the response time suggesting that an approximate procedure might still be viable. The membrane strain rate can be estimated remarkably accurately by calculating for the condition when half the kinetic energy in the system has been dissipated (which roughly corresponds to a condition of maximum strain rate). The associated yield stress is again taken as constant and the resulting equations of motion integrated very readily. Moreover, modal approximation techniques could be used in concert with this procedure resulting in a simple and useful engineering solution to a complex problem [10].

The essential results of this discussion on dynamic plastic structural response including rate sensitivity and large deformations is that with judicious application, it may be possible to effect a solution to the complex dynamic response of a structure by using an overall correction factor to account for rate sensitivity affects. The credibility of such an approach is discussed in greater detail in the next section of this chapter.

5.3. External crashworthiness

Crashworthiness is defined as the ability of a vehicle structure to provide maximum opportunity for survival of its occupants during impact. It includes the notion of providing a protective envelope within the occupant space which, if at all possible, will not be violated. We differentiate here between external and internal crashworthiness. The latter pertains to the environment inside the vehicle against which an occupant may impact and be injured during an accident. In this section we concentrate on the external crashworthiness problem or what we might term primary structural response.

Stimulated by a growing social consciousness, government regulations, and increasing product liability litigations, improved crashworthiness design has been a subject of concern to vehicle manufacturers from the early 1960s to date [11]. The essential problem presented is to design a structure which will result in a controlled energy absorption of the kinetic energy of vehicle during impact. It is well known that the kinetic energy of a body is as shown in eqn (5.5)

$$KE = WV^2/(2g). \tag{5.5}$$

This energy is directly proportional to the weight or mass of the body and varies as the square of its velocity; hence, a body going at 40 m.p.h. has four times the KE of one travelling at 20 m.p.h. For reference purposes, it might be useful to observe that the KE of a 3000 pound car going 30 m.p.h. is approximately 90 000 lbf.

Figure 5.8 provides a range of energy absorption and hence KE of various portions of a vehicle being impacted in different directions. From an external crashworthiness viewpoint, it is clear that energy absorbed from zero to 160 000 foot pounds are of significance from a design point of view. For the internal crashworthiness problem we are concerned primarily with the kinetic energy of the occupant and absorbing it in an effective non-injurious manner. Again, in Fig. 5.8 the human body KE which must be dissipated judiciously during an impact is also shown for purposes of comparison. Of course, these levels are much smaller being of the order of only a few thousand foot pounds.

It is also clear from Fig. 5.8 that one must account for the direction of the impact as well as the frequency of each directional event. For example, since half of all impacts are frontal [12] it is obviously of great importance that the associated structure in the forward part of the vehicle be well designed and of a long enough length for the weight and size of vehicle to enhance the prospects for occupant survival.

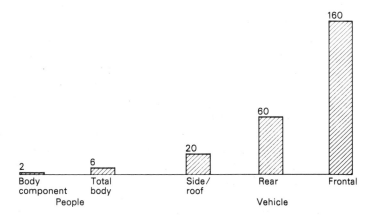

FIG. 5.8.   Energy-absorbing range (in K ft-lb) for vehicle impact.

Despite many advances in the last decade or so in computer capabilities and numerical methods, the technical problem of effecting a good crashworthy vehicle design based on analytical techniques is an extremely complicated task. Indeed, empirical or trial and error techniques are widely used for this reason. The technical problem presented is predicting the response of a structure which is undergoing very large deformations (of the order of a characteristic length of the structure), is being loaded well into the material nonlinear range, and has the final added complexity of the very nonlinear rate sensitive material property effects which were discussed in the previous section. When we stop to consider that most structures are designed to remain in the linear small deformation elastic region, we must conclude that the crashworthiness design problem is many orders of magnitude more complex than the traditional one.

It might be useful to examine a related area to demonstrate how controlled energy absorption does work if properly implemented. During the last decade and a half, the Federal Highway Administration in the United States has promoted a vigorous programme on the development of energy-absorbing systems on the highway, such as breakaway signposts, energy-absorbing steel or sand-filled barrels at gore areas, and energy-absorbing bridge rail systems [13–15]. These devices, which essentially absorb energy external to the vehicle, have worked remarkably effectively and are widely used throughout the United States. They do attest to the utility of good energy-absorbing concepts for vehicle-impact control (Fig. 5.9).

We could categorize the methods of quantitatively designing a vehicle for vehicle impact into three groups: (a) fully analytical predictive models, (b) a combination analytical and experimental approach, and (c) fully experimental or empirical trial and error. Great advances have been made in analytical techniques for predicting structural response including some plasticity effects. One of the most widely utilized techniques for computerized structural analysis, the finite element method, has been used to assess aircraft as well as vehicle crashworthiness response [16]. Such a fully deterministic approach is, however, very difficult, costly, and time consuming. It is necessary to run such mathematical models for extremely long periods because of the enormous deformation which takes place. Moreover, it is very difficult for such models to predict comprehensively the details of the crash response of a structure. Perhaps, as computer costs continue to decrease significantly, fully deterministic finite element models might be utilized to an increased extent.

At the other end of the spectrum, empirical or fully experimental approaches have been and continue to be used with some success. Many designs do evolve after a considerable series of full-scale tests. However, item (b), a combination of analytical and experimental holds considerable promise from a cost effectiveness point of view. Lumped mass models of a vehicle with nonlinear but massless springs whose physical properties are determined from experimental static data with these components can be of considerable utility [17]

FIG. 5.9.   Negligible damage to vehicle from impact with energy-absorbing empty oil barrel barrier system.

(Fig. 5.10(a)). Figure 5.10(b) provides component deformation characteristics of a 1974 Ford Pinto; these characteristics can be used as some of the K factors for the nonlinear springs for the limped mass model of Fig. 10(a).

One of the difficulties with a lumped mass model is that the nonlinear characteristics of the components such as in Fig. 5.10(b) do not include strain rate effects which can be quite significant for structural steels used in automobile construction. Considering a practical range of impact velocities of 20 to 50 mile h^{-1}, we can estimate using effective length plastic hinge concepts the order of the maximum strain rate during these impacts. It is not unreasonable to expect they would be proportional to the velocities which are only changing by a factor of 2.5. In view of the slow variation of strain rate for structural steel, Fig. 5.3, such a variation would not expect to be very noticeable as far as yield stress effects are concerned. Hence, an overall strain rate sensitivity correction factor should suffice. Indeed, Tomassoni has found [17] that the use of a constant strain rate sensitivity correction factor of 30 per cent over the static value provides very good correlation of Pinto crash versus impact velocity (Fig. 5.11). The shape of the force displacement curve for a constant strain rate sensitivity correction agrees fairly well with experiments

(a)

Typical car crash model

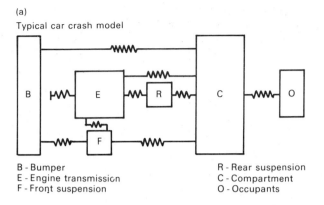

B - Bumper
E - Engine transmission
F - Front suspension

R - Rear suspension
C - Compartment
O - Occupants

Static crush data obtained from a
sub-compact size car front structure

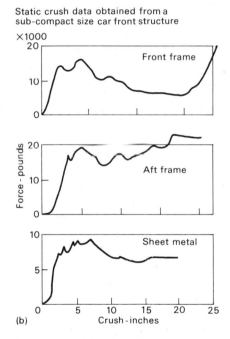

Fig. 5.10   (a) Lumped mass model of vehicle for frontal impact; (b) force-
displacement experimental curves for vehicle components.

relative to all other mathematical models which Tomassoni has attempted.
These results do tend to support the notion of an overall rate sensitivity
correction factor sufficing and using these with lumped mass models appear to
be the most useful from the point of view of cost effectiveness.

The same general technique should be usable for impact in the rear as well as

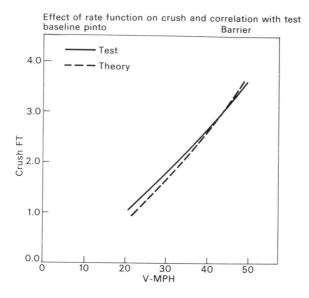

Fig. 5.11.   Pinto crash displacement versus velocity—tests correlated with lumped mass model and 30 per cent rate sensitivity correction factor [17].

side and roof. Specifically, the lumped mass model could be used in concert with static test data and an overall rate sensitivity correction factor applied which would be velocity independent. It is likely that a factor of 30 per cent over the static value should suffice for other directional impacts.

5.4. Internal vehicle design for body impact

Injuries related to vehicle impact are frequently characterized as second collision injuries. When a vehicle impacts an obstruction or is hit by another vehicle, during the brief 15 or 20 ms of the actual impact occupants do not move appreciably from their initial positions. However, because of the inertial forces which arise on them they do develop significant velocities relative to the car interior if they are not restrained by seat belts or other devices. For example, if the vehicle were to hit a rigid barrier at 30 m.p.h., a passenger would continue moving forward at about 25 m.p.h. after the vehicle interior had come to rest. Of course, it is for this reason that restraints such as seat belts would be so effective in minimizing or reducing injury levels from second collision impacts.

The kinetic energy which a human body or body component can have is relatively modest being up to a few thousand foot pounds (see Fig. 5.8). Regrettably, this kinetic energy is normally dissipated as a result of the impact forces acting on a body or body component during this second collision. In view of the fact that the deceleration distance is so small, being anywhere from a fraction of an inch to a few inches, the resulting impact forces can be very high, and indeed may also be fatal. A well designed vehicle is one which will minimize damage to people as a result of good energy management of the gross structure as well as the vehicle interior target areas.

The general interior design problem for impact safety could be described as follows: a given vehicle is to be designed to sustain some velocity level of impact in a particular direction and into a type of object (e.g. a rigid wall, a tree); determine, utilizing either mathematical models or experimental information, body component velocities and direction, and potential target areas. Experimentally determine the static target impedence that is the force deflection relationship for all potential targets as a result of impact with a simulated body segment. Find, using either analytical or experimental techniques, the force on the body as a result of contact with the target. Should this level of force be potentially injurious or not acceptable for some reason, the target must be redesigned to make it more accommodating to the impact by lowering the level of force which the body component would experience. The cycle is then repeated to examine whether the force level on the body is now acceptable. This procedure could be used for the design or analysis of any interior area which is a serious potential body impact point.

Occupant simulation models are mathematical models representative of the human body and its components and can be used to predict the motion of a certain percentage of humans in a given impact situation. These models are referred to in Chapter 6 by Dr Houston. For example, suppose a given vehicle had been impacted into a barrier and an acceleration guage away from the impact crushing portion of the vehicle gave a reading as is shown in Fig. 5.12. (With the lumped mass models discussed in the previous section, it should be possible to obtain an analytical approximation to this force–time response without doing a full-scale impact test). The pulse loading of Fig. 5.12 becomes an input to an occupant simulation model to determine the subsequent motion

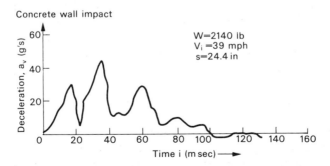

FIG. 5.12.   Vehicle impact force (in g) versus time.

of any representative portion of the human population to the impact sequence of the pulse shown. Such models could be used with confidence up to the point where impact occurs with some portion of the vehicle interior. Thereafter, because of the complications attendant on modelling such contact problems, the subsequent motion cannot normally be obtained with great confidence. However, if the force vs deflection characteristics of the target area are known (and can be modified if necessary to account for rate sensitivity effects), then the magnitude of the forces on the body components could be estimated. Parametric runs could of course be obtained for a given pulse shape as Fig. 5.12 with many different segments of the human population.

An enumeration of possible target areas inside a vehicle and associated body portion areas which may be injured by them is given in Table 5.2. Thorax-related injuries also include impact trauma to the heart and main arteries. Life-threatening injuries are those pertaining to the head, neck, and thorax.

TABLE 5.2

Car interior components and associated potential body injury area

Component	Probable impact injury area
Instrument panel (upper)	Skull/brain, shoulder, thorax, chest
Instrument panel (lower)	Patella, pelvic bone system, femoral shaft
Floor pan	Feet, ankles, tibia, fibula
Windscreen	Facial tissue, scalp, cervical column
Steering wheel	Facial bones, thorax, heart, descending aorta, lungs and other internal organs
A, B pillars, roof frame	Skull/brain
Doors	Femur, pelvic bones, thorax, shoulder
Seats	Cervical column, spinal

In the United States, the Federal Motor Vehicle Safety Standards (FMVSS) demand a rather limited form of protection against body contact with the interior of a vehicle. Indeed, they are based on the premise that a body will be restrained either by a comprehensive belt system or an air bag. The issue of installation of air bags or passive belts has been on the American scene for about a decade without substantial progress taking place. In 1983 all new cars in the United States should have a passive protection system. However, should that system turn out to be a belt which is disconnected by the occupant, then the hostile interior of a car may still cause significant damage to its occupants, as is the case today (FMVSS 201 and 208 are the governing regulations pertaining to occupant impact with the interior of a vehicle).

Mindful of the dangers attendant on occupant impact with the vehicle's interior, a number of preventative design procedures are possible. Sheet metal moulding should be placed around (a) and (b) pillars as well as roof frames. Head impact with these areas at low velocities can be fatal. Headform decelerations for normal impact into foam covered and sheet metal covered rigid structures are shown in Fig. 5.13 with data obtained from reference [18]. Clearly, the sheet-metal-covered rigid structure is better than the foam and the foam decidedly better than the bare rigid structure. Even the foam cover, it will be noted, shows decelerations as high as 160 gs at only 8 mile h^{-1} velocities. It has been shown that intrusions associated with side impacts cause significantly higher levels of injury [19].

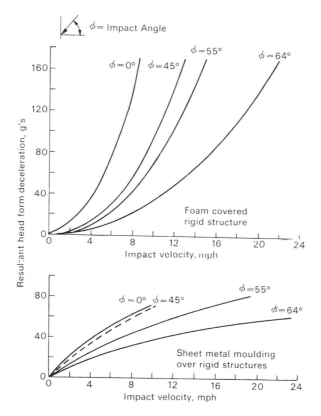

Fɪɢ. 5.13.   Resultant headform deceleration versus impact velocity for various impact angles.

To inhibit such intrusion, higher strength door latches as well as connections at the door-B pillar interface which enhance membrane action during the impact sequence should be very helpful. When a vehicle is impacted from behind, usually by another vehicle, front seated occupants can have their seats yield and allow them to fly rearward at a relatively high velocity causing head and/or neck injury when they strike the rear of their vehicle interior. FMVSS 207 requires a yield moment at the base of a seat back of only 3300 inch pounds;

as shown in Fig. 5.14, assuming an 80 pound effective force of the upper torso and seat back weight, a 20 g loading due to the gross rear-end impact itself, and a 15 inch moment arm, we find a yield moment at the base of the seat of 24 000 inch pounds is required. If we account for strain rate effects described in the previous section, a static yield moment of approximately 16 000 inch pounds should suffice to achieve this minimal level of injury protection. Should air bags be reliably implemented or some type of belt system be widely utilized, a dramatic decrease should obviously occur in the number of interior impact injuries.

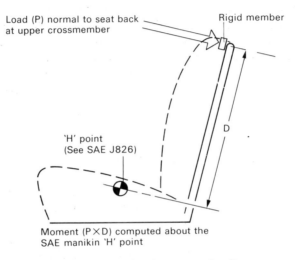

Load (P) normal to seat back at upper crossmember

Rigid member

D

'H' point (See SAE J826)

Moment (P×D) computed about the SAE manikin 'H' point

Fɪɢ. 5.14.   Simulated occupant loading.

The topic of human tolerance to impact injury is discussed for the head, thorax, and long bones in Chapters 1, 2, and 3 of this text. For completeness, Table 5.3 outlines the human tolerance injury criteria for the various segments of the human body.

Tᴀʙʟᴇ 5.3

Human injury tolerance criteria	
Brain (fatal)	HIC > 1500
Skull fracture	1200 lb -1 in^2
	3000 lb -2 in^2
Cervical damage	Moment > 30 lbf
Thorax (fracture)	> 1500 lb
Patella	1000 lb
Basic bone tissue	12000 psi

5.5. Summary

An overview of the general vehicle impact injury problem is presented. Special attention is paid to the problem of dynamic plastic structural response including especially rate sensitivity since most vehicle structures are made of materials which exhibit this behaviour. Simplified techniques are described to determine the response of structures made of rate-sensitive materials. These concepts are applied as the problem of external crashworthiness is addressed wherein the gross energy absorbed by a vehicle structure or portions thereof must be determined. Lumped mass models with nonlinear springs made of materials which do exhibit an overall rate sensitivity correction factor are most promising. The problem of impact with the interior of a vehicle by portions of a body are discussed. Use of occupant simulation models in concert with a limited amount of testing information offers a convenient testing approach to this design problem. A number of specific suggestions are made as to how to improve particular major component areas of the vehicle interior.

References

1 MANJOINE, M. J. (1944). Influence of rate of strain and temperature on yield stresses in mild steel. *J. appl. Mech.* **11**, 211.

2 COWPER, G. R., and SYMONDS, P. S. (1957). Strain hardening and strain rate effects in the impact loading of cantilever beams. *Tech. Rep.* **28**, Brown University, September.

3 PERRONE, NICHOLAS (1965). On a simplified method for solving impulsively loaded structures of rate sensitive materials. *J. appl. Mech.* **32**, 489–92.

4 HODGE, P. G. Jr. (1958). *Plastic analysis of structures.* McGraw-Hill, New York.

5 PERRONE, NICHOLAS (1971). Response of rate sensitive frames to impulsive load. *J. engng Mech. Div. ASCE* **97**, 49–62.

6 JONES, N., and WIERZBICKI, T. (1976). A study of the higher modal dynamic plastic response of beams. *Int. J. mech. Sci.* **18**, 135–47.

7 CHON, C. T., and MARTIN, J. B. (1976). A rationalization of mode approximations for dynamically loaded rigid plastic structures based on a simple model. *J. struct. Mech.* **4**, 1–31.

8 PERRONE, N., and EL-KASRAWY, T. (1968). Dynmic response of pulse loaded rate sensitive structures. *In. J. Solids Struct.* **4**, 517–30.

9 PERRONE, N. (1967). Impulsively loaded strain rate sensitive plates. *J. appl. Mech.* **34**, 380.

10 PERRONE, N., and BHADRA, P. (1979). A simplified method to account for plastic rate sensitivity with large deformations. *J. appl. Mech.* **46**.

11 Anon. (1976). Product liability—its growing business importance. *Automotive Engng* 30–6, October.
12 KIHLBERG, J. K. (1965). Driver and his right front passenger in automobile accidents. Cornell Aeronautical Laboratory Report No. V3-1823-R16.
13 Anon. (1971). Design of traffic safety barriers. Highway Research Board, Highway Research Record No. 343.
14 Anon. (1971). Design of sign supports, and structures. Highway Research Board, Highway Research Record No. 346.
15 PERRONE, N. (1972). Thick walled rings for energy absorbing bridge rail systems. Federal Highway Administration Report No. RD-73-49, December.
16 WINTER, R., CRONKITE, J. D., and PIFKO, A. B. (1979). Crash simulation of composite and aluminum helicopter fuselages using a finite elment program. American Institute of Aeronautics and Astronautics Special Symposium Vol. on Business Aircraft, 1979, pp. 233–40.
17 TOMASSONI, J. E. (1978). A study of the effect of strain rate on the automobile crash dynamic response. Proceedings of Aiaa/ASME 19 Structures, Structural Dynamics and Materials Conference.
18 Anon. (1978). Simplified method for simulating glancing blow impacts—motor vehicles. SAE-J136, 1978 SAE Handbook, pp. 34, 131–4, 133.
19 HARTEMAN, F. *et al.* (1976). Occupant protection in lateral impacts. Paper No. 760806, 20th STAPP Car Crash Conf. Proc. Society of Automotive Engineers, pp. 189–219, October.

Section II   Occupational tasks and environment

6. The mechanics of human body motion

(due to impulsive forces during occupational manoeuvres and jumping)

RONALD L. HUSTON and CHRIS E. PASSERELLO

6.1. Introduction

Accurate analyses of human body motions are relevant in a number of situations. For example it is important to know the precise sequence and time courses of motion of human limbs to enable an astronaut to reorient himself in space or to enable a diver to execute the precise number of somersaults and enter the water in a particular body orientation to avoid splash in a diving competition, or even to enable a gymnast to time the various movements of body segments accurately to execute a particular event. On the other hand, it is also relevant to determine the muscular effort required for certain occupational work functions to help provide prescriptions for optimization of human effort and the performance of man–machine systems.

Significant changes have occurred during the past decade in the methods of analysis of human body motion. New matrix methods and computer-oriented techniques are replacing the traditional approximate and intuitive approaches. Furthermore, new experimental procedures and techniques to study human body motion are being developed which greatly exceed those of only the recent past.

In 1967 Smith and Kane (1967) published a literature survey of the work done in human body mechanics up to 1966. They concluded that most of the studies were largely intuitive or they employed a human body model which was too simple to be quantitatively relevant. However, at about the same time (1966–70) a number of investigators began to seek and develop a more mathematical approach to the modelling and analysis of human body mechanics. In 1967 and 1968 Smith and Kane (1967, 1968) presented one of the first mathematical analyses of the dynamics of the human body in free fall. In 1970 Kane and Scher (1970) provided a more extensive analysis of this phenomenon. Passerello and Huston (1971a) also provided a mathematical analysis of human self-reorientation in space. This latter work was extended in 1971 to consider human body models in general force fields (1971b).

Roberts and Robbins (1969) examined mathematical human body models in crash simulation. Robbins and co-workers (Robbins, Bennett, Henke, and Alem 1970; Robbins, Snyder, McElhaney, and Roberts (1971) also provided experimental verification of mathematical crash simulation models. In addition to this a number of investigators such as Stewart Duffy, Soechting, Litchman, and Paslay (1971); Ewing and Thomas (1972), Liu (1971), Beckett and Pan (1971), Jacobs, Skorecki, and Charnley (1972), and Paul (1971) have studied particular aspects of human body motion such as arm movement, head movement, walking, etc. More recent work involving mathematical human body models includes studies by Kurzhals and Reynolds (1972), and Hewes and Glover (1972), and Kinzel, Hall, and Hillberry (1972). However, most of these analyses do not provide a general dynamical theory but rather are directed toward specific applications. Also, most of them employ the traditional methods of classical mechanics in analysing their mathematical model.

Young (1970) using Lagrange's equations, provided a more extensive analysis with particular emphasis on the reaction of a human body model to impulsive forces. Bartz (1971), using Newton's laws, presented a very general computer-simulation model of a crash victim. Passerello (1972), using Lagrange's form of d'Alembert's Principle, also proposed a very general computer-simulation model to study human body motion.

To some extent these more general approaches are based upon recent successes in the analysis of the dynamics of general systems of rigid bodies— the so-called 'general chain systems'. These analyses, which are generally computer oriented, usually also present some systematic procedure for organizing the complex geometry of the body components and their dynamics. Foremost among these analyses is the work of Hooker (1970) and Fleischer (1971) on multi-body satellites and the more general analyses of Chace and Bayazitoglu (1971) and Passerello (1972) and Passerello and Huston (1972) on general chain systems.

In this chapter we present a general procedure for studying the dynamics of human body motion. The analysis is based to a large extent upon the work of Passerello (1972) and Passerello and Huston (1972). It employs Lagrange's form of d'Alembert's Principle together with a matrix-tensor procedure for organizing the geometry. Illustrative examples are provided for a number of simple common motions.

The analysis is divided into six sections with the first section providing the preliminary considerations and background needed. The general dynamics of the human body is developed in the second and third sections. The fourth section presents a specialization to impulsive motions and the fifth section contains the illustrative motions demonstrating application of the analysis. The final section provides a summary and conclusions regarding the analysis.

The analysis is computer oriented resulting in equations which can be coded by the reader for computer solution. The illustrative examples are intended to demonstrate the range of applicability of the analysis.

6.2. Preliminary considerations

The model

In studying the mechanics of human body motion, it is of course necessary to introduce a mechanical model of the human body. To this end let us consider Hanavan's fifteen-body finite-segment model of the human body (Hanavan 1964) as shown in Fig. 6.1. The fifteen bodies $(B_j, j = 1, \ldots, 15)$ are connected

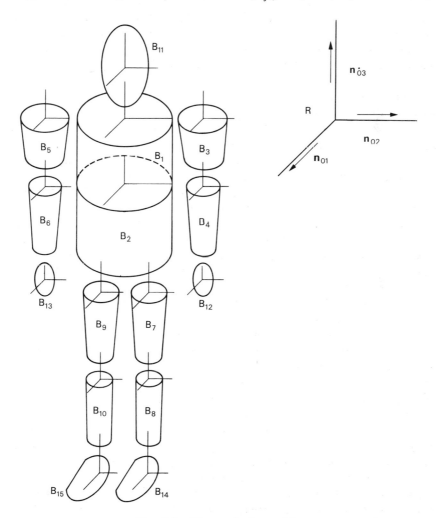

FIG. 6.1. Human body model.

to each other through hinges, ball-and-socket joints, and axially as with the human limb connections.

Following Hanavan (1964), B_1 is an elliptical cylinder representing the chest. B_{11} is an ellipsoid representing the head. It is attached by a ball-and-socket joint to B_1. B_3 and B_5 are frustrums of elliptical cones representing the upper arms. They are also attached by ball-and-socket joints to B_1. B_4 and B_6 are frustrums of elliptical cones representing the lower arms. They are attached by hinges to B_3 and B_5. B_{12} and B_{13} are spheres representing the hands. They are attached by ball-and-socket joints to B_4 and B_6. B_2 is an elliptical cylinder representing the lower torso. It shares a common axis with B_1. B_7 and B_9 are frustrums of elliptical cones representing the upper legs. They are attached by ball-and-socket joints to B_2. B_8 and B_{10} are frustrums of elliptical cones representing the lower legs. They are attached by hinges to B_7 and B_9. Finally, B_{14} and B_{15} are elliptical cones representing the feet. They are attached by hinges to B_8 and B_{10}.

The force field

Our first objective will be to obtain governing equations of motion for this system on rigid bodies which is now our model of the human body. This system of connected rigid bodies may be identified as a 'general chain system' such as that studied in Passerello (1972) and Passerello and Huston (1972). We shall consider the system to be subjected to an arbitrary force field. We shall seek to develop the governing equations so that (1) if the force field is specified the configuration and motion of the system are determined, (2) if the configuration and motion of the system are specified the force field is determined, and (3) if a combination of a portion of the force field and a portion of the configuration and motion is specified the remaining (unknown) portions are determined by the equations.

To begin this analysis it is helpful to consider first some preliminary ideas regarding the development of equations of motion and some ideas pertaining to the geometrical and kinematical relations between two adjoining rigid bodies.

Equations of motion

Consider actually writing down the equations of motion for the system shown in Fig. 6.1. For simplicity let us initially assume that all connections between adjoining bodies are ball-and-socket joints, i.e. we temporarily generalize the hinges to ball-and-socket joints. Conceptually then, the simplest approach is to use Newton's laws and write equations of motion for each individual body of the system. However, this approach has the disadvantage of introducing excessive unnecessary computation and analysis. For example, this approach would lead to $6 \times 15 = 90$ equations involving as unknowns non-working constraint forces between the bodies of the system. These non-working constraint forces are usually of no interest and thus they would need to be systematically eliminated from the 90 equations. In general there are $3(15 - 1)$ $= 42$ of these forces (three between each adjoining body). Hence this procedure would ultimately lead to $90 - 42 = 48$ equations of motion to be solved.

It is possible to avoid the computation associated with the elimination of these non-working constraint forces. Two methods are available of which the best known and most widely used is the method of Lagrange's equations. In this method the 15-body system is treated as a unit, the non-working constraint forces are automatically eliminated, and the 48 equations of motion are obtained directly. These equations can be written in the form

$$\frac{\mathrm{d}}{\mathrm{d}t} \frac{\partial K}{\partial \dot{x}_j} - \frac{\partial K}{\partial x_j} = F_j \qquad j = 1, \ldots, 48 \qquad (6.1)$$

where K is the kinetic energy of the system, the $x_j(j = 1, \ldots, 48)$ are the generalized co-ordinates of the system (one for each degree of freedom), and the $F_j(j = 1, \ldots, 48)$ are the generalized active forces acting on the system. If the externally applied forces acting on the system of bodies are replaced by an equivalent set of forces consisting of 15 forces and 15 couples acting on each of the 15 respective bodies of the system, then the generalized forces F_j can be written

$$F_j = \sum_{k=1}^{15} \left(\frac{\partial V^k}{\partial \dot{x}_j} \cdot \mathbf{F}_k + \frac{\partial \omega^k}{\partial \dot{x}_j} \cdot \mathbf{M}_k \right) \qquad j = 1, \ldots, 48 \qquad (6.2)$$

where $\mathbf{F}_k$ and $\mathbf{M}_k(k = 1, \ldots, 15)$ represent the 'equivalent' (Kane 1959) forces and couple torques acting on the respective bodies B_k and where $\mathbf{F}_k$ has its line of action passing through point G_k which is the mass centre of B_k. V_k and ω^k represent the velocity and angular velocity of G_k and B_k in an inertial reference frame R (see Fig. 6.1). (In some cases it may be of interest to consider internal 'working' moments between adjoining bodies. In these cases these moments can be included in $\mathbf{M}_k$.) The quantities $\partial V^k/\partial x_j$ and $\partial \omega^{-k}/\partial \dot{x}_j$ $(k = 1, \ldots, 15)$ are called 'partial rates of change of position' and 'orientation' respectively.

While providing a number of advantages such as those listed above, Lagrange's equation also lead to serious disadvantages. The principal disadvantage is that the computation of the derivatives in eqn (6.1) is extremely tedious and is actually unwieldy since there are 15 bodies. An alternative method, developed by Kane (1961, 1965, 1968; Kane and Wang 1965) in 1960, involves Lagrange's form of the d'Alembert Principle and is based upon the notion of generalized inertia forces; it retains the advantages of Lagrange's equations but avoids the differentiation problems. According to this principle we have

$$F_j + F_j^* = 0 \qquad j = 1, \ldots, 48 \qquad (6.3)$$

where the generalized active force F_j, is given by equation (6.2) and the generalized inertia force F_j^*, is given by

$$F_j^* = \sum_{k=1}^{15} \left(\mathbf{F}_k^* \cdot \frac{\partial \mathbf{V}^k}{\partial \dot{x}_j} + \mathbf{T}_k^* \cdot \frac{\partial \boldsymbol{\omega}^k}{\partial \dot{x}_j} \right). \tag{6.4}$$

F_k^* and T^* are the inertia force and torque respectively and are given by

$$\mathbf{F}_k^* = -m_k \mathbf{a}^k \tag{6.5}$$

and

$$\mathbf{T}_k^* = -\mathbf{I}_k \cdot \boldsymbol{\alpha}^k - \boldsymbol{\omega}^k \times (\mathbf{I}_k \cdot \boldsymbol{\omega}^k) \tag{6.6}$$

where m_k is the mass of B_k, $\mathbf{I}_k$ is the inertia dyadic of B_k relative to G_k, and $\boldsymbol{\alpha}^k$ is the angular acceleration of B_k in the inertial reference frame. From eqns (6.1) and (6.3) it can be seen that the 'generalized inertial force' F_j^* ($j = 1, \ldots, 48$) is given by

$$F_j^* = -\frac{\mathrm{d}}{\mathrm{d}t} \left(\frac{\partial K}{\partial \dot{x}_j} \right) + \left(\frac{\partial K}{\partial x_j} \right). \tag{6.7}$$

This method provides the same advantages as the method of Lagrange's equation (principally the elimination of the non-working constraint forces) together with the advantage of avoiding the differentiation required with Lagrange's equations. However, the function F_j^* (see eqn (6.6)) also introduces derivatives, but in this case they are fundamental vector quantities and can be calculated by vector multiplication. Given the necessary algorithms, a computer can be used to perform these calculations. Therefore in view of these advantages Kane's method is used here to develop the equations of motion for the human body model of Fig. 6.1.

Geometrical relations between two adjoining bodies—shifter matrices

We now consider two typical adjoining bodies of the system such as those shown in Fig. 6.2 where B_k and B_l are the bodies, and $\mathbf{n}_{ki}$ and $\mathbf{n}_{li}$ ($i = 1, 2, 3$) represent sets of mutually perpendicular unit vectors in B_k and B_l respectively. Our objective here is to develop convenient relations describing the relative orientation and the relative rate of change of orientation, i.e. the angular velocity of the two adjoining bodies.

Since the unit vector sets are fixed in the respective bodies the relative orientation of the bodies is determined by the relative orientation of the unit vector sets. We consider the matrix defined as

$$SKL_{im} = \mathbf{n}_{ki} \cdot \mathbf{n}_{lm} \qquad i, m = 1, 2, 3. \tag{6.8}$$

This 3×3 square matrix defines the relative orientation of the unit vector sets since it provides the scalar components of $\mathbf{n}_{ki}$ along $\mathbf{n}_{lm}$. This matrix also

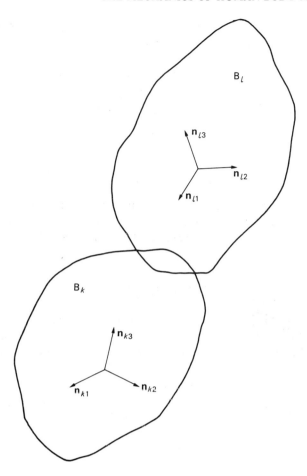

FIG. 6.2.   Two typical adjoining bodies.

provides the familiar transformation relation between the components of a vector referred to each set. That is suppose a vector $\mathbf{V}$ is expressed as

$$\mathbf{V} = V_i^{(k)} \mathbf{n}_{ki} = V_m^{(l)} \mathbf{n}_{lm} \qquad (6.9)$$

where following the summation convention there is a sum from 1 to 3 over the repeated subscripts. Then, $V_i^{(k)}$ and $V_j^{(l)}$ are related by the expressions

$$V_i^{(k)} = SKL_{im} V_m^{(l)}$$
$$V_i^{(l)} = SKL_{mi} V_m^{(k)} \qquad (6.10)$$

where again there is a sum over the repeated subscripts. The expressions of eqns (6.10) are obtained immediately by taking the dot products of eqn (6.9) with $\mathbf{n}_{ki}$ and $\mathbf{n}_{lm}$ respectively. The matrix SKL_{im} is thus seen to be the familiar transformation matrix encountered in elementary tensor analysis. It is frequently called the 'shifter' matrix [30] because of its shifting properties as displayed in eqns (6.10).

The shifter matrix also has the property of being an 'orthogonal' matrix, i.e.

$$SKL_{im} SKL_{km} = \delta_{ik}$$
$$SKL_{mi} SKL_{mk} = \delta_{ik} \tag{6.11}$$

where δ_{ik} is Kronecker's delta function defined as 1 for $i = k$ and 0 for $i \neq k$. In matrix notation these relations can be expressed as

$$(SKL)(SKL)^{\mathrm{T}} = (SKL)(SLK) = I \tag{6.12}$$

where the superscript T denotes the transpose and I is the identity matrix (elements δ_{ik}). Equations (6.11) and (6.12) follow immediately from the definition of SKL_{ij} of eqn (6.8). Finally, one other property which also follows immediately from eqn (6.8) is the 'chain rule', i.e. $K, L,$ and M refer to three sets of unit vectors and then

$$SKM = (SKL)(SLM) \tag{6.13}$$

The chain rule of eqn (6.13) together with the shifting property of eqn (6.10) provides for the transformation of the components of a vector referred to unit vectors of any body into components (of the same vector) referred to unit vectors of any other body. For example there are advantages in expressing a vector in terms of unit vectors fixed in an inertial reference frame because such unit vectors maintain constant orientation. Hence if SOK_{im} represents the shifter matrix between the unit vectors of body B_k and the unit vectors of the inertial reference frame, then the components $V_i^{(0)}$ of a vector referred to the inertial reference frame can be expressed in terms of the components $V_m^{(k)}$ referred to the unit vectors of B_k as (see eqn (6.10))

$$V_i^{(0)} = SOK_{im} V_m^{(k)}. \tag{6.14}$$

The shifter matrix SOK_{im} can be obtained by repeated application of the chain rule (eqn (6.13)). However, to use the chain rule it is necessary to know the shifter matrices between the respective adjoining bodies. In this regard Huston and Passerello (Huston and Passerello 1971; Passerello and Huston 1971) have developed a systematic scheme for obtaining these individual shifters. This scheme is briefly outlined below.

Configuration charts

Consider again the two adjoining bodies of Fig. 6.2. Introduce co-ordinate axes X_{ki} and X_{li} ($i = 1, 2, 3$) in bodies B_k and B_l respectively and let these axes be respectively parallel to the unit vector sets. Next, imagine the bodies B_k and B_l to be oriented relative to each other such that these axes are respectively parallel as shown in Fig. 6.3. This orientation when the respective axes are

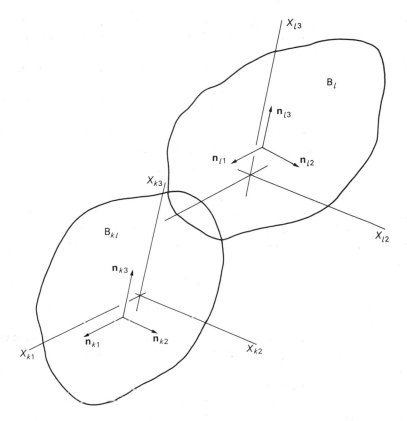

F$_{\mathrm{IG}}$. 6.3.   Reference configuration of two typical adjoining bodies.

parallel is called the 'reference configuration' between two bodies. Next, imagine three successive dextral rotations of B_l relative to B_k about the axes X_{l1}, X_{l2}, and X_{l3} through angles α_{kl}, β_{kl}, and γ_{kl} respectively (the subscripts on the angles refer to the bodies B_k and B_l). These three rotations bring B_l into 'general configuration' with respect to B_k as shown in Fig. 6.2.

   This process is schematically described by the configuration chart of Fig. 6.4. Configuration charts(Passerello and Huston 1971) provide a tabular represen-tation of the relation between sets of unit vectors. Each dot in the chart represents a unit vector indexed in the far left column and identified in the bottom row. The respective reference frames are listed in the top row. The two intermediate reference frames and their unit vectors are not named. These are the reference frames of the intermediate positions of B_l in the successive rotation process described above. The horizontal lines in the chart connect dots associated with common axes and the angle written beneath is the respective rotation angle about these axes. The inclined lines are used to

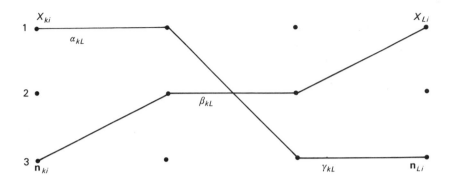

FIG. 6.4.   Configuration chart.

develop the relations between the unit vectors in adjacent columns. If two dots
are connected by an inclined line the corresponding unit vectors are related by
a positive sine term. If two dots in adjacent columns do not lie on any line the
corresponding unit vectors are related by a negative sine term. Unit vectors
corresponding to dots in a common row in adjacent columns are related by a
positive cosine term unless they are equal (as when the dots are connected by a
horizontal line). The argument of these trigonometric functions is the angle
between the respective columns.

As an illustration of this suppose that the unit vectors represented by the
second column of dots in Fig. 6.4 are denoted by N_{ki} ($i = 1, 2, 3$). Then, using
the instructions outlined above, the following relations between n_{ki} and N_{ki} are
obtained:

$$\mathbf{n}_{k1} = \mathbf{N}_{k1}$$
$$\mathbf{n}_{k2} = c\alpha_{kl}\mathbf{N}_{k2} - s\alpha_k\mathbf{N}_{k3}$$
$$\mathbf{n}_{k3} = s\alpha_{kl}\mathbf{N}_{k2} + c\alpha_{kl}\mathbf{N}_{k3} \qquad (6.15)$$
$$\mathbf{N}_{k2} = c\alpha_{kl}\mathbf{n}_{k2} + s\alpha_{kl}\mathbf{n}_{k3}$$
$$\mathbf{N}_{k3} = -s\alpha_{kl}\mathbf{n}_{k2} + c\alpha_{kl}\mathbf{n}_{k3}$$

where $s\alpha_{kl}$ and $c\alpha_{kl}$ are abbreviations for sin α_{kl} and cos α_{kl} respectively. It is easy
to verify these relations with a simple sketch of the two unit vector sets as
shown in Fig. 6.5. Note also that when $\alpha_{kl} = \beta_{kl} = \gamma_{kl}$ the respective unit
vectors are equal and the bodies are in reference configuration.

The configuration chart can now be conveniently used to determine the
shifter matrix SKL. Let αKL be the matrix defined as

$$\alpha KL = \begin{bmatrix} 1 & 0 & 0 \\ 0 & c\alpha_{kl} & -s\alpha_{kl} \\ 0 & s\alpha_{kl} & c\alpha_{kl} \end{bmatrix}. \qquad (6.16)$$

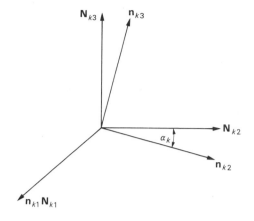

F$_{IG}$. 6.5.   Companion unit vector sets.

Then it is easily seen from eqns (6.15) that

$$\alpha K L_{im} = \mathbf{n}_{ki} \cdot \mathbf{N}_{km}. \tag{6.17}$$

Hence $\alpha K L$ is a shifter matrix between $\mathbf{n}_{ki}$ and $\mathbf{N}_{ki}$. Then by the chain rule of eqn (6.13)

$$SKL = (\alpha K L)(\beta K L)(\gamma K L) \tag{6.18}$$

where $\beta K L$ and $\gamma K L$ are the matrices

$$\beta K L = \begin{bmatrix} c\beta_{kl} & 0 & s\beta_{kl} \\ 0 & 1 & 0 \\ -s\beta_{kl} & 0 & c\beta_{kl} \end{bmatrix} \tag{6.19}$$

$$\gamma K L = \begin{bmatrix} c\gamma_{kl} & -s\gamma_{k} & 0 \\ s\gamma_{kl} & c\gamma_{kl} & 0 \\ 0 & 0 & 1 \end{bmatrix} \tag{6.20}$$

where the sine and cosine are again abbreviated. The shifter matrices $\alpha K L$, $\beta K L$, and $\gamma K L$ can be determined from the configuration chart by inspection by noting the following. Each matrix has the integer 1 and cosines on the diagonal. The integer occurs in the same row as the horizontal line of the configuration chart. The remaining elements in the row and column on the integer are zero. The other elements are $+$ sine and $-$ sine. The $+$ occurs in the lower row if the slope of the inclined line is positive.

The configuration chart of Fig. 6.4 is valid for any two typical adjoining bodies of the 15-body human body model of Fig. 6.1 if they are connected by ball-and-socket joints. If they are connected with hinges (as with the elbow between B_3 and B_4) the configuration chart is the same as in Fig. 6.4 except that two of the angles (e.g. α_{34} and β_{34}) are zero and their corresponding matrices in eqn (6.18) are simply identities. This means that all the configuration charts of adjoining bodies have the same form and hence all the shifter matrices between adjoining bodies have the same form. Therefore by using eqns (6.16), (6.18), (6.19), and (6.20) a computer subprogram can be written to compute all the shifter matrices between adjoining bodies. The chain rule of eqn (6.13) can then be used to compute any shifter matrix such as *SOK*.

Angular velocity

The shifter matrices determine the relative orientation of the bodies of the system in terms of the various rotation–orientation angles α, β, and γ. As mentioned above, it is also of interest to obtain expressions for the relative rate of change of orientation between the bodies, i.e. the relative angular velocities. To this end, consider again the two typical adjoining bodies B_k and B_l (Fig. 6.2). Recall that B_l is brought into general configuration relative to B_k by three successive dextral rotations about the axes X_{11}, X_{12}, and X_{13} through the angles α_{kl}, β_{kl}, and γ_{kl} respectively. Hence the addition formula for angular velocity (see for example Kane and Wang (1965) leads to the following expression for the angular velocity of B_l relative to B_k:

$$^k\omega^l = \dot{\alpha}_{kl}\,\mathbf{n}_{kl} + \dot{\beta}_{kl}\,\mathbf{N}_{k2} + \dot{\gamma}_l\,\mathbf{n}_{l3} \tag{6.21}$$

where the dots denote time differentiation. The vector $\mathbf{N}_{k2}$ is parallel to X_{l2} after the first rotation (through α_{kl}).

Although eqn (6.21) follows immediately from the addition formula and the manner of bringing B_l into general configuration relative to B_k, the equation can also be determined by inspection from the configuration chart (Fig. 6.4) by making the following observations. The relative angular velocity of the unit vectors associated with adjacent columns of dots (Fig. 6.4) is simple angular velocity (1967) directed along the common unit vector (horizontal line) with magnitude proportional to the derivative of the corresponding rotation angle (written under the horizontal line). Hence eqn (6.21) can be obtained by the sum of products of the orientation angle derivatives and the associated unit vectors corresponding to the horizontal lines of the configuration charts. Furthermore if $^k\omega^l$ is expressed in terms of $\mathbf{n}_{ki}$ as

$$^k\omega^l = {}^k\omega^l\,\mathbf{n}_{ki} \tag{6.22}$$

the components $^k\omega^l_i$ ($i = 1, 2, 3$) are given by

$$^k\omega^l_i = \dot{\alpha}_{kl}\delta_{il} + \dot{\beta}_{kl}\alpha K L_{i2} + \dot{\gamma}_{kl}\alpha K L_{im}\beta K L_{m3}. \tag{6.23}$$

The angular velocity of B_l in an inertial reference frame R can now be determined by repeated use of the addition formula for angular velocity, i.e.

$$\omega^l = {}^k\omega^l + \omega^k \tag{6.24}$$

and the components can all be referred to unit vectors $\mathbf{n}_{oi}$ ($i = 1, 2, 3$) fixed in R by multiplication of the appropriate shifter matrices.

Shifter derivatives

From the above discussion and from the earlier remarks it is seen to be convenient computationally to express all vectors in terms of $\mathbf{n}_{0i}$ ($i = 1, 2, 3$), the unit vectors of the inertial reference frame R. Also, this can be done conveniently through shifter matrices such as SOK (see eqn (6.14)).

It is sometimes necessary to differentiate the vector components which are referred to $\mathbf{n}_{oi}$. This means that shifters such as SOK will need to be differentiated. As mentioned above, however, these shifter derivatives can be obtained by a multiplication algorithm. To obtain this algorithm consider SOK to be given by

$$SOK_{im} = \mathbf{n}_{0i} \cdot \mathbf{n}_{km}. \tag{6.25}$$

Then since the $\bar{\mathbf{n}}_{0i}$ is fixed in R (and are therefore constant)

$$\frac{d}{dt}(SOK_{im}) = \mathbf{n}_{0i} \cdot \frac{{}^R d\mathbf{n}_{km}}{dt}. \tag{6.26}$$

However, since the $\mathbf{n}_{km}$ are fixed in B_k

$$\frac{{}^R d\mathbf{n}_{km}}{dt} = \omega^k \, X \, \mathbf{n}_{km}. \tag{6.27}$$

Hence $d(SOK_{im})/dt$ becomes

$$\frac{d}{dt}(SOK_{im}) = \mathbf{n}_{0i} \cdot \omega^k X \mathbf{n}_{km}$$

$$= -\omega^k X \mathbf{n}_{0i} \cdot \mathbf{n}_{km}$$

$$= -e_{iqn} \omega_n^k \mathbf{n}_{0q} \cdot \mathbf{n}_{km}$$

or

$$\frac{d}{dt}(SOK_{im}) = WOK_{iq} \, SOK_{qm} \tag{6.28}$$

where WOK_{im} is a matrix defined as

$$WOK_{iq} = -e_{iqn} \omega_n^k \tag{6.29}$$

where ω_n^k are the components of ω^k referred to n_{0n} and e_{iqn} is the permutation symbol which Erigen (1962) defined as

$$e_{iqn} = \begin{cases} 1 & i, q, n \quad \text{distinct and cyclic} \\ -1 & i, q, n \quad \text{distinct and anticyclic} \\ 0 & i, q, n \quad \text{not distinct.} \end{cases} \tag{6.30}$$

WOK_{iq} is simply the matrix whose dual vector [30] is $^0\boldsymbol{\omega}^k$. Eqn (6.28) then provides the desired multiplication algorithm.

Finally it is sometimes desirable to have derivatives of the matrices αKL, βKL, and γKL which are defined by equations (6.16), (6.19), and (6.20). These derivatives can be obtained directly from these definitions and are as follows:

$$\frac{d}{dt}(\alpha KL) = \dot{\alpha}_{kl} \begin{bmatrix} 0 & 0 & 0 \\ 0 & -s\alpha_{kl} & -c\alpha_{kl} \\ 0 & c\alpha_{kl} & -s\alpha_{kl} \end{bmatrix} \tag{6.31}$$

$$\frac{d}{dt}(\beta KL) = \dot{\beta}_{kl} \begin{bmatrix} -s\beta_{kl} & 0 & c\beta_{kl} \\ 0 & 0 & 0 \\ -c\beta_{kl} & 0 & -s\beta_{kl} \end{bmatrix} \tag{6.32}$$

$$\frac{d}{dt}(\gamma KL) = \dot{\gamma}_{kl} \begin{bmatrix} -s\gamma_{kl} & -c\gamma_{kl} & 0 \\ c\gamma_{kl} & -s\gamma_{kl} & 0 \\ 0 & 0 & 0 \end{bmatrix}. \tag{6.33}$$

Summary

To summarize then, the underlying principle of the analysis is to formulate the equations of motion so they can be adapted to programming on a digital computer. This is done by using Kane's dynamical equation (eqn (6.3)) and by expressing vector quantities in terms of unit vectors in an inertial reference frame. The corresponding component transformation such as eqn (6.14) is obtained through the shifter matrices which are in turn obtained from configuration charts (Fig. 6.4). The vector quantities are then easy to differentiate since the unit vectors in the inertial reference frame are constant. Furthermore, the necessary shifter derivatives can be obtained by multiplication (eqn (6.28)) and can thus be obtained using the computer.

These preliminary ideas provide a basis for the analysis given in the following sections.

6.2. Kinematics

Variable definition

Consider again the human body model shown in Fig. 6.1. As mentioned above, adjoining bodies are connected to each other through ball-and-socket joints except at the elbows, knees, ankles, and trunk where hinges are employed.

It is sometimes convenient to consider a hinge as a special case of a ball-and-socket joint with two of the rotation angles fixed at zero. For example, if the left elbow is considered as a ball-and-socket joint, three successive dextral rotations of the lower arm B_4 relative to the upper arm B_3 define three angles α_{34}, β_{34}, and γ_{34}. If now α_{34} and γ_{34} are fixed at zero, the ball-and-socket joint becomes a hinge with β_{34} being its rotation angle. Also, the shifter matrices α_{34} and γ_{34} become identity matrices. Hence,

$$S34 = \beta 34 \tag{6.34}$$

and the configuration chart of Fig. 6.4 takes the form shown in Fig. 6.6.

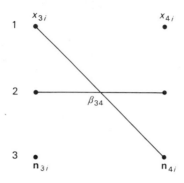

FIG. 6.6.   Configuration chart for a hinge.

Referring again to Fig. 6.1, if we consider the connections at the elbows, knees, ankles, and trunk to be hinges there are 34 degrees of freedom as opposed to the 48 mentioned earlier. (The seven hinges eliminate 14 degrees of freedom.) Let the variables $x_j(j = 1, \ldots, 34)$ represent these degrees of freedom and let them be defined as follows:

$x_1 = x$ translation of an arbitrary point in B_1 relative to the inertial
$x_2 = y$ reference frame R
$x_3 = z$

$x_4 = \alpha$ rotation of B_1 relative to the inertial reference frame R
$x_5 = \beta$
$x_6 = \gamma$

$x_7 = \alpha_{13}$ rotation of B_3 (upper left arm) relative to B_1
$x_8 = \beta_{13}$
$x_9 = \gamma_{13}$

$x_{10} = \beta_{34}$ rotation of B_4 relative to B_3 (left elbow)
$x_{11} = \alpha_{412}$ rotation of B_{12} (left hand)
$x_{12} = \beta_{412}$ relative to B_4 (lower left arm)
$x_{13} = \gamma_{412}$

$x_{14} = \alpha_{15}$ rotation of B_5 (upper right arm) relative to B_1
$x_{15} = \beta_{15}$
$x_{16} = \gamma_{15}$

$x_{17} = \beta_{56}$ rotation of B_6 relative to B_5 (right elbow)
$x_{18} = \alpha_{613}$ rotation of B_{13} (right hand) relative to B_6 (lower right arm)
$x_{19} = \beta_{613}$
$x_{20} = \gamma_{613}$

$x_{21} = \alpha_{111}$ rotation of B_{11} (head) relative to B_1
$x_{22} = \beta_{111}$
$x_{23} = \gamma_{111}$

$x_{24} = \gamma_{12}$ rotation of B_2 relative to B_1 trunk

$x_{25} = \alpha_{72}$ rotation of B_7 (upper left leg) relative to B_2
$x_{26} = \beta_{72}$
$x_{27} = \gamma_{72}$

$x_{28} = \beta_{78}$ rotation of B_8 relative to B_7 (left knee)

$x_{29} = \beta_{814}$ rotation of B_{14} relative to B_8 (left ankle)

$x_{30} = \alpha_{29}$ rotation of B_9 (upper right leg) relative to B_2
$x_{31} = \beta_{29}$
$x_{32} = \gamma_{29}$

$x_{33} = \beta_{910}$ rotation of B_{10} relative to B_9 (right knee)
$x_{34} = \beta_{1015}$ rotation of B_{15} relative to B_{10} (right ankle).

Angular velocity of the body segments

Equations (6.23) and (6.24) can be used to obtain expressions for the angular velocity of each of the various bodies of the model relative to the inertial reference frame R. For example, the angular velocity of B_1 in R is given directly from equation (6.23) as

$$\boldsymbol{\omega}^1 = (\dot{\alpha}\delta_{i1} + \dot{\beta}\alpha 01_{i2} + \dot{\gamma}\alpha 01_{im}\beta 01_{m3})\bar{\mathbf{n}}_{0i} \tag{6.35}$$

where, as before, there is a sum from 1 to 3 over the repeated indices. In view of the definitions listed above equation (6.35) can be rewritten as

$$\boldsymbol{\omega}^1 = (\dot{x}_4\delta_{i1} + \dot{x}_5\alpha 01_{i2} + \dot{x}_6\alpha 01_{im}\beta 01_{m3})\mathbf{n}_{0i}. \tag{6.36}$$

Using eqn (6.24) the angular velocity of B_2 in R can be written

$$\boldsymbol{\omega}^2 = {}^1\boldsymbol{\omega}^2 + \boldsymbol{\omega}^1 \tag{6.37}$$

where since B_2 is hinged to B_1, ${}^1\boldsymbol{\omega}^2$ is given by

$${}^1\boldsymbol{\omega}^2 = \dot{\gamma}_{12}\mathbf{n}_{13}$$
$$= \dot{\gamma}_{12}S01_{i3}\mathbf{n}_{0i}$$

or

$$^1\boldsymbol{\omega}^2 = \dot{x}_{24}\,S01_{i3}\,\mathbf{n}_{0i}. \tag{6.38}$$

Similarly, the angular velocity of the other bodies of the system can be expressed as follows:

$$\boldsymbol{\omega}^3 \;= \;^1\boldsymbol{\omega}^3 \;+\boldsymbol{\omega}^1 \tag{6.39}$$

$$\boldsymbol{\omega}^4 \;= \;^3\boldsymbol{\omega}^4 \;+\boldsymbol{\omega}^3 \tag{6.40}$$

$$\boldsymbol{\omega}^5 \;= \;^1\boldsymbol{\omega}^5 \;+\boldsymbol{\omega}^1 \tag{6.41}$$

$$\boldsymbol{\omega}^6 \;= \;^5\boldsymbol{\omega}^6 \;+\boldsymbol{\omega}^5 \tag{6.42}$$

$$\boldsymbol{\omega}^7 \;= \;^2\boldsymbol{\omega}^7 \;+\boldsymbol{\omega}^2 \tag{6.43}$$

$$\boldsymbol{\omega}^8 \;= \;^7\boldsymbol{\omega}^8 \;+\boldsymbol{\omega}^7 \tag{6.44}$$

$$\boldsymbol{\omega}^9 \;= \;^2\boldsymbol{\omega}^9 \;+\boldsymbol{\omega}^2 \tag{6.45}$$

$$\boldsymbol{\omega}^{10} = \;^9\boldsymbol{\omega}^{10} + \boldsymbol{\omega}^9 \tag{6.46}$$

$$\boldsymbol{\omega}^{11} = \;^1\boldsymbol{\omega}^{11} + \boldsymbol{\omega}^1 \tag{6.47}$$

$$\boldsymbol{\omega}^{12} = \;^4\boldsymbol{\omega}^{12} + \boldsymbol{\omega}^4 \tag{6.48}$$

$$\boldsymbol{\omega}^{13} = \;^6\boldsymbol{\omega}^{13} + \boldsymbol{\omega}^6 \tag{6.49}$$

$$\boldsymbol{\omega}^{14} = \;^0\boldsymbol{\omega}^{14} + \boldsymbol{\omega}^8 \tag{6.50}$$

$$\boldsymbol{\omega}^{15} = \;^{10}\boldsymbol{\omega}^{15} + \boldsymbol{\omega}^{10} \tag{6.51}$$

where the first term on the right-hand side of these equations has the form of equation (6.36) or equation (6.38).

In view of equations (6.36)–(6.51) it is seen that the angular velocities of the various bodies of the system can be compactly expressed as

$$\boldsymbol{\omega}^k = \omega_{ijk}\dot{x}_j\mathbf{n}_{0i} \tag{6.52}$$

where there is a sum from 1 to 34 on j and from 1 to 3 on i. The quantities ω_{ijk} $(i = 1, \ldots, 3, j = 1, \ldots, 34, k = 1, \ldots, 15)$ form a $3 \times 34 \times 15$ block matrix. They also form the coefficients of the partial rate of change or orientation vectors mentioned earlier, i.e.

$$\frac{\partial \boldsymbol{\omega}^k}{\partial \dot{x}^j} = \omega_{ijk}\mathbf{n}_{0i}. \tag{6.53}$$

It will be seen that the ω_{ijk} play an important role in the analysis which follows. The specific values of the elements of the ω_{ijk} matrix can be obtained from expressions such as equation (6.36). For example, from eqn (6.36) all ω_{ijl} are zero except that

$$\omega_{i41} = \delta_{i1} \tag{6.54}$$

$$\omega_{i51} = \alpha01_{i2} \tag{6.55}$$

$$\omega_{i61} = \alpha01_{im}\beta01_{m3}. \tag{6.56}$$

The complete listing of ω_{ijk} $(i = 1, \ldots, 3, j = 1, \ldots, 34, k = 1, \ldots, 15)$ is given in Appendix 6.1.

Angular acceleration of the body segments

The angular acceleration of each of the various bodies of the model can now be obtained simply by differentiating the angular velocity vectors above. Specifically, the angular acceleration of $\mathbf{B}_k$ $(k = 1, \ldots, 15)$ in the inertial reference frame R is

$$\boldsymbol{\alpha}^k = {}^R\mathrm{d}\,\boldsymbol{\omega}^k/\mathrm{d}t. \tag{6.57}$$

Then using eqn (6.52), $\boldsymbol{\alpha}^k$ can be written as

$$\boldsymbol{\alpha}^k = (\omega_{ijk}\dot{x}_j + \dot{\omega}_{ijk}\dot{x}_j)\mathbf{n}_{0i} \tag{6.58}$$

where, as before, there is a sum from 1 to 34 on j and from 1 to 3 on i. The quantities ω_{ijk} $(i = 1, \ldots, 3, j = 1, \ldots, 34, k = 1, \ldots, 15)$ can be obtained by simply differentiating the ω_{ijk} computed above and listed in Appendix 6.1. A glance at Appendix 6.1 shows that this in turn involves differentiating shifter matrices and $\alpha KL, \beta KL$, and γKL matrices. However, the algorithms of eqns (6.28), (6.31), (6.32), and (6.33) enable us to calculate these derivatives by simple matrix multiplication. For example, from eqns (6.54), (6.55), and (6.56) we see that all $\dot{\omega}_{ij1}$ are zero except that

$$\dot{\omega}_{i51} = \alpha\dot{0}1_{i2} \tag{6.59}$$

$$\dot{\omega}_{i61} = \alpha\dot{0}1_{im}\beta01_{m3} + \alpha01_{im}\beta\dot{0}1_{m3} \tag{6.60}$$

where $\alpha\dot{0}1_{ij}$ and $\beta\dot{0}1_{ij}$ are given by equations (6.31) and (6.32).

A complete listing of $\dot{\omega}_{ijk}x_j$ $(i = 1, \ldots, 3, k = 1, \ldots, 15, j$ summed $1, \ldots, 34)$ is given in Appendix 6.2.

Positions of the mass centres of the body segments

Figure 6.7 shows the model of Fig. 6.1 with additional symbols and vectors. Specifically the mass centre of each body $\mathbf{B}_k$ is $\mathbf{G}_k$ $(k = 1, \ldots, 15)$, $\mathbf{O}_k$ $(k = 1, \ldots, 15)$ is a reference point or co-ordinate origin for $\mathbf{B}_k$. $\mathbf{O}_k$ is located at the connecting point of $\mathbf{B}_k$ with its adjacent lower numbered body. (The reference point $\mathbf{O}_1$ may, however, be arbitrarily selected in $\mathbf{B}_1$ and usually this is taken to be at $\mathbf{G}_1$.) The vector $\mathbf{r}_k$ $(k = 1, \ldots, 15)$ locates $\mathbf{G}_k$ relative to $\mathbf{O}_k$. ($\mathbf{r}_k$ is thus fixed in $\mathbf{B}_k$.) The vector $\bar{\xi}_m$ is the position vector of $\mathbf{O}_m$ relative to $\mathbf{O}_l$ where $\mathbf{B}_l$ and $\mathbf{B}_m$ are adjacent connecting bodies with $l < m$. (ξ_m is thus fixed in $\mathbf{B}_l$, the adjacent lower numbered body.) Finally, O is the origin of R, the inertial frame, and the vector $\mathbf{R}$ is the position vector of $\mathbf{O}_1$ relative to O.

The position vector $\mathbf{P}_k$ of $\mathbf{G}_k$ $(k = 1, \ldots, 15)$ relative to O is now

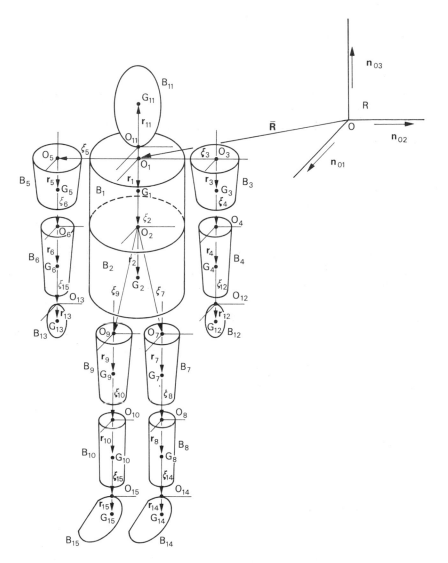

FIG. 6.7. Position vectors for the human body model.

determined. Specifically, for the 15 G_k these are

$$P_1 = R + r_1 \tag{6.61}$$
$$P_2 = R + \xi_2 + r_2 \tag{6.62}$$
$$P_3 = R = \xi + r_3 \tag{6.63}$$
$$P_4 = R + \xi_3 + \xi_4 + r_4 \tag{6.64}$$

$$\mathbf{P}_5 = \mathbf{R} + \boldsymbol{\xi}_5 + \mathbf{r}_5 \tag{6.65}$$

$$\mathbf{P}_6 = \mathbf{R} + \boldsymbol{\xi}_5 + \boldsymbol{\xi}_6 + \mathbf{r}_6 \tag{6.66}$$

$$\mathbf{P}_7 = \mathbf{R} + \boldsymbol{\xi}_2 + \boldsymbol{\xi}_7 + \mathbf{r}_7 \tag{6.67}$$

$$\mathbf{P}_8 = \mathbf{R} + \boldsymbol{\xi}_2 + \boldsymbol{\xi}_7 + \boldsymbol{\xi}_8 + \mathbf{r}_8 \tag{6.68}$$

$$\mathbf{P}_9 = \mathbf{R} + \boldsymbol{\xi}_2 + \boldsymbol{\xi}_9 + \mathbf{r}_9 \tag{6.69}$$

$$\mathbf{P}_{10} = \mathbf{R} + \boldsymbol{\xi}_2 + \boldsymbol{\xi}_9 + \boldsymbol{\xi}_{10} + \mathbf{r}_{10} \tag{6.70}$$

$$\mathbf{P}_{11} = \mathbf{R} + \boldsymbol{\xi}_{11} + \mathbf{r}_{11} \tag{6.71}$$

$$\mathbf{P}_{12} = \mathbf{R} + \boldsymbol{\xi}_3 + \boldsymbol{\xi}_4 + \boldsymbol{\xi}_{12} + \mathbf{r}_{12} \tag{6.72}$$

$$\mathbf{P}_{13} = \mathbf{R} + \boldsymbol{\xi}_5 + \boldsymbol{\xi}_6 + \boldsymbol{\xi}_{13} + \mathbf{r}_{13} \tag{6.73}$$

$$\mathbf{P}_{14} = \mathbf{R} + \boldsymbol{\xi}_2 + \boldsymbol{\xi}_7 + \boldsymbol{\xi}_8 + \boldsymbol{\xi}_{14} + \mathbf{r}_{14} \tag{6.74}$$

$$\mathbf{P}_{15} = \mathbf{R} + \boldsymbol{\xi}_2 + \boldsymbol{\xi}_9 + \boldsymbol{\xi}_{10} + \boldsymbol{\xi}_{15} + \mathbf{r}_{15} \tag{6.75}$$

These vectors can of course be expressed in terms of the unit vectors $\mathbf{n}_{0i}$ ($i = 1, 2, 3$) through use of the shifter matrices.

Velocities of the mass centres of the body segments.

The velocities of the mass centres of the body segments can now be obtained by differentiating the position vectors developed in the above subsection. Specifically the velocity of $\mathbf{G}_k$ ($k = 1, \ldots, 15$) in R is

$$\mathbf{V}^k = \frac{{}^R\mathrm{d}\mathbf{P}_k}{\mathrm{d}t} \tag{6.76}$$

Now since typically $\mathbf{P}_k$ ($k = 1, \ldots, 15$) involves the vectors $\boldsymbol{\xi}_l$ which are fixed in the adjacent lower numbered body and the vectors $\bar{\mathbf{r}}_j$ which are fixed in $\mathbf{B}_j$ ($j = 1, \ldots, 15$), the differentiation in eqn (6.76) can be obtained through vector multiplication. For example, the velocity $\mathbf{V}^{12}$ of the mass centre of the left hand is given by

$$\mathbf{V}^{12} = \frac{{}^R\mathrm{d}\mathbf{R}}{\mathrm{d}t} + \boldsymbol{\omega}^1 \times \boldsymbol{\xi}_3 + \boldsymbol{\omega}^3 \times \boldsymbol{\xi}_4$$
$$+ \boldsymbol{\omega}^4 \times \boldsymbol{\xi}_{12} + \boldsymbol{\omega}^{12} \times \mathbf{r}_{12} \tag{6.77}$$

In terms of $\mathbf{n}_{0i}$ ($i = 1, 2, 3$) this can be written as

$$\mathbf{V}^{12} = (\dot{x}_i + e_{ilm}\omega_{lj1}SO1_{mk}\xi_{3k}\dot{x}_j$$
$$+ e_{ilm}\omega_{lj3}SO3_{mk}\xi_{4k}\dot{x}_j + e_{ilm}\omega_{lj4}SO4_{mk}\xi_{12k}\dot{x}_j$$
$$+ e_{ilm}\omega_{lj12}SO12_{mk}r_{12k}\dot{x}_j)\mathbf{n}_{0i} \tag{6.78}$$

where all repeated indices are summed from 1 to 3 except j which is summed from 1 to 34. (It is worth noting that the same result could have been obtained by first expressing equation (6.72) in terms of $\mathbf{n}_{0i}$ and then differentiating, using eqn (6.28), to calculate the derivative of the shifter.)

In view of eqn (6.78) it is seen that in general the mass centre velocities can be expressed as

$$\mathbf{V}_k = V_{ijk}\dot{X}_j\mathbf{n}_{0i} \tag{6.79}$$

where there is a sum from 1 to 34 on j and a sum from 1 to 3 on i. The quantities V_{ijk} $(i = 1, \ldots, 3, j = 1, \ldots, 34, k = 1, \ldots, 15)$, like the ω_{ijk} given earlier, form a $3 \times 34 \times 15$ block matrix. They also form the coefficients of the partial rate of change of the position vectors mentioned above, i.e.

$$\frac{\partial \mathbf{V}^k}{\partial \dot{x}_j} = V_{ijk}\mathbf{n}_{0i} \tag{6.80}$$

Like the ω_{ijk}, it will be seen that the V_{ijk} play an important role in the analysis which follows. The specific values of the elements of the V_{ijk} matrix can be obtained from expressions such as in eqn (6.78). A complete listing of these elements is given in Appendix 6.3.

Accelerations of the mass centres of the body segments

The acceleration of the mass centres of the body segments can now be obtained simply by differentiating the velocity vectors above. Specifically, the acceleration of G_k $(k = 1, \ldots, 15)$ in R is

$$\mathbf{a}^k = (V_{ijk}\ddot{x}_j + \dot{V}_{ijk}\dot{x}_j)\mathbf{n}_{0i} \tag{6.81}$$

where, as before, there is a sum from 1 to 34 on j and from 1 to 3 on i. The quantities $\dot{V}_{ijk}$ $(i = 1, \ldots, 3, j = 1, \ldots, 34, k = 1, \ldots, 15)$ can be obtained by simply differentiating the V_{ijk} computed above and listed in Appendix 6.3. For example, from eqn (6.75) $\dot{V}_{ij12}\dot{x}_j$ is given by

$$
\begin{aligned}
\dot{V}_{ij12}\dot{x}_j = {} & e_{iPm}\dot{\omega}_{lj1}\dot{x}_j SO1_{mk}\xi_{3k} + e_{ilm}e_{mnq}\omega_{lj1}\dot{x}_j\omega_{mp1}\dot{x}_p \\
& \times SO1_{qk}\xi_{3k} + e_{ilm}\dot{\omega}_{lj3}\dot{x}_j SO3_{mk}\xi_{4k} \\
& + e_{ilm}e_{mnq}\omega_{lj3}\dot{x}_j\omega_{np3}\dot{x}_p SO3_{qk}\xi_{4k} \tag{6.81} \\
& + e_{ilm}\omega_{lj4}\dot{x}_j SO4_{mk}\xi_{12k} + e_{ilm}e_{mnq}\omega_{lj4}\dot{x}_j\omega_{np4}\dot{x}_p \\
& \times SO4_{qk}\xi_{12k} + e_{ilm}\dot{\omega}_{lj12}\dot{x}_j SO12_{mk}r_{12k} \\
& + e_{ilm}e_{mnq}\omega_{lj12}\dot{x}_j\omega_{np12}\dot{x}_p SO12_{qk}r_{12k} \tag{6.82}
\end{aligned}
$$

where there is a sum from 1 to 3 on $k, l, m, n,$ and q and a sum from 1 to 34 on j and p.

A complete listing of $\dot{V}_{ijk}\dot{x}_j$ $(i = 1, \ldots, 3, k = 1, \ldots, 15, j$ summed $1, \ldots, 34)$ is given in Appendix 6.4.

Summary

This concludes the section on kinematics. The principal results are the determination of the four matrices ω_{ijk}, V_{ijk}, $\dot{\omega}_{ijk}\dot{x}_j$, and $\dot{V}_{ijk}\dot{x}_j$. Furthermore, the elements of these matrices as listed in Appendices 6.1–6.4 can be evaluated through multiplication algorithms on a digital computer.

6.3. Equations of motion

We are now ready to write the dynamical equations of motion for our human body model (Fig. 6.1.). The equations of motion will be of the form of eqns (6.3), i.e.

$$F_j^* + F_j = 0 \qquad j = 1, \ldots, 34 \tag{6.83}$$

(note that here we are considering hinge joints at the trunk, elbows, knees, and ankles).

First, let us consider the generalized active force F_j ($j = 1, \ldots, 34$). From eqn (6.2) F_j is given as

$$F_j = \sum_{k=1}^{15} \left(\frac{\partial \mathbf{V}^k}{\partial \dot{x}_j} \cdot \mathbf{F}_k + \frac{\partial \boldsymbol{\omega}^k}{\partial \dot{x}_j} \cdot \mathbf{M}_k \right) \quad j = 1, \ldots, 34; \text{ no sum on } k \tag{6.84}$$

where $\mathbf{F}_k$ and $\mathbf{M}_k$ ($k = 1, \ldots, 15$) represent forces and torques which are equivalent to the applied forces acting on the respective bodies $\mathbf{B}_k$. $\mathbf{F}_k$ has its line of action passing through G_k and $\mathbf{M}_k$ is of the form

$$\mathbf{M}_k = \mathbf{M}_k^{\text{ext}} + \sum_L \mathbf{M}LK \qquad k = 1, \ldots, 15 \tag{6.85}$$

where $\mathbf{M}_k^{\text{ext}}$ is the torque due to external forces applied to $\mathbf{B}_k$ and $\mathbf{M}LK$ is a torque exerted on $\mathbf{B}_k$ by $\mathbf{B}_l$. $\mathbf{M}LK$ would normally arise owing to such forces exerted between $\mathbf{B}_k$ and $\mathbf{B}_l$. The sum in eqn (6.85) is to include all torques exerted on $\mathbf{B}_k$ by adjoining bodies. Hence in view of eqn (6.85) it is possible to write eqn (6.84) as

$$F_j = F_j^{\text{ext}} + F_j^{\text{int}} \quad j = 1, \ldots, 34 \tag{6.86}$$

where

$$\mathbf{F}^{\text{ext}} = \sum_{k=1}^{15} \frac{\partial \mathbf{V}^k}{\partial \dot{x}_j} \cdot \mathbf{F}_k + \frac{\partial \boldsymbol{\omega}^k}{\partial \dot{x}_j} \cdot \mathbf{M}_k^{\text{ext}} \quad j = 1, \ldots, 34 \tag{6.87}$$

and where

$$\mathbf{F}^{\text{int}} = \sum_{k=1}^{15} \left(\sum_L \frac{\partial \boldsymbol{\omega}^k}{\partial \dot{x}_j} \cdot \mathbf{M}LK \right) \quad j = 1, \ldots, 34. \tag{6.88}$$

Using eqns (6.53) and (6.80) eqn (6.87) can be rewritten as

$$F_j^{\text{ext}} = V_{ijk} F_{ki} + \omega_{ijk} M_{ki}^{\text{ext}} \quad j = 1, \ldots, 34 \tag{6.89}$$

where there is a sum from 1 to 3 on i and from 1 to 15 on k, and where F_{ki} and

M_{ki}^{ext} ($k = 1, \ldots, 15$, $i = 1, \ldots, 3$) are the $\mathbf{n}_{0i}$ components of $\mathbf{F}_k$ and $\mathbf{M}_k^{\text{ext}}$.

Next, to find a convenient form for F_j^{int} consider two adjacent bodies $\mathbf{B}_k$ and $\mathbf{B}_l$ where $k < l$ such as shown in Fig. 6.2. Then in general the generalized co-ordinates locating $\mathbf{B}_l$ relative to $\mathbf{B}_k$ are x_p, x_{p+1}, and x_{p+2} where $1 < p < 34$. The torque which $\mathbf{B}_k$ exerts on $\mathbf{B}_l$ is $\mathbf{M}KL$ and the torque which $\mathbf{B}_l$ exerts on $\mathbf{B}_k$ is $\mathbf{M}LK$. Note that owing to the law of action and reaction [26]

$$\mathbf{M}KL = -\mathbf{M}LK. \tag{6.90}$$

Now from eqn (6.24) we have

$$\omega^l = \omega^k + {}^k\omega^l \tag{6.91}$$

where from eqn (6.23) ${}^k\omega^l$ is

$${}^k\omega^l = (\dot{x}_p\delta_{il} + \dot{x}_{p+1}\alpha KL_{i2} + \dot{x}_{p+2}\alpha KL_{in}\beta KL_{n3})\mathbf{n}_{ki}. \tag{6.92}$$

From eqns (6.88), (6.90), and (6.91) we see that the contribution of $\mathbf{M}KL$ and $\mathbf{M}LK$ to F_j^{int} is calculated as

$$\frac{\partial\omega^k}{\partial\dot{x}_j}\cdot\mathbf{M}LK + \frac{\partial\omega^l}{\partial\dot{x}_j}\cdot\mathbf{M}LK = \frac{\partial^k\omega^l}{\partial\dot{x}_j}\cdot\mathbf{M}KL \tag{6.93}$$

Now, if $j \neq p, p+1$, or $p+2$, then $\partial^k\omega^l/\partial\dot{x}_j$ is zero and the contribution to F^{int} is zero. If $j = p, p+1$, or $p+2$, then from eqns (6.92) and (6.93) the contribution to F_j^{int} is

$$F_p^{\text{int}} = \mathbf{M}KL_1$$
$$F_{p+1}^{\text{int}} = \mathbf{M}KL_i\alpha KL_{i2} \tag{6.94}$$
$$F_{p+2}^{\text{int}} = \mathbf{M}KL_i\alpha KL_{in}\beta KL_{n3}$$

where $\mathbf{M}KL_i$ ($i = 1, \ldots, 3$) are the $\mathbf{n}_{ki}$ components of $\mathbf{M}KL$. Equations (6.94) are useful in determining muscle forces exerted between the bodies of the model.

Let us now consider the generalized inertia force F_j^* ($j = 1, \ldots, 34$) of eqn (6.83). From eqn (6.3) F_j^* can be written

$$F_j^* = \sum_{k=1}^{15}\left(\mathbf{F}_k^*\cdot\frac{\partial\mathbf{V}^k}{\partial\dot{x}_j} + \mathbf{T}_k^*\cdot\frac{\partial\omega^k}{\partial\dot{x}_j}\right) \tag{6.95}$$

where $\mathbf{F}_k^*$ and $\mathbf{T}_k^*$ ($k = 1, \ldots, 15$) are given by

$$\mathbf{F}_k^* = -m_k\mathbf{a}^k \quad \text{(no sum on } k) \tag{6.96}$$

and

$$\mathbf{T}_k^* = -\mathbf{I}^k\cdot\boldsymbol{\alpha}^k - \omega^k\times\mathbf{I}^k\cdot\omega^k \quad \text{(no sum on } k) \tag{6.97}$$

The inertia dyadic $\mathbf{I}^k$ ($k = 1, \ldots, 15$) can be written in the form

$$\mathbf{I}^k = I_{mnk}\mathbf{n}_{0m}\mathbf{n}_{0n} \tag{6.98}$$

where the elements I_{mnk} $(m, n = 1, \ldots, 3, k = 1, \ldots, 15)$ referred to $\bar{\mathbf{n}}_{0m}$ and $\bar{\mathbf{n}}_{0n}$ are related to the elements $I^{(k)}_{rsk}$ referred to $\bar{\mathbf{n}}_{kr}$ and $\bar{\mathbf{n}}_{ks}$ as follows:

$$I_{mnk} = SOK_{mr} SOK_{ns} I^{(k)}_{rsk} \tag{6.99}$$

Using eqns (6.97), (6.98), (6.52) and (6.58), $\mathbf{T}_k$ can be written as

$$\mathbf{T}^*_k = -\{I_{mnk}(\omega_{njk}\ddot{x}_j + \dot{\omega}_{njk}\dot{x}_j)$$
$$+ e_{iqm}\omega_{ipk}\omega_{njk}x_p\dot{x}_j I_{qnk}\}\mathbf{n}_{0m} \quad k = 1, \ldots, 15 \tag{6.100}$$

where there is no sum on k and $i, m, n,$ and q are summed from 1 to 3 and j and p are summed from 1 to 34. Using this result together with eqns (6.96), (6.81), (6.53), and (6.79) the generalized inertia force of eqn (6.95) can be written as

$$F^*_j = -\sum_{k=1}^{15} [m_k V_{ijk}(V_{ipk}\ddot{x}_p + \dot{V}_{ipk}\dot{x}_p)$$
$$+ \omega_{ijk}\{I_{ink}(\omega_{npk}\ddot{x}_p + \dot{\omega}_{npk}\dot{x}_p)\}$$
$$+ e_{mqi}\omega_{mpk}\omega_{nhk}\dot{x}_p\dot{x}_h I_{qnk}] \quad j = 1, \ldots, 34 \tag{6.101}$$

where there is a sum from 1 to 3 on $i, m, n,$ and q and from 1 to 34 on p and h.

Using this result together with eqn (6.86), the governing dynamical equations (6.83) can now be written in the following explicit form:

$$A_{jp}\ddot{x}_p = f_j + F^{\text{ext}}_j + F^{\text{int}}_j \quad j = 1, \ldots, 34 \tag{6.102}$$

where p is summed from 1 to 34 and where A_{jp} and f_j $(j, p = 1, \ldots, 34)$ are given by

$$A_{jp} = \sum_{k=1}^{15} m_k V_{ijk} V_{ipk} + I_{ink}\omega_{ijk}\omega_{npk} \tag{6.103}$$

and

$$f_j = -\sum_{k=1}^{15} (m_k V_{ijk} \dot{V}_{ipk}\dot{x}_p + I_{ink}\omega_{ijk}\dot{\omega}_{npk}\dot{x}_p$$
$$+ e_{mqi} I_{qnk}\omega_{ijk}\omega_{mpk}\omega_{nhk}\dot{x}_p\dot{x}_h) \tag{6.104}$$

with p and h summed from 1 to 34 and $i, m, n,$ and q summed from 1 to 3.

Equation (6.102) form a set of 34 second-order non-linear ordinary differential equations. These equations can be solved numerically for the 34 generalized co-ordinates x_p. If some (or all) of these co-ordinates are specified, eqns (6.102) reduce in part to algebraic equations for the unknown forces F^{ext}_j and F^{int}_j. In both cases all of the coefficients can be determined using the four matrices ω_{ijk}, V_{ijk}, $\dot{\omega}_{ijk}\dot{x}_j$, and $\dot{V}_{ijk}\dot{x}_j$ as listed in Appendices 6.1–6.4. Hence, the governing dynamical equations can be both generated and solved on a digital computer. These equations are specialized for impulsive forces in the next section.

6.4. Impulsive forces

One of the areas of greatest interest in human body mechanics is the response of the human body to impulsive forces. This is especially the case in the study of accident phenomena and accident reconstruction. It is also of interest in athletic studies.

An impulsive force is defined to be a force that becomes very large in magnitude but only over a short interval of time. For such forces it is possible to simplify and specialize the equations of motion of the previous section. Specifically, it is possible to specialize eqns (6.102) so that they provide direct information regarding the change in the first derivative of the generalized coordinates due to the action of an impulsive force. To this end it is helpful first to introduce and define some new quantities called 'generalized impulse' and 'generalized momentum' as follows. Let F_k and M_k^{ext} be equivalent (Kane 1959) to the externally applied impulsive forces acting on B_k ($k = 1, \ldots, 15$). Let MLK represent an impulsive torque exerted on B_k by B_l. Let the (short) time interval during which these forces and moments become large be $t_1 \leq t \leq t_2$. Then the generalized impulse I_j and the generalized momentum P_j ($j = 1, \ldots, 34$) are defined as (Kane 1959)

$$I_j = \sum_{k=1}^{15} \left\{ \frac{\partial V^k}{\partial \dot{x}_j} \cdot \int_{t_1}^{t_2} F_k \, dt + \frac{\partial \omega^k}{\partial \dot{x}_j} \cdot \int_{t_1}^{t_2} M_k^{ext} + \sum_L \int_{t_1}^{t_2} MLK \, dt \right\} \quad (6.105)$$

and

$$P_j(t) = \sum_{k=1}^{15} \left\{ m_k \frac{\partial V^k}{\partial \dot{x}_j} \cdot V^k(t) + \frac{\partial \omega^k}{\partial \dot{x}_k} \cdot I^k \cdot \omega^k(t) \right\} \quad (6.106)$$

where the sum in the last term of eqn (6.105) is to include all impulsive torques exerted on B_k by adjacent bodies.

Since the time interval $t_1 \leq t \leq t_2$ is 'short' the variables x_j ($j = 1, \ldots, 34$) remain approximately constant throughout the interval. However, the impulsive forces may lead to appreciable changes in the variables $\dot{x}_j$ ($j = 1, \ldots, 34$). Hence the partial rates of changes of position and orientation, $\partial V^k/\partial \dot{x}_j$ and $\partial \omega^k/\partial \dot{x}_j$ remain approximately constant through the interval since they depend only upon x_j, whereas the velocities and angular velocities V_k and ω^k may change appreciably through the interval since they depend upon $\dot{x}_j$. Therefore the generalized impulses I_j remain approximately constant through the interval while the generalized momenta P_j may change appreciably and are thus dependent on t through the interval.

Following the theory of impulse and momentum [29] it can be shown that

$$P_j(t_2) - P_j(t_1) = I_j \quad j = 1, \ldots, 34. \quad (6.107)$$

These equations are the governing equations for situations involving impulsive forces. They are sometimes called the 'generalized impulse–momentum equations'.

Comparing equations (6.105) with equations (6.84), (6.86), (6.87), (6.88), and (6.94) it is seen that, like F_j, I_j can be written in two parts as

$$I_j = I_j^{\text{ext}} + I_j^{\text{int}} \quad j = 1, \ldots, 34 \qquad (6.108)$$

where I_j^{ext} is given by

$$I_j^{\text{ext}} = \sum_{k=1}^{15} \left(\frac{\partial \mathbf{V}^k}{\partial \dot{x}_j} \cdot \int_{t_1}^{t_2} \mathbf{F}_k \, dt + \frac{\partial \omega^k}{\partial \dot{x}_j} \int_{t_1}^{t_2} \mathbf{M}_k^{\text{ext}} \, dt \right) \qquad (6.109)$$

and where the contribution to I_j^{int} by the impulsive torque between B_k and B_l is

$$I_p^{\text{int}} = \int_{t_1}^{t_2} MKL_i \, dt$$

$$I_{p+1}^{\text{int}} = \alpha KL_{i2} \int_{t_1}^{t_2} MKL_i \, dt \qquad (6.110)$$

$$I_{p+2}^{\text{int}} = \alpha KL_{in} \beta KL_{n3} \int_{t_1}^{t_2} MKL_i \, dt$$

where, as in eqn (6.94), x_p, x_{p+1}, and x_{p+2} are the generalized co-ordinates defining the relative orientation of B_k and B_l.

Similarly, comparing eqn (6.106) with eqns (6.95), (6.96), (6.97), and (6.101), it is seen that the generalized momenta P_j can be written as

$$P_j = \sum_{k=1}^{15} (m_k V_{ijk} V_{ipk} + I_{ink} \omega_{ijk} \omega_{npk}) \dot{x}_p \quad j = 1, \ldots, 34 \qquad (6.111)$$

where there is a sum from 1 to 3 on i and n and from 1 to 34 on p. Then by noting eqn (6.103) this can be rewritten as

$$P_j = A_{jp} \dot{x}_p \quad i = 1, \ldots, 34. \qquad (6.112)$$

Therefore by substituting eqns (6.108), and (6.112) into (6.107) the governing equations for the reaction to impulsive forces become

$$A_{jp} \Delta \dot{x}_p = I_j^{\text{ext}} + I_j^{\text{int}} \quad j = 1, \ldots, 34 \qquad (6.113)$$

where $\Delta \dot{x}_p$ is defined as

$$\Delta \dot{x}_p = \dot{x}_p(t_2) - \dot{x}_p(t_1) \qquad (6.114)$$

These equations show that it is possible to solve *algebraically* for the increments in the first derivatives of the generalized co-ordinates due to the action of impulsive or impact forces.

Illustrative examples using eqn (6.113) as well as the governing eqn (6.102) are presented in the next section.

6.5. Example motions and applications

General background

In this section we present some elementary applications of the governing dynamical equations which were derived in the two preceding sections. Specifically, we present applications of eqns (6.102) and (6.113) for the human body model of Fig. 6.1.

Physical data for the model for particular application are available in the work of Hanavan (1904). The physical data used herein, i.e. ξ_k, $\mathbf{r}_k$, m_k, and $\mathbf{I}^k$ ($k = 1, \ldots, 15$), are also obtained from Hanavan and are listed in Appendix 6.5.

The coefficients of the governing equations are computed using the algorithms and formulae listed in Appendices 6.1–6.4. These were evaluated numerically for the example motions studied on a digital computer. Finally the governing equations were solved numerically for these motions using a Runge–Kutta integration scheme on a high-speed digital computer.

Three specific classes of motions are considered. In the first class the model is subjected to a series of impulsive forces or sharp blows and the resulting velocity increments and the positions of the body segments at various instants are determined. In the second class a human-factors problem of a man attempting to use a hammer in a weightless environment is considered to determine the muscular effort required for the function. In the third class an analysis of simple jumping is made to determine the relative contributions of the legs, arms, and ankles to the motion. These examples are not meant to be exhaustive studies in themselves but instead they are presented as demonstrations of the versatility and wide range of applicability of the governing dynamical equations and of the potential of the method in providing guidelines for evaluation and possible equation of effort required for various occupational and functional motions.

The model subjected to impulsive forces

In this first example we employ the equations developed in the preceding section on impulsive forces. Specifically we use eqns (6.113) which need not be integrated but which may be solved algebraically, i.e. upon specification of the impulsive force the resulting velocity increments in the bodies of the model are obtained.

Two basic configurations were used, one for the model in a sitting position and the other for the model in a standing position. With the model in a sitting position an impulse of 167 lbf s was applied in several directions acting through a point located midway on a line connecting the hip joints. (This configuration could be used to study ejection seat problems or problems encountered in automobile crashes.) The results are presented in Table 6.1. $\Delta \dot{x}_j$ represents the incremental change in $\dot{x}_j$; $\Delta \dot{x}_j$ is thus a kind of velocity

TABLE 6.1

Response to impulsive forces (model in sitting position).

The impulse $\mathbf{I}$ is applied to a point midway between the hip joints with the model in a sitting position (see §6.2 for the notation of x_j).

j	x_j	$\mathbf{I} = 167\mathbf{n}_{03}$ lbf s $\Delta\dot{x}_j$	$\mathbf{I} = 167\mathbf{n}_{01}$ lbf s $\Delta\dot{x}_j$	$\mathbf{I} = 118\mathbf{n}_{01}$ $+118\mathbf{n}_{03}$ lbf s $\Delta\dot{x}_j$	$\mathbf{I} = 167\mathbf{n}_{02}$ lbf s $\Delta\dot{x}_j$	$\mathbf{I} = 118\mathbf{n}_{01}$ $+118\mathbf{n}_{02}$ lbf s $\Delta\dot{x}_j$
1	—	-0.0134^a	743^a	526^a	0.00180^a	526^a
2	—	0^a	0^a	0^a	677^a	479^a
3	—	511^a	0.0618^a	361^a	0^a	0.0437^a
4	—	0	0	0	-7.52	-5.31
5	—	-0.0101	25.0	17.6	0	17.6
6	—	0	0	0	0.00441	0.00309
7	—	0	0	0	-38.6	-27.3
8	—	0.00662	60.3	42.3	-0.00212	42.7
9	—	0	0	0	-0.0298	-0.0210
10	—	0.00455	-112.0	-79.4	0	-79.4
11	—	0	0	0	184.	130
12	—	-0.00248	62.5	44.2	0.00901	44.2
13	—	0	0	0	0.0311	0.0219
14	—	0	0	0	-38.6	-27.3
15	—	0.00662	60.3	42.7	0.00658	42.7
16	—	0	0	0	-0.0298	-0.0210
17	—	0.00457	-112	-79.4	-0.0121	-79.4
18	—	0	0	0	184	130
19	—	-0.00252	62.5	44.2	0.0206	44.2
20	—	0	0	0	0.0311	0.0219
21	—	0	0	0	95.5	67.5
22	—	0.0167	-130	-91.7	0	-91.7
23	—	0	0	0	-0.00307	-0.00215
24	—	0	0	0.00108	-0.0024	-17.5
25	—	0.0317	-2.64	-1.89	-1.38×10^4	-9.76×10^{-3}
26	$-90°$	33.0	-25.0	5.67	-1.50	-18.7
27	—	-0.0317	-2.63	-1.89	-1.38×10^4	-8.78×10^{-3}
28	$90°$	-33.0	29.8	-2.28	8.48	27.0
29	—	0.0806	-78.8	-55.6	18.5	-68.8
30	—	-0.0123	-2.37	-1.69	-1.30×10^4	-9.21×10^{-3}
31	$-90°$	33.0	-25.0	5.67	1.50	-16.6
32	—	-0.0122	-2.37	-1.69	-1.30×10^4	-9.22×10^{-3}
33	$90°$	-33.0	29.8	-2.28	-8.48	15.0
34	—	-0.08	-78.8	-55.6	18.5	-42.6

[a] These values are in in s^{-1}; the remainder are in rad s^{-1}.

increment. Special note should be taken of the velocities of the head for the five different cases tabulated. The most severe case results in the head sustaining an angular velocity change of 130 rad s^{-1} when the impulse is in the $\mathbf{n}_{01}$ direction (defined in Fig. 6.1). When the skull is subject to associated centrifugal inertia forces due to angular velocities of such orders of magnitude, the resulting tensile and shear stresses in the brain are of magnitudes that exceed the ultimate strengths of dura matter and of the arteries; the phenomenon of head injury is discussed elsewhere in this book.

Two further cases were considered with the model in a standing configuration: the first with an impulsive force applied to the head simulating a blow to the head, and the other with an impulsive force applied to the right knee simulating a side blow to the right leg. The results of these two motions are presented in Table 6.2.

TABLE 6.2

Response to impulsive forces (model in standing position)

In case (1) the impulse is applied to the head; in case (2) the impulse is applied to the right knee.

j	x_j	Case (1) $\mathbf{I} = -167\mathbf{n}_{01}$ lbf s^{-1} $\Delta \dot{x}_j$	Case (2) $\mathbf{I} = 167\mathbf{n}_{02}$ lbf s^{-1} $\Delta \dot{x}_j$
1	–	−369.0[a]	0[a]
2	–	0[a]	−39.2[a]
3	–	0[a]	0[a]
4	–	0	−23.0
5	–	−28.2	0
6	–	0	0
7	–	0	22.8
8	–	−17.3	0
9	–	0	0
10	–	59.9	0
11	–	0	0.873
12	–	−33.4	0
13	–	0	0
14	–	0	22.8
15	–	−17.3	0
16	–	0	0
17	–	59.9	0
18	–	0	0.87
19	–	−33.4	0
20	–	0	0
21	–	0	29.4
22	–	−487.0	0
23	–	0	0
24	–	0	0
25	–	0	41.7
26	–	36.0	0
27	–	0	0
28	–	−11.5	0
29	–	9.78	0
30	–	0	224.0
31	–	36.0	0
32	–	0	0
33	–	−11.5	0
34	–	9.78	0

[a] These values are in in s^{-1}; the remainder are in rad s^{-1}.

Hammering motion in space

In this second example the solution is obtained in two parts. First the model begins in an erect reference configuration. It then lifts its right arm and in a hammering motion it strikes an object. During this manoeuvre the arm motion is specified and the resulting angular displacement and translation of the chest

TABLE 6.3

Arm motion and impulse response

The impulse is applied to the right hand at $T = 2.425$ s.

j	x_j at 2.425 s	$\dot{x}_j$ at 2.425$^-$ s	$\Delta\dot{x}_j$	$\dot{x}_j$ at 2.425 s
		$\mathbf{I} = (167 \text{ lbf s})\mathbf{n}_{03}$		
1	-0.168[a]	-0.201[b]	4.85[b]	4.65[b]
2	-0.0564[a]	0.425[b]	-70.9[b]	-70.5[b]
3	-0.181[a]	1.07[b]	-49.5[b]	-48.4[b]
4	0.00481	-0.0345	8.08	8.05
5	0.00202	-0.0257	0.0964	0.0707
6	0.0842	0.233	5.11	4.88
7	–	–	-2.35	-2.35
8	–	–	-4.06	-4.06
9	–	–	-5.12	-5.12
10	–	–	4.32	4.32
11	–	–	-22.7	-22.7
12	–	–	-2.41	-2.41
13	–	–	-673	$-673.$
14	–	–	-0.237	-0.237
15	-0.0334	1.30	28.1	28.1
16	–	–	-0.112	-0.112
17	-1.52	4.45	-1380	-1376
18	–	–	59.3	59.3
19	–	–	-1.23×10^5	-1.23×10^5
20	–	–	7.72	7.72
21	–	–	-20.8	-20.8
22	–	–	-2.52	-2.52
23	–	–	-5.12	-5.12
24	–	–	-5.12	-5.12
25	–	–	-10.9	-10.9
26	–	–	-0.930	-0.930
27	–	–	0	0
28	–	–	0.228	0.228
29	–	–	-0.195	-0.195
30	–	–	-10.9	-10.9
31	–	–	-0.93	-0.93
32	–	–	0	0
33	–	–	0.228	0.228
34	–	–	-0.194	-0.194

[a] These values are in inches; the remainder are in radians.
[b] The values are in in s^{-1}; the remainder are in rad s^{-1}.

is calculated through the Runge–Kutta integration of the equations of motion. More specifically, x_j and $\dot{x}_j$ ($j = 1, \ldots, 6$) are determined with x_j and $\dot{x}_j$ ($j = 7, \ldots, 34$) specified. (All x_j, ($j = 7, \ldots, 34$) are specified as zero except x_{15} and x_{17}.) Next when $x_{15} + x_{17}$ is approximately $90°$ ($t = 2.435$ s), the right hand is struck with an impulsive force of 167 $\mathbf{n}_{03}$ lbf s. The resulting velocity changes are then calculated as in the first example above. Table 6.3 shows the time history of x_{15} and x_{17} prior to application of the impulsive force. The functions used in specifying x_{15} and x_{17} are of the form suggested by Smith and Kane [1, 2]:

$$f(t) = f_0 + (f_1 - f_0)\frac{t}{T} - \frac{1}{2\pi}\sin\left(\frac{2\pi t}{T}\right) \tag{6.115}$$

where T is the duration time of variable change, and f_0 and f_1 are the values of f for $t = 0$ and $t = T$. This function has the property of vanishing first and second derivatives for both $t = 0$ and $t = T$. Table 6.3 shows first the values of x_j and $\dot{x}_j$ ($j = 1, \ldots, 34$) just prior to the application of the impulse; next the values of the velocity increments $\Delta\dot{x}_j$ ($j = 1, \ldots, 34$) resulting from the impulsive force and finally the resulting velocities $\dot{x}_j$ ($j = 1, \ldots, 34$) just after the impulse.

Simple jumping

In this final example the motion of the model is specified and the forces required to produce the motion are determined. Four different cases of jumping motion are considered. In each case the time history of the normal force acting on the feet, the moment between the hip and upper leg, the moment between the lower and upper leg, an approximation of the work done by the leg muscles, and finally the height point O_1 reaches above its position in the reference configuration are computed. The four different cases considered are shown in Fig. 6.8. Case 1 is restricted to leg motion, case 2 is leg and arm motion, case 3 is leg, arm, and upper body motion, and case 4 is leg, arm, upper body, and ankle motion.

Figure 6.9 depicts the variables used in defining the four jumping motions. The functions used for these variables are as follows (see §6.2 for notation).

Case 1

$$x_{28} = x_{33} = +1.57 - 1.57\left\{\frac{t}{0.2} - \frac{1}{2\pi}\sin\left(\frac{2\pi t}{0.2}\right)\right\}$$

$$x_{26} = x_{31} = -\alpha = -\tan^{-1}\left(\frac{\xi_{14}}{\xi_8}\tan\tfrac{1}{2}x_{28}\right)$$

$$x_{29} = x_{34} = -x_{28} + \alpha$$

$$x_3 = \xi_7 + \xi_8\cos\alpha + \xi_{14}\cos(x_{28} - \alpha)$$

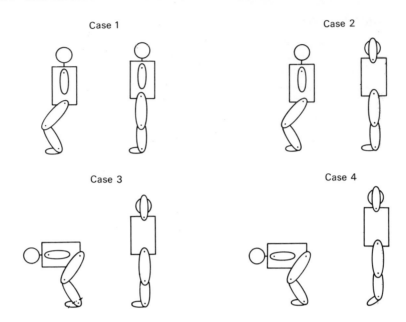

FIG. 6.8. Body segment configurations during jumping.

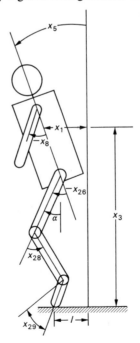

FIG. 6.9. Variables employed to define body segment configurations during jumping.

Case 2

$$x_{28} = x_{33} = 1.57 - 1.57 \left\{ \frac{t}{0.2} - \frac{1}{2\pi} \sin\left(\frac{2\pi t}{0.2}\right) \right\}$$

$$x_{26} = x_{31} = -\alpha = -\tan^{-1}\left(\frac{\xi_{14}}{\xi_8} \tan\tfrac{1}{2}x_{28}\right)$$

$$x_{29} = x_{34} = -x_{28} + \alpha$$

$$x_3 = \xi_7 + \xi_8 \cos\alpha + \xi_{14} \cos(x_{28} - \alpha)$$

$$x_8 = x_{15} = -3.14 \left\{ \frac{t}{0.2} - \frac{1}{2\pi} \sin\left(\frac{2\pi t}{0.2}\right) \right\}$$

Case 3

$$x_{28} = x_{33} = 1.57 - 1.57 \left\{ \frac{t}{0.2} - \frac{1}{2\pi} \sin\left(\frac{2\pi t}{0.2}\right) \right\}$$

$$x_5 = 1.57 - 1.57 \left\{ \frac{t}{0.2} - \frac{1}{2\pi} \sin\left(\frac{2\pi t}{0.2}\right) \right\}$$

$$\alpha = -\tan^{-1}\left(\frac{\xi_{14}}{\xi_8} \tan\tfrac{1}{2}x_{28}\right)$$

$$x_{26} = x_{31} = -(\alpha + x_5)$$

$$x_1 = \xi_7 \sin x_5$$

$$x_3 = \xi_7 \cos x_5 + \xi_8 \cos\alpha + \xi_{14} \cos(x_{28} - \alpha)$$

$$x_8 = x_{15} = -3.14 \left\{ \frac{t}{0.2} - \frac{1}{2\pi} \sin\left(\frac{2\pi t}{0.2}\right) \right\}$$

Case 4

$$x_{28} = x_{33} = 1.57 = 1.57 \left\{ \frac{t}{0.2} - \frac{1}{2\pi} \sin\left(\frac{2\pi t}{0.2}\right) \right\}$$

$$x_5 = 1.57 - 1.57 \left\{ \frac{t}{0.2} - \frac{1}{2\pi} \sin\left(\frac{2\pi t}{0.2}\right) \right\}$$

$$\alpha = -\tan^{-1}\left(\frac{\xi_{14}}{\xi_8} \tan\tfrac{1}{2}x_{28}\right)$$

$$x_{26} = x_{31} = -(\alpha + x_5)$$

$$x_{29} = x_{34} = -0.785 + 1.57 \left\{ \frac{t}{0.2} - \frac{1}{2\pi} \sin\left(\frac{2\pi t}{0.2}\right) \right\}$$

$$x_3 = \xi_7 \cos x_5 + \xi_8 \cos\alpha + \xi_{14} \cos(x_{28} - \alpha) + l \sin(x_{28} - \alpha + x_{29})$$

$$x_1 = \xi_7 \sin x_5 + l\{1 - \cos(x_{28} - \alpha + x_{29})\}$$

where l is the distance from the toe to the ankle joint, ξ_7, ξ_8, and ξ_{14} are the magnitudes of the vectors ξ_7, ξ_8, and ξ_{14}, as shown in Fig. 6.7.

Figures 6.10–6.12 are plots of the normal force at the foot (ground reaction), the moment at the hip, and the moment at the knee for the four motions. Table 6.4 shows the height reached by O_1, the work done by the moment at the hip and knee, and the velocity of the mass centre of the system when the feet leave the ground. In Fig. 6.10, when F_{normal} is zero, the feet leave the ground. Note that the maximum normal force occurs when the legs, arms, and upper body

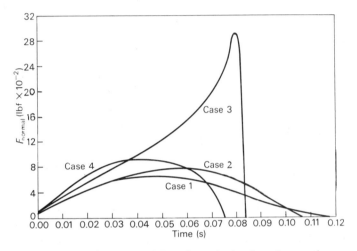

FIG. 6.10. Plots of normal foot force during jumping motions.

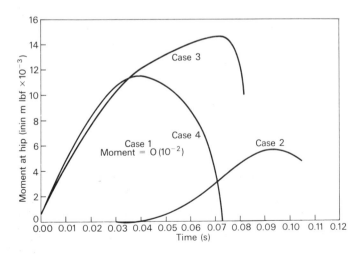

FIG. 6.11. Hip moment variations during jumping motions.

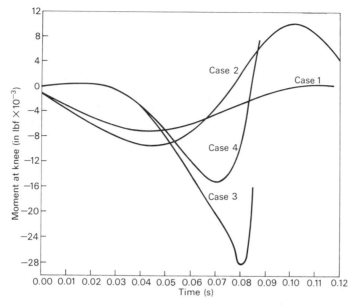

Fɪɢ. 6.12.   Knee moment variations during jumping motions.

(but not the ankles) are employed in the jump (Case 3). In Fig. 6.11 the hip moment, i.e. the moment between B_7 and B_2, is negligible for case 1. The maximum hip moment occurs when the legs, arms, and upper body (but not the ankles) are employed in the jump (Case 3). In Table 6.4 the work done by the upper leg muscles is calculated by simply integrating the product of the hip moment and the angle at the hips (x_{26} or β_{72}). Similarly the work by the lower leg muscles is the integrated product of the knee moment and the angle at the knees (x_{28} or β_{78}). From Table 6.4 it appears that the body motion is of primary importance in the modelled jumping motion.

TABLE 6.4

Height jumped and work done

Case	Time at zero normal force (s)	Height O_1 reaches above reference position (in)	Work of muscles in upper legs (in lbf)	Work of muscles in lower legs (in lbf)
1	0.116	3.4	6.61×10^{-3}	1.22×10^{4}
2	0.106	15	6.50×10^{3}	2.44×10^{4}
3	0.084	55	1.74×10^{5}	5.36×10^{5}
4	0.075	42	1.18×10^{4}	9.2×10^{3}

Summary

These examples illustrate the different kinds of problems which may be studied using the equations and methods developed in the sequel, i.e. given the forces applied to the system the resulting displacement and motion are determined (first example). Alternatively, given the displacement and motion the required forces are obtained (third example). Finally, given a combination of forces, displacement, and motion the remaining (unknown) displacement, motion, and forces are determined (second example).

Other illustrative motions using these same methods can be found in Passerello and Huston (1971); Huston and Passerello (1971); and Gallenstein and Huston (1972). The first of these pertains to human attitude control in space, the second to simple lifting and kicking, and the last to swimming motion.

6.6. Conclusions

We have presented an analysis and a general procedure for studying the mechanics of human body motion. This procedure and the resulting governing dynamical equations are applicable in a wide variety of situations, i.e. given any force field applied to the human body, or given any prescribed motion of the human body, or given any combination of these the governing dynamical equations lead to a determination of the unknown forces and motion.

This comprehensive nature of the analysis is due to a large extent to using Lagrange's form of d'Alembert's principle to develop the governing equations. Indeed, this principle essentially reduces the derivation of the governing equations to the evaluation of kinematical coefficients. In turn, simple algorithms for evaluating these coefficients are listed in a series of appendices. Therefore all of the numerical work beginning with the derivation of these coefficients and ending with the numerical solution of the governing equations can easily be performed on a high-speed digital computer.

The analysis is illustratively applied in three classes of motion: response to impact; the work of human factors in space; simple jumping. As mentioned in the preceding section these examples are not meant to be exhaustive but merely illustrative of the range of applicability of the method.

The areas of application are virtually unbounded; i.e. the analysis and procedure as presented herein are applicable in virtually every phase of human physical activity ranging from work in space to accident phenomena, to athletic and health studies, to problems of human factors, and beyond.

Nomenclature

$\mathbf{a}^k$	$(k = 1, \ldots, 15)$	acceleration of G_k in R (equation (6.80))
B_k	$(k = 1, \ldots, 15)$	bodies of the model (Fig. 6.1.)

F_j	$(j = 1, \ldots, 34)$	generalized active force (equation (6.2))
F_j^*	$(j = 1, \ldots, 34)$	generalized inertia force (equation (6.3))
$\mathbf{F}_k^*$	$(k = 1, \ldots, 15)$	inertia force on $\mathbf{B}_k$ (equation (6.4))
G_k	$(k = 1, \ldots, 15)$	mass centre of $\mathbf{B}_k$
$\mathbf{I}^k$	$(k = 1, \ldots, 15)$	inertia dyadic of $\mathbf{B}_k$ for G_k (equation (6.98))
m_k	$(k = 1, \ldots, 15)$	mass of $\mathbf{B}_k$
$\mathbf{n}_{ki}$	$(k = 1, \ldots, 15;$	
	$i = 1, \ldots, 3)$	unit vectors fixed in $\mathbf{B}_k$
O_k	$(k = 1, \ldots, 15)$	reference point of $\mathbf{B}_k$ (Fig. 6.7)
$\mathbf{P}_k$	$(k = 1, \ldots, 15)$	position vector of G_k in R (Fig. 6.7)
R		inertia reference frame
$\mathbf{r}_k$	$(k = 2, \ldots, 15)$	position vector of G_k relative to O_k (Fig. 6.7)
SKL_{im}	$(i, m = 1, \ldots, 3)$	shifter matrix (equation (6.8))
t		Time
$\mathbf{T}_k^*$	$(k = 1, \ldots, 15)$	inertia torque of $\mathbf{B}_k$ (equation (6.4))
$\mathbf{V}^k$	$(k = 1, \ldots, 15)$	velocity of G_k in R (equation (6.76))
x_j	$(j = 1, \ldots, 34)$	generalized co-ordinates (Table 6.1)
$\alpha KL, \beta KL, \gamma KL$		shifter matrices (equations (6.18)–(6.20))
α^k	$(k = 1, \ldots, 15)$	angular acceleration of $\mathbf{B}_k$ in R
ξ_k	$(k = 3, \ldots, 15)$	reference position vector of $\mathbf{B}_k$ (Fig. 6.7)
$^k\omega^l$		angular velocity of $\mathbf{B}_1$ relative to $\mathbf{B}_k$
ω^k	$(k = 1, \ldots, 15)$	angular velocity of $\mathbf{B}_k$ in R (equation (6.24))

Acknowledgments

The authors thank NASA and the University of Cincinnati Institute of Space Sciences for their support of a major portion of the research associated with this work. This support, which was granted under NASA Grant NGL 36-004-014, Multidisciplinary space related research in the physical, engineering, life and social sciences, is sincerely appreciated. The authors also acknowledge and sincerely appreciate support from the Office of Naval Research under Grant No. N0014-72-A-0027-0002.

Appendix 6.1

Below is a complete listing of *non-zero* ω_{ijk} $(i = 1, \ldots, 3; j = 1, \ldots, 34; k = 1, \ldots, 15)$ as defined in eqn (6.52):

$$\omega_{1\,4\,1} = 1$$
$$\omega_{i\,5\,1} = \alpha 01_{i2} \ (i = 1, 2, 3)$$
$$\omega_{i\,6\,1} = \alpha 01_{im}\beta 01_{m3} \ (i = 1, 2, 3)$$
$$\omega_{ijk} = \omega_{ij1} \ (i = 1, 2, 3, j = 4, 5, 6, k = 2, \ldots, 15)$$

$\omega_{i\,24\,2} = S01_{i3}\ (i = 1, 2, 3)$

$\omega_{i\,7\,3} = S01_{i1}\ (i = 1, 2, 3)$

$\omega_{i\,8\,3} = S01_{im}\alpha13_{m2}\ (i = 1, 2, 3)$

$\omega_{i\,9\,3} = S01_{im}\alpha13_{mn}\beta13_{n3}\ (i = 1, 2, 3)$

$\omega_{i\,j\,4} = \omega_{i\,j\,3}\ (i = 1, 2, 3, j = 4, \ldots, 9)$

$\omega_{i\,10\,4} = S03_{i2}\ (i = 1, 2, 3)$

$\omega_{i\,14\,5} = S01_{i1}\ (i = 1, 2, 3)$

$\omega_{i\,15\,5} = S01_{im}\alpha15_{m2}\ (i = 1, 2, 3)$

$\omega_{i\,16\,5} = S01_{im}\alpha15_{mn}\beta15_{n3}\ (i = 1, 2, 3)$

$\omega_{i\,j\,6} = \omega_{i\,j\,5}\ (i = 1, 2, 3, j = 4, 5, 6, 14, 15, 16)$

$\omega_{i\,17\,6} = S05_{i2}\ (i = 1, 2, 3)$

$\omega_{i\,24\,7} = \omega_{i\,24\,2}\ (i = 1, 2, 3)$

$\omega_{i\,25\,7} = S02_{i1}\ (i = 1, 2, 3)$

$\omega_{i\,26\,7} = S02_{im}\alpha27_{m2}\ (i = 1, 2, 3)$

$_{i\,27\,7} = S02_{im}\alpha27_{mn}\beta27_{n3}\ (i = 1, 2, 3)$

$\omega_{i\,j\,8} = \omega_{i\,j\,7}\ (i = 1, 2, 3, j = 4, 5, 6, 24, 25, 26, 27)$

$\omega_{i\,28\,8} = S07_{i2}\ (i = 1, 2, 3)$

$\omega_{i\,24\,9} = \omega_{i\,24\,2}\ (i = 1, 2, 3)$

$\omega_{i\,30\,9} = S02_{i1}\ (i = 1, 2, 3)$

$\omega_{i\,31\,9} = S02_{im}\alpha29_{m2}\ (i = 1, 2, 3)$

$\omega_{i\,32\,9} = S02_{im}\alpha29_{mn}\beta29_{n3}\ (i = 1, 2, 3)$

$\omega_{i\,j\,10} = \omega_{i\,j\,9}\ (i = 1, 2, 3, j = 4, 5, 6, 24, 30, 31, 32)$

$\omega_{i\,33\,10} = S09_{i2}\ (i = 1, 2, 3)$

$\omega_{i\,21\,11} = S01_{i1}\ (i = 1, 2, 3)$

$\omega_{i\,22\,11} = S01_{im}\alpha111_{m2}\ (i = 1, 2, 3)$

$\omega_{i\,23\,11} = S01_{im}\alpha111_{mn}\beta111_{n3}\ (i = 1, 2, 3)$

$\omega_{i\,j\,12} = \omega_{i\,j\,4}\ (i = 1, 2, 3, j = 7, 8, 9, 10)$

$\omega_{i\,11\,12} = S04_{i1}$

$\omega_{i\,13\,12} = S04_{im}\alpha412_{mn}\beta412_{n3}\ (i = 1, 2, 3)$

$\omega_{i\,j\,13} = \omega_{i\,j\,6}\ (i = 1, 2, 3, j = 14, 15, 16, 17)$

$\omega_{i\,18\,13} = S06_{i1}\ (i = 1, 2, 3)$

$\omega_{i\,19\,13} = S06_{im}\alpha613_{m2}\ i = 1, 2, 3)$

$\omega_{i\,20\,13} = S06_{im}\alpha613_{mn}\beta613_{n3}\ (i = 1, 2, 3)$

$\omega_{i\,j\,14} = \omega_{i\,j\,8}\ (i = 1, 2, 3, j = 24, 25, 26, 27, 28)$

$$\omega_{i\,29\,14} = S08_{i2} \ (i = 1, 2, 3)$$

$$\omega_{i\,j\,15} = \omega_{i\,j\,10} \ (i = 1, 2, 3, j = 24, 30, 31, 32\ 33)$$

$$\omega_{i\,34\,15} = S010_{i2} \ (i = 1, 2, 3)$$

$$(m, n \text{ summed } 1, \ldots, 3).$$

Appendix 6.2

Below is a complete listing of $\omega_{ijk}\dot{x}_j$ ($i = 1, \ldots, 3, k = 1, \ldots, 15, j$ summed $1, \ldots, 34$) (see equation (4.58)):

$$\dot{\omega}_{ij1}\dot{x}_j = \alpha\dot{0}1_{i2}\dot{x}_5 + (\alpha\dot{0}1_{im}\beta 01_{m3} + \alpha 01_{im}\beta\dot{0}1_{m3})\dot{x}_6$$

$$\dot{\omega}_{ij2}\dot{x}_j = \dot{\omega}_{ij1}\dot{x}_j + S\dot{0}1_{i3}\dot{x}_{24}$$

$$\dot{\omega}_{ij3}\dot{x}_j = \dot{\omega}_{ij1}\dot{x}_j + S01_{im}\{\alpha\dot{1}3_{m2}\dot{x}_8 + (\alpha\dot{1}3_{mn}\beta 13_{n3}$$
$$+ \alpha 13_{mn}\beta\dot{1}3_{n3})\dot{x}_9\} + S\dot{0}1_{im}\{\delta_{m1}\dot{x}_7$$
$$+ \alpha 13_{m2}\dot{x}_8 + \alpha 13_{mn}\beta 13_{n3}\dot{x}_9\}$$

$$\dot{\omega}_{ij4}\dot{x}_j = \dot{\omega}_{ij3}\dot{x}_j + S\dot{0}3_{i2}\dot{x}_{10}$$

$$\dot{\omega}_{ij4}x_j = \dot{\omega}_{ij3}\dot{x}_j + S\dot{0}3_{i2}x_{10}$$

$$\dot{\omega}_{ij5}\dot{x}_j = \dot{\omega}_{ij1}\dot{x}_j + S01_{im}\{\alpha\dot{1}5_{m2}\dot{x}_{15} + (\alpha 1\dot{5}_{mn}\beta 15_{n3}$$
$$+ \alpha 15_{mn}\beta\dot{1}5_{n3})\dot{x}_{16} + S\dot{0}1_{im}\{\delta_{m1}\dot{x}_{14}$$
$$+ \alpha 15_{m2}\dot{x}_{15} + \alpha 15_{mn}\beta 15_{n3}\dot{x}_{16}\}$$

$$\dot{\omega}_{ij6}\dot{x}_j = \dot{\omega}_{ij5}\dot{x}_j + S\dot{0}5_{i2}\dot{x}_{17}$$

$$\dot{\omega}_{ij7}\dot{x}_j = \dot{\omega}_{ij2}\dot{x}_j + S02_{im}\{\alpha\dot{2}7_{m2}\dot{x}_{26} + (\alpha 2\dot{7}_{mn}\beta 27_{n3}$$
$$+ \alpha 27_{mn}\beta\dot{2}7_{n3})\dot{x}_{27}\} + S\dot{0}2_{im}\{\delta_{m1}\dot{x}_{25}$$
$$+ \alpha 27_{m2}\dot{x}_{26} + \alpha 27_{mn}\beta 27_{n3}\dot{x}_{27}\}$$

$$\dot{\omega}_{ij8}\dot{x}_j = \dot{\omega}_{ij7}\dot{x}_j + S\dot{0}7_{i2}\dot{x}_{28}$$

$$\dot{\omega}_{ij9}\dot{x}_j = \dot{\omega}_{ij2}\dot{x}_j + S02_{im}\{\alpha\dot{2}9_{m2}\dot{x}_{31} + (\alpha 2\dot{9}_{mn}\beta 29_{n3}$$
$$+ \alpha 29_{mn}\beta\dot{2}9_{n3})\dot{x}_{32}\} + S\dot{0}2_{im}\{\delta_{m1}\dot{x}_{30}$$
$$+ \alpha 29_{m2}\dot{x}_{31} + \alpha 29_{mn}\beta 29_{n3}\dot{x}_{32}\}$$

$$\dot{\omega}_{ij10}\dot{x}_j = \dot{\omega}_{ij9}\dot{x}_j + S\dot{0}9_{i2}\dot{x}_{33}$$

$$\dot{\omega}_{ij11}\dot{x}_j = \dot{\omega}_{ij1}\dot{x}_j + S01_{im}\{\alpha 111_{m2}\dot{x}_{22} + (\dot{\alpha}111_{mn}\beta 111_{n3}$$
$$+ \alpha 111_{mn}\beta 111_{n3})\dot{x}_{23}\} + S\dot{0}1_{im}\{\delta_{m1}\dot{x}_{21}$$
$$+ \alpha 111_{m2}\dot{x}_{22} + \alpha 111_{mn}\beta 111_{n3}\dot{x}_{23}\}$$

$$\dot{\omega}_{ij12}\dot{x}_j = \dot{\omega}_{ij4}\dot{x}_j + S04_{im}\{\dot{\alpha}412_{m2}\dot{x}_{12} + (\alpha 412_{mn}\beta 412_{n3}$$
$$+ \alpha 412_{mn}\beta 4\dot{1}2_{n3})\dot{x}_{13}\} + S\dot{0}4_{im}\{\delta_{m1}\dot{x}_{11}$$
$$+ \alpha 412_{m2}\dot{x}_{12} + \alpha 412_{mn}\beta 412_{n3}\dot{x}_{13}\}$$

$$\dot{\omega}_{ij13}\dot{x}_j = \dot{\omega}_{ij6}\dot{x}_j + S06_{im}\{\alpha 6\dot{1}3_{m2}\dot{x}_{19} + (\alpha 6\dot{1}3_{mn}\beta 6\dot{1}3_{n3}$$
$$+ \alpha 6\dot{1}3_{mn}\beta 6\dot{1}3_{n3})\dot{x}_{20}\} + S\dot{0}6_{im}\{\delta_{m1}\dot{x}_{18}$$
$$+ \alpha 6\dot{1}3_{m2}\dot{x}_{19} + \alpha 6\dot{1}3_{mn}\beta 6\dot{1}3_{n3}\dot{x}_{20}\}$$

$$\dot{\omega}_{ij14}\dot{x}_j = \dot{\omega}_{ij8}\dot{x}_j + S\dot{0}8_{i2}\dot{x}_{29}$$

$$\dot{\omega}_{ij15}\dot{x}_j = \dot{\omega}_{ij10}\dot{x}_j + S0\dot{1}0_{i2}\dot{x}_{34}$$

$$(i = 1, \ldots, 3, m, n \text{ summed } 1, \ldots, 3).$$

Appendix 6.3

Below is a complete listing of the *non-zero* V_{ijk} ($i = 1, \ldots, 3, j = 1, \ldots, 34,$ $k = 1, \ldots, 15$) as defined in eqn (6.79):

$$V_{ij1} = \delta_{ij} + e_{ilm}\omega_{1j}S01_{mn}r_{1n}$$

$$V_{ij2} = \delta_{ij} + e_{ilm}\omega_{2jl}S02_{mn}r_{2n} + e_{ilm}\omega_{ijl}S01_{mn}\xi_{2n}$$

$$V_{ij3} = \delta_{ij} + e_{ilm}\omega_{3jl}S03_{mn}r_{3n} + e_{ilm}\omega_{1jl}S01_{mn}\xi_{3n}$$

$$V_{ij4} = \delta_{ij} + e_{ilm}\omega_{4jl}S04_{mn}r_{4n} + e_{ilm}\omega_{3jl}S03_{mn}\xi_{4n}$$
$$+ e_{ilm}\omega_{1jl}S01_{mn}\xi_{3n}$$

$$V_{ij5} = \delta_{ij} + e_{ilm}\omega_{5jl}S05_{mn}r_{5n} + e_{ilm}\omega_{ijl}S01_{mn}\xi_{5n}$$

$$V_{ij6} = \delta_{ij} + e_{ilm}\omega_{6jl}S06_{mn}r_{6n} + e_{ilm}\omega_{5jl}S05_{mn}\xi_{6n}$$
$$+ e_{ilm}\omega_{ijl}S01_{mn}\xi_{5n}$$

$$V_{ij7} = \delta_{ij} + e_{ilm}\omega_{7jl}S07_{mn}r_{7n} + e_{ilm}\omega_{2jl}S02_{mn}\xi_{7n}$$
$$+ e_{ilm}\omega_{ijl}S01_{mn}\xi_{2n}$$

$$V_{ij8} = \delta_{ij} + e_{ilm}\omega_{8jl}S08_{mn}r_{8n} + e_{ilm}\omega_{7jl}S07_{mn}\xi_{8n}$$
$$+ e_{ilm}\omega_{2jl}S02_{mn}\xi_{7n} + e_{ilm}\omega_{1jl}S01_{mn}\xi_{2n}$$

$$V_{ij9} = \delta_{ij} + e_{ilm}\omega_{9jl}S09_{mn}r_{9n}$$
$$+ e_{ilm}\omega_{2jl}S02_{mn}\xi_{9n} + e_{ilm}\omega_{1jl}S01_{mn}\xi_{2n}$$

$$V_{ij10} = \delta_{ij} + e_{lim}\omega_{10jl}S010_{mn}r_{10n} + e_{ilm}\omega_{9jl}S09_{mn}\xi_{10n}$$
$$+ e_{ilm}\omega_{2jl}S02_{mn}\xi_{9n} + e_{ilm}\omega_{1jl}S01_{mn}\xi_{2n}$$

$$V_{ij11} = \delta_{ij} + e_{ilm}\omega_{11jl}S011_{mn}r_{11n} + e_{ilm}\omega_{1jl}S111_{mn}\xi_{11n}$$

$$V_{ij12} = \delta_{ij} + e_{ilm}\omega_{12jl}S012_{mn}r_{12n} + e_{ilm}\omega_{4jl}S04_{mn}\xi_{12n}$$
$$+ e_{ilm}\omega_{3jl}S03_{mn}\xi_{4n} + e_{ilm}\omega_{1jl}S01_{mn}\xi_{3n}$$

$$V_{ij13} = \delta_{ij} + e_{ilm}\omega_{13jl}S013_{mn}r_{13n} + e_{ilm}\omega_{6jl}S06_{mn}\xi_{13n}$$
$$+ e_{ilm}\omega_{5jl}S05_{mn}\xi_{6n} + e_{ilm}\omega_{1jl}S01_{mn}\xi_{5n}$$

$$V_{ij14} = \delta_{ij} + e_{ilm}\omega_{14jl}S014_{mn}r_{14n} + e_{ilm}\omega_{8jl}S08_{mn}\xi_{14n}$$
$$+ e_{ilm}\omega_{7jl}S07_{mn}\xi_{8n} + e_{ilm}\omega_{2jl}S02_{mn}\xi_{7n}$$
$$+ e_{ilm}\omega_{1jl}S01_{mn}\xi_{2n}$$

$$V_{ij15} = \delta_{ij} + e_{ilm}\omega_{15\,jl}S015_{mn}r_{15n} + e_{ilm}\omega_{10\,jl}S010_{mn}\xi_{15n}$$
$$+ e_{ilm}\omega_{9\,jl}S09_{mn}\xi_{10n} + e_{ilm}\omega_{2\,jl}S02_{mn}\xi_{9n}$$
$$+ e_{ilm}\omega_{ijl}S01_{mn}\xi_{2n}$$

($\delta_{ij} = 0$ unless $i = j$ and there is a sum from 1 to 3 on l, m, and n).

Appendix 6.4

Below is a complete listing of $V_{ijk}x_j$ ($i = 1, \ldots, 3, k = 1, \ldots, 15, j$ summed $1, \ldots, 34$) (see eqn (6.81)):

$$\dot{V}_{ij1}\dot{x}_j = D_{1ik}r_{1k}$$
$$\dot{V}_{ij2}\dot{x}_j = D_{1ik}\xi_{2k} + D_{2ik}r_{2k}$$
$$\dot{V}_{ij3}\dot{x}_j = D_{1ik}\xi_{3k} + D_{3ik}r_{3k}$$
$$\dot{V}_{ij4}\dot{x}_j = D_{1ik}\xi_{3k} + D_{3ik}\xi_{4k} + D_{4ik}r_{4k}$$
$$\dot{V}_{ij5}\dot{x}_j = D_{1ik}\xi_{5k} + D_{5ik}r_{5k}$$
$$\dot{V}_{ij6}\dot{x}_j = D_{1ik}\xi_{5k} + D_{5ik}\xi_{6k} + D_{6ik}r_{6k}$$
$$\dot{V}_{ij7}\dot{x}_j = D_{1ik}\xi_{2k} + D_{2ik}\xi_{7k} + D_{7ik}r_{7k}$$
$$\dot{V}_{ij8}\dot{x}_j = D_{1ik}\xi_{2k} + D_{2ik}\xi_{7k} + D_{7ik}\xi_{8k} + D_{8ik}r_{8k}$$
$$\dot{V}_{ij9}\dot{x}_j = D_{1ik}\xi_{2k} + D_{2ik}\xi_{9k} + D_{9ik}r_{9k}$$
$$\dot{V}_{ij10}\dot{x}_j = D_{1ik}\xi_{2k} + D_{2ik}\xi_{9k} + D_{9ik}\xi_{10k} + D_{10ik}r_{10k}$$
$$\dot{V}_{ij11}\dot{x}_j = D_{1ik}\xi_{11k} + D_{11ik}r_{11k}$$
$$\dot{V}_{ij12}\dot{x}_j = D_{1ik}\xi_{3k} + D_{3ik}\xi_{14k} + D_{12ik}r_{12k}$$
$$\dot{V}_{ij13}\dot{x}_j = D_{1ik}\xi_{5k} + D_{5ik}\xi_{6k} + D_{6ik}\xi_{13k} + D_{1ik}r_{13k} + D_{4ik}\xi_{12k}$$
$$\dot{V}_{ij14}\dot{x}_j = D_{1ik}\xi_{2k} + D_{2ik}\xi_{7k} + D_{7ik}\xi_{8k} + D_{8ik}\xi_{14k} + D_{14ik}r_{14k}$$
$$\dot{V}_{ij15}\dot{x}_j = D_{1ik}\xi_{2k} + D_{2ik}\xi_{9k} + D_{9ik}\xi_{10k} + D_{10ik}\xi_{15k} + D_{15ik}r_{k5k}$$

where there is a sum from 1 to 3 on k and from 1 to 34 on j and where D_{Kik} ($K = 1, \ldots, 15$, $i, k - 1, \ldots, 3$) is defined as

$$D_{Kik} = e_{ilm}\dot{\omega}_{Kpl}\dot{x}_p SOK_{mk}$$
$$+ e_{irl}e_{lsq}\omega_{Kjr}\omega_{Kpq}\dot{x}_j\dot{x}_p SOK_{sk}$$

(l, m, q, r, s summed from 1 to 3, j, p summed from 1 to 34, no sum on K).

Appendix 6.5

Below is a listing of quantities ξ_k, $\mathbf{r}_k$, m_k, and $\mathbf{I}^k$ ($k = 1, \ldots, 15$) which provide specific physical data for the human body model of Figs. 6.1 and 6.6. The data

were obtained from Hanavan (1964) and is used in the example motions of
§ 6.5.

$$\xi_1 = 0$$
$$\xi_2 = 0$$
$$\xi_3 = (8.35\,\mathbf{n}_{12} + 1.85\,\mathbf{n}_{13})\text{ in}$$
$$\xi_4 = -11.7\,\mathbf{n}_{33}\text{ in}$$
$$\xi_5 = (-8.35\,\mathbf{n}_{12} + 1.85\,\mathbf{n}_{13})\text{ in}$$
$$\xi_6 = -11.7\,\mathbf{n}_{53}\text{ in}$$
$$\xi_7 = (3.09\,\mathbf{n}_{22} - 16.75\,\mathbf{n}_{23})\text{ in}$$
$$\xi_8 = -18.6\,\mathbf{n}_{73}\text{ in}$$
$$\xi_9 = (-3.09\,\mathbf{n}_{22} - 16.75\,\mathbf{n}_{23})\text{ in}$$
$$\xi_{10} = -18.6\,\mathbf{n}_{93}\text{ in}$$
$$\xi_{11} = 3.95\,\mathbf{n}_{13}\text{ in}$$
$$\xi_{12} = -11.2\,\mathbf{n}_{43}\text{ in}$$
$$\xi_{13} = -11.2\,\mathbf{n}_{63}\text{ in}$$
$$\xi_{14} = -15.6\,\mathbf{n}_{83}\text{ in}$$
$$\xi_{15} = -15.6\,\mathbf{n}_{103}\text{ in}$$
$$\mathbf{r}_1 = 0$$
$$\mathbf{r}_2 = -12.05\,\mathbf{n}_{23}\text{ in}$$
$$\mathbf{r}_3 = -4.46\,\mathbf{n}_{33}\text{ in}$$
$$\mathbf{r}_4 = -4.75\,\mathbf{n}_{43}\text{ in}$$
$$\mathbf{r}_5 = -4.46\,\mathbf{n}_{53}\text{ in}$$
$$\mathbf{r}_6 = -4.75\,\mathbf{n}_{63}\text{ in}$$
$$\mathbf{r}_7 = -9.87\,\mathbf{n}_{73}\text{ in}$$
$$\mathbf{r}_8 = -6.62\,\mathbf{n}_{83}\text{ in}$$
$$\mathbf{r}_9 = -9.87\,\mathbf{n}_{93}\text{ in}$$
$$\mathbf{r}_{10} = -6.62\,\mathbf{n}_{103}\text{ in}$$
$$\mathbf{r}_{11} = -6.35\,\mathbf{n}_{113}\text{ in}$$
$$\mathbf{r}_{12} = -1.89\,\mathbf{n}_{213}\text{ in}$$
$$\mathbf{r}_{13} = -1.89\,\mathbf{n}_{113}\text{ in}$$
$$\mathbf{r}_{14} = -1.55\,\mathbf{n}_{153}\text{ in}$$
$$\mathbf{r}_{15} = -1.55\,\mathbf{n}_{153}\text{ in}$$

$m_1 = 0.750$ slugs	$m_6 = 0.104$ slugs	$m_{11} = 0.433$ slugs
$m_2 = 1.832$ slugs	$m_7 = 0.555$ slugs	$m_{12} = 0.0391$ slugs

$m_3 = 0.178$ slugs $\qquad m_8 = 0.278$ slugs $\qquad m_{13} = 0.0391$ slugs

$m_4 = 0.104$ slugs $\qquad m_9 = 0.555$ slugs $\qquad m_{14} = 0.0798$ slugs

$m_5 = 0.178$ slugs $\qquad m_{10} = 0.278$ slugs $\qquad m_{15} = 0.0798$ slugs

$$I_1 = (11.23\,\mathbf{n}_{11}\mathbf{n}_{11} \quad + 7.740\,\mathbf{n}_{12}\mathbf{n}_{12} \quad + 11.16\,\mathbf{n}_{13}\mathbf{n}_{13}) \quad \text{slugs in}^2$$

$$I_2 = (61.56\,\mathbf{n}_{21}\mathbf{n}_{21} \quad + 49.24\,\mathbf{n}_{22}\mathbf{n}_{22} \quad + 30.66\,\mathbf{n}_{23}\mathbf{n}_{23}) \quad \text{slugs in}^2$$

$$I_3 = (2.82\,\mathbf{n}_{31}\mathbf{n}_{31} \quad + 2.82\,\mathbf{n}_{31}\mathbf{n}_{31} \quad + 0.00827\,\mathbf{n}_{33}\mathbf{n}_{33}) \quad \text{slugs in}^2$$

$$I_4 = (1.03\,\mathbf{n}_{41}\mathbf{n}_{41} \quad + 1.03\,\mathbf{n}_{42}\mathbf{n}_{42} \quad + 0.00283\,\mathbf{n}_{43}\mathbf{n}_{43}) \quad \text{slugs in}^2$$

$$I_5 = (2.816\,\mathbf{n}_{51}\mathbf{n}_{51} \quad + 2.816\,\mathbf{n}_{52}\mathbf{n}_{52} \quad + 0.00827\,\mathbf{n}_{53}\mathbf{n}_{53}) \quad \text{slugs in}^2$$

$$I_6 = (1.03\,\mathbf{n}_{61}\mathbf{n}_{61} \quad + 1.03\,\mathbf{n}_{62}\mathbf{n}_{62} \quad + 0.00283\,\mathbf{n}_{63}\mathbf{n}_{63}) \quad \text{slugs in}^2$$

$$I_7 = (10.16\,\mathbf{n}_{71}\mathbf{n}_{71} \quad + 10.16\,\mathbf{n}_{72}\mathbf{n}_{72} \quad + 0.065\,\mathbf{n}_{73}\mathbf{n}_{73}) \quad \text{slugs in}^2$$

$$I_8 = (5.34\,\mathbf{n}_{81}\mathbf{n}_{81} \quad + 5.34\,\mathbf{n}_{82}\mathbf{n}_{82} \quad + 0.0129\,\mathbf{n}_{83}\mathbf{n}_{83}) \quad \text{slugs in}^2$$

$$I_9 = (10.16\,\mathbf{n}_{91}\mathbf{n}_{91} \quad + 10.16\,\mathbf{n}_{92}\mathbf{n}_{92} \quad + 0.065\,\mathbf{n}_{93}\mathbf{n}_{93}) \quad \text{slugs in}^2$$

$$I_{10} = (5.34\,\mathbf{n}_{10}\mathbf{n}_{10} \quad + 5.34\,\mathbf{n}_{102}\mathbf{n}_{102} \quad + 0.0129\,\mathbf{n}_{103}\mathbf{n}_{103}) \text{slugs in}^2$$

$$I_{11} = (4.64\,\mathbf{n}_{111}\mathbf{n}_{111} \quad + 4.64\,\mathbf{n}_{112}\mathbf{n}_{112} \quad + 2.30\,\mathbf{n}_{113}\mathbf{n}_{113}) \quad \text{slugs in}^2$$

$$I_{12} = (0.056\,\mathbf{n}_{121}\mathbf{n}_{121} \quad + 0.056\,\mathbf{n}_{122}\mathbf{n}_{122} \quad + 0.056\,\mathbf{n}_{123}\mathbf{n}_{123}) \quad \text{slugs in}^2$$

$$I_{13} = (0.056\,\mathbf{n}_{131}\mathbf{n}_{131} \quad + 0.056\,\mathbf{n}_{132}\mathbf{n}_{132} \quad + 0.056\,\mathbf{n}_{133}\mathbf{n}_{133}) \quad \text{slugs in}^2$$

$$I_{14} = (0.0016\,\mathbf{n}_{141}\mathbf{n}_{141} \quad + 0.158\,\mathbf{n}_{142}\mathbf{n}_{142} \quad + 0.158\,\mathbf{n}_{143}\mathbf{n}_{143}) \quad \text{slugs in}^2$$

$$I_{15} = (0.0016\,\mathbf{n}_{151}\mathbf{n}_{151} \quad + 0.158\,\mathbf{n}_{152}\mathbf{n}_{152} \quad + 0.158\,\mathbf{n}_{153}\mathbf{n}_{153}) \quad \text{slugs in}^2$$

References

1 SMITH, P. G., and KANE, T. R. (1967). The reorientation of a human body in free fall. *Tech. Rep. 171*, Stanford University, Stanford, Ca.

2 SMITH, P. G., and KANE, T. R. (1968). On the dynamics of the human body in free fall, *J. appl. Mech.* **35**, 167.

3 KANE, T. R., and SCHER, M. P. (1970). Human self-rotation by means of limb movements. *J. Biomech.* **3**, 39–49.

4 PASSERELLO, C. E., and HUSTON, R. L. (1971). Human attitude control. *J. Biomech.* **4**, 95–102.

5 HUSTON, R. L., and PASSERELLO, C. E. (1971). On the dynamics of a human body model. *J. Biomech.* **4**, 369–78.

6 ROBERTS, V. L., and ROBBINS, D. H. (1969). Multidimensional mathematical modelling of occupant dynamics under crash conditions. *Int. Automotive Engineering Congr., Detroit, 1969*, Pap. no. 690248. Soc. Automotive Engineers, Warrendale, Pa.

7 ROBBINS, D. H., BENNETT, R. O., HENKE, A. W., and ALEM, N. N. (1970). Prediction of mathematical models compared with impact sled test results using anthropometric dummies. *Proc. 14th Stapp Car Crash Conf.*, Pap.

no. 700907. Soc. Automotive Engineers, Warrendale, Pa.

8 ROBBINS, D. H., SNYDER, R. G., McELHANEY, J. H., and ROBERTS, V. L. (1971). A comparison between human kinematics and the predictions of mathematical crash victim simulators. *Proc. 15th Stapp Car Crash Conf., 1971*, Pap. no. 710849. Soc. Automotive Engineers, Warrendale, Pa.

9 STEWART, P. A., DUFFY, J., SOECHTING, J., LITCHMAN, H., and PASLAY, P. R. (1971). Control of the human forearm during abrupt acceleration. *Symp. on Biodynamic Models and Their Applications, AMRL, WPAFB, Ohio, 1971*, pp. 193–210. WPAFB, Dayton, Ohio.

10 EWING, C. L., and THOMAS, D. J. (1972). Human head and neck response to impact acceleration. *Army–Navy Joint Rep.* Naval Aerospace Medical Research Laboratory, US Army Aeromedical Research Laboratory, New Orleans.

11 LIU, Y. K. (1971). The biomechanics of spinal and head impact: problems of mathematical simulation. *Symp. on Biodynamic Models and their Applications, AMRL, WPAFB, Ohio, 1971*, pp. 701–36. WPAFB, Dayton, Ohio.

12 BECKETT, R. E., and PAN, K. C. (1971). Analysis of gait using a minimum energy approach. *Symp. on Biodynamic Models and Their Applications, AMRL, WPAFB, Ohio, 1971*, pp. 823–42. WPAFB, Dayton, Ohio.

13 JACOBS, N. A., SKORECKI, J., and CHARNLEY, J. (1972). Analysis of the vertical component of force in normal and pathological gait. *J. Biomech.* **5**, 11–34.

14 PAUL, J. P. Forces transmitted by joints in the human body. *Proc. Inst. Mech. Engrs.* **181**, 8–15.

15 KURZHALS, P. R., and REYNOLDS, R. B. (1972). Development of a dynamic analytical model of man on board a manned spacecraft, Appendix B. *NASA Tech. Note TN D-6584.*

16 HEWES, D. E., and GLOVER, K. E. (1972). Development of Skylab experiment T020 employing a foot-controlled manoeuvering unit. *NASA Tech. Note TND-6674.*

17 KINZEL, G. L., HALL, JR. A. S., and HILLBERRY, B. M. (1972). Measurement of the total motion between two body segments. I. Analytical development. **5**, 93–106.

18 YOUNG, R. D. (1970). A three-dimensional mathematical model of an automobile passenger. *Texas Transportation Inst. Res. Rep. 140–2.*

19 BARTZ, J. A. (1971). A three-dimensional computer simulation of a motor vehicle crash victim. *Cornell Aeronautical Lab. Rep. DOT HS-800 574.*

20 PASSERELLO, C. E. (1972). On the dynamics of general chain systems. *Ph.D. Diss.* University of Cincinnati.

21 HOOKER, W. W. (1970). A set of r dynamical attitude equations for an arbitrary n-body satellite having r rotational degrees of freedom *AIAA J.* **8**, 1205–7.

22 FLEISCHER, G. E. (1971). Multi-rigid-body attitude dynamics simulator. *JPL Tech. Rep. 32–1516.*

23 CHACE, M. A., and BAYAZITOGLU, Y. O. (1971). Development and application of a generalized d'Alembert force for multifreedom mechanical systems. **93**, 317–27.

24 PASSERELLO, C. E., and HUSTON, R. L. (1972). An analysis of general chain systems. *NASA Contract. Rep. CR-127924.*

25 HANAVAN, E. P. (1964). A mathematical model of the human body. *AMRL Tech. Rep. AMRL-TR-64-102, WPAFB, Ohio.*

26 KANE, T. R. (1959). *Analytical elements of mechanics, Vol. 1.* Academic Press, New York.

27 KANE, T. R. (1961). Dynamics of nonholonomic systems. *J. appl. Mech.* **28**, 574–8.

28 KANE, T. R., and WANG, C. F. (1965). On the derivation of equations of motion. *J. Soc. ind. appl. Math.* **13**, 487–92.

29 KANE, T. R. (1968). *Dynamics.* Holt, Rinehart, and Winston, New York.

30 ERIGEN, A. C. (1962). *Nonlinear theory of continuous media*, p. 447. McGraw-Hill, New York.

31 GALLENSTEIN, J., and HUSTON, R. L. (1973). Analysis of swimming motions. *Hum. Factors* **15**, 91–8.

7. Low-back stresses during load lifting

Don B. Chaffin

7.1. Introduction

Contrary to what many people state, manual material handling has not been replaced by automation in modern industries. Heavy loads are still being lifted by people both in their employment and leisure time activities. Often the body position allowed by the immediate physical environment or by the size of the object being lifted does not allocate the stresses on the body to those components that are best capable of coping with them. One of the parts of the body that is often the most highly stressed during a lifting act is the low back, specifically the lower lumbar segments of the spinal column and their associated muscles and ligaments. Unfortunately, all too often these stresses can result in 'low-back pain'. The prevention of low-back problems is quickly becoming the major concern of industrial physicians who are responsible for the selection of workers for jobs having manual materials handling requirements. Its prevention is also now concerning those engineers who are responsible for designing materials-handling tasks so that they are not injurious to the general population. The basis for this concern in modern industry is substantiated by some of the statistics that follow.

Estimates by Hult [1] of the proportion of compensatable medical claims that are low back in origin range from about 15 per cent for all United States industry (based on US National Safety Council statistics) to as high as 30 per cent for certain industries studied in Sweden. Snook and Ciriello [2] report that the incidence rate of low-back pain appears to be increasing faster than the rates for other types of injuries in the United States. As an example they cite the statistics for the state of Wisconsin wherein compensatable back injuries increased from 7.7 per cent of all claims in 1938 to 19.1 per cent in 1965. Troup and Chapman [3] estimate that 30 million working days were lost in Great Britain during 1968 as a result of low-back pain. Hult [1] estimates that approximately two million working days are lost in Sweden for similar reasons. In the fiscal year 1971 one estimate based on the US Department of Labor and Industries for the State of Washington indicates that over half a million days were lost from work as a result of the more serious compensatable type of back injuries alone.

As an estimate of the severity of the problem, the author reviewed the number of days lost for each on-the-job low-back case reported during a period of one year by a large US electronics manufacturing industry and found a mean of four days per case. The distribution was heavily skewed, however, so that the mode was closer to two days per case. The average time lost per case for the more serious compensatable back problems has been reported by the Department of Labor and Industries in the State of Washington to average more than 125 days. The average in-patient hospital stay in the United States for patients diagnosed as having disc-related problems is about 13 weeks according to Holtz and Keys [4]. Rowe [5] reported that the average time lost for all employees at Eastman Kodak as a result of low-back complaints was second only to upper respiratory ailments. In this same regard, it has been well documented that the length of incapacitation due to low-back problems is much greater (three to four times greater according to Magora and Taustein [6]) if the person is engaged in heavy labour.

It should also be recognized that low-back problems are not only an ailment of older persons. Rather, the onset of symptoms occurs most often in persons between 20 and 35 years old according to Hult [1], with remission intervals of from three months to three years based on Rowe's findings [5]. Hult [1] confirms a similar remission trend. Nachemson [7] states that the frequency of repeated episodes appears to peak while in the 40s. He also summarizes studies by Hirsch, Horal, and Hult and concludes that from 70 to 80 per cent of the world's population will have suffered from this malady at some time in their working lives.

It must therefore be concluded that low-back pain is a major source of incapacitation, suffering, and cost to the world today. In fact, in eleven states in the United States spinal-disc-related problems are ranked as the major reasons for the high number of days of care necessary per patient in the hospital [1]. Finally, low-back pain should be characterized as striking younger persons and is recurrent in nature, though between episodes the person is often completely symptom free.

7.2. Biomechanical modelling of load-lifting activities

It is evident from papers by Tichauer [8], Roaf [9], and Davis, Troup, and Burnhard [10] that the estimation of stresses on various parts of the musculoskeletal system during lifting activities will require a complex methodology which takes account of such factors as (1) instantaneous positions and accelerations of the extremities and trunk, (2) geometry changes in the spine, (3) strength variations within different muscle groups and people, and (4) effects of the abdominal pressure reflex during lifting.

It is the intent now to describe one of the simpler computerized bio-mechanical models which has been used to estimate the forces and torques that are created at six major articulations of the body (i.e. wrist, elbow, shoulder, hip, knee, and ankle) as well as at the L_5/S_1 (lumbosacral) spinal disc of a person who is performing a weight-lifting activity. This particular model has been chosen as an example of the types of models that are now practical because of the relative ease of use and computational speed of today's digital computers.

It is over 80 years since Braune and Fischer published their data regarding the mass distribution for the various body segments. Since then fundamental extensions by Dempster [11] and Drillis and Contini [12] have resulted in better estimates of (1) the location of the mass centre of gravity for various body segments, (2) the link lengths of body segments, and (3) the magnitudes of the moments of inertia of the various body segments.

However, it was not until the widespread use of the high-speed digital computer that data of this type could be easily used in developing analytical models to study the mechanics of the human body. The digital computer has provided a computational capacity which, in turn, has fostered the develop-ment of several different types of biomechanical models. Some of these models have been formulated to determine the location of the whole-body centre of gravity when the body is placed in various configurations [13] or to estimate the effects of sudden forces acting on the body masses, occurring in a vehicle-crash situation [14].

Another type of computerized biomechanical model primarily estimates the forces and torques at various articulations of the body during voluntary actions (e.g. lifting, running, or throwing). An early example is a two-link model of the arm developed by Pearson, McGinley, and Butzel [15]. The intent of this model was to compute the forces and torques at the elbow and shoulder during a sagittal plane motion of the arm–forearm–hand aggregate. Figure 7.1 is an illustration of the mechanical analogue of the arm used in this type of model. Stroboscopic photographs of the various arm motions of interest are taken to determine the instantaneous positions, velocities, and accelerations of the arm segments. These 'activity' data along with the anthropometric dimensions of the segment lengths and weights provide enough input information to compute the stress levels (in forces and torques) at the elbow and shoulder, and thus provide a means of achieving a better understanding of both the complex muscle actions required for control of the arm and the resulting strain at the articulations.

An extension of the Pearson arm model was developed by Plagenhoef [16]. Again, high-speed photographic data are used to describe the body configur-ations during the relevant task. This spatial information combined with additional anthropometric dimensions regarding the length of the arm, trunk, and leg segments, and the total weight of the subject provides adequate

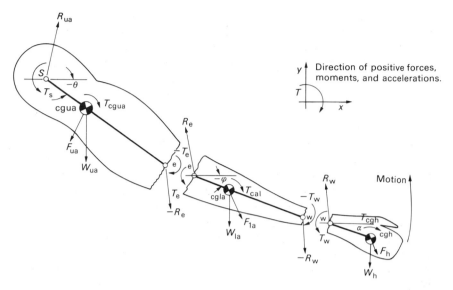

FIG. 7.1.   Three-segment human arm for analysis of sagittal plane movements: R, reactive forces; T, reactive torques; F, inertial forces; W, segment weights. (From ref. 21.)

information to compute the forces and torques at the elbow, shoulder, hip, and knee during various physical activities.

7.3.   The present SSP model—background

Recent efforts have been made to extend the earlier biomechanical model of Plagenhoef [16] to include (1) an estimate of the stresses in the lumbosacral disc, (2) the addition of external loads on the hands (e.g. in a materials handling task, and (3) an evaluation of the effects of various muscle group strengths on the performance of the person being simulated. In accomplishing this several different computerized biomechanical models have evolved. The following is a description of one of these models, along with a presentation of a few of the more interesting results that have been obtained from using the model to study various whole-body lifting activities.

The model has been referred to as the *Static Sagittal Plane (SSP) model*, by the research group at the University of Michigan which has used it over the last 12 years. As the name implies this particular model has been developed to evaluate various 'static' situations, such as when one is holding a weight or pushing or pulling on a non-moving object with both hands. In addition to these applications the model can be used to analyse the more normal 'slow' moves used in heavy materials handling by formulating the input data to describe a sequence of static positions with very small changes in each successive position. In making this type of pseudo-dynamic analysis it must be assumed that the effects of acceleration and momentum are negligible.†

The SSP model is also restricted to symmetric sagittal plane activities; thus a rotation or lateral deviation of the body cannot be analysed. A complete relaxation of this latter restriction was accomplished by Schanne [18] in the form of a three-dimensional model. Two of the greatest problems in this more advanced model are (1) difficulty in obtaining good spatial data describing the position of each body segment in three-dimensional space in other than a laboratory setting and (2) difficulty in intuitively understanding the complex vector representations of forces and torques that result from a three-dimensional analysis of human motion. Computer-generated graphical displays are in fact necessary even to begin to reduce this latter problem so as to allow the user to interpret the model outputs.

7.4. A description of the SSP model for load-lifting simulations

The static sagittal plane model develops the estimates of the forces and torques at each of the major articulations of the extremities in much the same manner as the Pearson and Plagenhoef biomechanical models. In essence this requires that the body be treated as a series of seven solid links which are articulated at the ankles, knees, hips, shoulders, elbows, and wrists as shown in Fig. 7.2.‡ Each of the links in the model is considered to have a mass whose estimate is based on the proportionality constants presented by Dempster and Gaughcan [22] and Clauser McConville, and Young [23]. The distribution of the mass within each link is based on the data of Dempster [11]. The link lengths are established from over-the-body measurements with the reference landmarks described by Dempster [11]. Specifically, the body measurements needed as input data are body stature, body weight, centre of gravity of the hand to wrist distance, lower arm length, lower leg length, foot length, and elbow height when standing erect. From these the link lengths (i.e. the straight line distances between the articulation points of rotation) are estimated based on the empirical relationships developed by Dempster and co-workers [22, 24]. Table 7.1 presents these data for a sample of US working personnel obtained from ref. 25. Additional estimates of the mass distributions and mass centre-of-gravity locations are given in Appendix 7.1.

A task being analysed by the SSP model is described by two types of data. First, any external force that may be exerted against the hands is measured and entered into the program as a vector acting at the centre of gravity of the

† A dynamic version of this same model which includes estimates of acceleration and momentum has been developed by Fisher [17] but is much more difficult to use.
‡ The computational techniques of the model have been described previously and in the interest of conserving space are not presented here. It is suggested that if interested the reader refers to Plagenhoef [19], Pearson *et al.* [15], Williams and Lissner [20], or Chaffin [21].

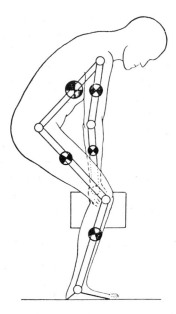

FIG. 7.2.   Linkage representation used to estimate torques and forces at major articulations during weight lifting.

TABLE 7.1

Estimated population link lengths (in cm)

Links	Males				Females			
	Percentiles smaller than each dimension							
	5%	50%	95%	Ref.	5%	50%	95%	Ref.
Hand (wrist-to-hand centre of gravity)	6.7	7.0	7.4	11	6.1	6.4	6.7	26
Forearm	25.7	27.2	28.5	11	23.1	24.2	25.6	26
Upper arm	28.6	30.2	32.0	11	26.7	27.7	28.7	11,26
Trunk erect (shoulder to hip centres)	45.6	48.7	51.7	27	42.2	45.2	48.3	27
Thigh (upper leg)	40.5	43.7	46.0	11	39.1	40.9	42.7	11,27
Shank (lower leg)	38.0	40.9	43.9	11	34.0	36.4	38.9	27
Foot (overall)	24.8	26.7	28.6	11	22.0	23.9	25.9	26
Stature	161.5	173.5	189.5	4	149.9	159.8	170.4	4

Except for the data from ref. 11 the link lengths are estimated from over-body anthropometry and adjusted by the methods of Dempster *et al.* [24].

hands. For example, if a person is holding a 10 kp box it is entered into the program as a 10 kp force acting downward (i.e. a force acting in respect to some defined reference axis). The second type of information required to describe the activity is the position of the body. To obtain these data it is necessary to measure the articulation angles from either a lateral photograph of a person who is in the position of interest or an articulated drawing-board body template placed in the 'Task' position.

The above data are sufficient for the model to compute the torques and forces at each of the six major articulations of the body. As an example of the insight that can be gained from the model the effects of various body positions on the effects at various joints in the body due to gravity forces acting on the body masses of a 50 percentile man are clearly indicated by examining the articulation torques in Fig. 7.3.

The need to include in the model a specific type of spinal stress predictor during lifting was based on the following studies. First these studies [7, 29–31] disclosed that compressive forces on the spinal column could cause the disc cartilage end-plates to suffer small microfractures. Furthermore, degeneration of the intervertebral discs and resulting herniations have been attributed by both Perey [31] and Gordon [32] to the compressive forces to which they are subjected in daily activities. Based on epidemiological data both Rowe and Morris, cited in ref. 33, have stated that 70–80 per cent of all chronic low-back pain is discogenic. Thus, it was believed that an estimate of the compressive forces created at the various disc end-plate interfaces during specific physical tasks would provide an important contribution towards the further understanding of low-back injury mechanisms as related to physical activities.

The particular estimations of compressing and shearing forces at the lumbosacral disc were chosen for inclusion in the model based on statistics regarding back disorders. These disclosed that between 85 and 95 per cent of all serious back injuries (i.e. disc herniations) occur with relatively equal frequencies at the L_4L_5 and L_5/S_1 levels [29, 34, 35]. Since on a gross basis the L_5/S_1 disc is more stressed in lifting activities (because it usually has a larger moment arm to the load in the hands), it was chosen to represent the lumbar stresses during lifting.

To perform the analysis of the lumbar spinal stress the following concepts were included in the SSP model. First the geometry of an average erect spinal column was developed from the dimensions given by Fick [36], Lanier [37], and Chaffin, Schutz, and Snyder [38]. The dimensions of this average male column were then proportionally scaled in the model, based on the hip-to-shoulder distance, to enable the study of smaller or larger individuals. The superior surface of the sacrum was estimated to be at an angle of 40° from the horizontal when standing erect [39].

The curvature change for the column during sagittal rotation of the hips was assumed from the data of Dempster [11] which disclosed that for the first 27°

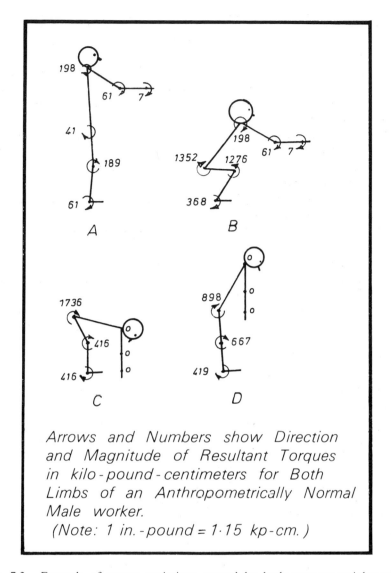

Arrows and Numbers show Direction
and Magnitude of Resultant Torques
in kilo-pound-centimeters for Both
Limbs of an Anthropometrically Normal
Male worker.
(Note: 1 in.-pound = 1·15 kp-cm.)

FIG. 7.3.   Example of torque variations caused by body-segment weights when
assuming various positions.

of trunk flexion the pelvis does not usually rotate forward (i.e. the rotation is in
the lumbar spine) and for each additional degree of trunk rotation the pelvis
contributes about two-thirds of a degree. Also it was assumed that 22 per cent
of the lumbar rotation occurs at the L_5/S_1 disc based on data obtained by
Davis et al. [10], Allbrock and Uganda [40], Lindahl [41], Rolander [42], and

Chaffin *et al.* [38]. An assumption is also made that when squatting the thighs rotate the pelvis two-thirds of a degree in a posterior direction for each degree of flexion of the thighs after the first 13° of flexion.

The contribution of intra-abdominal pressure (created when one tenses the diaphragm and transverse and oblique muscles of the abdomen during lifting) in relieving compression on the lumbar spine was estimated from the data of Morris, Lucas, and Bresler [30]. The procedure required correlating their abdominal pressure data with the torque and angles estimated at the hip of the subjects used in their experiments. The least squared error expression developed from this procedure is

$$\text{Abdominal pressure} = 10^{-4}\{0.6516 - 0.005\,447 \text{ (hip angle)}\}$$
$$\times \{(\text{hip torque})^{1.8}\}$$

where (1) the abdominal pressure is in mm Hg with a maximum limit of 150 mm Hg, (2) the hip angle is in degrees from the erect or longitudinal axis of the legs, and (3) the hip torque is in kp cm.† This expression was found to have a correlation coefficient of 0.73. The error was attributed to (1) not knowing the exact position of the trunk during each test, thus causing an unknown variation in both the assumed hip angles and torques, and (2) not being able to assess the time rate of force application.

This latter factor is important in that recent experiments by Asmussen and Poulsen [43] disclosed that probably the method of lifting (i.e. quick jerk or slow sustained pull) can significantly affect the abdominal pressure reflex.

The amount of force created by the abdominal pressure was estimated by assuming the following three conditions which are similar to those proposed by Morris *et al.* [30]: (1) a diaphragm area of 465 cm^2 upon which the abdominal pressure can act, (2) the line of action of the force acts parallel to the line of action of the normal compressing forces on the lower lumbar spine, and (3) the force acts through a finite moment arm distance from the centre of the L_5/S_1 disc. In formulating the SSP model the moment arm of this force has been assumed to vary as the sine of the angle at the hips: for the erect position it has a moment arm of 6.7 cm and for a 90° hip angle position (bent over) it has a moment arm of 14.9 cm to the L_5/S_1 disc centre.

Additional compressive forces on the lumbar spine due to the abdominal muscles were assumed to be negligible since Bartelink [44] disclosed that the rectus abdominis which could mechanically cause a spinal compressive force, was relatively inactive during lifting activities. The abdominal pressure was therefore attributed to the oblique and transverse muscles which are not well

† The unit 'kilopond' (kp) is used as a measure of force throughout this chapter to be consistent with prior reports and to avoid large numbers (i.e. a kilopond is assumed to be equivalent to 9.8×10^5 dyn which would be consistent with the commonly used centimetre unit of length or in the American engineering system to about 2.21 lbf).

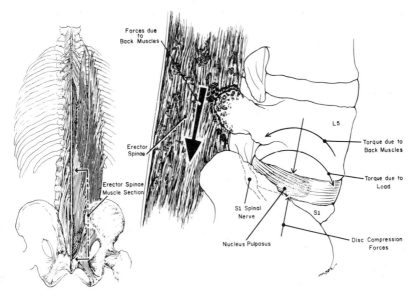

Fɪɢ. 7.4.   Forces and torques operating during load lifting.

positioned to assist or hinder directly in sagittal plane flexion or extension of the trunk.

The link of action of the muscles of the lower lumbar back were assumed to act parallel to the normal compressing force on the vertebral-discs interface and with a moment arm of 5.0 cm as illustrated in Fig. 7.4.† The estimate of the magnitude of the muscle force required to maintain a particular trunk position against the gravitational forces that act on the body masses and any mass being held in the hands is accomplished by dividing the estimated torque due to the body weight and load (after the abdominal pressure effect has been subtracted) at the centre of the L_5/S_1 disc by the 5 cm moment arm assumed for the back muscles. This logic assumes that the bending strength of the column is only due to the spinal muscles, which at this point in time is probably reasonable. It also assumes that the articulating facets do not resist compression in the lumbar column during a lifting act. This assumption is being tested by Belytschko et al. [46] in research of the intrinsic mechanical functions of the column. One recent investigation of this by Kraus, Robertson, and Farfan [47] disclosed that the lumbar facets are primarily for resisting shearing forces when the spine is flexed, thus supporting this latter assumption. With a straight spine, however, the facets may be capable of bearing some compression load, though more research on this is warranted.

† This is an average moment arm which is based on values published by Bartclink [44], Munchinger [45], Perey [31], and Thieme [39] as well as on cadaver measurements.

7.5. Some numerical examples

To assist further in understanding the operation of the SSP model various lifting situations have been simulated. The results of these simulations are now described.

Case 1. Lifting a 100 lbf (45 kp) load close to the body

For this case we assume a common lifting posture as illustrated in Fig. 7.5. Here the torso is tilted (30° from vertical). With average male anthropometrics assumed, the resulting torque at the hips is 2056 kp cm due to the weight BW of the trunk, neck, head and arms and the 100 lbf load FH held in the hands. The abdominal pressure P_{abdom} is predicted based on this hip torque and the angle between the trunk and thighs by the empirical relation:

$$P_{abdom} = 10^{-4}\,(0.6516 - 0.005\,447\,(70°)\,(2056)^{1.8}$$
$$P_{abdom} = 24.8\,mmHg = 33\,120\,dyn\,cm^{-2} = 0.0337\,kp\,cm^{-2}$$

which gives

$$F_{abdom} = (P_{abdom})\,(diaphragm\ area)$$
$$= F_{abdom} = (0.038)\,(465) = 16\,kp.$$

Thus assuming static equilibrium conditions at the L_5/S_1 disc gives

$$T_{L_5/S_1} = 0 = T_{body\ weight} + T_{hand\ force} - T_{abdom\ pres} - T_{musc}$$
$$= (BW \times B) + (FH \times H) - (F_{abdom} \times A) - (F_{musc} \times M)$$

and

$$F_{L_5/S_1} = 0 = \sin \alpha\,BW + \sin \alpha\,FH - F_{abdom} + F_{musc} - F_{comp}.$$

Using male 50 percentile anthropometric data and the position indicated in Fig. 7.5, a 45 kp load in the hands results in the following computations:

$$F_{musc} = \frac{(BW \times B) + (FH \times H) - (F_{abdom} \times A)}{M}$$
$$= \frac{(36 \times 20) + (45 \times 28.5) - (16 \times 14.6)}{5}$$
$$= 354\,kp.$$

The compressive force on the superior surface of the sacrum is estimated for $\alpha = 66°$ (based on pelvic rotation as described earlier) by solving the static force equilibrium equation

$$F_{comp} = (\sin 66° \times 36) + (\sin 66° \times 45) - 16 + 354 = 412\,kp.$$

It should be noted in the above four force components that the erector

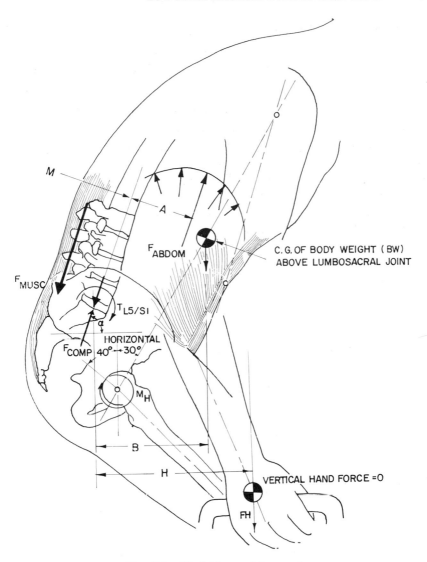

F<small>IG</small>. 7.5.   Variables used in case 1.

spinae muscles are the primary source of compression force on the disc during a volitional lifting activity. By stooping over further (tilting the torso towards horizontal) the effect of the body weight and load held in the hands is shifted still further from a compression force to a shearing force which is primarily resisted by the lumbar facets and their associated ligaments which could cause or aggravate facetogenic problems. This will be further illustrated with other numerical examples later.

The major point is that the erector spinae muscles are essential to the stability of the column but their actions create high compression forces within the column. Studies of cadaver spinal column strengths [31, 48, 49] indicate that such forces could cause microfractures of the cartilage end-plates, perhaps then initiating or aggravating disc degeneration.

The role of abdominal pressure in relieving the compression load on the spinal column during lifting has been depicted earlier [30] as a major source of column support. Data obtained by Asmussen and Poulsen [43] would appear to limit the general effectiveness of this reflex mechanism. A maximum limit of 150 mmHg appears to be possible with highly trained individuals although 90 mm is probably a more reasonable limit for the normal population. It should be noted though that this spinal relief may not be possible when a person is carrying a load for a sustained period since it depends on maintaining a counteracting thoracic pressure on the diaphragm by holding one's breath. When assumed to be a well-developed reflex in an individual, it appears that it can relieve approximately 15–20 per cent of the lumbar spinal column compression during normal weight-lifting actions.

Case 2. Lifting variable loads close to body

As a more general case let us assume the body position shown earlier in Fig. 7.5 but vary the load from zero to 125 lbf (i.e. 0–56.7 kp). The predicted L_5/S_1 compression forces for each load resulting from this procedure are shown in Fig. 7.6. What is clearly depicted is that even with moderate loads held reasonably close to the body high compression forces are created at the lumbosacral level. Perhaps this is why one recent epidemiological study [50] has shown that some people had an increased incidence rate of low-back pain when lifting only 16 kp in their jobs.

Case 3. Lifting loads with two different postures

It is often stated that lifting with the legs straight and the back stooped over is much more stressful on the low back than lifting with a near-vertical back, using the legs for the major lifting action. Unfortunately, this rule, though widely quoted, has not been subjected to critical analysis. One reason for suspecting that the rule may be wrong is that observations of people lifting loads by two independent researchers [51, 52] have shown that people who have been repeatedly instructed to lift in this manner do not follow these instructions. Rather they more often lift with their backs, holding their legs in a nearly erect manner.

The numerical example that follows depicts two alternative postures of lifting an object which is 15 in in front of the ankles and 15 in above the floor. The static load on the hands was 30 lbf (13.6 kp) with a dynamic component added to represent the initial load accelerations (at about 100 ms into the

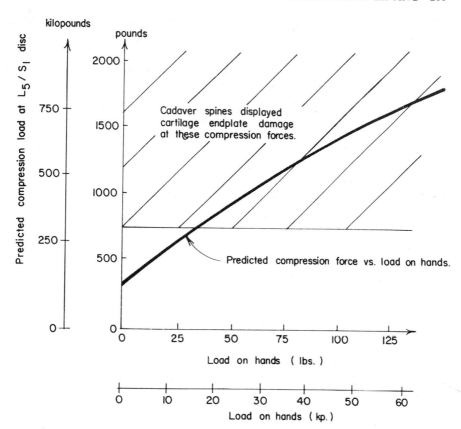

FIG. 7.6. Predicted L_5/S_1 compression loads for torso–arm configurations depicted in Fig. 7.5.

lifting action).† What occurs in dynamic load lifting when the load is horizontally forward of the feet is that a person normally lifts the load in a direction towards the torso, rather than only vertically as was assumed in cases 1 and 2. The two postures assumed in the model are shown in Fig. 7.7 with the predicted L_5/S_1 compression forces shown under each figure.

What clearly must be concluded from these simulations is that the back-stooped load-lifting posture reduces the compressive loading on the L_5/S_1 disc below that predicted when lifting with a more nearly vertical back. The reason for the larger compression forces when lifting with the back near vertical is that (1) the load moment arm is increased and (2) the vertical component of both the body weight and the hand forces add more directly to the compressive forces on the more vertical spine. Of course the shearing forces on

† The dynamic component was determined by an empirical study of load lifting by Park [52].

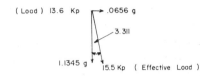

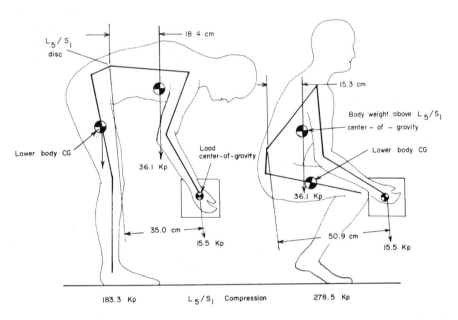

Fig. 7.7.   Low-back compression associated with two different lifting postures (load placed 15 in from the ankle and 15 in from the floor) with an assumed static load of 30 lbf (13.6 kp). (From ref. 53.)

the L_5/S_1 disc are greater when lifting with the back horizontal. This may be a problem for individuals who have pathological or anatomical changes of the articulating facets since these provide the primary resistance to shearing forces in the lumbar spine.

In short, it appears that people lift loads with their backs rather than their legs. In so doing they minimize the energy necessary to move the body–load mass combination as described by both Brown [51] and Park [52]. Though it was commonly believed that 'back lifting' was more stressful than 'leg lifting', this does not appear to be so in regard to compression loading of the column especially when lifting loads that are larger than can be brought between the knees or loads which are located horizontally out from the feet. If a person can maintain the torso in a completely erect position while lifting, which because of limited shoulder strength and arm reach requires a small object in close to the body, then the compressive force is minimized on the L_5/S_1 disc for moderate load lifting. A comparison of the resulting compression forces at the lumbosacral disc due to lifting of loads small enough to pass between the legs (with similar H values) is shown in Fig. 7.8 for both a completely erect torso lift and for a lift shown earlier in case 1 (Fig. 7.5). The difference in compression forces between these two postures is primarily due to the added torso moment when in the stooped position. Of note is that if the person stooped all the way over, instead of just a slight crouch as depicted in Fig. 7.8, the compression force function would lie below the top curve because the load and body weight components acting on the L_5/S_1 disc are now shifted to a shearing effect.

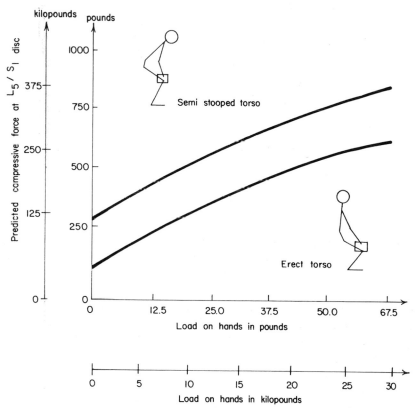

FIG. 7.8. Comparison of the lifting of small loads which can be lifted between the legs by an average sized man.

7.6. Applications of occupational biomechanics

The preceding biomechanical concepts have had a direct application in two arenas. First these concepts provide a rational basis for instruction as to safe ways to lift heavy loads. Secondly they have provided a basis for the development of load-lifting guides for both the National Institute for Occupational Safety and Health and the Federal Trade Commission. Both of these aspects are discussed below.

7.6.1. A biomechanical recommendation for safe lifting instructions

To ensure that people minimize the compression loading of the lumbar column when lifting loads care must be taken to instruct them on the basic biomechanics of load lifting. The following general points should be considered in this regard.

(1) Loads should always be brought in as close to the body as possible to minimize H before lifting (see Fig. 7.5).

(2) Provided the load is compact enough to permit lifting it between the legs, keep the torso near vertical and lift with the legs. This posture will significantly reduce the resulting compressive load on the lumbar spine as compared with lifting with the torso flexed forward from vertical (as depicted in Fig. 7.8).

(3) If a load is too large to go between the legs or is horizontally located more than about 15 in away from the front of the body when picking it up or setting it down, a lift wherein the torso is nearly horizontal better allows the load to be moved so that H is minimized which then minimizes the compression loading on the spinal column (as presented in Fig. 7.7). When performing a back lift of this type, care must be taken to avoid a hyperflexed back as this places further stress on the posterior ligaments and posterior annulus of the disc. The low back should be flattened to pre-tension the erector spinae muscles and possibly distribute some compressive load to the facets.

(4) Loads should always be moved slowly and in a well-controlled (i.e. pre-planned) manner to minimize the effects of accelerations and unco-ordinated muscle recruitment which can greatly increase the resulting peak compression loads on the spinal column.

7.6.2. A biomechanical recommendation for maximum acceptable loads to be lifted

Both NIOSH (National Institute for Occupational Safety and Health) and the FTC (Federal Trade Commission) are developing federal guidelines as to the maximum load that can safely be lifted by some assumed proportion of the adult population in the United States. The preceding biomechanical analysis has provided one criterion for such policies.

One of the major factors which such biomechanical concepts has stressed in specifying maximum allowable loads to be lifted is that the specification must be directly linked to the H distance or bulkiness of the object to be lifted. This is necessary not only to minimize the L_5/S_1 compression but also to minimize the muscle strength requirements at other joints while lifting. One proposed specification developed from such biomechanical modelling is depicted in Fig. 7.9. The recommendations reflect the great variability that exists in the population to lift heavy loads. Accepting that a 362 kp limit is reasonably tolerated by most adults, though older individuals may be at higher risk, the lower 'action level' is recommended. This limit requires that individuals lifting loads exceeding the specification be medically evaluated and carefully instructed for such tasks if performed in industry. This would also approximate the weight of products that could be sold with the label 'portable'. The latter problem arises in that most warranties on products labeled 'portable' today require the purchaser to return the product for service rather than allowing home service.

The upper 'permissible limit' recommendation is just that for industrial

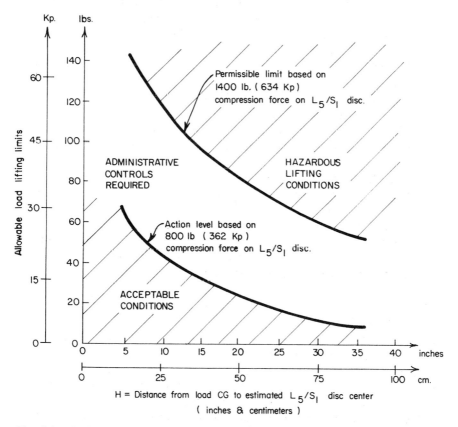

FIG. 7.9. Recommended weight and size limits for occasional lifts (less than once every 5 min). See text for assumptions.

tasks. It assumes that 634 kp dictates hazardous lifting conditions for most people and should not be permitted in job design and specification.

At the time of the writing of this chapter a great deal of debate regarding the assumptions and interpretations of these limits is underway in the agencies involved. To develop these guidelines strong biomechanical assumptions were involved. Slow motions with static equivalent conditions were assumed though it is known that some common lifting motions can impart significantly higher forces on the body with rapidly changing force directions. Since lifting motions are never standardized, however, it is difficult to estimate the dynamic effects on normal motions. As the object lifted approaches a person's lifting muscle strength, the person tends to slow the motion and hence minimizes the dynamic effects, thus giving some justification for the static assumptions.

Average anthropometry was assumed for both men and women. Varying body weight and link length parameters can change the values by as much as ± 10 per cent of those plotted. In some specific design situations this may be critical.

The lifts modelled assumed that both hands and the object centre of gravity were the same H distance from the L_5/S_1 disc and were equally loaded and that they were located between the floor and normal table height above the floor, i.e. they were low lifts. Clearly other types of lifts (e.g. above the waist, one handed, or twisting lifts to the side) create different body reactions and limitations than considered in these guidelines. Nevertheless, the guidelines do provide a rational basis upon which many of these other considerations can be added as both the technical knowledge and practitioner sophistication improves in the future.

7.7. Summary of load and posture effects on compression forces of lumbar discs

It should be clear from the preceding that moderate loads can produce high compression forces in the lumbar spinal discs. Both epidemiological and cadaver data indicate that compression forces in the ranges depicted here produce structural failures in the weight-bearing cartilage end-plates. The resulting microfractures and scarring are believed by some researchers to be a major factor in accelerating the natural aging and degeneration of the spinal discs.

The recommendations are based on a concern for minimizing compression forces incurred by the spinal column when lifting loads. It should be evident that much more research is needed to understand the pathobiomechanics of load lifting better. Simple biomechanical models of the type described are able to assist one in determining how the many anatomical, postural, and load-related variables combine to affect the stresses in common load-lifting situations. Controlled laboratory and field studies are now needed to determine the validity of the predictions of such models. In addition, as more data are developed regarding the physiology of load lifting they need to be incorporated into the development of better models. It is only through such combined efforts of both the physical and life scientists that the reduction and control of low-back pain will be forthcoming.

Appendix 7.1

Reference tables of the locations population centres of gravity and body mass distributions are given below.

Distances to segment centres of gravity (CG) (cm)

	Percentile smaller than given distance						
	Male			Female			Percentage link length [11]
Dimension	5%	50%	95%	5%	50%	95%	(%)
Wrist to hand CG	6.7	7.0	7.4	6.1	6.4	6.7	–
Elbow to lower arm CG	11.0	11.7	12.3	9.9	10.4	11.0	43.0
Shoulder to upper arm CG	12.5	13.2	14.0	11.6	12.1	12.5	43.6
Hip to trunk, neck, head CG	18.1	19.3	22.5	16.7	17.9	19.1	39.6[a]
Knee to upper leg CG	23.0	24.8	26.1	22.2	23.2	24.2	56.7
Ankle to lower lmg CG	21.5	23.2	24.9	19.3	20.6	22.1	56.7
Heel to foot CG	10.6	11.4	12.3	9.4	10.3	11.1	42.9

All the dimensions are based on link lengths presented in Table 5.5 using the Dempster percentage [11] of link length estimates.

[a] When in erect posture.

Estimates of mass distributions (kp and lbf)

Link	Percentile lighter than given amount						Percentage of of body weight[a] (%)
	Male			Female			
	5%	50%	95%	5%	50%	95%	
Hand	0.4 (0.8)	0.4 (1.0)	0.6 (1.3)	0.3 (0.6)	0.4 (0.8)	0.5 (1.2)	0.6
Forearm	0.9 (2.0)	1.2 (2.7)	1.6 (3.5)	0.7 (1.6)	1.0 (2.2)	1.4 (3.2)	1.6
Upper arm	1.6 (3.5)	2.1 (4.7)	2.8 (6.1)	1.3 (2.9)	1.7 (3.8)	2.5 (5.6)	2.8
Head, neck, and trunk	33.0 (72.8)	43.4 (95.9)	56.8 (125.4)	27.2 (60.1)	35.8 (79.2)	52.1 (115.0)	57.8
Arms, head, and torso above L_5/S_1 disc[b]	27.2 (60.0)	35.8 (79.0)	46.8 (103.3)	22.4 (49.5)	29.5 (65.2)	42.9 (94.7)	47.6
Upper leg	5.7 (12.5)	7.4 (16.4)	9.7 (21.5)	4.7 (10.3)	6.2 (13.6)	8.9 (19.7)	9.9
Lower leg	2.6 (5.8)	3.4 (7.6)	4.5 (9.9)	2.2 (4.8)	2.8 (6.3)	4.2 (9.2)	4.6
Foot	0.7 (1.7)	1.0 (2.3)	1.4 (3.0)	0.7 (1.5)	0.9 (1.9)	1.3 (2.8)	1.4
Body weight[c]	57.1 (126)	75.2 (166)	98.3 (217)	47.1 (104)	62.1 (137)	90.1 (199)	

[a] Estimates are from Dempster [11] as corrected for supposed fluid loss by Clauser *et al.* [23]
[b] Based on ref. 30.
[c] Based on ref. 28.

References

1 HULT, L. (1954). Cervical, dorsal and lumbar spinal syndromes. *Acta orthop. scand.* Suppl. 17.
2 SNOOK, S. H., and CIRIELLO, V. M. (1972). Low back pain in industry. *Am. Soc. Safety Engnrs J.* **17**, 17.
3 TROUP, J. D. G., and CHAPMAN, A. E. (1969). The strength of the flexor and extensor muscles of the trunk. *J. Biomech.* **2**, 49.
4 HOLTZ, C. L., and KEYS, M. A. (1971). Regional differences in the use of hospital days. *Prof. Activ. Stud. Rep. CPHA* **9**, 4.
5 ROWE, L. M. (1971). Low back disabilities in industry: updated position. *J. occup. Med.* **13**, 476.
6 MAGORA, A., and TAUSTEIN, I. (1969). An investigation of the problem of sick-leave in the patient suffering from low back pain. *Ind. Med.* **38**, 80.
7 NACHEMSON, A. L. (1971). Low back pain: its etiology and treatment. *Clin. Med.* **18**, 18–24.
8 TICHAUER, E. R. (1965). The biomechanics of the arm–back aggregate

under industrial working conditions. *ASME Rep. No. 65-WA/HUF-1.*

9 ROAF, R. (1960). A study of the mechanics of spinal injuries. *J. bone jt. Surg. Br. Vol.* **42**, 810–32.

10 DAVIS, P. R., TROUP, J. D. G., and BURNHARD, J. H. (1965). Movements of the thoracic and lumbar spine when lifting: A Chronocyclophotographic study. *J. anat. Lond.* **99**, 13–26.

11 DEMPSTER, W. T. (1955). Space requirements of the seated operator. *WADC Tech. Rep. No. 55–159.*

12 DRILLIS, R., and CONTINI, R. (1966). Body segment parameters. *Tech. Rep. No. 1166.03.* New York University, New York.

13 HANAVAN, E. P. (1964). A mathematical model of the human body. *A.D. Rep. No. 608–463.*

14 ROBBINS, D. H., BENNETT, R. O., and BOWMAN, B. M. (1973). The MVMA 2-dimensional crash victim simulation. *Rep. No. UM-HSRI-BI-73–5.* HSRI, Ann Arbor, Michigan.

15 PEARSON, J. R., MCGINLEY, D. R., and BUTZEL, L. M. (1961). Dynamic analysis of the upper extremity for planar motions. *Rep. No. 04468-1-T.* University of Michigan, Ann Arbor, Mich.

16 PLAGENHOEF, S. C. (1966). Methods for obtaining kinetic data to analyze human motions. *Res. Q.* **37**, 1.

17 FISHER, B. (1967). *A biomechanical model for the analysis of dynamic activities. M.S. Thesis.* University of Michigan, Ann Arbor, Mich.

18 SCHANNE, F. J. (1972). *A three-dimensional hand force capability model for a seated person,* Vols. I, II. *Doctoral Diss.,* University of Michigan, Ann Arbor, Mich.

19 PLAGENHOEF, S. C. (1968). Computer programs for obtaining kinetic data on human movements. *J. Biomech.* **1**, 221–34.

20 WILLIAMS, M., and LISSNER, H. R. (1962). *Biomechanics of human motion.* W. B. Saunders, Philadelphia.

21 CHAFFIN, D. B. (1967). *The development of a prediction model for the metabolic energy expended during arm activities,* pp. 213–36. *Ph.D. Thesis,* University of Michigan, University Microfilms Inc., Ann Arbor, Mich.

22 DEMPSTER, W. T., and GAUGHRAN, G. R. L. (1967). Properties of body segments based on size and weight. *Am. J. Anat.* **120**, 33–54.

23 CLAUSER, C. E., MCCONVILLE, J. T., and YOUNG, J. W. (1970). *Weight, volume, and center of mass of segments of the human body. AMRL Tech Rep.* 69–70.

24 DEMPSTER, W. T., SHERR, L. A., and PRIEST, J. G. (1964). Conversion scales estimating humeral and femoral lengths and the lengths of functional segments. *Hum. Biol.* **36**, 246–61.

25 CHAFFIN, D. B. (1972). Some effects of physical exertions. *Tech. Rep.* Industrial Engineering Department, University of Michigan, Ann Arbor, Mich.

26 CHURCHILL, E., and BERNHARDI, B. (1957). WAF training body dimensions. *WADC Tech. Rep. 57–197.* Wright Field, Ohio.

27 MORGAN, C.T., COOK, J. S., CHAPANIS, A., and LUND, M. (1963). *Human engineering guide to equipment design.* McGraw-Hill, New York.

28 National Health Survey (1965). *Weight, height, and selected body dimensions of adults. PHS Publ. 1000—Ser. 11, no. 8.* US Government Printing Office, Washington, DC.

29 ARMSTRONG, J. R. (1965). *Lumbar disc lesions,* pp. 42–45. Williams and Wilkins, Baltimore. Md.

30 MORRIS, J. M., LUCAS, D. B., and BRESLER, B. (1961). Role of the trunk in stability of the spine. *J. bone jt. surg., Am. Vol.* **43**, 327.

31 PEREY, O. (1957). Fracture of the vertebral end plate in the lumbar spine. *Acta orthop. scand.* Suppl. 25.

32 GORDON, E. E. (1961). Natural history of the intervertebral disc. *Arch. phys. Med.* **42**, 750–63.

33 BADGER, D. W., DUKES-DOBOS, F. N., and CHAFFIN, D. B. (1972). Prevention of low back injury in the industrial work force. *NIOSH Symp. Rep.,* pp. 1–3. NIOSH Ergonomics Branch, Cincinnati, Ohio.

34 KRUSEN, F., ELLWOD, C. M., and KOTTKE, F. J. (1965). *Handbook of physical medicine and rehabilitation.* W. B. Saunders, Philadelphia.

35 SMITH, A. DeF., DEERY, E. M., and HAGMAN, G. L. (1944). Herniations of the nucleus pulposus—a study of 100 cases treated by operation. *J. bone jt. Surg.* **26**, 821.

36 FICK, R. (1904). *Handbuch der Anatomic und Mechanik der Gelenke.* Von Gustav Fisher, Jena.

37 LANIER, R. R. (1939). Presacral vertebrae of white and negro males. *Am. J. Phys. Med.* **25**, 343–420.

38 CHAFFIN, D. B., SCHUTZ, R. K., and SNYDER, R. G. (1972). A prediction model of human volitional mobility. *Automotive Engineering Congr., Detroit, Mich., January 1972,* Paper 720002. Soc. Automotive Engineers, New York.

39 THIEME, F. P. (1950). *Lumbar breakdown caused by erect posture in man. Anthropometric Paper 4.* University of Michigan, Ann Arbor, Mich.

40 ALLBROCK, D., and UGANDA, K. (1957). Movement of the lumbar spinal column. *J. bone jt. Surg., Br. Vol.* **39**, 339–45.

41 LINDAHL, O. (1966). Determination of the sagittal mobility of the lumbar spine. *Acta orthop. scand.* **37**, 241–54.

42 ROLANDER, S. D. (1966). Motion of the lumbar spine with special reference to the stabilizing effect of posterior fusion. *Acta orthop. scand.* Suppl. 90.

43 ASMUSSEN, E., and POULSEN, E. (1968). On the role of the intra-abdominal pressure in relieving the back. *Commun.-Dan. natn. Ass. Infant. Paral.,* **28**, 1–11.

44 BARTELINK, D. L. (1957). The role of abdominal pressure in relieving the

pressure on the lumbar intervertebral discs. *J. bone jt. Surg., Br. Vol.* **39**, 718–25.

45 MUNCHINGER, R. (1962). Manual lifting and carrying. *Int. Occup. Safety Health Inform. Sheet*, **3**.

46 BELYTSCHKO, T., ANDRIACCHI, T., SCHULTZ, A., and GALANTE, J. (1973). Analog studies of forces in the human spine: computational techniques. *J. Biomech.* **6**(4), 361–72.

47 KRAUS, H., ROBERTSON, G. H., and FARFAN, H. F. (1973). On the mechanics of weight lifting, *New England Bioengineering Conf. Proc., University of Vermont, April 1973*. New England Bioengineering Society.

48 EVANS, F. G., and LISSNER, H. R. (1965). Studies on the energy absorbing capacity of human lumbar intervertebral discs. *Proc. 7th Stapp Car Crash Conf., Springfield, Ill.* Soc. Automotive Engineers, New York.

49 SONODA, T. (1962). Studies on the strength for compression, tension, and torsion of the human vertebral column. *J. Kyoto Prefect. Med. Univ.*, **71**, 659–702.

50 CHAFFIN, D. B., and PARK, K. S. (1973). A longitudinal study of low-back pain as associated with occupational weight lifting factors. *Am. Ind. Hyg. Ass. J.*

51 BROWN, J. R., *Lifting as an industrial hazard*. Labour Safety Council of Ontario, Ontario Department of Labour, Toronto.

52 PARK, K. S. (1973). A computerized simulation model of postures during manual materials handling. Ph.D. Thesis. University of Michigan, Ann Arbor, Mich.

53 PARK, K. S., and CHAFFIN, D. B. (1974). A biomechanical evaluation of two methods of load lifting. *AIIE Trans.*, **6**(2), 105–13.

8. Human body vibration, with applications to industrial environments, ambulance design, cardiovascular monitoring, and cardiac assist techniques

K. V. Frolov, B. A. Potemkin, E. Rohl, H. Wolff,
A. Bhattacharya, and D. N. Ghista

8.1. Vibration as an environmental factor imposed by man

Introduction

During the process of technical and industrial evolution, man has consciously or unconsciously changed his environment. His wish to mechanize heavy manual labour as well as the desire to travel rapidly over long distances has resulted in the creation of effective machines and high-speed transportation equipment; however, as technology has developed the problem of protecting man from the harmful influence of vibrational loads generated by various machines, mechanisms, and automatic production lines has become acute.

Indeed, the problem of the interaction of man with his environment has become the theme of the century. The study of the influence of vibration on man, in order to create effective means of vibration protection, is a part of the overall struggle for the quality of the environment of our planet, for improvement of life, and for protection of the natural riches around us. This modern problem has both social and economic aspects.

Classification of vibrations

There is no one single type of vibration in nature. The concept of vibration actually covers a great variety of physical phenomena and processes. It is not surprising therefore that a universal means has never been achieved for protecting man from vibration and from the noise which it generates. A deterministic physical process of vibration can be characterized mathematically. A classical example is found in stable harmonic vibrations described by the function:

$$u(t) = u_0 \cos \omega t \tag{8.1}$$

where u_0 is the amplitude of the vibrations, ω is the angular frequency, and t is time.

The amplitude and frequency of such vibrations remain constant with time. Although this classical form of vibration is actually only an idealization of more complex oscillating processes in actual machines, it is sometimes used in laboratory tests and also in calculating the simplest forms of oscillations.

We must frequently deal with more complex vibrations which are usually represented as the sum of a finite or infinite number of harmonic components:

$$u(t) = \sum_{i=1}^{N} u_i \cos(\omega_i t + \phi_i) \qquad (8.2)$$

where ϕ_i is the phase of the *ith* harmonic.

Generally speaking aperiodic vibrations vary much more widely than do periodic vibrations. Important in this group are the so-called attenuating harmonic and quasi-harmonic unstable vibrations with continually changing frequency. Such vibrations arise, for example, during acceleration and braking of mechanisms with rotating elements. Oscillations following an attenuating sine wave can be mathematically represented by the expression

$$u(t) = u_0 e^{-\delta t} \sin(\omega t + \phi) \qquad (8.3)$$

where $u_0, \delta, \omega,$ and ϕ are constant quantities characterizing the initial amplitude, attenuation, frequency, and phase shift of the oscillations respectively.

Quasi-harmonic vibrations with continually increasing frequency can be described by the expression

$$u(t) = u_0 \sin(\omega t + \lambda t^2 + \phi) \qquad (8.4)$$

where λ is the rate of change of the frequency.

Periodically repeated impact pulses, which arise when comparatively great forces are applied briefly, also constitute frequently encountered vibration loads. Impact effects can quite arbitrarily be divided into impulse-type and limitation-type effects. The form of impulse effect is rather arbitrary, but in all cases the following expressions are correct:

$$u(t) \neq 0 \quad \text{where } t < \tau$$
$$u(t) = 0 \quad \text{where } t > \tau$$

Limitation motions are characterized by the fact that as time passes they approach a certain constant limiting value. This can be represented mathematically, for example, as follows:

$$u(t) = (u_1 e^{\alpha t} - u_2 e^{-\alpha t}) e^{-\beta t} \qquad (8.5)$$

where $u_1, u_2, \alpha,$ and β are real numbers and $\beta > \alpha$. Examples of these idealized forms of vibration are shown in Fig. 8.1.

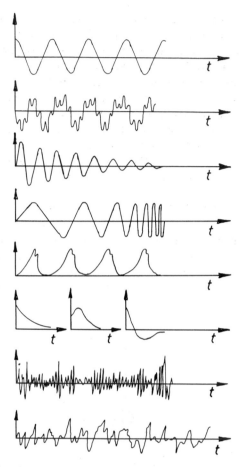

Fig. 8.1. Idealized forms of vibration processes.

Strictly speaking, in nature there are not and cannot be any 'purely' deterministic processes. Therefore, actual vibrations can be considered to be only approximately deterministic. Consequently, the second class of vibrations consists of random vibrations, the behaviour of which cannot be described as a regular function of time. At each moment the parameters of this type of vibration may take on some quantitative value from the area of possible values. Mathematical description of random vibrations utilizes statistical characteristics, the sense of the use of which lies in the resultant transition from random functions to deterministic functions defining the mean estimates of the random vibrations.

In practice we quite frequently encounter random vibrations for which the statistical characteristics do not change over the time interval analysed as time is shifted, i.e. as t is replaced by $t + a$ where a is an arbitrary quantity. These can be thought of as random oscillations about a certain mean value, and the source of the oscillations has a nature which does not change with time. These random vibrations are called stationary vibrations, in contrast to non-stationary vibrations which include all other vibrations not satisfying the above condition. Depending on the range (or band) of frequencies contained in the random vibrations, they are arbitrarily divided into narrow-band and wide-band vibrations. In the theory of random vibrations an important role is played by the so-called Gaussian random processes which are distinguished from non-Gaussian processes in that they are fully defined by a single statistical characteristic—the correlation function.

Parameters of a vibration stimulus and units of measurement

When we study three-dimensional vibrations of the human body we generally distinguish the primary axes of motion. The orientation of the axes of the system of co-ordinates thus introduced and its connection with the human body is shown in Fig. 8.2. In accordance with these symbols, vibrations are arbitrarily decomposed into vertical z, longitudinal x, and transverse y vibrations.

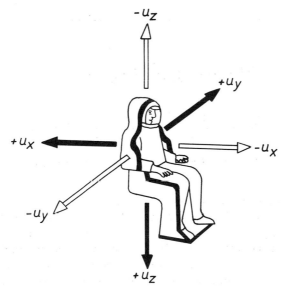

FIG. 8.2. Primary axes of vibrational motion of the human body.

To provide a quantitative description of a vibrational motion along one of the axes x, y, or z, we most frequently use three main parameters: displacement u, velocity $\dot{u}$, and acceleration $\ddot{u}$. We should note that estimates of these parameters can be developed by several methods.

The instantaneous value of a parameter being studied corresponds to the value of the parameter at a fixed instant in time.

The peak value of a parameter is defined as the greatest deviation of its instantaneous values in some given direction from the zero level

$$u_{\text{peak}} = |u_{\text{max}}|. \tag{8.6}$$

The peak value of displacement represents the maximum deflection of an oscillating point, which is important in determining the clearance between oscillating bodies. The peak value $\ddot{u}_{\text{peak}}$ of acceleration is used to determine the maximum forces of inertia.

The mean value of a vibration parameter over time T is calculated as the arithmetic mean of the instantaneous values (ignoring the sign)

$$u_{\text{m}} = \frac{1}{T} \int_{t_0}^{t_0 + T} |u(t)| \, dt \tag{8.7}$$

and is used to estimate the overall intensity of vibrations.

The effective value of the vibrations

$$u_{\text{e}} = \left\{ \frac{1}{T} \int_{t_0}^{t_0 + T} u^2(t) \, dt \right\}^{1/2} \tag{8.8}$$

has a definite physical sense in the case of the vibration velocity $\dot{u}_{\text{e}}$ since the energy of oscillations is generally proportional to the square of the vibration velocity. The expression under the square root in the last formula is called the mean square value of the vibration parameter: $u_{\text{m.sq.}} = u_{\text{e}}^2$.

Vibrations of complex shape are sometimes characterized by the following quantities: the form factor $k_{\text{f}} = u_{\text{e}}/u_{\text{m}}$ and the amplitude factor $k_{\text{a}} = u_{\text{peak}}/u_{\text{e}}$.

For monoharmonic vibrations, the estimates described above take on the following values:

$$u = u_0 \cos (\omega t + \phi)$$
$$u_{\text{peak}} = u_0$$
$$u_{\text{m}} = (2/\pi) u_0 = 0.637 \, u_0$$
$$u_{\text{e}} = u_0 / \sqrt{2} = 0.707 \, u_0$$
$$u_{\text{m.sq.}} = u_0^2/2$$
$$k_{\text{f}} = \pi/2 \sqrt{2} = 1.111$$
$$k_{\text{a}} = \sqrt{2} = 1.41.$$

For monoharmonic vibrations the amplitudes of the acceleration, velocity, and displacement are quite simply interrelated:

$$\ddot{u}_{\text{peak}} = \omega \, \dot{u}_{\text{peak}}$$
$$\dot{u}_{\text{peak}} = \omega \, u_{\text{peak}}$$

where $\omega = 2\pi f$ is the angular frequency of the vibrations, measured in rad s^{-1} (1 rad $= 57.295°$).

Vibrational displacements are measured in metres, velocities in m s^{-1}, and accelerations in m s^{-2}. It is quite common practice to measure accelerations in units of the acceleration of the gravity of the earth: $g = 9.807$ m s^{-2}. Vibration acceleration, measured in g units $(g = \ddot{u}_0/G = (2\pi f)^2\, u_0/G)$ is frequently called the overload. In the practice of vibration measurements dimensionless units of measurement of vibration relative to a certain initial level are used. They are called decibels and are defined by the following expression:

$$L = 20 \log \frac{u}{u_0} \left[\mathrm{db} \right].$$

The initial level of vibration acceleration generally used is $\ddot{u}_{e0} = 0.31 \times 10^{-4}$ m s^{-1}. This value is produced from the relationship between sound pressure at the threshold of audibility and the oscillating acceleration in a flat sound wave.

Sources of vibration in industry and domestic life

Vibrations refer primarily to perturbing forces which vary in their nature and mode of action. Vibration sources might be irregularities of roads or fields, gusts of wind, pressure pulsations in turbulent layers of the atmosphere or a body of water, unevenness or imbalance of the rotation of machine and engine parts, friction and microscopic impacts between the working parts of machine tools, gaps in bearings and gears, oscillations of the rotors of electric machines under the influence of magnetic fields, acoustical loads resulting from the exhaust streams of jet engines and airplane propellers, and pulsations of pressure in pneumatic and hydraulic machines. An example of road irregularities and its effect on ambulance-transported patients will be discussed in some detail at the end of this chapter.

The random vibrations of a wheel rolling over a road with an uneven surface are to a great extent transmitted through to the floor of the cabin and the seats in the vehicle. Figure 8.3 shows the spectral planes of the vertical vibration accelerations of a ZIL-130 truck as it drives over a dirt road at various speeds. Figure 8.4 shows the spectral characteristics for an automobile.

In addition to the vertical oscillations, a passenger sitting in a motor vehicle also experiences longitudinal and transverse oscillations. Tests have shown that the horizontal accelerations in the body of a motor vehicle are less than the vertical accelerations; for trucks the mean square value of longitudinal oscillations is $0.4\,g$ and of transverse oscillation is $0.3\,g$ in the 0–15 Hz frequency band. The longitudinal oscillations in the 0–5 Hz band amount to 30–50 per cent of the vertical oscillations. Figures 8.5–8.7 show the characteristics of vertical oscillations of a tractor, a grader, and a streetcar. Whereas it is

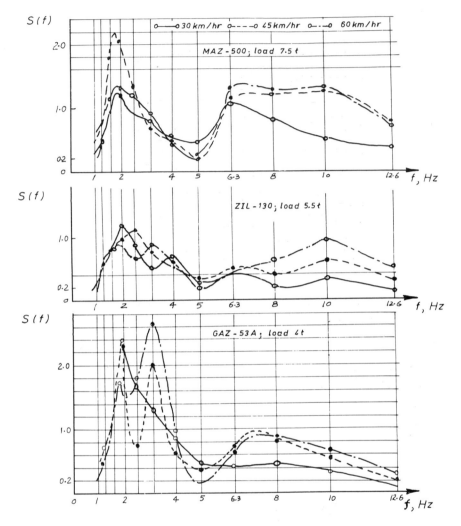

FIG. 8.3.   Spectral planes of the vertical vibration accelerations of a ZIL-130 truck as it drives over a dirt road at varying speeds: solid curves, 30 km h^{-1}; broken curve, 45 km h^{-1}; chain curve, 60 km h^{-1}.

possible to design the suspension system to prevent the passenger from being subject to large-amplitude vertical oscillations in the frequency range that is harmful to the subject's physiological system, it is quite difficult to protect passengers from horizontal oscillations by changing the design of the seat since the back must not be too soft because it is used as a support for the passengers.

The vibrations which act on passengers in boats have been less fully studied. Obviously the characteristics of the vibrations in this case depend greatly on

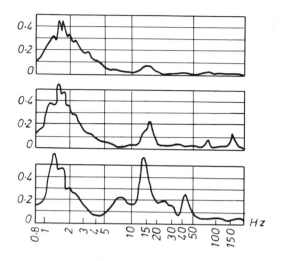

Fig. 8.4.   Spectral characteristics of an automobile.

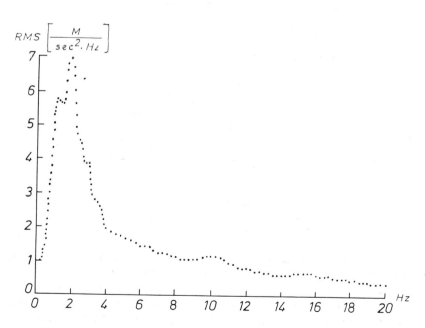

Fig. 8.5.   Spectral density of the vertical oscillations of a tractor.

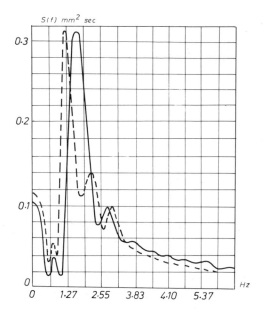

FIG. 8.6.   Spectral densities of the vertical accelerations of a KTM-5M streetcar: solid curve, rigid track, $V = 50 \text{ km h}^{-1}$; broken curve, soft track, $V = \text{km h}^{-1}$.

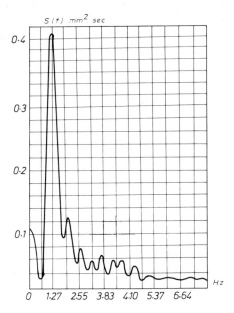

FIG. 8.7.   Spectral density of vertical oscillations of the body of an MTV - 82 streetcar: rigid track, $V = 20 \text{ km h}^{-1}$.

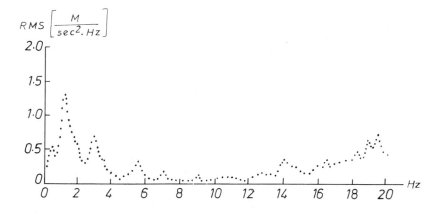

F_IG_. 8.8.   Spectral density of vibrations on the cabin floor at a ship helm with a swell
on the water.

the design of the ship and the state of the surface of the water. Figure 8.8 shows
the spectral density of vibrations on the cabin floor at the helm with a swell on
the water.

Aircraft, particularly helicopters, generate rather intensive vibrations in the
directions of all the primary axes. Helicopters characteristically vibrate in the
10–70 Hz range with amplitudes ranging from 0.4 mm for 70 Hz to 2.4 mm for
10 Hz. Table 8.1 presents data on the frequencies of the primary sources of
perturbing forces—the rotors and propellers of Soviet passenger aircraft—as

T_ABLE_ 8.1

Type of aircraft	Type of engine	Rotating speed			Defining vibration frequencies in cabin (Hz)
		Engine rotor	Propeller shaft	Propeller blades	
TU-104	Jet	60–79	–	–	20–40; 60–100
TU-124	Fan jet	128–147	–	–	4–25; 120–140; 160–180
TU-134	Fan jet	176–193 95–110	–	–	4–25; 95–110; 140–160
TU-114	Turboprop	166–192 137.5	12	48; 96	12; 48; 96; 144
IL-62	Fan jet	26–86	–	–	4–12; 80; 120
IL-18	Turboprop	50–117 205	18	72	1.5–10; 18; 72; 144
AN-10	Turboprop	205	18	72	18; 72; 144
AN-24	Turboprop	252	21	84	20; 80; 160
YaK-40	Fan jet	144–176 188–291	–	–	11–20; 60–100

well as values defining the vibration frequencies of cabin floors measured in flight [1]. For engines with variable rotor or propeller speed the limiting frequencies corresponding to the rotating speed at idle and the maximum speed are shown; for jet engines the numerator shows the rotating speed of the low-pressure stage and the denominator shows the rotating speed of the high-pressure stage.

The level of vibrations and their frequency spectrum in the cabin depend primarily on the type of engine and the arrangement of the engines on the aircraft as well as on the elasticity of the engine supports, the structural elements of the aircraft, and the equipment in the cabin. Figure 8.9 shows the relative spectra of the amplitudes of vibration motions measured in the floor panels of the cabins of three types of aircraft: version (a) for four turboprop engines in the wings, far from the fuselage; version B—two jet engines in the center wing section, near the fuselage; version C—four fan jets on the tail portion of the fuselage. Zones I, II, and III represent the forward, middle, and after portions of the passenger cabin.

Production workers in many branches of industry are constantly exposed to vibration of various spectra and levels. In factories and at working locations in the machine building, textile, construction, and other industries, human operators generally work standing up. Vibrations are transmitted to the body through the floor. Figure 8.10 shows the characteristics of accelerations measured on the floors of a construction material plant and a thermal electric power plant.

When working with mechanized hand tools vibration and impulse effects are transmitted to the body primarily through the arms. Figure 8.11 shows the levels of vibration on the handles of a pneumatic drill and an electric knife. In this case the nature of the effect depends essentially on the operating mode of the tool and the properties of the rock or product being worked, while the degree of the effect is closely related to the working position of the operator which is determined by the angles of the joints, the muscular force expended by the operator, the force of the grip on the tool etc.

Vibrations have also begun to reach into people's homes. Housework in the modern home is largely mechanized and frequently quite reminiscent of production processes. The well-known architect Le Corbusier is justified when he calls the modern apartment a 'living machine'. In the modern apartment we find an ever-increasing number of domestic machines, instruments, and mechanisms generating intensive vibrations and noise.

Furthermore, the tendency in construction is to increase the height and decrease the weight of structures, resulting in the fact that the tall flexible buildings of today oscillate significantly in the wind. In the late 1960s the problem of decreasing the deflections of tall buildings became more pressing for the construction industry than the reduction of stresses in structures [2]. Low-frequency oscillations of buildings resulting from lateral wind pressure

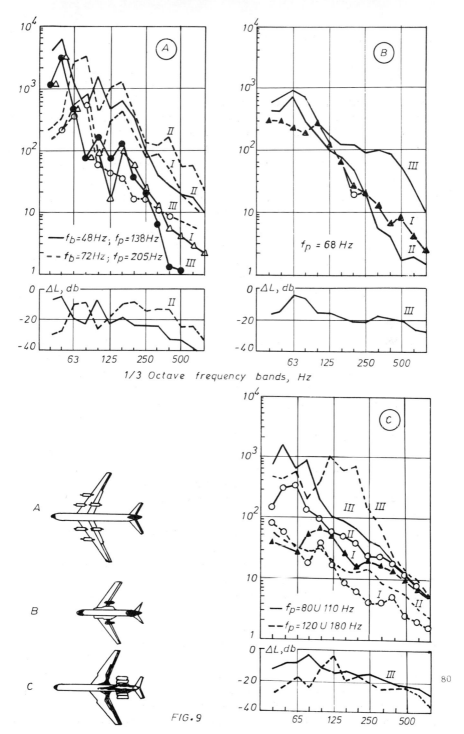

FIG. 8.9. Relative spectra of amplitudes of vibrational motions measured in the floor panels of cabins of three types of aircraft.

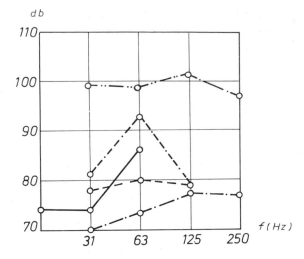

FIG. 8.10. Level of acceleration on the floor of a construction material plant: —— panelling shop (1 m from vibrator); -------- sand grate (2 m from grate); -.-.-.-. sand grate (3 m from grate); —·—·—·— compressor shop; -..-..-..wind test stand.

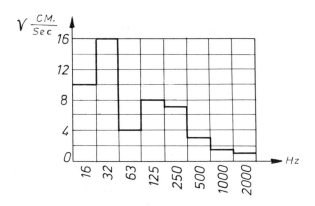

FIG. 8.11. Levels of vibration on the handles of a pneumatic drill and vibrating knife.

have a significant influence on the comfort of the inhabitants of the upper stories. Analysis of the oscillations of a 55-storey building in New York [3] showed that during strong north-east winds the inhabitants of the upper floors could not write so that workers employed in firms on the upper floors were regularly given time off on days with such winds. In this same work the author notes that the inhabitants of the upper.storeys of most skyscrapers in New York experience unpleasant sensations caused by the systematic rocking of the buildings.

Considering the spectral composition of vibrations, working positions, and the paths of propagation of vibrations through the human body, we can classify the types of vibration effect as shown in Table 8.2.

TABLE 8.2

Working position and path of propagation of vibrations	Source of vibration	Nature of vibration	Frequency range (Hz)
(1) Seated position (through seat and control organs)	Motor transport	Random	0–30
	Construction machines	Narrow-band random	0–3
	Rail transport	Random	0–6; 12–20
	Aviation	Wide-band random	20–500
	Water transport	Random	0–30
(2) Standing position (through floor)	Textile industry mechanisms	Polyharmonic and random	5–20
	Machine tool industry mechanisms	Random and periodic impulse	5–20
(3) Standing position (through control organs)	Riveting machines	Polyharmonic and random	20–200
	Electric saw	Polyharmonic	80–200
	Electric nut driver	Polyharmonic	13
	Reinforced concrete moulding	Polyharmonic	40–50
	Pneumatic tramper	Polyharmonic	11–15
	Gasoline-powered saw	Polyharmonic	83–87
	Drill	Random	10–2000
	Pick hammer	Random	10–2000

Influence of vibrations on the human body, on operator efficiency, and in medical applications

The role of man as the operator of technical systems is ever increasing. It is therefore quite important to study the influence of various factors on the efficiency of an operator as a link in the man–machine system. The vibrations of machines, acting on man, can reduce the productivity and quality of labour significantly. The results of many studies [4–8] indicate the unfavourable influence of vibrations on the functions of the visual analyser of a human operator, the inaccuracy of performance of tracking tasks, etc. In ref. 9 it is noted that the Gemini astronauts, when subjected to vibration at a frequency of 50 Hz, could not read the indications of their instruments since the eyeballs vibrate at this frequency and the eyes are literally covered by a film.

We have performed special investigations designed to evaluate the influence of vertical low-frequency vibrations over periods of 2–4 h on the efficiency of human subjects. Analysis of the results of our studies has shown that by the beginning of the second hour of exposure to the vibrations the mean square error of operators performing compensatory tracking of a random signal had increased by 1.5 times in comparison to the initial level of error. Furthermore, periodic changes in the quality of tracking appeared with the passage of time. The slowing of the motor reaction of our human subjects was particularly clearly seen in the resonant mode (vibration frequency 4.75 Hz) 3 h after the beginning of a session.

When exposed to harmonic vibrations with frequencies of 3–8 Hz with a mean square level of acceleration of 1.5 m s^{-2}, the visual acuity of operators remained practically unchanged over a 2 h session; however, when the vibration level was increased to 4.5 m s^{-2} at frequencies of 4.5 and 8 Hz, visual acuity dropped by 17 per cent. This deterioration was observed throughout the entire session which lasted 2 h.

Figure 8.12 shows curves of equal resolving capacity of the eyes under the influence of vibrations. The curve corresponding to unity shows the resolving capacity of the eye at rest; the other curves illustrate the deterioration in resolving capacity of the eye under the influence of vibrations. In Fig. 8.12 we can see three zones of increased sensitivity of the eyes to vibration: around 5 Hz, 14–30 Hz (except for the 22–26 Hz area), and 60–70 Hz (higher frequencies were not studied). At frequencies of 4–14 Hz, 22–26 Hz, and over 30 Hz increased 'interference stability' of vision was observed. We can assume

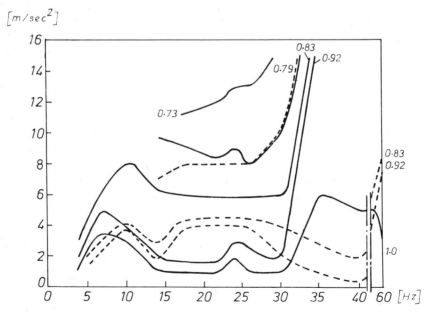

FIG. 8.12. Curves of equal resolving capacity of the eye: solid curves, without additional tie; broken curves, with additional tie.

that for the 22–26 Hz zone this phenomenon is a result of adaptation of the organs of vision to vertical vibrations at the natural oscillating frequency of the head (20–30 Hz). At the end of a vibration session the functional condition of the operator was rapidly restored (in no less than 5 min) to its initial level.

Vibrations in the resonant mode (4–5 Hz) have a greater influence on the functional state of a human subject than vibrations in other modes, which are manifested both as a significant reduction in the speed and accuracy and in a significant deterioration of the functional status after the vibrations are turned off. In addition, vibrations cause a sensation of discomfort, irritation, nausea, and other unpleasant phenomena. When applied briefly, the subjects complain of dysphoria, pain in the stomach and back, headache, general fatigue, difficulty in breathing, itching, deafness, etc.

Under the influence of vibration the mechanical reaction of the human body, manifested as displacement of mobile structures and deformation and bending of parts of the body, can cause the following effects: (1) changes and possible disruption of the normal course of processes both in individual organs of the body and in molecular or cellular structures; (2) pinching of the tissues and blood and lymph vessels; (3) resonance and standing waves in the blood vessels; (4) extension and compression of nerve tissues; (5) heating resulting from friction.

Vibrations also cause (1) dynamic loads on the skeleton, (2) possible damage to the tissues as a result of counteracting forces from supports, and (3) changes in the compliance and pressure of the perivascular tissue and structures in the thorax, particularly those adjacent to a support and a body fixation system ('passive pump' of peripheral blood vessels and lungs). These phenomena may be accompanied by the following main physiological reactions by the body: (1) stimulation of vascular and muscular mechanoreceptors; (2) phase shifts in the central and peripheral intravascular pressure which are capable of changing the filling of the heart and its output as well as the peripheral blood flow; (3) phase influences on the resistance of the vessels due to periodic fluctuations in pressure in the arterioles and veinules resulting in instantaneous reactions (the Beilies effect) and a change in capillary pressure; (4) possible intermittent vascular spasms with subsequent ischemia or stasis.

Vibrations also influence anaphase, i.e. the stage of cell division during which the chromosomes begin to split. Thus in experiments with mice subjected to vibrations characteristic of a rocket engine after one day the percentage of anaphase formations in the spinal column reached 9.79 per cent whereas in a control group of animals it was only 2.61 per cent.

Vibrations cause the earliest and most significant changes in the cardiovascular and nervous–muscular systems. Vibrations acting on the peripheral nerve endings cause changes of various types in their sensitivity. The reactions of the central nervous system are manifested as loss of equilibrium of nervous processes. It has been established that the cortical areas of the brain are quite sensitive to vibrational stimulus. According to data in ref. 9 vibrations cause depression of the alpha rhythm on the electroencephalogram followed by elevation of this rhythm with longer exposure to vibration.

It has been shown in a number of works [10, 11] that vibration at frequencies of 10–70 Hz causes disruption of the static and dynamic co-ordination of motion and has an unfavourable influence on the neuromuscular system. Many investigators [12] have observed changes in the pulse frequency and arterial pressure under the influence of vibration. A dependence of the change in the respiratory function on the amplitude of motion or acceleration of vibrations at frequencies of 4–5 Hz has been noted.

We have performed experimental investigations [13] of the simultaneous influence of vibrations and static load on the muscles. In the experiments we measured excitation in the neuromuscular system by means of electromyograms (EMG), muscular tonus, vibrasensitivity, pulse frequency, electrocardiograms (ECG), and arterial pressure.

The greatest influence of vibration was observed in the condition of the cardiovascular system. A stable reliable increase in the frequency of heart contractions was noted under the influence of random vibrations following 20 min of the experiment, while sinusoidal vibrations with a frequency of 5 Hz caused a change in pulse frequency after 30 min. During the period of the tests we also observed a change in the arterial pressure. The maximum systolic pressure generally dropped and the minimal pressure increased. Vibrations and measured physical load also caused a significant change in the nervous–muscular apparatus. The tone of the muscles of the back and neck increased significantly. The greatest changes, as in the functions of the cardiovascular system, were observed under the influence of random vibrations.

Analysis of the peculiarities of the changes of all the functions that we have studied shows the difference in the nature of the influence of sinusoidal as opposed to random vibrations rather clearly. Sinusoidal vibrations caused changes in peripheral circulation and the nervous–muscular apparatus which were restored within 20 min. After random vibrations we noted an increase in the tone of the peripheral vessels combined with changes in the nervous–muscular apparatus of inhibitory type. The increase in electrical activity of the muscles along with a reduction of skin sensitivity, muscular endurance, and productivity (elongation of chronaxy) indicates complex interactions in the central nervous system in response to the influence of random vibration.

Vibrations cause a change in the morphological composition of the blood which is characterized by a reduction in the number of erythrocytes and the percentage of haemoglobin [14,15]. Many investigators relate these changes in the composition of the blood to changes in the central nervous system.

The changes detected under the influence of vibration in the condition of the endocrine system [16] indicate significant irritation of various elements of the endocrine system by vibration. The long-term influence of vibration can cause stable irreversible changes in metabolic processes in the human body. It should be noted that the general clinical indicators (ECG, electroencephalograms (EEG), frequency of respiration, pulse, blood pressure, etc.) have a comparatively short period of after-effect and rather rapidly return to the normal level following interruption of the vibration. However, with a test subject in good general health with no visible deviations, latent disruptions in the internal medium of the organism may arise, particularly in the metabolism of biologically active substances (acetylcholin, catecholamines, histamine, serotonin, etc.). Those functions of the organism which depend significantly on the humoral mechanisms of regulation, particularly those whose regulation involves the autonomic innervation, are comparatively resistant to vibration. However, if changes do occur they are relatively stable, i.e. the after-effect period is long.

Studies performed [17] indicate changes under the influence of vibration in the activity of the enzyme diaminoxidase (DO) as well as changes in the ratio of histamine to serotonin. Histamine has extremely high activity and a broad spectrum of action. It participates in a number of important physiological processes: it increases the permeability of blood vessel walls, causes contraction of smooth muscle fibres, participates in the regulation of the microcirculation of the blood, and stimulates the secretion of gastric juice. In various functional states of the organism resulting from extreme stimuli the changes in the content of histamine are accompanied by changes in the activity of enzymes participating in its formation and breakdown. An excess content of histamine in the organism results in vegetative disorders in the functioning of its systems and in the development of allergic states. Serotonin is closely related to histamine and noradrenalin, indicating the functional relationship between serotonin and the hypothalamus–hypophysis–adrenal regulatory system.

The results of investigations have shown that when vibrations at frequencies of 2–10 Hz with amplitudes of 0.2–0.8 mm act on the body the dynamics of the histamine:DO and histamine:serotonin ratios reflect a long-term after-effect, while the changes in general clinical indicators are quite transient. For clarity Fig. 8.13 shows the correlation relationships between the level of histamine and the DO activity. Here the enzyme has shown its significance as a 'balancer' reflecting the capability of the regulatory systems of the body for adaptive behaviour.

Researchers have described various forms of vibration sickness which developed under the systematic influence of the vibration stimulus over a period of years. In spite of the tremendous quantity of work dedicated to the study of the influence of vibration on the human body, the pathogenesis and mechanism of development of vibration disease remain insufficiently clear.

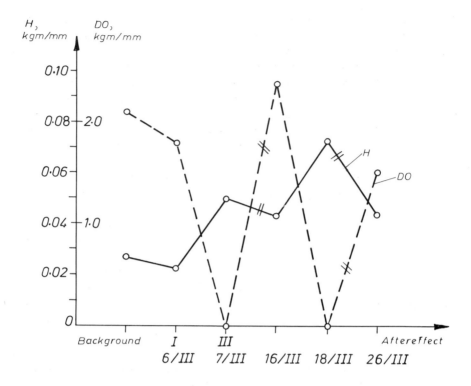

F<small>IG</small>. 8.13.   Content of histamine and DO activity in the blood of persons exposed to vibration.

Complex clinical and hygienic investigations have allowed us to reveal three main forms of vibration pathology: (1) peripheral—from the influence of local vibrations on the arms of the workers; (2) cerebral—from the primary influence of overall vibrations; (3) cerebral–peripheral—from the equal influence of both general and local vibrations.

The clinical symptoms of peripheral pathology vary significantly depending on the structure, power, and frequency characteristics of the vibrating tool. The primary indicators of the disease are attacks of pale cold fingers and paresthesia and pain in the distal segments of the arms at rest and at night. Trophic and sensory disorders of the distal type and hypertrophy of the skeletal muscles and less frequently of the shoulder girdle are noted.

The cerebral form of vibration disease is characterized in its initial stages by general cerebral vascular and cortical–subcortical meso-disencephalic neurodynamic disorders; in later stages organic brain damage develops. With this form of the disease peripheral vegetovascular and sensory disorders are also observed, but they are of secondary significance.

We should also note the fact that vibrations can have a useful influence on the human body. General vertical vibrations have been successfully used for the removal of stones of the ureter since 1965. Horizontal vibrations facilitate the passage of stones from the kidneys. In ref. 18 results are reported of the use of vibration therapy as an effective method of dropping stones in the ureter and in the diagnosis of certain forms of urolithiasis. Studies of the respiration, pulse, arterial pressure, and biopotentials of the ureter have allowed parameters of vibration to be found which combine the absence of any harmful influence on the body with the maximum therapeutic effect. Analysis of the resonant frequencies of the organs in the lower abdomen, as well as concrements of the ureter, have allowed the most effective parameters of sine wave vibrations to be found, i.e. frequency 10–15 Hz and amplitude 2 mm.

A vibration stimulus, strictly measured as to frequency and exposure, has been successfully used to treat the peripheral nervous system, skeletomotor apparatus, non-specific diseases of the lungs, the gastrointestinal tract, and gynaecological and other diseases. The therapeutic nature of moderate doses of audiofrequency mechanical oscillations results from the fact that they act as stimulators to the protective mechanisms developed by the body itself in the process of evolution. As a result of the influence of vibration of low intensity and brief duration a complex of protective and adaptive mechanisms comes into play in the body; however, more intensive longer acting vibration suppresses the protective reaction of the body. One well-known basis for an explanation of the mechanism of action of vibration as a function of the initial condition of the organism is the study of N. Yc. Vvedensky who found that the intensity of parabiosis can be decreased by weak stimuli used in small gradually increasing doses.

Finally, we draw the following conclusions.

(1) Vibration can be looked upon as an environmental stimulus acting on the human body during daily life both at work and at home.

(2) The vibrations are quite complex in nature and varied as to the form of stimulus which must be described using several parameters or groups of parameters.

(3) The selection of any given group of parameters to describe vibrations is rather arbitrary which complicates the investigation of the influence of vibrations on the human body and also leads to difficulties in comparing the results of different investigations.

(4) A tremendous amount of factual material has been accumulated concerning the harmful and useful influence of vibration on human health and man's working capacity. However, no single theory has yet been developed to form a foundation for objective criteria for evaluating the influence of vibration on the human body.

8.2. Biodynamic characteristics of the components of the human body

A systems approach to the evaluation of the dynamic characteristics of the components of a human body subject to vibration

The reactions of the body to vibrations represented by a generalized oscillatory input $u(x, y, z, t)$ can be arbitrarily represented by a simplified block diagram as shown in Fig. 8.14.

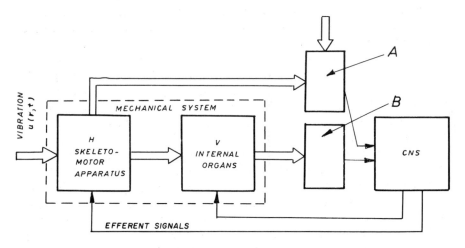

Fig. 8.14.   Block diagram of the human-body response to vibration.

This diagram includes three elements: the mechanical system (including the body's connective tissue, fluid components, skeleto-motor apparatus and internal organs), a group of receptors converting mechanical oscillations to electrical impulses, and the central nervous system regulating the processes in the body. The skeleto-motor apparatus consists of passive and active parts. The passive part consists of the skeleton which, serving as the support for the entire body, also represents a system of levers articulated to each other. The active part of the motor system consists of organs called muscles which connect the skeletal components and, owing to their capability for active contraction, move the levers of the passive system.

The oscillations of the elements of the mechanical system are received by receptors which can react to various stimuli. The receptors are quite varied structurally. They include comparatively simply constructed nerve endings, highly differentiated special formations (sensors), and individual elements of the complex sense organs. Usually, in response to a stimulus, a receptor generates a sequence of electrical impulses (afferent signals) by means of highly complex electrical and chemical processes which, when they reach a certain intensity, pass along the nerve fibres to the cerebral cortex. It has been established that within certain limits the number of afferent impulses is directly proportional to the logarithm of the intensity of the stimulus.

The receptors are divided into exteroceptors, proprioceptors, and interoceptors. The exteroceptors are located on the outer surface of the body and receive stimuli from the environment. The proprioceptors are located in the muscles and joints and perceive the contractions and extensions of the musculature and the positions of the joints, i.e. signal the position and movements of the body. The interoceptors are located in the internal organs and perceive changes in the internal environment and the condition of the visceral sphere of the organs.

Each receptor has its own threshold of stimulation, i.e. the minimum intensity of a stimulus sufficient to cause excitation. Each type of receptor is adapted to its own qualitatively specific stimulus, e.g. sound waves for the ear etc. The specific stimulus for a given receptor, to which it has become adapted in the process of evolution, is called its adequate stimulus. We should note that mechanical oscillations are a particular type of stimulus since they cause activity even in receptors for which they are not adequate. For example, mechanical stimulus of the retina causes the sensation of flashes of light.

In assessing the effects of vibration on man, we can distinguish the following basic levels of analysis of phenomena occurring in the human body: (1) the subcellular level; (2) the cellular level; (3) the tissue level; (4) the level of organs and systems; (5) the level of the entire organism. However, in studying any one of these levels we must always first determine the mechanical characteristics of the biological components which participate in the oscillating process.

In the particular case when the influence of unidirectional vibrations is being analysed and it is considered that the system is linear, the operators defining the dynamic characteristics of the mechanical system are the transfer functions and input mechanical impedance function. These functions have frequency characteristics and are simple in form:

$$H(j\omega) = \frac{\sum_i A_i(j\omega)^i}{\sum_k B_k(j\omega)^k}. \tag{8.9}$$

It should be noted that for a linear system the transfer function is invariant to the selection of kinematic and force parameters of the vibrations. Thus the transfer functions calculated from motion, velocity, acceleration, or force should coincide. If, however, a non-linear system is selected as the object of study we must use different mathematical approaches which lead to more complex relationships and operators.

Furthermore, the receptors themselves, which can be looked upon as sensors converting mechanical energy to electrical signals, have their own dynamic characteristics. Figure 8.15 is a schematic diagram of a mechanoreceptor—a

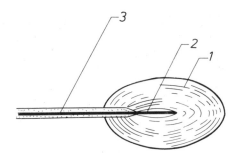

FIG. 8.15.   A schematic diagram of a mechanoreceptor

Pacinian corpuscle. It is bulb shaped, 1 mm long, and 0.5 mm thick. We can draw an analogy between this receptor cell and a microphone. However, the frequency characteristics of the Pacinian corpuscles, in contrast to those of a microphone, have never been completely studied.

Thus the mechanical reaction in a study of the effects of vibration must be looked upon as the primary reaction of the biological system and analysis of the mechanical characteristics should be performed quite carefully.

Model of the human body under the influence of general vibrations

Let us now briefly present the experimentally confirmed characteristics of the mechanical reactions of the human body to vibration.

Below 2 Hz the human body acts as a solid body. In the 2–100 Hz band the mechanical energy propagates through the body in the form of waves, the length of which is significantly greater than the dimensions of the body. Based on this the model should be one of an oscillating system with concentrated parameters and several degrees of freedom. The basic resonant frequencies of vertical oscillation of the human body lie in the 4–6 Hz area in the sitting position and in the areas of 5 and 12–15 Hz in the standing position. At frequencies over 100 Hz the human body acts as a more complex system with distributed parameters; the mechanical energy may propagate in the form of shear waves, surface waves or compressive waves. The type of waves propagating through the body depends to a significant extent on the frequency and conditions of transmission of vibrations.

As the frequency of the exciting oscillations increases, even smaller parts of the body become involved in the oscillating process and the zone of mechanical effect of the vibration is ever more localized.

The mechanical properties of the human body depend on the direction of the vibrations applied. When vibrations propagate in the transverse direction the physical reaction differs significantly from the reaction to vertical vibrations. The basic resonant frequencies of the human body subjected to horizontal vibrations lie in the 1–3 Hz frequency band.

The non-linearity of the elastic and damping properties of the human body is manifested in the form of the dependence of mechanical reactions of the human body on the amplitude of external vibrations. The dynamic characteristics of the human body also change as functions of the position of the body. Further, when the vibrations are applied over a long period of time smooth changes occur in the dynamic characteristcs of the human body, indicating their instability. Constant activity of the muscles can significantly influence the measured dynamic characteristics.

During the last 15 years the human body has been modelled as an elastic system in a number of ways [19–25]. One of the basic structural elements of the human body is the curved spinal column. The spine consists of segments called vertebrae. Although they share a common type of structure the vertebrae have significantly different shapes in the various areas of the spine. The combination of the solid bone vertebrae connected by elastic discs and bound by a system of fibrous ligaments forms a strong and flexible column allowing significant movement due to the intervertebral joints. The results of experimental studies [26] performed on cadavers in order to determine the dynamic characteristics of the spinal column are shown in Fig. 8.16.

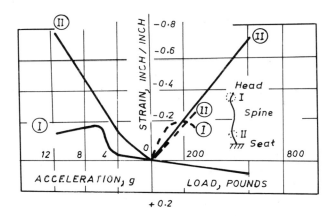

Fig. 8.16.   Dynamic characteristics of a spinal column under loads applied to the head (broken curve) and shoulder (solid curve).

The mechanical characteristics of individual vertebrae and intervertebral discs have also been studied [27]. Figure 8.17 shows a stress–strain graph for the first lumbar intervertebral disc together with the body of the twelfth thoracic and first lumbar vertebrae. In experiments involving extension it has been found that a load of 198–248 kgf is sufficient to burst a disc. The ruptures were generally found to take place at the point of connection of a disc with a vertebra.

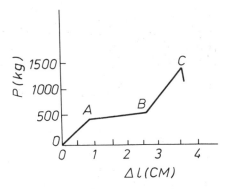

FIG. 8.17. Compression load versus deflection for the first lumbar intervertebral disc together with the body of the twelfth thoracic and first lumbar vertebrae.

For the spine and the supported structures it is convenient to utilize an elastic rod with masses attached to it as shown in Fig. 8.18. Figure 6.19 shows the positions of the human body and the corresponding curvatures of the lumbar portion of the spine. Thus for a man seated on a chair in three different positions as shown in Fig. 8.20 we obtained the amplitude–frequency characteristics shown in Fig. 8.21. Then the analytic approximation of the experimental frequency characteristics yielded the expressions for the transfer functions (for the human body) between accelerations measured the head of the subject and at the seat of the chair, expressed as a function of frequency. Matching of the analytical solutions [28] of the three-, two-, and one-mass mechanical models (Fig. 8.21) with the experimental data also enabled determination of the model parameters.

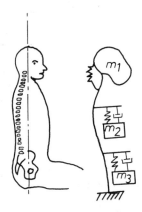

FIG. 8.18. Schematic diagram of the spinal column and the associated simplistic human body—vibrational model.

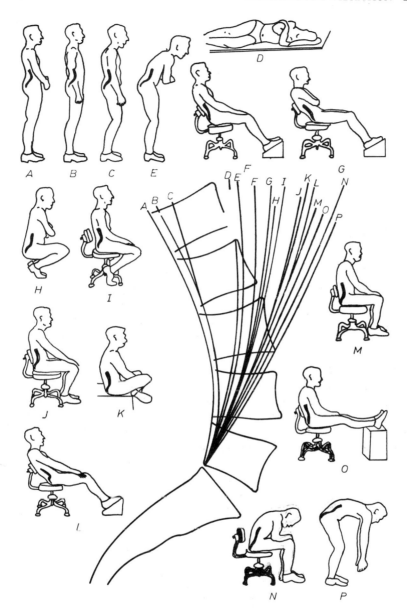

FIG. 8.19. Various aspects of lumbar curvature as a function of position. The figures were obtained from X-rays taken of test subjects. The normal position (position B) corresponds to the position lying on one side; the angles formed by the body and femur and the femur and tibia are about 135°. Note the significant variations in the lumbosacral lordosis from position A to position P.

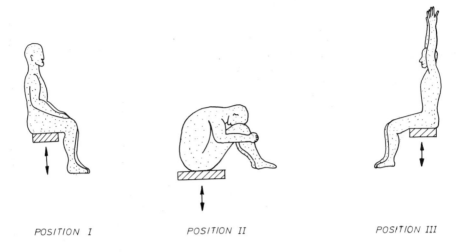

POSITION I   POSITION II   POSITION III

FIG. 8.20.   Seated man in various postures subject to vibrations through the seat.

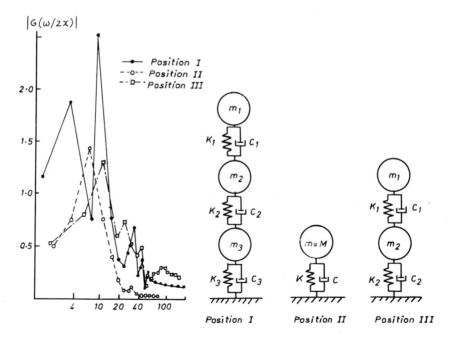

FIG. 8.21.   Amplitude–frequency characteristics and biodynamic models of the three seated positions shown in Fig. 6.20: solid curve, position I; broken curve, position II; chain curve, position III.

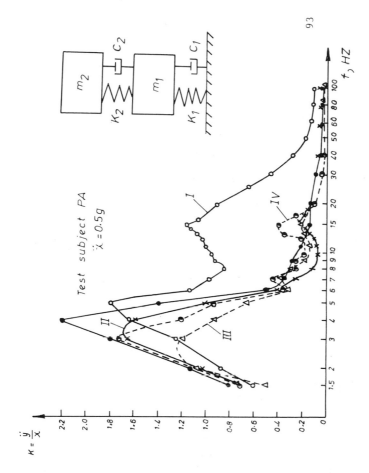

Fig. 8.22. Acceleration amplitude ratio (ratio of head acceleration to that applied at the seat) versus frequency for different orientations of the test subject.

Often a human operator must work standing up while being exposed to vibration. In ref. 29 the dynamic characteristics of the body in the standing position with various angles of the knee joint are presented. Figure 8.22 shows the amplitude–frequency dependences for this case and for the corresponding model. Figures 8.23 and 8.24 indicate how the parameters of the model change as the bending angle is changed. The values of the parameters expressed in

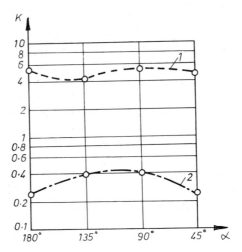

FIG. 8.23.   Variations of the model parameter with the bending angle.

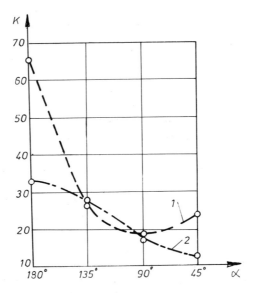

FIG. 8.24.   Variations of the model parameter with the bending angle.

terms of the orientation angle (defined in Fig. 8.22), are as follows:

$$k_1 = 16\alpha^2 - 44.5\alpha + 49.4$$
$$k_2 = 5.4\alpha^3 + 31.8\alpha^2 - 46.6\alpha + 33.4 \qquad (8.10)$$
$$c_1 = 0.945\alpha^3 - 5\alpha^2 - 7.65\alpha + 1.8$$
$$c_2 = 0.0735\alpha^2 + 0.268\alpha + 0.16.$$

Models of the human arm under the influence of local vibration

Studies of the oscillating properties of the human arm developed intensively in the middle of the twentieth century owing to the widespread use in industry of mechanized (vibration-inducing) hand tools. These initial studies [30, 31] investigated the elastic and damping parameters of the human arm and the propagation of oscillations along the arm. In ref. 30 it was established that when vibrations act through the hand the amplitude of oscillations of the elbow joint is greater than the amplitude of oscillations of the hand. In later studies [32–34] of the human arm subjected to vibration the concept of mechanical impedance came to be widely used.

The next stage in the study of the arm as a mechanical system was the development of the Dieckmann model in the form of an oscillating system with two degrees of freedom [33]. The input impedance of this model has a maximum at frequencies below 5 Hz and in the 35–40 Hz area, the effective mass of the arm is approximately 1 kg, and the compliance of the palmar tissue is estimated as $2 \times 10^8 \, \text{cm dyn}^{-1}$.

Investigations [23] have shown that the amplitude–frequency characteristics of the human arm change significantly with various degrees of muscular stress. An increase in stress results in an increase in the level of vibrations directly measured at various points on the human arm. The resonant frequencies of the system also increase which is explained by the increase in equivalent rigidity of the entire system.

Vasil'yev [34] developed an experimental method of investigating the dependence of the input mechanical impedance of the system on the stress in the muscles of the arm and on the working position. First the frequency characteristics of the free oscillating system of an electrodynamic vibrator (with the handle of a pick hammer attached to it) were estimated. The test subject then imitated various working modes, grasping the handle with his arm straight or bent and in various working positions. Comparison of the frequency characteristics of the free vibrator and the vibrator held by the arms of the test subject allowed the oscillating properties of the arms to be calculated.

Figure 8.25 shows typical frequency characteristics of a free vibrator and a vibrator–arm system. As we can see from the figure, the nature of the curves differs for straight and bent arms. The characteristic of the straight arm has one maximum which consequently can be modelled by a system with one

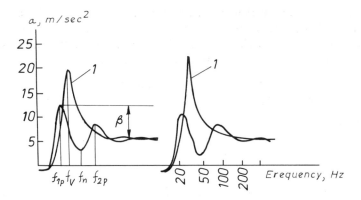

FIG. 8.25.   Acceleration amplitude versus frequency curves for the vibrator–arm system (for arms straight and bent).

degree of freedom (Fig. 8.26). The values of parameters k_p and R_p of this mechanical model can be determined by the formulas

$$k_p = 4\pi^2 m_v f_v^2 \frac{\beta^2 - 1}{\beta^2} - k_v \tag{8.11}$$

$$R_p = 4\pi m_v f_v \frac{[(\beta^2 - 1)\{\beta - (\beta^2 - 1)^{1/2}\}]^{1/2}}{2\beta^3} R_v$$

where m_v, k_v, and R_v are the mass, rigidity, and damping factor of the moving portion of the vibrator, f_v is the resonant frequency of the vibrator–straight arm system, and β is the ratio of amplitudes of acceleration of the moving portion of the vibrator–straight arm system at resonance to the amplitude at high frequencies.

The input mechanical impedance of the straight arm is determined by the formula

$$Z_{sa} = R_p + j\frac{k_p}{\omega}. \tag{8.12}$$

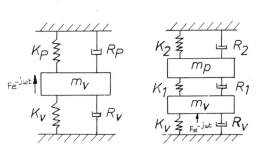

FIG. 8.26.   Model of systems with one and two degrees of freedom for the vibrational characteristics of straight and bent arms respectively.

The bent arm was studied, with the elbow joint angles of $22°$, $45°$, and $60°$. The frequency dependences for the bent arm had a form characteristic of an oscillating system with two degrees of freedom (Fig. 8.26). On solving the equation of motion of the vibrator–bent arm system for

$$m_v \ddot{u}_1 = R_v(\ddot{u}_1 - \ddot{u}_2) + k_v u_1 = Fe^{-j\omega t} \tag{8.13}$$
$$m_p \ddot{u}_2 + R_v \dot{u}_2 + k_2 u_2 + R_1(\dot{u}_2 - \dot{u}_1) + k_1(u_2 - u_1) = 0 \tag{8.14}$$

we obtain

$$k_1 = 4\pi^2 f_{2p}^2 m_v \frac{\beta^2 - 1}{\beta^2} - k_v$$

$$k_2 = 4\pi^2 f_{1p}^2 (m_v + m_p) \left\{ 1 - \frac{m_v^2}{\beta_1^2 - (m_v + m_p)^2} \right\} - k_v$$

$$R_1 = 4\pi f_{2p} m_v \left[\frac{(\beta^2 - 1)\{\beta - (\beta^2 - 1)^{1/2}\}}{2\beta^3} \right]^{1/2} - R_v$$

$$R_2 = 4\pi f_{1p}(m_v + m_v) \left[\frac{(\beta_1^2 - 1)\{\beta_1 - (\beta_1^2 - 1)^{1/2}\}}{2\beta_1^3} \right]^{1/2} - R_v$$

$$m_p = \frac{k_1}{4\pi^2 f_n^2 \left\{ 1 - \dfrac{f_{1p}^2 f_{2p}^2 - f_n^2 f_2^2}{f_p^2(f_{1p}^2 + f_{2p}^2 - f_n^2 - f_p^2)} \right\}^{-1}} \tag{8.15}$$

where f_{1p}, f_{2p}, f_n, and f_v are the co-ordinates of the maxima and minima on the frequency characteristics (Fig. 8.25), and β_1 is the ratio of the amplitude of acceleration at the resonant frequency f_{1p} to the amplitude of acceleration at high frequencies where

$$\beta_1 = \beta \frac{m_v + m_p}{m_v}. \tag{8.16}$$

Thus the input mechanical impedance of the bent arm is given by the formula

$$z_{ba} = \frac{(R_1 + jk_1/\omega)\{R_2 + j(k_2/\omega - \omega m_p)\}}{(R_1 + R_2) + j\{(k_1 + k_2)/\omega - \omega m_p\}}. \tag{8.17}$$

To allow the results to be used in practice, the degrees of stress of the muscles of the arm were distinguished as follows: weak, 0–200 N; moderate, 200–500 N; strong, 500 N or more. The mean values of the components of the input impedance of the arm are presented in Table 8.3.

Regulation of the dynamic characteristics of the human body under the long-term influence of a vibration stimulus

When vibrations are applied over an extended time, a significant change occurs in the dynamic properties of the human body. The changes are so significant

TABLE 8.3

Mean values of components of input impedance of the arm

Muscular Stress	Mechanical parameters				
	$10^5\,\mathrm{N\,m^{-1}}$	$10^4\,\mathrm{N\,m^{-1}}$		$\mathrm{N\,s\,m^{-1}}$	$\mathrm{N\,s\,m^{-1}}$
	Angle of elbow joint 0° (straight arm)				
Weak		26		140	
Moderate		7		240	
Strong		20		350	
	Angle of elbow joint 20°				
Weak	1	2.8	2.1	350	300
Moderate	2	3.5	3.1	430	400
Strong	3.1	4.1	3.4	630	540
	Angle of elbow joint 45°				
Weak	1.9	2.7	3.1	600	530
Moderate	2.7	3.3	3.6	660	560
Strong	3.9	3.8	4.0	710	670
	Angle of elbow joint 60°				
Weak	2.6	3	3.4	630	450
Moderate	3.2	3.4	3.6	690	450
Strong	4.1	5.2	4.1	820	520

that to ignore this fact may lead to basic errors, for example in the development of means for vibration protection.

In our studies [35–37] test subjects were subjected to monoharmonic vibrations during sessions lasting from 2 to 4 h. The amplitude–frequency characteristics of the body were measured before and after the sessions. The intensity and frequency of vibrations were constant during each session but changed from session to session. A frequency range of 2–7 Hz was covered, and the following conclusions were drawn.

(1) The amplitude of vibrations of the human body increases during the course of a vibration session, while the amplitude of vibration accelerations decreases.

(2) The rate of change of the amplitude of oscillations of the human body depends on the frequency of the vibration stimulus. The maximum absolute value of acceleration is reached at the frequency of the human body.

The phenomenon noted in point 1 can be explained from the standpoint of the theory of oscillations if we consider that the parameters of the ligaments of the skeleto-muscular system are non-stationary, while the non-linearity of the elastic ligaments is 'soft' in nature. During the long-term application of vibrations, the elastic ligaments become 'softer'. According to the equation of motion for the system, which can be written as

$$m\ddot{\Delta} + \delta\dot{\Delta} + c\Delta - \gamma\Delta^3 = -m\ddot{u} \tag{8.18}$$

where the Δ are the relative vibration motions, the phenomenon described in point (1) can be expressed mathematically by the condition

$$\min_{R} \max_{t} ||R(\Delta,\dot{\Delta})|| = I_0 \tag{8.19}$$

i.e. the maximum forces in the mechanical couplings of the system must be minimal.

With respect to point (2) we note, in Fig. 8.27, the dependence of the change in the relative resonant frequency ω/ω_0 of the human body under the influence of vibrations of frequency p. The left branch of the graph, in the shaded area, corresponds to the trans-resonant mode where $p > \omega_0$, while the right branch corresponds to the sub-resonant mode, i.e. $p < \omega_0$.

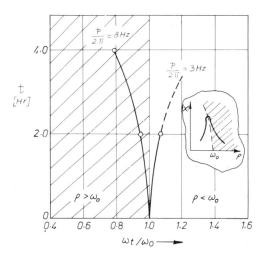

FIG. 8.27.   Dependence of the change in the relative resonant frequency ω/ω_0 of the human body under the influence of vibrations of frequency p. The left and right branches of the graph correspond to the trans-resonant and sub-resonant modes respectively.

It is noted that the natural frequency of the human body under long-term exposure to vibration changes quite independently of any effort of the subject in such a way as to 'detune' as far away from the frequency of the vibration stimulus as possible. This can be described mathematically by the simple inequality

$$\frac{d}{dt}\{|\omega_t(t) - p|\} > 0. \tag{8.20}$$

This behaviour of the biological system is quite 'intelligent' from the standpoint of mechanics. We know, in agreement with the results of the theory of oscillations, that in developing a vibration protection system the designer attempts to place the natural frequency of the protected object, installed on shock absorbers, as far as possible from the frequency of the vibration stimulus.

The results described above show that certain factual data on the dynamic properties of the system studied cannot be explained by representing the human body as a passive mechanical system. In accordance with these ideas we can supplement the passive mechanical models of the human body and represent it by a system such as that shown in Fig. 8.28. Here the vector V characterizes the effect of regulation which is performed by the biological system; each component corresponds to a specific regulation mechanism and the next resultant effect produces certain changes in the dynamic properties of the mechanical system. Thus the behaviour of a biomechanical system under long-term vibration is organized so as to minimize the influence of the unfavourable factor in the environment by for example a local change of position. This hypothesis is also confirmed by the results of ref. 40, which revealed search activity by the muscles.

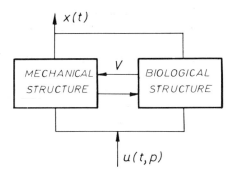

FIG. 8.28.   Conceptual modelling of the human body response to long-term vibration.

8.3. Optimized design of a low vibration ambulance

We now apply these ideas to a particular example, that of an ambulance and attendants transporting a patient along a rough road. The reason for the development of a new concept for rescue vehicles (or ambulances) is the high vibration stress of patient and attendants in the vehicles used at present. In rescue service, vehicles are used in which the patient is placed, attended to, and transported on a stretcher which is fixed to a stretcher-frame which is adjustable in height. Since no specially manufactured rescue vehicles are as yet available, small mass-produced trucks of adequate size and a gross vehicle weight of 30 000–40 000 N are generally modified for this use. Such adapted ambulances are small in size and therefore do not permit appropriate treatment of the patient.

Undue vibrations arise owing to inadequate design of the suspension system and to the poor coupling between the truck suspension, the stretcher frame, and the stretcher. Conventional stretcher-frames are susceptible to vibration and provide poor damping. The stretcher itself forms an oscillating system owing to the flexibility of its beams and the covering material. The natural frequencies of these components are sometimes almost in resonance with the vehicle's natural body frequency (which is shifted from its normal values of 3–1.5 Hz to higher values of 3–4 Hz because of the high internal friction of the leaf springs). As a result the patient is not isolated from the vibrations of the bodywork and in fact these vibrations are even amplified. This effect occurs precisely in the 3–6 Hz frequency range where human beings are considered to be extremely sensitive.

Human sensitivity to vibration depends on numerous factors, and considerable research on the classification of vibration exposure of human beings in the lying position still needs to be done. However, in the Federal Republic of Germany there exists, as a provisional criterion, the intensity of perception K or the K value (defined in Appendix 8.1) which is defined in the *VDI-Richlinie 2057* [41] as being an objectively measurable factor for the evaluation of the vibrations perceptible by the human being in lying or sitting position.

Figure 8.29 shows the vibration characteristics, as a function of the frequency, in the form of curves of constant intensity of perception as described in the relevant norms. These curves of constant intensity of perception were determined by means of vibration tests with human beings.

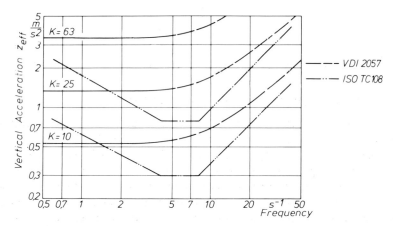

FIG. 8.29. Curves of constant intensity perception.

For this purpose the test persons were exposed to defined vertical vibration accelerations at a specific frequency and then asked to find out the same intensity as perceived by this reference vibration at other frequencies. Thus the dependence of human perception on the vibration frequency was determined.

The current requirements for the suspension of ambulances dictate that 'a suspension system should be achieved wherein the mechanical vibrations do not exceed the range of K values between 10 and 25 on a seat or a litter being loaded with 75 kg'.

Suspension characteristics of presently used ambulances

As a typical example of ambulances, a vehicle (vehicle A) of 34 900 N gross vehicle weight with the live rear axle suspended on leaf springs and with dual tyres was chosen. Vehicle B has a gross vehicle weight of 26 000 N and a lighter live rear axle on parabolic springs and single tyres. For comparative tests a conventional ambulance based on a vehicle of the pick-up truck category (D) and a middle-class passenger car (E) were also included.

The tests were conducted at different driving speeds in the range 20–70 km h^{-1}, and the K value of the patient was established by measuring the vertical acceleration below the centre of gravity of the human body. The location of the measuring point for the K value of the attending personnel was on the vehicle floor.

The results of these comparative measures for the patient and attendant are presented in Fig. 8.30. It is seen that sufficient driving comfort is achieved in a conventional ambulance (vehicle D) on roadway 1 (of average surface quality);

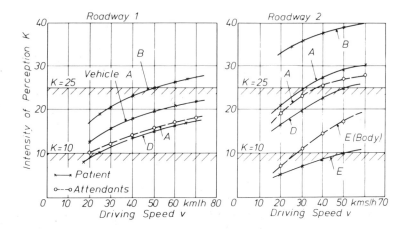

F$_{IG}$. 8.30.   Vibration exposure of the patient (solid curve) and attendant (broken curve) in mass-produced ambulances on normal roads (roadway 1) and on poorly surfaced roads (roadway 2).

however, on roadway 2 the limit of vibration exposure $K = 25$, fixed as the norm, is exceeded at speeds above $50\,\mathrm{km\,h^{-1}}$. The type B rescue vehicle exceeds this limiting value even on roadway 1 at a speed of $50\,\mathrm{km\,h^{-1}}$, whereas vehicle A still satisfies the norm. On roadway 2, however, neither vehicle can fulfill the norm even at low speeds, thus pointing the need for a new optimized ambulance suspension system design.

Analogue computer simulation system for design optimization

For the optimization of ambulance design parameters, a mechanical–electrical analogue of the vehicle suspension system is required. In order to afford a measure for comparison of the analogue simulation system Fig. 8.31 shows the vibration system of stretcher frame, stretcher, and man. Figure 8.32 shows the entire vehicle on a test bench where both harmonic and random activation (or forcing) functions were given to the vehicle to simulate road and surface unevenness of actual test stretches. Frequency analyses were carried out by monitoring the vibration acceleration amplitudes as functions of the frequency. The vibration stresses of the patient and attendant were also established.

The mechanical simulation of the ambulance system, employed for optimization on the analogue computer, is shown in Fig. 8.33. The differential equations describing this mechanical system are represented in Appendix 6.2 and presented on the electrical analogue circuit (of Fig. 8.34). Bench tests (described above) carried out in parallel served as proof of the conformity of the simulation system and the actual vehicle with regard to their vibrational behaviour. The simulation system consists of the following components: road

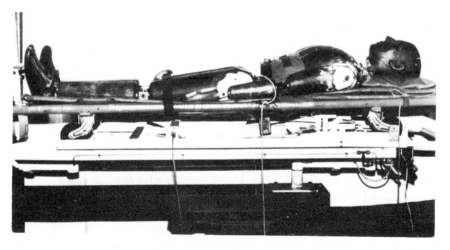

FIG. 8.31. Stretcher frame and stretcher dummy on a vibration test bench. The tests were carried out with a harmonic sine activator.

FIG. 8.32.   Vehicle study on a servohydraulic vibration test bench.

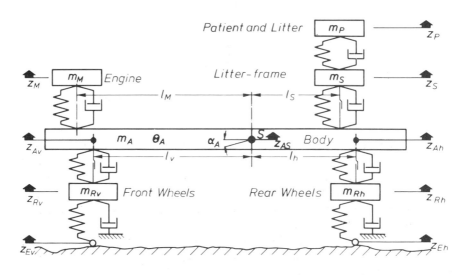

FIG. 8.33.   Mechanical simulation system for the ambulance vehicle.

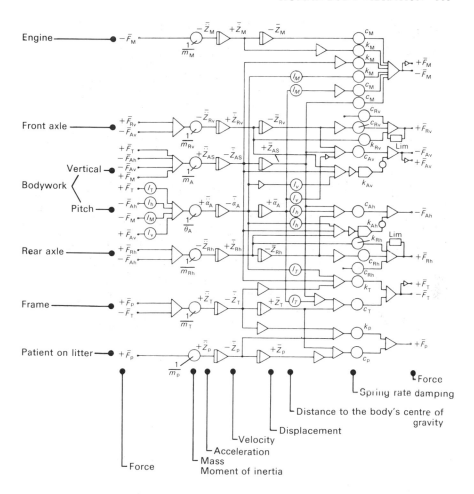

FIG. 8.34. Analogue circuit of the ambulance's mechanical simulation system.

surface, vehicle, stretcher frame, stretcher with patient, and the vibration evaluation for the sitting (attendant) and lying (patient) man. The profile of surface unevenness of a roadway was determined by measurement of the vehicle acceleration of a special wheel scanning the road surface. The double integration of this signal was employed to activate both axles of the vehicle's mechanical simulation system.

The analogue computer simulation system has the advantage of being amenable to having its parameters changed and optimized for testing and checking in a simple and efficient fashion. Figure 8.35 provides a comparison of the vibration stress of the driving test with that achieved on the computer to

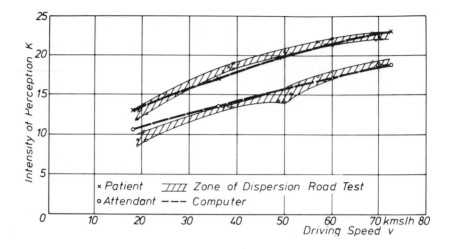

Fig. 8.35. Comparison of *K* values for vehicle A obtained in an actual driving test with those obtained on the computer: x patient; o attendant; broken curve, computer simulation; shaded area, zone of dispersion road test.

confirm the good matching obtained between the responses of the actual vehicle and the analogue vehicle-simulation system.

Development of an improved design based on the optimization results

A low bodywork (or bodyframe) natural frequency is an important influence on the vibration stress of the patient and attending personnel. The realization of a low natural frequency is not possible with the structures of the presently used rescue vehicle.

In order to achieve the wide spring track required for an adequate roll stiffness, an independent front suspension is required, which by virtue of its smaller mass leads to a reduction of the dynamic wheel loads (and hence improved active safety of the vehicle and driving stability). On the rear axle space must be gained for the wider spring track, at least by means of single tyres. This can be realized by an alteration of the axle load distribution by moving the rear axle further to the rear of the vehicle. It should thus be possible to place the patient in the space between the axles, which is ideal from the comfort viewpoint.

A reduction in the level of the bodywork centre of gravity (at unchanged ground clearance) requires the elimination of the drive shaft, i.e. a change to a front-wheel drive. As a result, the height of entrance to the vehicle can be reduced considerably, which simplifies moving the patient in and out of the ambulance. Also, a lighter rear axle can be realized.

In order to maintain a constant natural frequency of the bodywork for various equipment and different loadings a gas spring (as commonly used in

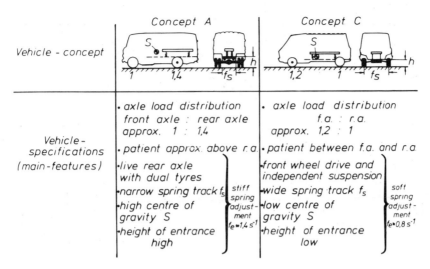

Vehicle - concept	Concept A	Concept C

Vehicle - specifications (main-features)	• axle load distribution front axle : rear axle approx. 1 : 1,4 • patient approx. above r.a. •live rear axle with dual tyres •narrow spring track f_s •high centre of gravity S •height of entrance high	stiff spring adjust- ment $f_e = 1,4\,s^{-1}$	• axle load distribution f.a. : r.a. approx. 1,2 : 1 • patient between f.a. and r.a. •front wheel drive and independent suspension •wide spring track f_s •low centre of gravity S •height of entrance low	soft spring adjust- ment $f_e = 0,8\,s^{-1}$

FIG. 8.36. Comparison of different ambulance concepts.

passenger cars, for example) can be incorporated in parallel with the steel spring.

In Fig. 8.36 the newly developed ambulance concept (C) is compared, with respect to its most important characteristics, to the current concept (A, B).

Resulting solution and comparison with the present design

Figure 8.37 shows the vibration intensity variations for the new optimized ambulance concept in comparison with the presently used ambulance (or

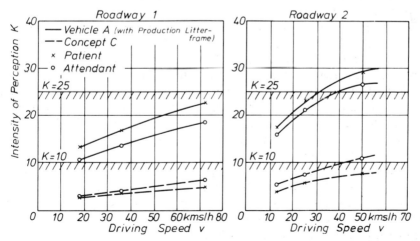

FIG. 8.37. Comparison of K values for vehicle A with a serial stretcher frame (solid curve) and the optimized concept C (Fig. 6.34) (broken curve): x patient; o attendant.

rescue vehicle). This comparison demonstrates that the vibration stress of patient and attendants is reduced considerably by the new concept. On the present ambulance with a serial stretcher frame, K values of 19 and 16 were measured for the patient and the attending personnel respectively on a fairly normal test stretch 1 at a driving speed of $15 \, \text{km h}^{-1}$. On the other hand, in the ambulance designed according to the new concept (delineated earlier), K values of 4 and 5 are measured for the driver and attendants respectively. On the rough test stretch 2, the reduction in vibration stress is also considerable with the new design. For instance at a driving speed of $50 \, \text{km h}^{-1}$, values of $K = 8$ for the patient and $K = 11$ for the attending personnel were achieved with the new concept vehicle compared with their values of $K = 29$ and 27, respectively for the presently used ambulance. Thus the new ambulance concept adequately meets the earlier specified criteria for patient comfort and operator convenience.

General conclusions

A lumped-parameter representation and a parametric study of the system on an analogue computer indicate that an improvement in the driving comfort for the patient can be achieved by conventional standard vehicles, provided that one makes optimal use of low-friction body springs and a well-tuned stretcher frame. However, vibrational stress to both the patient and the attending personnel can be further reduced by the use of a new non-standard vehicle design characterized by a front-wheel drive, an independent suspension, a wide wheel base, and a vehicle possessing a low centre of gravity.

8.4. Cardiovascular response to whole-body oscillation and its therapeutic implications

Considerable efforts have been made since 1940 to understand the biodynamic response to oscillatory forces generated during tractor driving and high-speed air flight [42–45]. Therefore much of the information was used to better the seat design of high-speed vehicles in order to minimize the transmitted oscillatory forces to the human body. However, under severe driving conditions and extreme military and aerospace environment, high-level oscillatory forces are still transmitted through the musculo-skeletal system and finally to the cardiovascular system (CV). In the 1960s, studies [46–48] were initiated to include the response of the CV system of animals and man to the stress of vibration or oscillation. These early studies, as well as most of the more recent ones, viewed whole-body oscillation (WBO) as an environmental hazard.

In 1969, researchers began to investigate the use of WBO for its possible beneficial effects such as a non-invasive technique for cardiac assistance [49], exercise substitutes [50], a countermeasure for CV deconditioning resulting from long-term immobilization (e.g. paraplegic) or space flight [50], and as a passive CV stresser (for subjects who are not capable of performing conventional upright exercise, such as the treadmill test, etc.) to evaluate the performance of the CV system [51].

In this chapter, the terms vibration and oscillation are used interchangeably. Efforts to evaluate the potential of WBO as a therapeutic and/or diagnostic tool require understanding the integrated nature of the physiological and biomechanical responses to the oscillatory forcing function. Most of the previous studies [52–54] included little information concerning quantification of the oscillatory forcing function to the CV system. As a result CV responses were correlated with vibration parameters of little or no significance. Only recently, Bhattacharya *et al.* [55] made a concerted effort to define and measure oscillation or vibration parameters which best describe oscillation-induced CV responses in awake animals. Results from their study will be presented here to provide the framework in which the potential of WBO as a therapeutic tool can be discussed.

Results of animal experiments with WBO

An important aspect of any study in which perturbation is introduced into a system is the identification and quantification of the forcing function. In the study by Bhattacharya *et al.* [55] the forcing function to the CV system was WBO. In addition, an effective vibration study should also include (1) instrumentation insensitive to artefact but capable of maintaining accurate physiological measurements and (2) an animal preparation capable of tolerating the required instrumentation and still capable of providing realistic responses, e.g. the awake chronically instrumented animal. The research effort reported by Bhattacharya [50, 55] was designed to include all of the above-mentioned requirements.

In their study the response of the CV system to vibration, as measured by the mean aortic pressure flow (MAF), which is defined as cardiac output minus coronary flow, were studied in six chronically instrumented awake canines. The chronically implanted preparation was used, since it provided the most normal subject (awake with an intact physiological system) as well as the opportunity for comparisons of repeated tests on the same subject. This technique also provided enhanced transducer stability, which is especially important for the oscillatory environment. In this study, canines were restrained with spines vertical and exposed to non-synchronous (not synchronized with the cardiac cycle) WBO for a constant acceleration amplitude of $1\,g$ (frequency range 2–12 Hz; oscillating table displacement waveform, sinusoidal; vibrated along g_z axis; heart non-paced). The major finding was the evidence of a linear relationship between the changes in MAF (from no vibration case) and the peak net transmitted force (from the

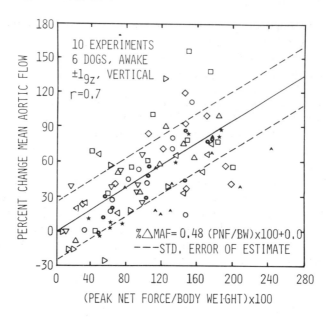

F<small>IG</small>. 8.38.   Percentage change in MAF versus (peak force/body weight).

oscillating table to the posterior of the subject) normalized by body weight (see Fig. 8.38).

In another study Bhattacharya [50] and McCutcheon *et al.* [56] subjected chronically instrumented anaesthetized canines to vibration exposure similar to those of the awake canine. A comparison between overall responses of awake and anaesthetized animals indicated that the anaesthetized group response was similar to that of awake canines but 68 per cent lower. In both the awake and anaesthetized cases, however, the pre-exposure state of stroke volume SV and mean heart rate MHR were important determinants of the response mechanisms. Variation in the control state (no vibration case) associated with the effects of the morphine, the chloralose–urethane anaesthetic (for the anaesthetized canine study), or the level of excitement in the awake case played an important role in determining the subject's response. In animals with elevated control MHR (usually above 150 beats min^{-1}) increases in MAF were produced as a result of SV increases. Animals with low initial MHR (usually less than 150 beats min^{-1}) raised MAF through heart rate increase only.

While the stress level as measured by peak net transmitted force (PNF) was of major importance, it was not the exclusive biomechanical forcing function for the relationship between the percentage change in MAF and the logarithm of the ratio of mean heart rate f_h to the vibration frequency f_t also demonstrated a linear relationship. In the anaesthetized study the percentage changes in MAF and log (f_h/f_t) also showed a linear relationship. Based upon the above discussion, vibration-induced CV (e.g. MAF) responses can, in general, best be described by the following functional equation:

$$Q = Q(F, f_T, f_R) \qquad (8.21)$$

where Q is the percentage change in MAF with vibration from the no vibration case, $F = (PNF/\text{body weight})(BW) \times 100$ is the dimensionless peak net transmitted force, $f_T = f_t/f_h$ is the dimensionless frequency and $f_R = f_t/f_r$ is the dimensionless body resonant frequency where f_r is the whole-body resonant frequency.

The term f_R highlights the importance of non-linear effects associated with oscillation frequencies near the whole-body resonant frequency. The existence of proportionality between the percentage MAF and log $(1/f_t)$ in both awake and anaesthetized studies suggested that the time relationship between cardiac and oscillatory cycles influenced the physiological responses. Such results indicate interesting fluid-mechanical implications of a CV system subjected to oscillatory forces. Consider for example a longitudinal tube with a pulsatile source filled with fluid and exposed to externally applied oscillation. Obviously, when these two oscillations (one from the pulsating source and the other due to externally induced oscillations) have different frequencies, a beat pattern is introduced in the pressure and flow waveforms of the fluid. Evidence of such beat patterns in the pressure and flow waveforms were seen in both animal [46] and analogue models [57] when subjected to WBO. The frequency of the beat pattern and the characteristics of the pressure and flow waveforms within the envelope were studied in detail by Knapp [57]. In his studies, results from an investigation of the responses of the CV system to heart synchronous (synchronous with the heart beat) and non-synchronous vibration using an analogue computer model of the hydraulic aspect of the CV system were compared with results obtained using chronically instrumented animals exposed to the same vibration protocol.

Knapp [57] also showed that, when vibration frequency was a multiple of the heart rate and the two frequencies were synchronized (beat by beat), the beat pattern disappeared (Fig. 8.39). These results suggest that non-synchronous oscillation could be used to survey the CV responses for all of the possible time relationships between the heart and oscillation cycles. Therefore once a certain CV response is chosen from such a survey it can be maintained by the heart synchronous technique as described by Bhattacharya et al. [58]. This capability of heart synchronous WBO to 'dial in' the desired CV responses indicates the potential of the technique as a therapeutic tool.

Therapeutic application of WBO: discussion

From the available data in the literature subjects exercising are usually confronted with peak acceleration levels as high as 20 g [59, 60]. The

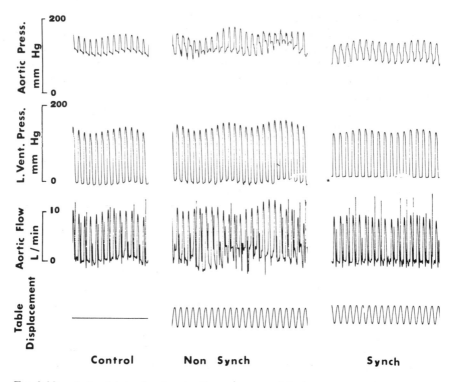

F IG. 8.39.   Animal data showing the disappearance of the beat pattern when the heart
and vibration cycles are synchronized. The asterisk indicates a recorder artefact.

predominant frequency components of the force transmitted during physical
activities such as running and jumping are about 1–3 Hz which is close to the
resonant frequency of the whole body. Since the transmitted forces measured
during WBO exposure are not unlike those measured on subjects exercising
[59–63], it is worthwhile to compare the corresponding CV responses.

Bhattacharya et al. [54], in their experiments on animals, have shown that for
non-synchronous oscillation the maximum increase in MAF was about 150
per cent from the no oscillation case. The corresponding changes in MHR and
SV were of the same order, depending upon whether the subject increased his
MAF mainly via MHR or via SV. In another study Bhattacharya et al. [58]
also showed that on the average non-synchronous WBO increased the mean
coronary flow (MCF) by 36 per cent (heart paced), the myocardial oxygen
consumption (MVO_2) by 42 per cent, and the pulse pressure by about 40 per
cent.

Heart synchronous WBO exposure produced positive and negative percentage changes in CV responses in most instances [64], and it provided the capability to maintain a desired response. For example on the average the range

of percentage changes was 11 per cent to 39 per cent in MAF, -4 per cent to 48 per cent in MCF, and -11 per cent to 63 per cent in MVO_2. From a comparison of the data of Bhattacharya and co-workers [50, 54, 58, 64] and of other workers [47, 52, 53, 57, 65, 66] with those of mild–moderate exercise [67–70] it can be said that the WBO can produce comparable CV responses. However, the physiological pathways responsible for CV responses induced by active exercise may not be the same as those responsible during WBO exposure. A detailed discussion of relative physiological pathways during exercise and WBO is given in ref. 50.

The potential of vibration as an exercise substitute provides a basis for possible therapeutic applications. One such application is the use of vibration as a cardiovascular reconditioning technique. The cardiovascular deconditioning phenomena associated with bed rest and more recently with weightlessness has become a subject of greater interest because of the advent of long-term manned space flights. From the standpoint of the labyrinth and proprioceptors, the most conspicuous aspect of the space environment is prolonged weightlessness [71–74]. Such a gravity-free environment induces cardiovascular deconditioning as reflected in orthostatic intolerance. To date, no established hypothesis has been validated to describe the mechanisms responsible for the cardiovascular deconditioning phenomenon. However, the occurrence of a purported 'disuse atrophy' or lack of stimuli to gravity sensitive mechanoreceptors during long-term space flights is one mechanism that can explain such cardiovascular deconditioning [75].

Previous investigators [76, 77] have indicated that while mild exercise in the weightlessness environment does help, it is not adequate as a countermeasure for cardiovascular deconditioning. However, it was hypothesized [77, 78] that severe exercise might facilitate cardiovascular conditioning although it might not be operationally feasible to realize such conditions of severe exercise during long-term space missions. Therefore, the need for a technique which will simulate the physiological mechanisms associated with severe exercise is obvious. A technique which could crowd the daily mechanoreceptor stimuli resulting from musculo-skeletal stress and the displacement of internal body organs into parts of an hour would be most desirable for long-term scientific space missions.

From the results of studies on WBO [47, 50, 52, 53, 54, 57, 58, 64–66] it can be implied that vibration stimulates mechanoreceptors (a stimulus which is greatly diminished with space flight and bedrest) and also increases MAF, MCF, and MVO_2. Moreover, the magnitude of the changes in the cardiovascular variables could be enhanced by increasing the amplitude of the input force [55]. The acceleration amplitudes of 0.75–1 g used in the studies by Bhattacharya and co-workers [50, 55] were not thought to be excessive. Further enhancement in cardiovascular responses can be achieved by ensuring that the vibration frequency is near the whole-body resonance frequency [65]. If used properly, WBO may well be a promising technique for the prevention of cardiovascular deconditioning. At present Bhattacharya *et al.* [51] are conducting experiments in this direction and results will be presented at a later date.

Another possible therapeutic application of WBO is its use as a non-invasive cardiac assist technique. Non-invasive assistance to the circulation could play a vital role in sustaining cardiovascular function during critical periods of cardiac failure. Assistance given to the heart during this critical time can decrease cardiac energy requirements, improve organ perfusion and enhance the patient's chance for survival. Efforts to use heart synchronous vibration for therapeutic application date back to the late 1960s when Arntzenius *et al.* [78] argued on the basis of ballistocardiographic results that proper synchronization of vibration with the heart cycle could enhance the blood volume ejection without a concomitant increase in left ventricular work. However, the number of studies since that time have been limited and results have often been inconsistent. A discussion of the results from the work of Bhattacharya and co-workers [50, 58, 64] which are relevant to cardiac assistance follows.

The main purpose of an assist procedure is to reduce the input energy to the myocardium (VMO_2) and still provide an adequate amount of blood volume and perfusion for the maintenance of the proper function of the total organism. In other words internal efficiency (or performance of myocardium, i.e. input energy to myocardium relative to the contractile effort or tension generated) and external efficiency (or performance of heart relative to MVO_2) must be elevated with the help of the assist procedure. However, when changes in efficiency are interpreted, caution must be exercised. For example an increase in efficiency may not necessarily indicate assistance because exercising subjects with normal hearts also increases efficiency, and it is a well-established fact that exercise is a stressful situation to the heart and the rest of the organism.

For a thorough evaluation of a cardiac assist technique, a set of experiments would have to be designed in which two groups of chronically instrumented animals with experimentally induced myocardial ischemia are used. One group would be exposed to the assist technique and the other allowed to follow the natural response to the induced stress of ischemia. Left ventricular efficiency could be measured in both groups and compared as a function of the survival rate between the assisted and unassisted groups. For these groups, if the application of an assist device on the normal heart decreases the energy requirements of the myocardium without impeding the ejection of adequate blood volume and perfusion pressure or increases the useful work done during systole without increasing the energy required by the myocardium, one might be able to imply that such a device has potential for providing circulatory assistance.

In 1975 Bhattacharya *et al.* [64] performed experiments on normal canines to determine the extent to which synchronization of the time relationship between the input force and the cardiac cycle could modify cardiac function. They restrained tranquillized canines with spines horizontal and exposed them to heart synchronous sinusoidal WBO at a constant acceleration of 0.75 g (the ratio of oscillation frequency to heart rate was unity, vibrated along g_z axis). For their study, MVO_2 showed substantial changes as a function of the various applied force types. The extreme values of changes from control in MVO_2 for the experiments ranged from -32 per cent to 111 per cent and the mean values ranged from -11 per cent to 63 per cent. The most promising force type for cardiac assistance appeared to be type III where the peak negative force delivered during the isovolumetric contraction period induced, on the average, a twofold increase in cardiac work (CW) relative to the increase in MVO_2. This response of MVO_2 and CW due to force type III in η_{LV} observed in their study. It was also observed that for this force type the peak time derivative of the left ventricular pressure (dP/dt) response corresponded to relatively less response in MVO_2 compared with that used for the other force types. The investigators implied that a better internal efficiency resulted during the application of force type III.

Can it then be implied that use of force type III on a normal dog will actually assist the heart in time of failure? A definite answer to this question can only be given when the technique is applied to animals with depressed hearts as discussed above. However, further insight may be achieved by reviewing the work of other investigators [79–85] who have applied intra-aortic balloon pumping and other counterpulsation techniques on dogs with normal and depressed hearts and measured changes in MVO_2, η_{LV}, MCF, SV, and other cardiovascular variables. A detailed comparison between the different assist techniques is presented by Bhattacharya [50]. His findings showed that the changes in η_{LV} resulting from assist exposure applied to animals with depressed hearts were greater in magnitude than those values for the normal animals. Such observations indicate that the responses of normal animals to assist techniques may be conservative. This was substantiated by the study of Knapp *et al.* [86] where they applied heart synchronous WBO on experimentally induced myocardially ischemic canines and found a relatively enhanced response compared with normal dogs.

The potential of WBO as a therapeutic tool, as discussed in this section, is certainly obvious. However, future studies are required to establish the clinical usefulness of this technique. This means that several years of basic studies are needed to investigate the various side effects and basic physiological mechanisms involved when a subject is exposed to WBO. Furthermore, application of WBO as an assist device and/or reconditioning tool for paraplegic patients requires conservative investigations in the areas of (1) criteria for candidate selection for WBO therapy and (2) WBO exposure periods. However, use of WBO as an exercise substitute for normal subjects should be a relatively safe area to begin evaluation of this technique.

Appendix 8.1. Severity index K

Vehicles expose human beings to mechanical vibrations which can interfere with their comfort and in extreme cases with their safety and health. As a criterion to rate the severity of exposure there exists the intensity of perception K, which can be determined for sinusoidal vibrations by using the formula

$$K = \frac{\alpha}{\{1+(\omega/\omega_0)^2\}^{1/2}} \ddot{z}_{\text{eff}}$$

where ω is the cyclic frequency, $\ddot{z}$ is the effective acceleration given by

$$\ddot{z}_{\text{eff}} = \left(\frac{1}{T}\int_0^T z^2 \, dt\right)^{1/2}$$

and $\omega_0 = 62.83\,\text{s}^{-1}$ and $\alpha = 18\,\text{s}^2\,\text{m}^{-1}$ are constants.

For random vibrations, which consist of numerous sinusoidal vibrations, the K value can be determined by summation of the K values of the particular sine vibrations:

$$K_{\text{total}} = \left(\sum_{i=1}^n K_i^2\right)^{1/2}.$$

Appendix 8.2. Governing equations of motion of the ambulance simulation

Using the abbreviations given in Fig. 6.33 the governing equations of motion of the rescue-vehicle simulation system are as follows:

$$m_p \ddot{z}_p - F_p = 0$$
$$m_T \ddot{z}_T + F_p - F_T = 0$$
$$m_A \ddot{z}_{AS} + F_T - F_{Ah} + F_M - F_{Av} = 0$$
$$\theta_A \alpha_A + F_T l_T - F_{Ah} l_h - F_M l_M + F_{Av} l_v = 0$$
$$m_{Rh} \ddot{z}_{Rh} + F_{Ah} - F_{Rh} = 0$$
$$m_{Rv} \ddot{z}_{Rv} + F_{Av} - F_{Rv} = 0$$
$$m_M \ddot{z}_M - F_M = 0$$

and the corresponding equations of forces are

$$F_p = c_p(z_T - z_p) + k_p(\dot{z}_T - \dot{z}_p)$$
$$F_T = c_T(z_{AS} + l_T \alpha_A - z_T) + k_T(\dot{z}_{AS} + l_T \dot{\alpha}_A - \dot{z}_T)$$
$$F_{Ah} = c_{Ah}(z_{Rh} - z_{AS} - l_h \alpha_A) + k_{Ah}(\dot{z}_{Rh} - \dot{z}_{AS} - l_h \dot{\alpha}_A)$$

$$F_{Av} = c_{Av}(z_{Rv} - z_{AS} + l_v\alpha_A) + k_{Av}(\dot{z}_{Rv} - \dot{z}_{AS} + l_v\dot{\alpha}_A)$$
$$F_M = c_M(z_{AS} - l_M\alpha_A - z_M) + k_M(\dot{z}_{AS} - l_M\dot{\alpha}_A - \dot{z}_M)$$
$$F_{Rh} = c_{Rh}(z_{Eh} - z_{Rh}) - k_{Rh}\dot{z}_{Rh}$$
$$F_{Rv} = c_{Rv}(z_{Ev} - z_{Rv}) - k_{Rv}\dot{z}_{Rv}.$$

where F is the force, c is the spring rate, k is the damping factor, l is the distance to the body's centre of gravity, m is the mass, z is the vertical displacement, α is the pitch angle and θ is the moment of inertia.

References

1 MUNIN, A. G., and KVITKA, V. YE. (eds.) (1973). *Aviatsionnaya akustika (Aviation acoustics)*. Mashinostroyeniye Press, Moscow.

2 RUDERMAN, J. (1965). High rise steel office buildings in the U.S. *Struct. Eng,* **43** (1), 23–33.

3 FELD, J. (1968). *Construction failure*, p. 151. Wiley, New York.

4 LANGE, K. O., and COERMAN, R. R. (1962). Visual acuity under vibration. *Hum. Factors* **4** (5), 291–300.

5 DENNIS, J. P. (1965). The effect of whole-body vibration on a visual performance task. *Ergonomics* **8** (2), 193–205.

6 VYKUKAL, H. C., and DOLKAS, C. B. (1966). Effects of combined linear and vibratory accelerations on human dynamics and pilot performance capabilities. *Life in Spacecraft, Proc. 17th Int. Astronautical Congr., Madrid, 1966*, p. 107. Dunod Editeur, Paris; Gordon and Breach Inc., NY; PWN–Polish Scientific Publishers, Warsaw.

7 SHARP, M. (1971). *Chelovek v. kosmose (Man in space)*, p. 32. Mir Press, Moscow.

8 BYNDAS, L. A., GLUKHAREVE, K. K., POTEMKIN, B. A., SAFRONOV, YU. G., SIRENKO, V. N., FROLOV, K. V. (1973). Estimation of the functional condition of a human operator exposed to vibration. In *Vibrozashchita cheloveka operatora i voprosy modelirovaniya (Vibration protection of a human operator and problems of modeling)*. Nauka Press, Moscow.

9 ANDREYEVA-GALANINA, YE. TS. OROGICHINA, E. A., and ARTAMONOVA, V. G. (1961). *Vibratsionnaya bolezn' (Vibration sickness)*. Medgiz Press, Leningrad.

10 GOLDMAN, D. E., and GIERKE, H. E. (1961). Effects of shock and vibration on man, *Shock and vibration handbook*, Vol. 3. McGraw-Hill, New York.

11 LOECKLE, W. E. (1950). The physiological effects of mechanical vibration, *German aviation medicine World War II* Vol. 2, pp. 716–22. U.S. Air Forces, Washington.

12 CLARAC, J., and KRESMAN, R. (1967). Conduite des transports en commun et lombalgies. *Arch. Mal. Prof.* **28** (3), 412–13.

13 BUTKOVSKAYA, Z. M., KADYSKINA, YE. N., POTEMKIN, B. A., and FROLOV, K. V. (1973). Study of the influence of vibrations of machines and tools on the arms of a human operator with various initial conditions. In *Vibroizolyatsiya mashin i vibrozashchita cheloveka-operatora (Vibration insulation of machines and vibration protection of the human operator)*, pp. 5–16. Nauka Press, Moscow.

14 VERSHCHEVSKIY, I. YA., BORSCHCHEUSKY, I. YA., EMELYANOV, M. D., KORESHKOV, A. A., MARKAVIN, S. S., PETROV, YU. P., and TERENTYEV, V. G. (1963). *Obshchaya vibratsiya i yeye vliyaniye na organizm cheloveka (General vibration and its influence on the human organism)*, p. 156. Medgiz Press, Moscow.

15 MEGEL, H., WOZMAK, N. U., and SUN, L. (1962). Effects on rats of exposure to heat and vibration. *J. appl. Physiol.* **17** (5), 759–62.

16 SACKLER, A. M., and WELTMAN, A. S. (1966). Effects of vibration on the endocrine system of male and female rats, *Aerosp. Med.* **37** (2), 158–66.

17 VAYSFEL'D, I. L., IL'ICHEVA, R. F., KOKAROUTZEV, V. V., PANOVOKO, G. YA., KHAZON, I. M., *et al.* (1972). Histamine and serotinin metabolism when vibration is applied to the body of a human operator. In *Vliyaniye vibratsiy razlichnykh spektrov na organizm cheloveka i problemy vibroz-ashchity (The influence of vibration of various spectra on the human body and problems of vibration protection)*, pp. 132–136. Nauka Press, Moscow.

18 GOLUBCHIKOV, V. A. (1972). The mechanism of the influence of general vertical vibration on the urinary tract in man. In *Vliyaniye vibratsiy razlichnykh spektrov na organizm cheloveka i problemy vibrozashchity. (The influence of vibration of various spectra on the human body and problems of vibration protection)*, pp. 184–8. Nauka Press, Moscow.

19 VON GIERKE, H. E. (1964). Biodynamic response of the human body. *Appl. Mech. Rev.* **17** (2), 951–8.

20 LIPPERT, S. (ed.) (1963). *Human vibration research*. Pergamon Press, New York.

21 VON GIERKE, H. E. (1971). Biodynamic models and their applications. *J. acoust. Soc. Am.* **50** (6), 1284–7.

22 POTEMKIN, B. A., and FROLOV, K. V. (1971). Model representation of the biomechanical system of a human operator with random vibrations. *Dokl. Akad. Nauk SSSR* **197** (6), 1284–7.

23 POTEMKIN, B. A., and FROLOV, K. V. (1972). Experimental study of the reaction of a human operator to vibration. In *Nelineynyye kolebaniya i perekhodnyye protsessy v mashinakh (Nonlinear oscillations and transient processes in machines)*, pp. 67–73. Nauka, Moscow.

24 GRIGANOV, A. S., and FROLOV, K. V. (1968). The problems of estimation of the influence of vibration of machines and tools on the organism of a human operator. *Kolebaniya i ustoychivost' priborov mashin i elementoy sistem upravleniya (Oscillations and stability of instruments, machines and*

elements of control systems), pp. 20–9. Nauka Press, Moscow.

25 POTEMKIN, B. A. (1968). Some problems of the influence of machine vibration on the work of a human operator. In *Kolebaniya i ustoychivost' priborov mashin i elementov sistem upravleniya (Oscillations and stability of instruments, machines and elements of control systems)*, pp. 30–5. Nauka Press, Moscow.

26 WHITTMANN, T. M., and PHILLIPS, N. S. (1969). Human body nonlinearity. In *Mechanical impedance analysis. Am. Soc. Mech. Eng. Publ. 69BHF-2.*

27 OBYSOV, A. S. (1971). *Nadeznnost' biologicheskikh tkaney (The reliability of biological tissue)*, pp. 37–48. Meditsina Press, Moscow.

28 POTEMKIN, B. A., and FROLOV, K. V. (1973). Construction of a dynamic model of the body of a human operator subjected to wideband random vibration. In *Vibroizolyatsiya mashin i vibrozashchita cheloveka operatora (Vibration insulation of machines and vibration protection of the human operator)*, pp. 17–30. Nauka Press, Moscow.

29 PANOVKO, G. YA., POTEMKIN, B. A., and FROLOV, K. V. (1972). Determination of the parameters of models of the body of a human operator exposed to vibration and impact. *Mashinovedenie*, No. 3, 31–6.

30 VON BEKESY, G. (1939). Uber die Vibrationsempfindung. *Akust Zn.* 4, 317.

31 KUHN, F., and SCHEFTLER, H. (1954). Beim gebrauch von druckluft schlagwerkzengen Einwickende Krafte. *Int. Z. angew. Physiol. Einschl. Arbeitsphysiol.* 15, 277.

32 KUH, F. (1953). Uber die Mechanische Impedanz des Menschen bei der Arbeit mit Presstufthammer. *Int. Z. angew. Physiol. Einschl. Arbeitsphysiol.* 15, 79.

33 DIECKMANN, D. (1958). Din Mechnisches Modell fur das Schwingserrgte Hand–Arm System des Menschen. *Int. Z. angew. Physiol. Einschl. Arbeitsphysiol.* 17, 125.

34 VASIL'YEV, YU. M. (1972). Problems of the dynamics of a system consisting of a human operator and an impact tool. In *Nelineynyye kolebaniya i perekhodnyye protsessy v mashinakh (Nonlinear oscillations and transient processes in machines)*, pp. 74–88. Nauka Press, Moscow.

35 GLUKHAREV, K. K., POTEMKIN, B. A., and FROLOV, K. V. (1973). Peculiarities of the biodynamics of the human body with vibrations. In *Vibrozashchita cheloveka operatora i voprosy modelirovaniya* (Vibration protection of the human operator and problems of modelling), pp. 22–8. Nauka Press, Moscow.

36 GLUKHAREV, K. K., POTEMKIN, B. A., and SIRENKO, V. N. (1972). The nonlinear and unstable characteristics of the human body. *Mashinovendenie*, No. 4, 9–14.

37 GLUKHAREV, K. K., POTEMKIN, B. A., and FROLOV, K. V. (1972). Nonlinearity and instability as manifestations of the regulation of dynamic properties of the human body. *Vliyaniye vibratsiy razlichnykh*

spektrov na organizm cheloveka i problemy vibrozashchity (The influence of vibration of various spectra on the human body and problems of vibration protection), pp. 46–50. Nauka Press, Moscow.

38 GLUKHAREV, K. K., POTEMKIN, B. A., and FROLOV, K. V. (1972). The construction of a simple mechanical model of the human body exposed to harmonic vibrations. *Konf. poKilebaniyam Mekhanischeskikh Sistem (Conf. on Oscillations of Mechanical Systems), Abstracts.* Naukova Dumka Press, Kiev.

39 POTEMKIN, B. A., ROTENBERG, R. V., SAFARISHVILI, G. A., SIRENKO, V. N., and FROLOV, K. V. (1973). Unstable dynamic characteristics of the human body with horizontal oscillation. *Vliyaniye vibratsiy razlichnykh spektrov na organizm cheloveka i problemy vibrozashchity (The influence of vibration of various spectra on the human body and problems of vibration protection*), pp. 25–9. Nauka Press, Moscow.

40 AIZERMAN, M. A., and ANDREEVA, E. A. (1970). *Issledovaniye protsessov upravleniya myshechnoy aktivnost'yu (Study of processes of the control of muscular activity*), pp. 5–49. Nauka Press, Moscow.

41 *VDI 2057: Beurteilung der Einwirkung mechanischer Schwingungen auf den Menschen NDI 1962.*

42 CLAYBERG, H. D. (1949). Pathologic physiology of truck and car driving. *Mil. Surg.* **105**, 299–311.

43 RAAB, W. (1932). Alimentare faktoren in der Entstehung von Arterio-sklerose und Hypertonie. *Med. Klin. (Munich)* **28**, 487–521.

44 ASHE, W. F. (1961). Physiological and pathological effects of mechanical vibration on animals and man. *NIH Rep. No. 862–4, Grant No. OH-6.*

45 FISHBEIN, W. I., and SALTER, L. C. (1950). The relationship between truck and tractor driving and disorders of the spine and support structures. *Ind. Med.* **19**, 444–5.

46 HOOVER, G. N., ASHE, W. F., DINES, J. H., and FRASER, T. M. (1961). Vibration studies, III. Blood pressure response to whole body vibration in anaesthetized dogs. *Arch. Environ. Health* **3**, 426–32.

47 HOOD, W. B., MURRAY, R. H., URSCHEL, C. W., BROWERS, J. A., and CLARK, J. G. (1966). Cardiopulmonary effects of whole body vibration in man. *J. appl. Physiol.* **21**, 1725.

48 HOOKS, L. E., NAREN, R. M., and BENSON, J. J. (1969). A momentum integral solution for pulsatile flow in a rigid tube with and without longitudinal vibration. *Proc. Eng. Sci. Med.*

49 ARNTZENIUS, A. C., KOOPS, J., RODRIGO, F. A., ELSBACH, H., and VAN BRUMMELEN, A. G. W. (1970). Circulatory effects of body acceleration given synchronously with heart beat (BASH). *Proc. 2nd World Congr. Ballistocardiography and Cardiovascular Dynamics, Oporto, 1969. Bibl. Cardiol.* **26**, 180.

50 BHATTACHARYA, A. (1975). Modification of cardiac function by heart

synchronous whole body vibration applied to awake chronically instrumented canines. Ph.D. Thesis. University of Kentucky.

51 BHATTACHARYA, A., KNAPP, C. F., and McCUTCHEON, E. P. (1975). Personal communication.

52 HOOD, W. B., JR. and HIGGINS, L. S. (1965). Circulatory and respiratory effects of whole body vibration in anaesthetized dogs. *J. Appl. Physiol.* **20** (6), 1157–62.

53 CLARK, J. G., WILLIAMS, J. D., HOOD, W. B., JR., and MURRAY, R. H. (1967). Initial cardiovascular response to low frequency, whole body vibration in humans and animals. *J. aerosp. Med.* **38**, 464–7.

54 DINES, J. H., SUTPHEN, J. H., ROBERTS, L. B., and ASHE, W. F. (1965). Intravascular pressure measurements during vibration. *Arch. Environ. Health* **11**, 323–6.

55 BHATTACHARYA, A., KNAPP, C. F., McCUTCHEON, E. P., and EDWARDS, R. G. (1977). Parameters for assessing vibration induced cardiovascular in awake, chronically instrumented canines exposed to whole body vibration. *J. appl. Physiol.* **42** (5), 682–9.

56 McCUTCHEON, E. P., KNAPP, C. F., WIBEL, F., and BHATTACHARYA, A. (1976). Efferent neural components of vibration induced cardiovascular responses in dogs (Abstract). *American Physiological Society and Biomedical Engineering Society Meetung, Philadelphia, 1976; Physiologist* **19** (3), 288.

57 KNAPP, C. F. (1974). Models of the cardiovascular system under whole body vibration stress. In *Vibration and Combined Stresses in Advanced Systems, AGARD Conf. Proc. 145.*

58 BHATTACHARYA, A., KNAPP, C. F., McCUTCHEON, E. P., and EVANS, J. M. (1975). Comparison of cardiac responses to nonsynchronous and heart synchronous whole body vibration. *Proc. 28th Annu. Conf. on Engineering in Medicine and Biology (ACEMB) Meeting, New Orleans, 1975* (Abstract). Alliance for Engineering in Medicine and Biology Md.

59 BHATTACHARYA, A., McCUTCHEON, E. P., GREENLEAF, J. E., and SHVARTZ, E. (1976). Comparison of mechanophysiological responses during running and jumping in man. *Physiologist* **19** (3), 128.

60 GROSSMAN, O. (1975). *Erschitterungsmessungen beim Trampolin und Minitrampspringen.* Eidg. Technische Hochschule, Labor fur Biomechanik, Zurich.

61 CAVAGNA, G. A. (1970). Elastic bounce of the body. *J. appl. Physiol.* **29**(3), 279.

62 CAVAGNA, G. A., SAIBENE, F. P., and MARGARIA, R. (1964). Mechanical work in running. *J. appl. Physiol.* **19**, 249.

63 BHATTACHARYA, A., and EVANS, J. (1974). Preliminary study. Wenner-Gren Research Laboratory, University of Kentucky.

64 BHATTACHARYA, A., KNAPP, C. F., McCUTCHEON, E. P., and EVANS, J. M.

(1976). Sensitivity of cardiac responses to the time relationship between vibration and cardiac cycles. *29th ACEMB Meeting, Boston, Mass.* (Abstract). Alliance for Engineering in Medicine and Biology, Maryland.

65 EDWARDS, R. G., McCUTCHEON, E. P., and KNAPP, C. F. (1972). Cardiovascular changes produced by brief whole body vibration of animals. *J. appl. Physiol.* **32**, 386–90.

66 KNAPP, C. F. (1974). The causes of decrements in aircrew performance physiological changes produced by vibration and other environmental stresses. *Prog. Rep. (AFOSR Contract No. F44620-74-C-0012).* Wenner-Gren Research Lab., University of Kentucky.

67 KOUHRI, E. M., GREGG, D. E., and RAYFORD, C. R. (1965). Effect of exercise on cardiac output, left coronary flow and myocardial metabolism in the unanesthetized dog. *Circ. Res.* **17**, 427.

68 VATNER, S. F., HIGGINS, C. B., FRANKLIN, D., and BRAUNWALD, E. (1972). Role of tachycardia in mediating the coronary hemodynamic response to severe exercise. *J. appl. Physiol.* **32** (3), 380.

69 MESSER, J. V., WAGMAN, R. J., LEVINE, H. J., NEILL, W. A., KRASNOW, N., and GORLIN, R. (1962). Patterns of human myocardial oxygen extraction during rest and exercise. *J. clin. Invest.,* **41** (4), 725–42.

70 VAN CITTERS, R. L., and FRANKLIN, D. L. (1969). Cardiovascular performance of Alaska sled dogs during exercise. *Circ. Res.* **24**, 33.

71 GLENN, J. H. (1962). Pilot's flight report. *Proc. IASNASA National Meeting on Manned Space Flight,* pp. 296–309. New York Institute of Aerospace Sciences, New York.

72 Mercury Project summary including results of the fourth manned orbital flight. *NASA Spec. Publ.* 45, 1963.

73 KASYAN, I. I., KOPANEV, V. I., and YAZDOVSKY, V. K. (1966). Reactions of cosmonauts to weightlessness. *Problems of space biology* (ed. N. M. Sisakyan). *NASA Tech. Transl.* **368**, pp. 260–77.

74 CRAMER, R. L. (1962). Response of mammalian gravity receptors to sustained tilt. *Aerospace Med.* **33**, 663.

75 GHISTA, D. N., BHATTACHARYA, A., SANDLER, H., and McCUTCHEON, E. P. (1976). Indices characterizing deconditioning following bedrest using LBNP. *Proc. 47th Ann. Meeting of the Aerospace Medical Association*, Bel Harbour, Florida, May 1976, p. 224. Aerospace Medical Association, Washington, DC.

76 JOHNSON, R. L., NICOGOSSIAN, A., BERGMAN, S. A., JR., and HOFFLER, G. W. (1974). Lower body negative pressure. *The Second Manned Skylab Mission Abstract, 45th Aerospace Medical Association Meeting, Washington, May 1974.* Aerospace Medical Association, Washington DC.

77 RUMMEL, J. A., MICHEL, E. L., SAWIN, C. F., and BUDERER, M. C. (1975). Skylab experiment M171: results of the second manned mission. *Aviat. space environ. Med.* **46**, 679.

78 ARNTZENIUS, A. C., LAIRD, J. D., NOORDERGRAFF, A., VERDOUW, P. D., and HUISMAN, P. H. (1972). Body acceleration synchronous with the heartbeat (BASH). *Prog. Rep. Proc. 15th Annual Meeting of the Ballistocardiography Research Society, Atlantic City, 1971. Bibl. Cardiol.* 29, 19.

79 JOHANSEN, K. H., DELARIA, G. A., and BERNSTEIN, E. F. (1973). Effects of external counterpulsation in reduction of the myocardial ischemia following coronary artery occlusion. *Am. soc. artific. int. organs* 19, 419–23.

80 URSCHEL, C., EBER, L., MATLOFF, J., CARPENTER, R., and SENNENBLICK, E. (1970). Alteration of the mechanical performance of the ventricle by intra-aortic balloon counterpulsation. *Am. J. Cardiol.* 25, 546–51.

81 BROWN, B. G., GUNDEL, W. D., MCGINNIS, G. E., *et al.* (1968). Improved intra-aortic balloon diastolic augmentation with a double balloon catheter in the ascending thoracic aorta. *Am. thorac. Surg.* 6, 127–37.

82 SPONITZ, H. M., COVELL, J. W., ROSS, J. R., and BRAUNWALD, E. (1969). Left ventricular mechanics and oxygen consumption during arterial counterpulsation. *Am. J. Physiol.* 217 (5), 1352–8.

83 FEOLA, M. NORMAN, N. A., HAIDERER, O., and KENNEDY, J. H. (1969). Assisted circulation: experimental intra-aortic balloon pumping. *Artificial Heart Program Conf., June, 1969,* pp. 637–57.

84 WEBER, E. T., and JANICKI, J. S. (1973). Coronary collateral flow and intra-aortic balloon counterpulsation. *Trans. Am. Soc. artific. int. organs.* 19, 395–401.

85 MEIJNE, N. G., KLOPPER, P. J., and BULTERIJS, A. H. K. (1973). The combined effect of a balloon catheter and synchronized partial bypass in the failing heart: an experimental study. *J. Cardiol. Surg.* 14, 261–70.

86 KNAPP, C. F., MCCUTCHEON, E. P., BHATTACHARYA, A., and EVANS, J. M. (1974). Study of heart synchronous whole body acceleration for circulatory assistance. *Rep. No. NOL-HT-3-2928-1.* National Heart and Lung Institute Contract No. NO1-HT-3-2928.

9. Vibration exposure in the industrial–occupational environment

DEVENDRA P. GARG

9.1. Introduction

The human body is routinely subjected to a variety of vibrating environments for example during travel in vehicles on land, water, or air, while operating heavy earth-moving machinery in agriculture or industry; or while operating hand-held power tools such as chain saws, grinders, or pneumatic jack hammers and drills [1, 2]. A clear understanding of human response to vibration is important since exposure to levels of mechanical vibration can significantly interfere with human task performance, comfort, safety, and efficiency.

Basically, three types of vibration exposure have been identified by the International Organization for Standardization [3]. The first type includes vibration transmitted simultaneously to the whole body surface or a substantial portion of it. This occurs when the body is located in a vibrating medium, such as when high intensity sound in air or water excites resonances of the body. The second type consists of vibrations transmitted to the body as a whole through the supporting surface, such as through the feet of a person standing in the vicinity of working machinery; through the buttocks of a seated passenger or machine operator; or through the supported area of a reclining person seated in a moving vehicle or a vibrating building. Finally, the third type of exposure is the one where vibrations are applied to specific parts of the body such as head, arms, and hand, through vibrating handles, pedals, head-rests, or a variety of hand-held powered tools and appliances.

Recognizing the importance of vibration exposure in the industrial occupational setting, this chapter will deal primarily with the second and third categories mentioned above. Specifically first the modelling and measuring approach of standing humans subjected to vibratory inputs such as those emanating from transportation vehicle or industrial shop floors will be presented. This will be followed by a discussion of vibration exposure to seated subjects such as tractor and earth-moving machinery operators. Vibrations resulting from hand-held powered tools and their effect on hand and arm will be given next, and protective measures and systems will be presented, where applicable, to mitigate the undesirable influence of vibration exposure.

9.2. Effect of vibration

A thorough understanding of the biological effect of vibrations on the human body involves a study of the effects of periodic mechanical forces or displacements as they are applied to the human body. Of particular concern are those forces and displacements which are capable of inflicting internal bodily damage as a result of working in conditions where there is a dangerous level of resonance or vibration [4]. Extensive literature exists on the subject of human response to vibrations [5–34]. The criteria generally used in research on this subject include preservation of comfort, efficiency of performance, and safety for health. Human response to vibrations, as defined by almost any criteria, is a functional characteristic of vibration transmitted to the body. These characteristics depend upon certain variables which may be classified as extrinsic or intrinsic [35]. Examples of typical extrinsic variables include frequency, amplitude, time history, direction, and point of application of vibration; whereas variables such as body size, body weight, body posture, and body tension are examples of intrinsic variables.

Each of the extrinsic and intrinsic variables affects the vibrating response of the human body in a variety of ways. For example, the human body is normally modelled by a multiple degree of freedom system using lumped-parameters of masses, springs, and dampers. The models exhibit several resonant frequencies, the dominant and whole body resonance frequency being close to 6 Hz. However, for a seated subject in the vertical mode the resonance frequency is within approximately an octave of 6 Hz, i.e. 3 Hz to 12 Hz. In this range the human body often amplifies the input vibrations. The effect of change in amplitude cannot be straightforwardly calculated since the human body itself is nonlinear. Also, the time history of vibration significantly affects the response. Whereas data exist on human body response for inputs of constant amplitude, frequency, and duration, not a great deal is known about the behaviour of the human body in complex situations where both vibration amplitude and frequency vary with time. Furthermore, it is possible for a subject to oscillate in any of the six degrees of freedom, namely, vertical, lateral, and longitudinal, roll, pitch, and yaw. Most of the response information available on whole body vibration is confined to the vertical direction only. Entry points for vibration may include feet, hands, seat, and back depending upon the contact of specific body part with source of excitation. Transmission of vibration may be affected by intermediate agents such as seat, cushions, harness, and gloves. Using such media judiciously it is possible appropriately to decrease body vibrations to ensure safety [36–37], comfort [38], and quality of performance [39–41].

In March 1979, the Committee on Hearing, Bioacoustics, and Biomechanics of the National Academy of Sciences studied the effects of whole body vibration on health [42]. The working group concluded that 'clinical health effects from prolonged occupational exposures to whole-body vibration have not yet been clearly established. This appears to indicate that for most common exposures there is little or no direct risk to health. On the other hand, because it is possible for such vibration to be of a magnitude or quality that can produce annoyance and pain and that can lead to functional alterations such as muscular weakness, high blood pressure, fatigue, and decreased nerve conduction velocity, sober consideration must be given to the possibilities for such clinical health effects and to safe limits for exposures. Even though all presently known effects are assumed to be reversible, they none the less may have an effect on the productivity and efficiency of the factory worker and perhaps on his well-being. Chronic effects on the musculo-skeletal system and the urinary tract have been suspected to be produced by some more severe exposures. A balanced research programme on the effects on human health of chronic exposure to whole-body vibration needs to be developed in order to permit the setting of standards for vibration.' Specifically, the following recommendations were made.

1. Field studies should be continued for high-dose populations such as truck drivers, bus drivers, heavy equipment operators, foundry workers, farm tractor drivers and commercial helicopter pilots. These studies must include environmental exposure data as well as epidemiological evidence of increases in morbidity. A description of the vibration environment which is as complete as possible is essential to the establishment of a meaningful vibration standard through the establishment of correlations of various exposure parameters with specific detrimental conditions. In this regard, one must be sure that workers in their off-duty time are not getting additional dosages of vibration. Epidemiological data should include absenteeism and job turnover rates as well as failure of physiological systems such as the skeletal, cardiovascular, or gastrointestinal systems.

2. Simultaneously, clinical studies should be conducted toward identifying any effects in the musculoskeletal system (i.e., insidious arthritis, lumbo-sacral strains, sciatica, lumbar disc syndrome, and facet syndrome), cardiovascular system (hypertension, coronary artery disease, obstructive syndromes, and vasospastic syndromes), gastrointestinal system and genitourinary system. Other disease conditions that may be aggravated would be spinal compression fractures, Schmorl's nodes, ankylosing spondylitis, spondylolisthesis, spondylosis, Scheurman's disease, supply spondylosis, detached retina, prostatitis, and ulcers of the stomach and small intestines.

3. Laboratory studies to parallel and explain findings under 1 and 2 should be oriented toward investigation of the physiological responses. Work should concentrate on the abdominal organs and their response to whole-body vibration. In order of importance these organs would be the gastrointestinal tract, endocrines (especially with respect to the pregnant subject), kidney, and liver. Other parts of the body are felt to be adequately covered by current laboratory research programmes. The purpose of the laboratory analysis should be to (a) attempt to define damage mechanisms for those pathological conditions identified as vibration-related by the field and clinical studies; (b) study vibration effects in isolation from other environmental stressors; and

(c) study the interrelationship between vibration effects and the effects of other environmental stressors. Other environmental stressors should include realistic stimuli such as heat, noise, and noxious fumes.

The whole spectrum from low- to high-frequency vibration needs to be considered. Repetitive impacts of various pulse shapes must be studied as part of the vibration continuum. Exposure time must be another experimental parameter. The adequacy of various environmental descriptors such as long-time averages, weighted rms-values, crest factors and the effects of individual peak exposures must be investigated. The preponderance of whole-body vibration exposure is in the vertical direction with the subject either seated or standing. Other axes of vibration appear of minor importance for workers in industry.

Although the identification of an entity which might be called vibration disease may be difficult, a decrease in the efficiency, production, and well-being of workers alone would justify research leading to vibration standards. It may be that there is no vibration disease as such from whole-body vibration comparable to the well-established vibration disease from hand-transmitted vibration, but that vibration simply acts to exacerbate existing disease states in individuals. Clarification of this point is highly desirable.

A variety of complex factors are involved in determining the human response to vibrations. In addition, there is paucity of consistent quantitative data concerning human perception of vibration and reaction to it. Several methods of rating the severity of exposure and defining limits of exposure based upon laboratory or field data have been developed in the past for specific applications. As such none of these methods can be considered to be applicable in all situations, and consequently, none has been universally accepted. There are many discrepancies in the reported research studies, primarily due to factors such as subjective human opinions, variation in subject posture, environment, and experiment conditions. The NAS report quoted above appropriately suggests the areas of study and research to develop a better understanding of human vibration response.

9.3. Vibration exposure while standing

In the case of standing human subjects, such as people operating machine tools in a factory, or riding in ground-traversing machines, vibrations are transmitted from feet to head progressing through various organs of the body. As those vibratory inputs vary over a range of frequencies, different parts of the human body pass through a resonant state at various frequencies. A knowledge of these resonance frequencies can be extremely useful in a variety of ways, such as for arriving at better designs of vehicle suspension systems for maximizing comfort and safety, for designing orthopaedic aids to replicate natural functions, or simply to provide a pleasant and protective environment for work and leisure.

An approach to develop a clear understanding of human body dynamics which has been found to be extremely effective is via lumped-parameter modelling and computer simulation. An appropriately designed human body model can be validated by comparing its responses to various inputs with those obtained experimentally by applying corresponding inputs directly to the human subjects. An appropriate model is one which resembles human anatomy and reproduces measured human responses including magnitude and location of resonance peaks.

9.3.1. Background study

A preliminary study directed toward the development of a lumped-parameter model was initiated by the author at the Massachusetts Institute of Technology. The experimental tests were performed using the instrumentation facilities of the MIT Draper Laboratories at Bedford, Mass. The setup included an electromechanical shaker, with a precision variable oscillator, power amplifier, high-power DC electromagnet, an AC armature and the associated shaker baseplate. With the human subject standing on the shake table, the driving frequency was varied over the range of interest. One vibration pick-up was attached to the shake-table baseplate and the other pick-up was attached to the head of the subject. Input to the feet and output from the head pick up were simultaneously recorded on a dual-beam Tektronix oscilloscope. At each frequency a photographic record of the input and output traces was made using a Polaroid camera attached to the oscilloscope screen.

The attenuation and phase angle plots were obtained from the input–output traces. The shaker facility had a frequency of operation ranging from 5 to 2000 Hz, with the restriction that the acceleration of the plate did not exceed 50 g. The amplitude of vibration was constrained on the basis of available data related to human physiological changes and tolerance to discomfort due to vibrations. The driving frequency for these experiments ranged from 5 to 40 Hz.

Based on the results of these experiments several lumped-parameter models, in increasing order of complexity, were advanced [43]. Analytical transfer functions were derived for each of these models. Three predominant peaks were found in each frequency response plot. This observation suggested a three-mass, three-spring, and three-damper model. It was hypothesized that each peak indicated the resonance of a specific subsystem of the body. For example, one peak would correspond to the resonance of the torso, the second to that of the internal organs, and the third to the head. The resonances of these body segments were actually experienced by the subjects as evidenced by their comments recorded during the tests. Analog and digital computer simulations of the derived transfer functions were also carried out.

9.3.2. Experimental setup

Facilities for the experiments conducted at Duke University [44] were designed to provide sinusoidal inputs and outputs over a wide range of frequencies and to operate within the boundaries of human tolerance. The standing human vibration transmissibility measuring setup is schematically depicted in Fig. 9.1. The test table was driven by a hydraulic exciter. A five-to-one lever arm ratio was used for the platform on which the test subject stood. At each frequency, the ratio of X (head), the displacement amplitude at the head, to X(feet), the displacement amplitude at the feet, was measured from sinusoidal traces on a dual-beam storage oscilloscope and plotted on graph paper. The oscilloscope traces were sampled until curves, free of such effects as fidgeting, were obtained. During experiments the subjects stood with their feet on a platform attached to the shake table and positioned between two sets of side angles. Before the application of input each subject placed his or her feet in a comfortable position with heels pressed against the angle section at the back. The side angles were then repositioned alongside the feet and the straps were tightened. This arrangement allowed for varied feet size and decreased the effect of fidgeting. An accelerometer was attached to the back side of heel angle as indicated in Fig. 9.1.

A second accelerometer on the human body was located at the head of the subject. This was attained by attaching the accelerometer first to a plexiglass frame and then strapping the frame to the head via a plastic hat. This scheme appropriately accommodated the accelerometer assembly with head dimensions varying from subject to subject. The entire arrangement proved to be light-weight and comfortable also for the test subjects. The curved plexiglass frame on which the accelerometer was mounted pressed against the head of the

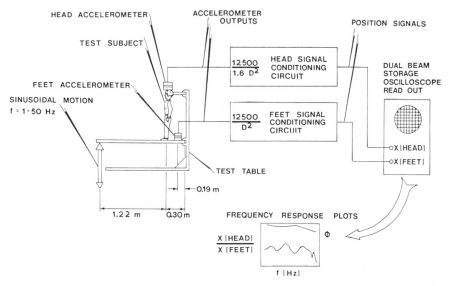

FIG. 9.1.   Schematic representation of test setup for whole body vibration studies.

test subject, thus providing a positive contact with the head at all times during the experiment.

The direct output of the two accelerometers initially contained higher-order harmonics. Two signal conditioning circuits were designed using a MiniAC analog computer (manufactured by EAI Associates). Each conditioning circuit consisted of four sequential stages. Over these stages it was attempted to filter out self-induced motions such as fidgeting and swallowing. Essentially, each conditioning circuit acted as a band-pass filter with double integration. Using two integration steps on the analog computer the acceleration signal was converted into a displacement signal. The factor of 1.6 difference between the two conditioning circuits was used to account for the placement of feet accelerometer.

Twelve subjects (eight male and four female) were tested and experimental plots were produced for each subject. In a typical experiment, a test subject was instructed to stand on the test table. With the subject in position, the test hat and feet restraint systems were adjusted. All subjects were provided with a switch located near their right hand to interrupt the experiment any time they felt uncomfortable or had a need to stop. In no case was the use of this switch found necessary by the subjects.

Test frequency was varied over a 1 to 50 Hz range with feet displacement ranging from a value of 0.0508 to 0.5842 mm (0.002 to 0.023 in). The upper limit was dictated by the restriction imposed by the operation of the actuator. Signals from the head and foot accelerometers were passed through two conditioning circuits to obtain position-like signals. For the two signals, X(HEAD) and X(FEET), using traces and records of these signals the magnitude ratio X(HEAD)/X(FEET) and the corresponding phase difference between them were derived for each frequency. Figure 9.2 shows typical test results for two of the twelve subjects who participated in the study. It was noted that in spite of a minor scatter of the data points, the two curves in their general trends were essentially similar. In particular, there were five noticeable peaks in the magnitude portion. These peaks successively occurred around 2 Hz, 6 Hz, 20 Hz, 33 Hz, and 43 Hz. The last peak was due to a combined translation and rotation of the head during vibrations.

9.3.3. Discussion of results

Test data were obtained for the twelve subjects as discussed in the previous section. A graphical process was used to arrive at an average curve representative of human body response under vibratory inputs. The average was computed by first superimposing the twelve test data curves shifted in frequency such that each extreme appeared at average location of extremes in frequency for each peak and valley for the curve. A smooth frequency-response plot was drawn by passing it through the median of superimposed test results. Figure 9.3 shows the average frequency response plot thus

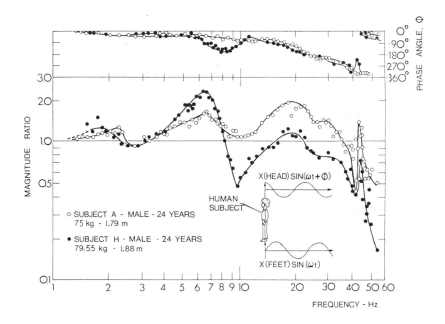

FIG. 9.2.   Typical experimental test data for two subjects.

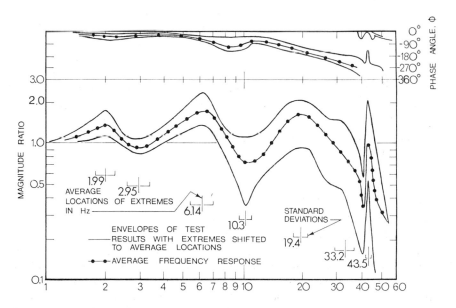

FIG. 9.3.   Envelopes of test responses, average frequency response plot, and mean and standard deviation for each peak and valley.

obtained. Envelopes of response curves for the twelve subjects and the mean and standard deviation for each peak and valley are also shown in the figure.

A lumped-parameter model [45] matching human anatomy and having transmissibility characteristics similar to ones obtained in the average test response was formulated. Values of body parameters were obtained from literature [46–63] where possible. Figure 9.4 represents the model configuration. The model description essentially consisted of three sets of parameters: mass distribution, spring stiffness, and damping coefficients. Linear parametric variations were assumed in the model. The head mass was augmented with an additional spring-mass-damper system to account for a rotational and translational mode of head vibration. Detailed parametric values used in the model are given in reference [1].

Figure 9.5 shows a comparison of the frequency response plots in both magnitude and phase angle of the matching model with the average test response for the twelve subjects participating in this study. The standard

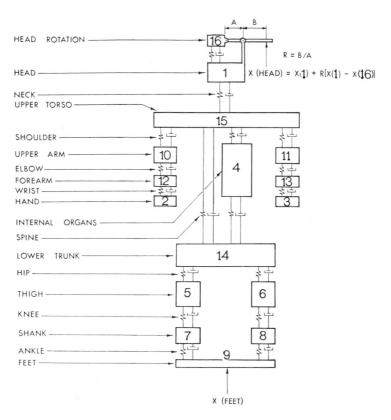

FIG. 9.4.   Proposed lumped—parameter vertical mode human body vibration model.

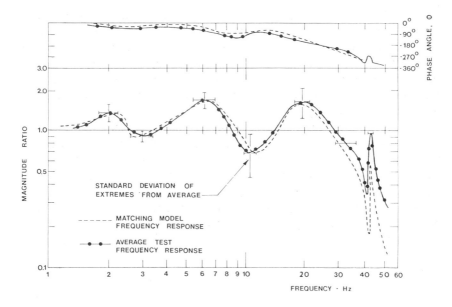

FIG. 9.5.   Average test response of human subjects as compared to the frequency
response plot for the lumped-parameter model of Fig. 9.4.

deviation bars in Fig. 9.5 are indicative of the variations at peaks and valleys of
the average test response. It is evident that the model response closely matches
the average test response in both magnitude and frequency.

A review of results obtained in this research effort indicated that it is feasible
to measure human vibration transmissibility over a frequency range of 1 Hz to
50 Hz. Also, it is possible to derive a linear lumped-parameter model which
accounts for both human anatomy and matches the average test frequency
response in both magnitude and phase angle. Standing human subjects under
vibratory input conditions exhibit four predominant peaks at frequencies in
the neighbourhood of approximately 2, 6, 20, and 33 Hz. Resonance close to
43 Hz is due to head rotation about some line parallel to a line through the ears.
The proposed model can be utilized for predicting human body response to
other types of input and combinations thereof [64]. Also, the parametric data
such as obtained in this study can be used for arriving at a better design of
prosthetics. Information on these resonance frequencies can be very prudent in
the design of a relatively risk free environment isolated from excessive
vibrations thus providing the needed protection from possible injury.

In order to reduce the transmission of vibration to shop floor, resulting from
rotating machinery, commercially available vibration isolators are normally
used. Special efforts have to be made for isolation when it is not possible to
mount the machinery directly upon isolators. In those cases, a relatively heavy

and rigid concrete block is supported on isolators and machinery is mounted on the block. The isolators applied for supporting the blocks may use materials such as felt, cork, rubber, steel spring, elastomers, or they may employ pneumatic damping. Figure 9.6 shows a typical arrangement to attain a rigid and massive supporting structure using a concrete block [65]. The block may be positioned above the floor level. Commercially available isolators may be used effectively to provide the necessary protection. Height adjustment capability facilitates leveling of the block. The entire arrangement, in conjunction with a judicious use of isolators, significantly reduces the level of vibrations experienced by operators while standing on industrial shop floors.

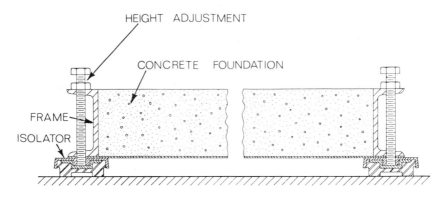

FIG. 9.6. Foundation supported on vibration isolators for placement above floor level [65].

9.4. Vibration exposure while seated

In a seated position, such as that adopted during the use of modern tractors and heavy earth-moving machinery, the human operator is subjected to a variety of stresses and environmental discomforts. Much of the discomfort results from an inappropriate seat suspension system and vibratory environment leading to fatigue and deterioration in sensitivity and rate of reaction to inputs. It has been shown [66] that increased output could be achieved by reducing operator discomfort. Also, among a group of tractor drivers [67], approximately 72 per cent were found to have some form of spinal deformation compared to a corresponding figure of 14 per cent for the population as a whole.

9.4.1. Typical studies

Huang and Suggs [68] studied fieldwork conditions inorder to determine the vibrational characteristics of farm tractors, and used laboratory simulations to investigate physiological response and performance deterioration of tractor operators under variable temperature conditions, and when subjected to a range of sinusoidal vibrations in the vertical, longitudinal, and transverse directions. A general four-wheel tractor with a metal seat mounted on a single-leaf spring suspension was chosen for the study. Vibration levels were measured on the seat and chassis for ploughing, discing, and riding across furrows without implements. Field tests indicated that the discing of ploughed ground generated most severe vibration for a tractor operator. Also, single-leaf spring suspension magnified the vibration thereby degrading the ride characteristics of the tractor over a large part of the frequency range. Of the three directions the vertical vibrations resulted in the highest acceleration and frequency distribution ranges.

It was also found during the study that higher vibrations in vertical and longitudinal directions resulted in an increased ventilation rate. The heart rate and ventilation rate were high in the neighbourhood of 1 g acceleration level but they were slightly lower at 1.5 g and 2 g accelerations for transverse oscillations. Vibrations transmitted through the steering wheels had a significant effect on ventilation rate and body acceleration. The waist acceleration was considerably larger than chest acceleration for both the hands-on and hands-off positions. For longitudinal acceleration, both waist and chest accelerations tended to increase as the imposed acceleration was increased. However, they had a tendency to decrease at higher imposed accelerations for transverse vibration. Also, the subjects experienced an increase in heart rate and slight decrease in oxygen consumption with an increase in temperature from 60° to 90 °F.

Efforts have been in progress at the National Institute of Agricultural Engineering (NIAE) Silsoe, England, to improve the performance characteristics of tractors, and to reduce the potential risks to tractor operators from the high level of vibration and consequent spinal damage or abnormality in the lumbar region. It has been shown [69] that vibration leads to induced discomfort and loss of control in a number of agricultural tasks. The NIAE research emphasized in part, the development of an experimental tractor on which complete driver's cab was suspended from the tractor chasis along the three axes of vertical, roll and pitch directions [70]. Parallel links guided the vertical motion of the cab controlled by rubber torsion spring elements, mounted on the rear pedestal. On the forward ends of the links was carried a large Hooke's joint, the bearings of which were replaced by rubber torsion springs. This Hooke's joint carried the cab. Separate pairs of hydraulic dampers were provided for each degree of freedom for the suspension. The natural frequencies of the suspensions were 0.8 Hz, 0.6 Hz, and 0.5 Hz in vertical, pitch, and roll respectively.

Measurements were made to study the ride isolation characteristics of the suspended cab tractor while towing a loaded 2-wheel unbalanced trailer at 22 km/h over a consolidated farm track. The cab suspension was designed to give the lowest natural frequencies practicable. Predominant peak frequencies of 1.0 to 3.0 Hz were encountered for both free and locked suspension configurations in vertical, longitudinal, lateral, roll, and pitch motions. With suspension in operation, an improvement in the ride quality of the tractor was noted as evidenced by a reduction in vertical and longitudinal root-mean-squared acceleration by 25 and 31 per cent respectively as compared to the case of locked suspension. There was a slight reduction in lateral acceleration as well. The suspension, in effect, altered the frequency spectra of vibrations in a favourable manner from the viewpoint of operator's vibration sensitivity.

A number of recent investigations have dealt with dynamic lumped-parameter modelling of tractor-suspension-operator systems. Such studies can be of use in identifying frequencies of harmful vibrations and can assist in the selection of suspension parameters such that the response of seated human subjects [71–77] are minimized in the frequency range of interest. The knowledge is important since it has been observed that mechanical stimuli of impacts and vibrations such as experienced by operators of tractors, trucks, and earth-moving machinery can lead to vasoconstriction of arterioles, herniated discs, lumbosacral pain, and traumatic fibrositis. According to Patil *et al.* [78] tractor operators are generally exposed to high intensity vibration levels in the 0.5 to 11 Hz range for extended periods of time, which they are not physically equipped to tolerate. Acceleration levels in conventional tractors are of the order of 0.5 g to 1.5 g in the frequency range of 2 to 7 Hz, and standard seats with different suspension parameters give rise to amplitude ratios of 2.5 to 4.5. With a suitable choice of system parameters, it was feasible to reduce the amplitude ratio to 1.55 and acceleration level to 0.9 g.

Significant improvements were obtained with a new type of seat suspension designed by the Patil & Ghista group [79], designated as PPG suspension illustrated in Fig. 9.7, consisting of a compression spring and a leaf spring in parallel with a dashpot acting in opposite direction to each other. The tractor was idealized by the seat–chassis–body and tyre masses lumped together. The parameters of the tractor were based on the work by Mathews [80]. The operator was modelled via lumped-masses of head, back, torso, thorax, diaphragm, abdomen, and pelvis connected by spring and dashpots, representing the elastic and damping properties of connective tissues between the segments. Inputs to the model consisted of sinusoidal vertical vibrations applied to tractor tyres, and suddenly applied disturbing force to the tyres to simulate sudden obstruction encountered by the tractor.

A CSMP simulation on IBM370/155 computer yielded the displacement and acceleration of body parts at different input frequencies. It was shown using the computed steady-state and transient responses that the PPG suspension located at the plane of centre of gravity of tractor with an optimal

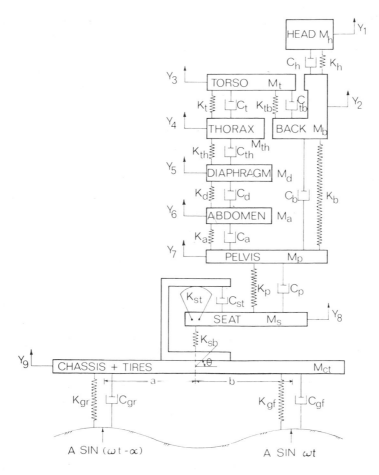

FIG. 9.7.   Occupant—tractor model with a 'Patil-Palanichamy-Ghista' suspension system [79].

set of parameters, reduced the amplitude ratios of body parts to 0.0076, thereby effectively isolating the tractor occupant from the harmful vibrations, and reduced the acceleration of body parts to levels well below the 8-hour reduced comfort boundary curves (Fig. 9.8) proposed by ISO [3], thus providing best riding comfort to the tractor operator. The suspension also significantly reduced the relative displacements between body parts.

Frolov [28] presented experimental results exhibiting dynamic characteristics of human body under random vibrations as influenced by the working position. The experimental amplitude-frequency characteristics of a seated subject in an erect position with the random input applied at the seat and

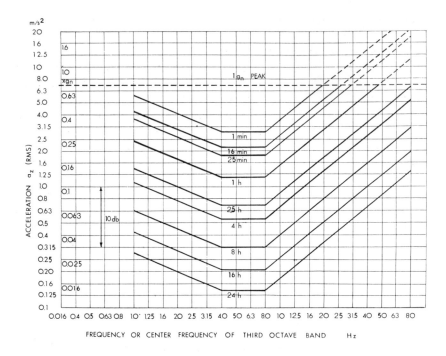

FIG. 9.8.   Longitudinal (a_z) Acceleration limits as a function of frequency and exposure time. To obtain 'exposure limits' multiply acceleration values by 2(6 db higher). To obtain 'reduced comfort boundary' divide acceleration values by 3.15 (10 db lower).

output measured at the head showed three dominant resonance peaks with natural frequencies at 3.9 Hz, 11.9 Hz and 35.8 Hz. Other peaks, though not so pronounced, occurred at 45 Hz and 60 Hz. In a variation of the above seated position, wherein the operator raised and bent his knees with his head resting between them, showed one dominant peak at 8 Hz and another small peak at 30 Hz. Finally, seated in an erect position with his hands raised, the experimental plot showed two dominant resonance peaks at 12.7 Hz and 25 Hz. Other smaller peaks occurred at approximately 45, 60, 80, and 95 Hz. In each case the subject vibrating in a seated position acted as a low-pass filter. The filter parameters changed with the actual configuration of the seated subject.

9.4.2. Vibration reduction for seated subjects

It is important that seated operators should be protected from potentially damaging vibrations. To some extent, such protection can be achieved by reducing the applied forces, or by manipulating body postures and supports.

Cushions may not always be effective. In some cases, they may even amplify vibrations at whole body vibration frequency [20]. At higher frequencies, however, the resonance may be attenuated. Therefore, cushions are used mostly for static comfort. Seat suspension systems generally consist of hydraulic shock absorbers, coil and leaf springs, torsion bars, or some combination of these. All of these systems have been effectively used for motion in a linear direction. However, when designed for the seat to pivot around a centre of rotation, these systems frequently provide undesirable pitching motion and operator fatigue. Hydraulic shock absorbers provide the most acceptable performance level from the point of view of fatigue. Figure 9.9 illustrates the vibration-induced stress characteristics for various types of seat design [81]. The data represent an average of over ten tests, each of fifteen minutes duration, on five subjects exposed to vibration characteristics of tractor operation. The superiority of seat designs which restrict seat vibrations to linear compared to pivoting motions is clearly evident from the data.

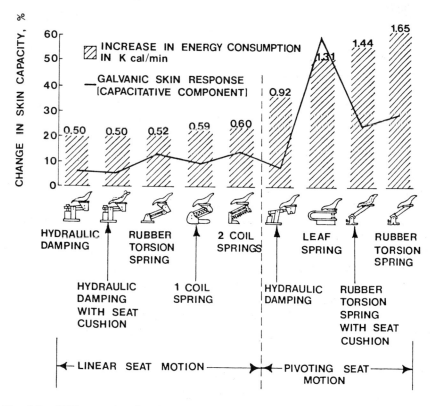

F IG. 9.9.   Difference in vibration-induced stress on a subject as a function of seat design [81].

Active vibration isolation [82–85] can be achieved for seated subjects in occupational environments, for example, while operating machine tools, tractors, and earth-moving machinery, or while piloting an aircraft. In order to define the degree of vibration isolation required for a given application, it is necessary to determine both the nature and the level of expected excitation and the allowable response. An active electrohydraulic vibration isolation system [86] based upon sensing the dynamic response of isolated payload and employing automatic feedback control techniques was used to produce actuator relative motions that opposed and cancelled the accelerations of the support structure. The motion of the actuator was determined by the fluid flow to and from a servovalve which in turn was controlled by a command signal from a servoamplifier. Sensor signals proportional to payload acceleration and the relative displacement between the payload and the vibration source were fed back to the servoamplifier, with appropriate linear and nonlinear compensation networks employed to provide proper stability margins, degree of isolation, and displacement control. Applications of these concepts for vibration isolation were shown in ref. [85] for the cases of jet and helicopter pilot seats.

A discussion of kinematic type of active vibration-isolation devices was presented in a recent paper by Frolov *et al.* [87]. The objective was to maintain a constant static displacement between the isolated subject and a vibrating base. Examples of desirable application include automobile and bus suspensions, high-speed railway passenger transport, multi-axis truck-trailer systems, and operators seated in human-operator cabin systems. The isolation characteristics of an active vibro-isolation device were to be achieved by using signals from accelerometer installed on a vibrating base. This accelerometer provided a stable vibrating base signal and allowed high values of gain in the control circuit which led to instability. When the accelerometer was located on a mass to be isolated, its signal decreased with vibro-isolation improvement. Using the proposed scheme it was possible to adjust mass, damping, and elastic system characteristics for providing it with optimal properties.

9.5. Vibration of hand and arm

Vibrations arising from hand-held power tools, vibrating machinery or vibrating workpieces normally used in production and industrial environments often can be very intense and harmful for operators. Typical examples of occupational environments susceptible to such injuries include manufacturing, mining, and construction industries when handling pneumatic and electrical hand tools, rock drills, pedestal grinders, polishers, and in forestry work using chain saws for snedding and felling of trees [88–91]. The vibrations are usually transmitted through the hand and arm to shoulders and other parts of the human body resulting in discomfort and reduced efficiency. A variety of disorders can result from a continued and habitual use of many vibrating tools. These include a variety of diseases affecting the blood vessels, nerves, bones, joints, muscles or connective tissues of the hand and forearm. For example [92], in certain cases, prolonged vibration of hand and arm can lead to inflammation of muscles and tendons, disruption of peripheral circulation and nerve injury, skeletal changes such as cyst and fissure formation, pain, fatigue, numbness and cyanosis, and vibration-induced white fingers.

9.5.1. Vibration-induced white fingers

Dr Maurice Raynaud, in his MD thesis [93], written in 1862, identified a condition in which one or more fingers may become pale and cold for a duration lasting from a few minutes to several hours. The skin of the affected part may assume a dead white colour, and the cutaneous sensibility may become blunted, and finally annihilated. This phenomenon is clinically known as the Primary Raynaud's Disease. In the initial stages the attacks may be intermittent and of short duration. However, the condition becomes progressively worse, leading to blue cyanotic fingers, the skin becoming atrophic and ulcerated, and finally gangrenous [94]. The cause of these changes was assumed to be a periodic spasm of the arterioles of the fingers. A similar condition is frequently encountered by users of hand-held power-operated tools. It was suggested by Agate [95] in 1949 that the association of white finger with vibration in industrial processes should be termed as 'Raynaud's Phenomenon of Occupational Origin' to distinguish it from 'Primary Raynaud's Disease'. The Industrial Injuries Advisory Council of Great Britain in 1970 adopted the description 'Vibration Induced White Finger' [96–97], and the related syndrome including the wrists, elbows, and shoulders, the 'Vibration Syndrome' [98–100].

The account of connection between occupational activities and the occurrence of Raynaud's phenomenon in the hands of pneumatic tool miners [101] was given in the year 1911. Subsequently, studies were conducted on mine drill operators from limestone quarries [102], fetters in iron foundries [103], riveters [104], operators carrying hand-held rotating tools [105, 106], cutting burrs [107], and pedestal grinder operators [108, 109]. Instances of Raynaud's phenomenon of occupational origin [110] were found in many cases. Grounds [111] in 1964 reported a connection between chain-saw operators and Raynaud's phenomenon for the first time. The study was conducted in Tasmania where 91 per cent of the sawmen examined were found to have symptoms, but the disease was not severe enough to discontinue their professional activity.

In Japan [112] an investigation was carried out in 1966 to relate the incidence of Raynaud's phenomenon in drillers, limestone quarry miners, and chain-saw operators. It was in this study that an objective measurement procedure through the use of finger pads was proposed and threshold vibration levels were identified and compared between exposed and control group of workers. An extensive research effort was undertaken for Scandinavian workers [113] in which over 450 tree fellers were studied and a strong correlation was found between vibration-induced white-finger condition and the duration of exposure resulting from hand-held powered chainsaw. Following this work, in 1970 a method was developed [114] for determining a temporary shift in the threshold of vibratory sensation as the exposure was changed from mild to intense.

It has been shown that the finger-tip blood-flow in response to cold is reduced in the case of vibratory tool users whether or not they show clinical evidence of Raynaud's phenomenon [101]. Repetitive vibration exposure possibly lowers the threshold for the reflex spasmodic contraction to which even normal arteries are subject, when traumatized or cooled. In normal vessels exposed to low temperatures, dilation occurs following the contraction. For operators subjected to vibratory exposure, there is a marked delay, and even failure, of this subsequent phase of cold dilation. A breakdown of local biochemical regulation, possibly due to an injury from vibrations to digital vessels walls and nerve endings, may be responsible for this delay.

In the case of vibratory tool workers affected by Raynaud's phenomenon the cessation of tool usuage rarely results in complete recovery. This is due to underlying structural changes. Whether this is primarily in the digital nerves or in the vessels themselves has not been adequately established. Lewis [115] carried out a detailed study of digital vessels in patients with Raynaud's disease, related to non-occupational causes. It was shown that a strong correlation existed between the severity and duration of symptoms, and the structural changes in digital vessels. His histological observations were restricted to arteries alone, and not to cutaneous nerves or nerve endings.

More recently, a study [116] was undertaken by Dupuis, to investigate the effect of vibration frequencies, acceleration amplitudes and elbow angles during vibration exposure of the hand–arm system. Specifically the impact of changes in these variables was studied on subjective perception, muscle activity, and peripheral blood circulation. The frequencies ranged from 8 Hz to 500 Hz, and the elbow angles ranged from 60° to 180° in 30° steps with the forearm kept horizontal in all cases. Biomechanical investigations showed a resonance of the wrist, and with smaller amplifications, of the elbow at low vibration frequencies of 10 to 20 Hz. Resonance in hand–arm system introduced extraordinary levels of mechanical strain of the body tissue, high subjective reaction, and changes of physiological functions. Equal sensation curves showed a frequency-dependent increase in acceleration. The application of static grip force—even without vibration—caused a significant reduction of skin temperature, which under vibration stress remained at the same lowered level. Also, vibration stress led to a significant decrease of the finger pulse amplitude.

Another recent study [117] identified three major disorders associated with the use of vibrating tools, namely the Raynaud Syndrome, osteoarthropathy and neuropathy. Raynaud Syndrome has been discussed earlier. Osteoarthropathy is a disease of the bones and the joints consisting of a degeneration process at the wrist, elbow, and shoulder. In neuropathy, the patient experiences a disorder of the peripheral nerves of the upper limits causing a feeling of pins and needles, and a reduction of sensation to touch, pin pricks, and temperature. The study was a part of an ergonomics research effort carried out on the risks and disorders present in an iron foundry. One hundred and sixty seven people whose work was related to the jobs of cast finishing and moulding were investigated. They included both those still engaged in this work, and subjects who had done this work at some time in the past. Neurological conclusions with regard to this study indicated that neuropathy was caused by hand-held vibrating tools. A clear neurological symptomatology was never present before two years of exposure; for periods of exposure up to three years, severe neuropathies were not observed, and finally a relevant clinico-functional recovery was observed if the subjects were removed from work attendance before a three-year period, whereas for higher-exposure periods reversibility was insufficient or absent.

Two extensive studies were carried out for the Forestry Commission in England to investigate the prevalence of Raynaud's phenomenon in chain-saw operators [118, 119]. The population consisted of workers who had used the saw 5.5 to 6.5 hours per day, five days per week, for a period of five to fourteen years. Of this population, the progress of 56 workers was monitored on an annual basis for four years. The prevalence of vibration induced white finger condition was found to be 90 per cent when first seen in December 1969 which rose to 95 per cent in January 1971. The prevalence rates for the control group were 6 per cent and 6.6 per cent respectively. The degree of disease severity [120] was graded according to that shown in Table 9.1.

In the four annual surveys, the same 46 workers were seen, the vibration-induced white fingers assessed by stage and by degrees within each stage. From 1961 to 1970, all workers were handling saws with high acceleration values on both handles (200–400m s^{-2}). From 1971 onward, all saws used were equipped with antivibration handles (i.e. handles with acceleration levels of 60 m s^{-2} each). A comparison of conditions over a period of time indicated significant improvement with the use of antivibration handles.

The prevalence of cold-induced white fingers was reported by Hamilton [102] in a finding related to a survey of stonecutters using pneumatic hammers. It was found that men using air hammers commonly suffered from a disturbance in the circulation of hands, consisting of spasmodic contraction of the blood vessels of certain fingers, making them blanched, shrunken, and numb. These attacks came under the influence of cold. The fingers affected were numb and clumsy while the vascular spasm persisted. There were no

TABLE 9.1

Stages of Raynaud's phenomenon [120]

Stage	Condition of digit	Work and social interference
0	No blanching of digits	No complaints
0_T	Intermittent tingling	No interference with activities
0_N	Intermittent numbness	
1	Blanching of one or more finger tips with or without tingling or numbness	No interference with activities
2	Blanching of one or more complete fingers with numbness usually confined to winter	Slight interference with home and social activities. No interference at work
3	Extensive blanching. Usually all fingers bilateral. Frequent episodes summer as well as winter	Definite interference at work, at home, and with social activities. Restriction of hobbies
4	Extensive blanching. All fingers. Frequent episodes summer and winter	Occupation change to avoid further vibration exposure because of severity of signs and symptoms

serious secondary effects following these attacks. The condition was most marked in those branches of stonework where the air hammer was most continuously used and was absent in the case of those stonecutters who did not use the air hammer. Another study [121] dealt with workers using chipping hammers and grinders in a modern foundry. It was observed that the risk of developing white-finger numbness with or without swollen hands, loss of grip, and painful shoulders and elbows was greatest among users of chipping hammers in contrast with other employees using grinders or no vibrating tools or equipment. Also, inverse variations were observed: i.e. the shorter the latent period, the higher the risk of developing Raynaud's syndrome, within a short follow-up period of sixteen months.

Bevenzi *et al.* [122] recently reported the results of a study conducted between September 1977 and July 1978 dealing with 169 shipyard caulkers who used pneumatic hand-held tools in their work. In an eight-hour shift, the total exposure time was five to six hours and the operators used only leather gloves. The investigations included a measurement of vibration spectra using accelerometers mounted on tool handles along three orthogonal axes. In addition, medical investigations were undertaken of the 169 caulkers and a control group of 60 workers not exposed to hand–arm vibrations such as electricians and welders. The study showed that approximately one-third of the shipyard caulkers tested were suffering from Raynaud's phenomenon of occupational origin. In terms of Taylor's classification (Table 9.1) their vibration disease was at stage 1 or 2. Among the 39 caulkers who used the chipping hammer for more than 3.5 hours, no fewer than 17 (43.6 per cent) of the caulkers were affected. Of those using it for less than 2 hours, only 22.4 per cent were affected. A moderate prevalence of typical bone cyst (31.3 per cent) and elbow extoses (10 per cent) was found among all caulkers.

9.5.2. Guidelines for hand and arm vibration

Studies cited in the previous section indicate that the vibration exposure to hand and arm in occupational environments presents possible hazards. Based upon limited data available [123–127] and on experience with current exposure conditions [128–131] the International Organization for Standards [132] has proposed tentative limits for safe exposure as a guidance to protect the majority of workers against serious health impairment and to assist in the development of hand-held tools with reduced risk of producing vibration-related disorders. The Standard specifies general methods for measuring and reporting hand-transmitted vibration exposure in three orthogonal axes for the one-third octave bands having centre frequencies between 8 and 1000 Hz. The exposure limits are expressed in terms of one-third octave band and octave band, vibration acceleration or velocity, daily exposure time, intermittency of exposure and direction of vibration relative to hand. The limits apply to periodic as well as to random or non-periodic vibrations.

The directions of vibration transmitted to the hand are recommended by the Standards to be measured and reported in appropriate directions of an orthogonal coordinate system, as specified in Fig. 9.10, having its origin in the head of the third metacarpal bone. The subsubscript 'h' refers to 'hand' to avoid a possible conflict with terminology used in biodynamics to define whole-body vibration. The primary quantity used for describing the magnitude of vibration is acceleration, expressed as root-mean-square (rms) value in metres per second squared. The measurements in the three axes are to be made at the surface of the hand where the energy enters the body. If the hand of the person is in direct contact with the vibrating surface of the hand grip, the transducer should be fastened to the vibrating structure. The size, shape, and mounting of the transducer should not have significant influence on the transfer of vibration to the hand.

Maximum acceptable levels of daily four to eight hour continuous exposure (correction factor 1) are given in Table 9.2 for third-octave bands and Table 9.3 for octave bands. Whenever possible, the Standards make a recommendation to use the third-octave band limit, which can be more stringent for a discrete frequency spectrum. The same limits apply separately for the vibration acceleration in the three coordinate axes. If the exposure is for less than 4 hours per day, or is intermittent, correction factors shown in Table 9.4 should be used for adjusting the exposure limits of Tables 9.2 and 9.3. If the continuous exposure time is shorter than four hours, the four and eight hour exposure limits can be raised by the correction factors of Table 9.4. This is illustrated in Figures 9.11 and 9.12 for third-octave and octave bands, respectively. For exposure of thirty minutes and shorter, the maximum

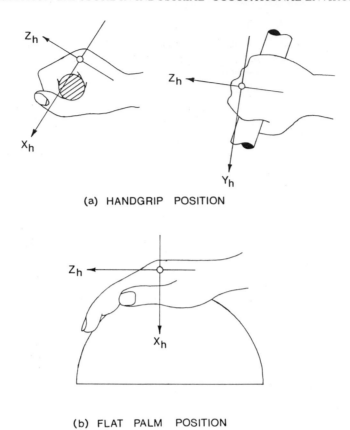

(a) HANDGRIP POSITION

(b) FLAT PALM POSITION

FIG. 9.10. Co-ordinate system for the hand [132]. The Z(hand)-axis is defined by the longitudinal axis of the third metacarpal bone. The X-axis projects backwards from the origin when the hand is in the normal anatomical position (palm facing forwards). The Y-axis passes through the origin and is perpendicular to the X-axis.

permissible exposure is at accelerations five times higher than those for four to eight hours of continuous exposure. Similarly, if the operation is such that vibration exposure periods are regularly followed by periodic rest periods without vibration, the correction factors for regularly interrupted vibration exposures of Table 9.4 should be applied.

It must be emphasized that the information and guidelines discussed above are all tentative at the present time and reflect a consensus based upon data available from both practical experience and laboratory experimentation of human responses. They cannot be taken to delineate completely safe exposure ranges in which vibration diseases cannot occur. The exposure limits presented will require further interpretation with respect to their application to specific operational exposure conditions and with respect to the definition

TABLE 9.2

Exposure guidelines for hand-transmitted vibration for 4 to 8 h uninterrupted or not-regularly-interrupted daily exposure [132].

Third Octave Band

Frequency (centre frequency of third octave band) Hz	Maximum rms value of vibration acceleration in each axis m/s^2	Maximum rms value of vibration velocity in each axis m/s
6.4	0.8	0.016
8	0.8	0.016
10	0.8	0.013
12.5	0.8	0.010
16	0.8	0.008
20	1	0.008
25	1.3	0.008
31.5	1.6	0.008
40	2	0.008
50	2.5	0.008
63	3.2	0.008
80	4	0.008
100	5	0.008
125	6.3	0.008
160	8	0.008
200	10	0.008
250	12.5	0.008
315	16	0.008
400	20	0.008
500	25	0.008
630	31.5	0.008
800	40	0.008
1000	50	0.008

Correction factor = 1

TABLE 9.3

Exposure guidelines for hand-transmitted vibration for 4 to 8 h uninterrupted or not-regularly-interrupted daily exposure [132]

Octave band

Frequency (centre frequency of octave band) Hz	Maximum rms value of vibration acceleration in each axis m/s^2	Maximum rms value of vibration velocity in each axis m/s
8	1.4	0.027
16	1.4	0.014
31.5	2.7	0.014
63	5.4	0.014
125	10.7	0.014
250	21.3	0.014
500	42.5	0.014
1000	85	0.014

Correction factor = 1

TABLE 9.4

Tentative correction factors for uninterrupted or not-regularly-interrupted and regularly-interrupted vibration exposure during daily shift (8 hour) [132]

VIBRATION EXPOSURE

Exposure time during daily shift	Uninterrupted or not regularly interrupted	Regularly-interrupted				
		Duration of recurrent time interval without vibration (in minutes per working hour)				
		Up to 10	More than 10 up to 20	More than 20 up to 30	More than 30 up to 40	More than 40
Up to 30 min	5	5	–	–	–	–
More than 30 min up to 1 h	4	4	–	–	–	–
More than 1 h up to 2 h	3	3	3	4	5	5
More than 2h up to 4 h	2	2	2	3	4	5
More than 4 h up to 8h	1	1	1	2	3	4

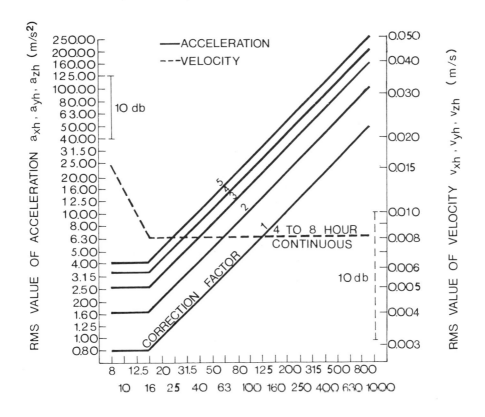

CENTER FREQUENCY OF THIRD OCTAVE BANDS IN Hz

FIG. 9.11.   Vibration exposure guidelines for hand (third-octave bands) [132].

and standardized measurement of the exposures produced by specific tools or machinery.

9.5.3. Reduction of hand–arm vibration

Vibration spectra resulting from machines and tools may be of constant acceleration level types or they may be frequency dependent [133]. The former relates to a constant energy per unit bandwidth, and results when the source of vibration is an impact or impulse, or a series of impacts or impulses such as in the case of pneumatic and electrical hammers, chisels, drills, pedestal grinders, and internal combustion engines. Sinusoidal sources of vibration excitation on the other hand produce frequency dependent spectra in which the fundamental frequency equals the sinusoidal repetition rate. Examples of sources of such vibrations include oscillations resulting from magnetic

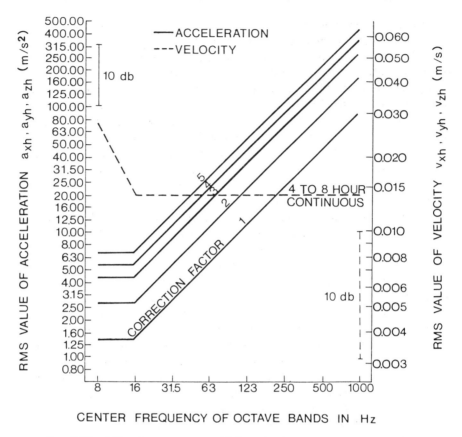

Fɪɢ. 9.12.   Vibration exposure guidelines for hands (octave bands) [132].

excitation or from forces arising from out-of-balance rotating machinery, such as in the case of a chain-saw. The vibration measurement is generally made as acceleration using accelerometers. Of course, conversion of measurement to velocity or displacement can be easily accomplished. Accelerometers are usually preferred over velocity or displacement transducers [134] due to their low cost, stability with time and temperature, small size and light weight (usually 1 to 30 g). In addition, their useful frequency range, with suitable amplification spans from approximately 0.5 Hz to 50 kHz which virtually covers the entire range of human sensitivity to vibrations. Usually accelerometers are used in conjunction with low-pass filters. Some configurations of chain saws can give rise to excessive vibrations. For example, single-cylinder gasoline-engine saws have a configuration in which the centre of gravity of rotating parts moves almost circularly, causing unbalance thereby vibrations [135]. Use of opposed-firing twin-cylinder engine has been made as a possible solution to eliminate the unbalance; however, the increased weight and width of the two-cylinder saw makes such a configuration unattractive. The solution that appears to have merit is to insulate the saw from vibration as is done in outboard motors and automobile engines.

Reynold and Soedel [136] analysed the vibration mechanism of chain-saw using two sets of tests. The first set included a narrow-band vibration analysis of a typical chain-saw during its operation. The second set involved a mechanical impedance analysis of the same chain-saw. In these tests the saw was excited with an electrodynamic shaker using an input of constant amplitude and variable frequency. For the corresponding frequencies, the vibration response of the saw was measured. From this information the broad-band resonance condition existing in the chain-saw body was derived. It was observed by the authors that the resonance conditions in the body reinforced the mechanism that caused the saw to vibrate, thus increasing the risk of potential injury.

The authors concluded that there were primarily two types of forcing function which caused the chain saw to vibrate. One type included shaking forces, composed of the fundamental shaking force and its harmonics, and the forces caused by the unbalanced masses of the clutch and the flywheel. The other type consisted of the pulses on the chain-saw caused by the exploding gas–air mixtures of the internal combustion engine. Also, the resonant structural characteristics of the chain-saw contributed significantly to the vibration amplitudes of the saw at frequencies above the fundamental forcing frequency of the saw. The tests showed that the influence of the cutting action of the saw on the vibration response was minimal.

One technique to reduce the perception of vibration is to isolate the hands from vibrations by using special rubber or foam backed gloves [137–140]. However, this arrangement has an effect on only a small part of vibration spectrum—normally over 800 Hz frequency and a few microns of amplitude. A more attractive option is to introduce rubber blocks between the handles and the body of the saw [141–143]. Figure 9.13 shows schematically the location of rubber blocks used for isolation purposes [135]. Two potential

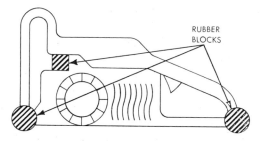

RUBBER
BLOCKS

FIG. 9.13.   Use of rubber blocks for vibration isolation of chain-saw handles.

problems with this design are that the rubber blocks are very sensitive to gasoline, oil, and heat; and that the shore rating of rubber blocks varies from one delivery to another. Significant improvements can be made by connecting the two handles, thereby increasing the weight of the handle-bar system, and by using softer rubber insulator blocks [140]. With an increase in weight, the resonance frequencies can be lowered below the principal frequency of the engine. The insulating rubber blocks are normally preloaded.

The attenuation of vibration transmission from the tool to the hands of the operator via covering the vibrating handle with sponge rubber or other resilient material was investigated by Suggs [144]. Figure 9.14 schematically illustrates this concept and gives a corresponding representation for the purpose of analytical modelling. Impedance measurements were made with the subject holding a vibrating handle with or without isolating cover sequentially. A change in impedance results from the cushion action of the handle cover. Foam rubber pipe insulation in two different thicknesses was wrapped around the tool handle. It was found that with resilient handle surfaces, the isolation of upper frequencies was possible. Additional studies on hand–arm–tool grip were carried out. It was shown that the mechanical impedance increased with an increase in gripping force. A tighter grip physically increases the effective stiffness of the tissue, and hence the force transmitted.

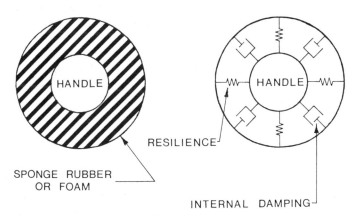

FIG. 9.14. Schematic representation of tool handle covering for vibration isolation.

9.6. Hand–arm system modelling

Models of hand–arm systems are important since an analytical framework can be developed to predict the resonance frequencies of hand–arm–tool system [145–149]. This information is of use to propose limitations on hazardous hand vibrations produced during the tools usage. Reynolds and

Soedel [150], Agate and Druett [130], Miwa [131] and Abrams and Suggs [151] have studied the response characteristics of hand–arm systems, in addition to several other researchers. Typically, a model for hand–arm systems, indicative of the vibratory response, in conjunction with a model of the hand-held power tools, yields important information leading to appropriate design of such tools.

A system of six accelerometers and seven-channel FM tape-recorders was used by Reynolds and Soedel with a 10 ft tape loop to determine the vibration characteristics of chain-saws [150]. Acceleration signals were measured and recorded in three orthogonal directions at both top and back handles. The recorded acceleration signals were then transferred to a constant bandwidth, narrow-band spectral analyser to obtain their auto power spectrums. These tests were very useful in understanding the broad-band response characteristics of the chain-saw at the handle.

A series of analytical and experimental studies of hand–arm vibration systems using cadaver limbs and fresh embalmed pig limbs was carried out by Suggs [144]. In a laboratory environment, bone mounting of accelerometers was used with shoulder end secured in an anchorage stand and a vibration input was applied at the distal end using an electrodynamic shaker. Bone was considered to be the basic transmitter of vibration, since it represents the most rigid component of the arm, with the surrounding tissue serving as an absorber or amplifier depending upon relative resiliencies. Incisions were made into the limb at the proximal and distal ends of the ulna and radius, and at the distal end of the humerus, and bones were drilled and tapped to allow for the mounting of the accelerometers. Using a driving point mechanical impedance technique it was possible to model and obtain the dynamic response of hand–arm system. The results indicated that the hand–arm system could be thought of as a low-pass mechanical filter which progressively attenuated higher frequencies.

9.7. Conclusions

The exposure to and attenuation of vibrations in an industrial and occupational context is of paramount importance from both biodynamic and engineering viewpoints. Transient and steady-state response data using human subjects have been found to be of much use in analysis and proper design of hand-held powered tools, automobiles, mass-transit vehicle suspension systems, and passenger elevators. Other typical areas of concern include manually-operated and controlled machinery, aerospace vehicles, ground-traversing devices and orthopaedic aids.

This chapter has considered the problem of human exposure to vibrations—while standing, such as on an industrial shop-floor in the vicinity of operating machinery; while seated, such as during a tractor, bulldozer, rough-terrain vehicle or bus operation; and while operating a powered hand-held tool, such as a chain-saw, pneumatic rock drill, riveter, grinder, or chipper. First, the effects of vibration exposure on health are discussed. Then for standing subjects, a measuring and modelling approach is presented for variation in input frequency applied to the feet. A sixteen-mass lumped-parameter model is advanced which reproduces measured dynamic response peaks at various resonance frequencies and closely conforms to the experimental response curves at the remaining frequencies.

Exposure to vibrations of seated subjects is discussed next. A survey of typical recent studies related to seated tractor operators is presented. Techniques for vibration attenuation via both passive and active suspensions are indicated. This is followed by a discussion of the occupational hand–arm vibration exposure problem in the United States and elsewhere [152,153]. Proposed guidelines from a draft ISO document entitled 'Principles for the Measurement and Evaluation of Human Exposure to Vibration Transmitted to the Hand' are given wherein tentative exposure limits and their frequency dependence are identified. Methods in use for the reduction or isolation of vibrations are enumerated. Finally, a brief discussion of studies relating to hand–arm system modelling is presented. It is apparent that extensive epidemiological, biological, medical, and engineering studies are needed which will lead to proper designs and working conditions with a view to protect the worker from excessive and harmful levels of vibration exposure in occupational and industrial environments.

References

1 GARG, D. P., and ROSS, M. A. (1976). Vertical mode human body vibration transmissibility. *IEEE Transactions on SMC* Vol. SMC–6, No. 2, pp. 102–12.

2 FROLOV, K. V. (1979). Modern problems of vibrations in the systems—man–machine environment. *International Symposium on Man Under Vibration: Suffering and Protection*, Udine, Italy. Centre International des Science Mecanique, Udine, Italy.

3 *Guide to the evaluation of human exposure to whole-body vibration*, ISO–2631, International Standards Organization, Geneva, Switzerland.

4 MORGAN, C. *et al.* (ed.), (1963). Human engineering guide to equipment design, Chapter 8, McGraw Hill, New York.

5 PRADKO, F., LEE, R., and KALUZA, Y. (1966). Theory of human vibration response. *ASME Paper No. 66-WA/BHF-15*.

6 COERMANN, R. R. (1962). The mechanical impedance of human body in sitting and standing position at low frequencies. *Hum. Factors* **4**, 227–53.

7 GARG, D. P. (1973). Analytical modeling of the human body subjected to

vibratory inputs. 8th Annual Meeting of the Association for the Advancement of Medical Instrumentation, Washington, DC, March 1973, Abstract in *Medical Instrumentation* **7**, 55.

8 GUIGNARD, J. C. (1965). Vibration. In *The textbook of aviation physiology* (ed. R. Gillies). Pergamon Press, New York.

9 SUGGS, C. W., and ABRAMS, C. F., JR. (1973). Simulation of whole body dynamics. *Proc. 5th Annual Southeastern Symp. System Theory*, March 1973, pp. 176–80. North Carolina State University, Raleigh, NC and Duke University, Durham, NC.

10 MACDUFF, J. N. (1969). Transient testing of man. *Sound Vibrat.* **3**, 16–21.

11 WEISE, E. B., and PRIMIANO, F. P., JR. (1966). The motion of the human center of mass and its relationship to mechanical impedance. *Hum. Factors* **8**, 399–405.

12 CHANEY, R. E., and BEAUPEURT, J. E. (1964). Subjective relation to whole-body vibration. *Tech. Rep. No. D3-6474*. The Boeing Company, Nichita, Kan.

13 PRADKO, F., LEE, A., and GREENE, J. D. (1965). Human vibration response theory. *ASME Paper No. 65-WA/HUF-19*.

14 GOLDMAN, D. E., and von GIERKE, H. E. (1960). Effects of shock and vibration on man. *Report No. 60–3*. Naval Medical Research Institute, Bethesda, Md.

15 STECH, E. L., and PAYNE, P. R. (1966). Dynamic models of the human body. *Report No. TR66-157*. Aerospace Medical Research Laboratory, Dayton, Ohio.

16 VON GIERKE, H. E. (Chairman) (1978). *Symposium on Biodynamic Models and their Applications*. Aviation Space and Environmental Medicine, Spec. Vol. **49**, No. 1, pp. 109–348.

17 MAGID, E. B., COERMANN, R. R., and ZIEGENRUECKER, G. H. (1960). Human tolerance to whole-body sinusoidal vibration. *J. Aerosp. Med.* **31**, 915–24.

18 VON GIERKE, H. E. (1964). Biodynamic response of the human body. *Appl. Mech. Rev.* 17, 951–8.

19 HANES, R. M. (1970). Human sensitivity to whole-body vibration in urban transportation systems: a literature review, *Report No. APL/JHU-TPR 004*, Johns Hopkins University, Baltimore, MD.

20 LOVESEY, E. J. (1970). The multi-axis vibration environment and man. *Appl. Ergonom.* **1**, 258–61.

21 HUANG, B. K. (1972). Digital simulation analysis of biophysical systems. *IEEE Trans. Biomed. Engng.* **19**, 128–39.

22 JONES, A. J., and SAUNDERS, D. J. (1974). A scale of human reaction to whole-body vertical sinusoidal vibration. *J. Sound Vibrat.* **35**, 367–75.

23 LANGE, W. (1974). A review of biomechanical models for evaluation of vibration stress. *AGARD-CP-145*.

24 LIPPERT, S. (1963). *Human vibration research*. Pergamon Press, New York.

25 VON GIERKE, H. E. (1971). Biodynamical models and their applications. *J. acoust. Soc. Am.* **50**, 1397–1413.

26 BRAUNOHLER, W. (1975). Effects of vibration on the musculoskeletal system. *Proceedings of the AGARD Conference*, Paper No. 145.

27 SANDOVER, J. (1978). Modelling human responses to vibration. *Aviat. Space envir. Med.* **49**, 335–39.

28 ASME (1970). *Dynamic response of biomechanical systems*. Special Publication. ASME, New York.

29 OSBORNE, D. J., and HUMPHREYS, D. A. (1976). Individual variability in human response to whole body vibration. *Ergonomics* **19**, 719–26.

30 BENEVOLENSKAYA, N. P., BENEVOLENSKAYA, N. P., BASOVA, T. T., and LYSENKO, L. L. (1979). Man–machine–object being worked–environment—system and vibration. *Int. Symp. Man Under Vibration: Suffering and Protection*, Udine, Italy. Centre International des Science Mecaniques, Udine.

31 SJOFLOT, L., and SUGGS, C. W. (1973). Human reactions to whole-body transverse angular vibrations compared to linear vertical vibrations. *Ergonomics* **16**, 455–68.

32 LEWIS, C. H., and GRIFFIN, M. J. (1978). Predicting the effects of dual frequency vertical vibration on continuous manual control performance. *Ergonomics* **21**, 637–50.

33 SHOENBERGER, R. W., and HARRIS, C. S. (1976). Psychophysical assessment of whole-body vibration. *Hum. Factors* **13**, 41–50.

34 FOTHERGILL, L. C., and GRIFFIN, M. J. (1977). The evaluation of discomfort produced by multiple frequency whole-body vibration. *Ergonomics* **20**, 263–76.

35 GRIFFIN, M. (1974). Some problems associated with the formulation of human response to vibration. In *The vibration syndrome* (ed. W. Taylor) pp. 13–23. Academic Press, London.

36 WASSERMAN, D. E., and BADGER, D. W. (1973). Vibration and the worker's health and safety. *Report No. 77*. United States Department of Health, Education and Welfare, Washington, DC and National Institute for Occupational Safety and Health, Cincinnati, Ohio.

37 British Standards Institution (1973). *Guide to the safety aspects of human vibration experiments*. Publication No. DD23. British Standards Institution, London.

38 PRADKO, F., and LEE, R. A. (1966). Vibration comfort criteria. *SAE Publication No. 660139*. Society of Automotive Engineers, New York.

39 LEWIS, C. H., and GRIFFIN, M. J. (1976). The effects of vibration on manual control performance. *Ergonomics* **19**, 203–16.

40 LANGE, K. O., and COERMANN, R. R. (1962). Visual acuity under vibrations. *Hum. Factors*, **4**, 291–300.

41 DENNIS, J. P. (1965). The effect of whole body vibrations on a visual performance task. *Ergonomics* **8**, 192–205.

42 National Research Council (1979). *The effects of whole body vibration on health.* Committee on Hearing, Bioacoustics and Biomechanics, National Research Council, NAS, Washington, DC.

43 GARG, D. P., and BENNETT, T. (1971). Transfer function and dynamic simulation of human beings. *Proc. 24th Annual Conf. Eng Med. Biol.* **13**, 302.

44 GARG, D. P. (1979). Dynamic modeling and vibratory response of human subjects in heave mode. *Int. Symp. Man Under Vibration: Suffering and Protection*, Udine, Italy. Centre International des Science Mecaniques, Udine.

45 GARG, D. P. (1978). Human body modeling under vibratory environments. *Proc. Nat. Systems Conf.* Ludhiana, India, pp. 21–5. Punjab Agricultural University, Ludhiana, India.

46 CARTER, D. R., and HAYES, W. C. (1976). Bone compressive strength. *Science, N.Y.* **194**, 1174–7.

47 CURREY, J. (1970). The mechanical properties of bone. *Clin. Orthop. Rel. Res.* **73**, 210–31.

48 YAMADA, H. (1970). *Strength of biological materials.* Williams and Wilkins, Baltimore, MD.

49 MARKOLF, K. (1970). Stiffness and damping characteristics of the thoraco-lumbar spine. *Proceedings of the Workshop on Bioengineering Approaches to Problems of the Spine.* National Institute of Health, Washington, DC.

50 PRASAD, P., and KING, A. (1974). An experimentally validated dynamic model of the spine. *J. appl. Mech.* **41**, 546–50.

51 WALKER, L. B., HARRIS, E. H., and PONTIUS, U. R. (1973). Mass, volume center of mass, and mass moment of inertia of head and neck of human body. Paper No. 730985, *Proc. 17th Stapp Car Crash Conf.* pp. 525–37. Society of Automotive Engineers, New York.

52 ADVANI, S., HUSTON, J., POWELL, W., and COOK, W. (1975). Human head–neck dynamic response: analytical models and experimental data. In *Aircraft crashworthiness* (eds. K. Saczalsi) pp. 197–212. University Press of Virginia.

53 BOWMAN, B., and ROBBINS, D. (1972). Parameter study of biomechanical qualities in analytical neck models. *Proc. 16th Stapp Car Crash Conf.* Paper No. 720957, pp. 14–43. Society of Automotive Engineers, New York.

54 CLOSE, R. (1972). Dynamical properties of mammalian skeletal muscles. *Physiol. Rev.* **52**, 129–97.

55 CULVER, C. C., NEATHERY, R. F., and MERTZ, H. J. (1972). Mechanical necks and human-like responses. *Proc. 16th Stapp Car Crash Conf.* Paper No. 720959, pp. 61–75. Society of Automotive Engineers, New York.

56 MOFFATT, C. A., HARRIS, E. G., and HALSAM, E. T. (1969). An experimental and analytical study of the dynamic properties of the human leg. *J. Biomech.* **2**, 373–87.

57 VON GIERKE, H. (1973). Dynamic characteristics of the human body. In *Perspectives in biomedical engineering* (ed. R. M. Kenedi) MacMillan, New York.

58 CLAUSER, C., MCCONVILLE, J., and YOUNG, J. (1969). Weight, volume, and center of mass of segments of the human body. *AMRL Report No. AMRL-TR-69-70.* Aerospace Medical Research Laboratory, WPAFB, Dayton, Ohio.

59 DRILLIS, R., CONTINI, R., and BLUESTEIN, M. (1964). Body segment parameters—a survey of measurement techniques. *Artif. Limbs* **8**, 44–66.

60 HIRSCH, A. E., and WHITE, L. A. (1965). Mechanical stiffness of man's lower limbs. *ASME Paper No. 65-WA/HUF-4.*

61 VIRGIN, W. J. (1951). Experimental investigation into the physical properties of the intervertebral disc. *J. Bone Jt Surg.* **33B**, 607–10.

62 COERMANN, R. R., ZIEGENRUECKER, G. H., WITTWER, A. L., and VON GIERKE, H. H. (1960). Passive dynamic mechanical properties of the human thorax–abdomen system and of the whole body system. *Aerospace Med.* **31**, 443–55.

63 BECKER, E. B. (1972). Measurement of mass distribution parameters of anatomical segments. *Proc. 16th Stapp Car Crash Conf.* Society of Automotive Engineers, New York.

64 GARG, D. (1978). Heave mode modeling and measurement of human body vibration transmissibility. *Proc. 28th Annual Conf. Engng Med. Biol.* p. 163.

65 CREDE, C. E. (1961). Application and design of isolators. In *Shock and vibration handbook* (ed. C. E. Harris and C. E. Crede) Vol. 2, pp. 32–1–32–53. McGraw-Hill, New York.

66 MANBY, T. C. D. (1964). Comfort and safety on tractors. National Institute of Agricultural Engineering, *Annual Report,* 1963–64.

67 ROSEGGER, R., and ROSEGGER, S. (1960). Health effects of tractor driving. *J. agric. Engng Res.* **5**, 241–75.

68 HUANG, B. K., and SUGGS, C. W. (1967). Vibration studies of tractor operators. *Trans. Soc. Automotive Engrs* **10**, 478–82.

69 DUPUIS, H., and CHRIST, F. (1966). Study of the risk of spinal damage to tractor drivers. Max-Planck Institute Report, West Germany.

70 DALE, A. K. (1979). The NIAE suspended cab tractor. *Int. Symp. Man Under Vibration: Suffering and Protection,* Udine, Italy. Centre International des Science Mechanique, Udine, Italy.

71 PARSONS, K. C., and GRIFFIN, M. J. (1978). Effect of rotational vibration

in roll and pitch axes on the discomfort of seated subjects. *Ergonomics* **21**, 615–25.

72 HILTON, D. J., and MORAN, P. (1975). Experiments in reducing tractor ride vibration with a cab suspension. *J. Agric. Engng Res.* **20**, 4–14.

73 GRIFFIN, M. (1975). Vertical vibration of seated subjects: effects of posture, vibration level, and frequency. *Aviat. Space Environ. Med.* **46**, 269–76.

74 STIKELEATHER, L. F., HA, G. O., and RADKET, A. O. (1972). Study of vehicle vibration spectra as related to seating dynamics. *Society of Automotive Engineers Publication No.* 720001.

75 HAACK, M. (1955). Tractor seat suspension for easy riding. *Soc. Automotive Engineers Trans.* pp. 452–69.

76 MUKSIAN, R., and NASH, C. D. (1974). A model for the response of seated humans to sinusoidal displacements of the seat. *J. Biomech.* **7**, 209–15.

77 MENDEL, M. J., and LOWRY, R. D. (1962). One-minute tolerance in man to vertical sinusoidal vibrations in the sitting position. *Report No. AMRL-TDR-62-121.* Wright–Patterson Air Force Base, Dayton, Ohio.

78 PATIL, M. K., PALANICHAMY, M. S., and GHISTA, D. N. (1978). Man–tractor system dynamics: towards a better suspension system for human ride comfort. *J. Biomech.* **11**, 397–406.

79 PATIL, M. K., PALANICHAMY, M. S., and GHISTA, D. N. (1979). Minimization of tractor-occupant's traumatic vibrational response by means of the PPG tractor seat suspension. *Int. Symp. Man Under Vibration: Suffering and Protection*, Udine, Italy. Centre International des Science Mechanique, Udine, Italy.

80 MATHEWS, J. (1973). The measurement of tractor ride comfort. *Society of Automotive Engnrs Paper No.* 730795.

81 GOLDMAN, D. E., and VON GIERKE, H. E. (1961). Effects of shock and vibration on man. In *Shock and vibration handbook* (ed. C. E. Harris and C. E. Crede) Vol. 3, pp. 44–1–44–50. McGraw Hill, New York.

82 SCHULTZ, O. (1978). Active multivariable vibration isolation for a helicopter by decoupling and frequency-domain methods. *7th Int. Congr.* International Federation of Automatic Control, Helsinki.

83 KSIAZEK, M. (1978). Determination of optimum vibroisolation systems for a sitting human operator. *Aviat. Space Environ. Med.* **49**, 257–61.

84 NASA (1972). *Evaluation, design, fabrication and testing of two electrohydraulic isolation systems for helicopter environments*, NASA Langley Research Center, Contract No. NAS1–10103, Hampton, Va.

85 SHUBERT, D. W., and RUZICKA, J. E. (1969). Theoretical and experimental investigation of electrohydraulic vibration isolation systems, Journal of Engineering for Industry, Series B, *Trans. ASME*, **91**, 981–90.

86 CALCATERRA, P. C. (1972). Active vibration isolation of aircraft seating. *Sound Vibrat.* **6**, 18–23.

87 FROLOV, K. V., SINJOV, A. V., SOLOVJOV, V. S., and SAFRONOV, J. G.

(1979). Kinematic type of active vibrational device. Presented at the *International Symposium on Man Under Vibration: Suffering and Protection*, Udine, Italy. Centre International des Sciences Mecanique, Udine, Italy.

88 ASANOVA, T. P. (1976). Vibration disease among workers using portable power tools in Finnish shipyards. In *Vibration and Work: Proceedings of the Finnish–Soviet–Scandinavian Vibration Symposium* (ed. O. Korhonen), pp. 52–62. Institute of Occupational Health, Helsinki.

89 BUTKOVSKAYA, Z. M., KADYSKINA, Y. N., POTEMKIN, B. A., and FROLOV, K. V. (1973). Study of the influence of vibrations of machines and tools on the arms of a human operator with various initial conditions. In *Vibration insulation of machines and vibration protection of the human operator*, pp. 5–16. Nauka Press, Moscow.

90 LOUDA, L., and LUKAS, E. (1975). Hygienic aspects of occupational hand–arm vibration. *Proceedings of the International Occupational Hand–Arm Vibration Conference*, pp. 60–6. National Institute of Occupational Safety and Health, Cincinnati, Ohio.

91 BENEVOLENSKAYA, N. P., BASOVA, T., and VANAG, G. A. (1978). *Problems of labor protection during riveting operations*. Nauka, Moscow.

92 LIDSTRÖM, I. (1974). Medical aspects of recent vibration work in Sweden. In *The vibration syndrome* (ed. W. Taylor) pp. 187–93. Academic Press, London.

93 RAYNAUD, M. (1862). Local asphyxia and symmetrical gangrene of the extremities. M.D. Thesis, Rignoux, Paris.

94 Taylor, W. and Pelmear, P. (ed.) (1975). *Vibration white finger in industry*. Academic Press, London.

95 AGATE, J. (1949). An outbreak of cases of Raynaud's disease of occupational origin. *Br. J. ind. Med.* **6**, 144–63.

96 TAYLOR, W., PELMEAR, P., and PEARSON, J. (1975). Vibration-induced white finger epidemiology. In *Vibration white finger in industry* (ed. W. Taylor and P. Pelmear) pp. 1–14.

97 JAMES, P., and GALLOWAY, R. (1974). Brachial arteriography in vibration induced white finger. In *The vibration syndrome* (ed. W. Taylor) pp. 195–203. Academic Press, London.

98 STEWART, A., and GODA, D. (1970). Vibration syndrome. *Br. J. ind. Med.* **27**, 19–27.

99 TAYLOR, W. (ed.) (1974). *The vibration syndrome*. Academic Press, London.

100 MCCALLUM, R. I. (1971). Vibration syndrome. *Br. J. Ind. Med.* **28**, 90–3.

101 WALTON, K. W. (1974). The pathology of Raynaud's phenomenon of occupational origin. In *The vibration syndrome* (ed. W. Taylor) pp. 109–19. Academic Press, London.

102 HAMILTON, A. (1918). A study of spastic anaemia in the hands of stone

cutters. U.S. Department of Labor Statistics, *Bulletin No. 236*, Washington, DC.

103 SEYRING, M. (1930). Maladies from work with compressed air drills. *Arch. Gewerbehyg* (Berlin) **1**, 359–75.

104 HUNT, J. H. (1936). Raynaud's phenomenon in workmen using vibratory instruments. *Proc. R. Soc. Med.* **30**, 171–7.

105 LIDSTRÖM, I. M. (1975). Vibration injury in rock drillers, chislers, and grinders. *Proceedings of the International Occupational Hand-Arm Vibration Conference*, NIOSH, Cincinnati, Ohio, pp. 77–83.

106 TEISINGER, J., and LOUDA, L. (1972). Vascular disease, disorders resulting from vibrating tools. *J. occup. Med.* **14**, 129–33.

107 TELFORD, E. D., McCANN, M. B., and McCORMACK, D. H. (1945). Dead hand in users of vibrating tools. *Lancet* **ii**, 359–61.

108 AGATE, J., and TOMBLESON, J. (1946). Raynaud's phenomenon in grinders of small metal castings. *Br. J. ind. Med.* **3**, 167.

109 PELMEAR, P. L., TAYLOR, W., and PEARSON, J. (1974). Raynaud's phenomenon in pedestal grinders. In *The Vibration Syndrome* (ed. W. Taylor) pp. 141–7. Academic Press, London.

110 ASHE, W. F., COOK, W. T., and OLD, J. W. (1962). Raynaud's phenomenon of occupational origin. *Arch. environ. Hlth* **5**, 333–43.

111 GROUNDS, M. (1964). Raynaud phenomenon in the users of chain saws. *Med. J. Aust.* **1**, 270–2.

112 MIURA, T., KIMURA, K., TOMINAGA, Y., and KIMOTSUKI, K. (1966). *On Raynaud's phenomenon of occupational origin due to vibrating tools—its incidence in Japan.* Report No. 65 of the Institute for Science and Labor, Tokyo, Japan. pp. 1–11.

113 HELLSTROM, B., and LANGE ANDERSON, K. (1970). Vibration injuries in Norwegian forest workers. *Br. J. ind. Med.* **29**, 255–63.

114 BJERKE, N., KYLIN, B., and LIDSTRÖM, I. M. (1970). Changes in the vibratory sensation threshold after exposure to powerful vibration. *Work-Environment-Health* **7**, 1–7.

115 LEWIS, T. (1938). The pathological changes in the arteries supplying the fingers in warm handed people and in cases of so-called Raynaud's disease. *Clin. Sci.* **3**, 287.

116 DUPUIS, H., and JANSEN, G. (1979). Immediate effects of vibration transmitted to the hand. *International Symposium on Man Under Vibration: Suffering and Protection*, Udine, Italy. Centre International des Science Mecanique, Udine, Italy.

117 GILIOLI, R., TOMASINI, M., BULGHERONI, C., FILIPPINI, G., and GRIECO, A. (1979). Effects of vibrating tools on the peripheral vessels and the peripheral nervous system in workers of an iron foundry—preventive suggestions. *International Symposium on Man Under Vibration: Suffering*

and Protection, Udine, Italy. Centre International des Science Mecanique, Udine, Italy.

118 TAYLOR, W., PEARSON, J., KELL, R. L., and KEIGHLEY, G. D. (1971). Vibration syndrome in forestry commission chain saw operators. *Br. J. ind. Med.* **28**, 83–90.

119 TAYLOR, W., PELMEAR, P., and PEARSON, J. (1975). A longitudinal study of Raynaud phenomenon in chain saw operators. In *Vibration white finger in industry* (ed. W. Taylor and P. Pelmear) pp. 15–20. Academic Press, London.

120 TAYLOR, W., PELMEAR, P. L., and PEARSON, J. (1974). Raynaud's phenomenon in forestry chain saw operators. In *The vibration syndrome* (ed. W. Taylor) pp. 122–39. Academic Press, London.

121 LENODIA, D. (1975). Ecological elements of vibration (chipper's) syndrome. *Proceedings of the International Occupational Hand–Arm Vibration Conference*, National Institute of Occupational Safety and Health, Cincinnati, Ohio, pp. 89–96.

122 BOVENZI, M., PETRONI, L., and DI MARINO, F. (1979). Epidemiological survey of shipyard workers exposed to hand–arm vibration. In *International Symposium on Man Under Vibration: Suffering and Protection*, Udine, Italy. Centre International des Science Mecanique, Udine, Italy.

123 TICHAUER, E. R. (1975). Thermography in the diagnosis of work stress due to vibrating implements. *Proceedings of the International Occupational Hand–Arm Vibration Conference*, National Institute of Occupational Safety and Health, Cincinnati, Ohio, pp. 160–8.

124 HEMPSTOCK, T., and O'CONNOR, D. (1975). The measurement of hand–arm vibration. In *Vibration white finger in industry* (ed. W. Taylor and P. Pelmear) pp. 111–22. Academic Press, London.

125 REYNOLDS, D. D. (1975). Hand–arm vibration: a review of 3 years' research. *Proceedings of the International Occupational Hand-Arm Vibration Conference*, National Institute of Occupational Safety and Health, Cincinnati, Ohio, pp. 99–128.

126 KRAUSE, P., ORBAN, A., PANZKE, K. J., and POPOV, K. (1979). Critical assessment of common methods to determine vibrational stress of hand–arm system. Paper presented at the *International symposium on man under vibration: suffering and protection*, Udine, Italy. Centre International des Science Mecanique, Udine, Italy.

127 KITCHENER, R. (1975). The measurement of hand–arm vibration in industry, *Proceedings of the International Occupational Hand–Arm Vibration Conference*, National Institute of Occupational Safety and Health, Cincinnati, Ohio, pp. 153–9.

128 KULMIN, T., WIIKERI, M., and SUMARI, P. (1973). Radiological changes in carpal and metacarpal bones and phalanges caused by chain saw vibration. *Br. J. Ind. Med.* **33**, 71–3.

129 KEIGHLEY, G. D. (1975). Hand–arm vibration: British Standards Institution draft for development. In *Vibration white finger in industry* (ed. W. Taylor and P. Pelmear) pp. 135–48. Academic Press, London.

130 AGATE, J., and DRUETT, H. (1946). A method for studying vibrations transmitted to the hands. *Br. J. Med.* **3**, 159–66.

131 MIWA, T. (1968). Evaluation methods for vibration effect. Part IV: measurement of vibration greatness for whole body and hand in vertical and horizontal directions. *Ind. Hlth* **6**, 1–10.

132 *Guide for the measurement and the evaluation of human exposure to vibration transmitted to the hand.* International Standard Organization Draft Proposal No. 5349, ISO TC 108/SC4/WG3, (1978).

133 ACTON, W. (1974). Aspects of field vibration measurement. In *The vibration syndrome* (ed. W. Taylor) pp. 25–33. Academic Press, London.

134 KUEHN, J. (1974). Vibration measurement. In *The vibration syndrome* (ed. W. Taylor) pp. 71–83. Academic Press, London.

135 NÄSLUND, U. (1974). Design problems in the reduction of vibration in chain saws. In *The vibration syndrome* (ed. W. Taylor) pp. 61–70. Academic Press, London.

136 REYNOLDS, D., and SOEDEL, W. (1974). Vibration testing of chain saws. In *The vibration syndrome* (ed. W. Taylor) pp. 91–107. Academic Press, London.

137 Brammer, A. J. (1975). Influence of hand-grip on the vibration amplitude of chain-saw handles. *Proceedings of the International Occupational Hand Arm Vibration Conference*, National Institute of Occupational Safety and Health, Cincinnati, Ohio, pp. 179–86.

138 MIWA, R. (1964). Studies on hand protectors for portable vibrating tools, Part I: Measurements of the attenuation effect of porous elastic materials. *Ind. Hlth* **2**, pp. 95–105.

139 AXELSSON, S. A. (1975). Progress in solving the problem of hand–arm vibration for chain saw operators in Sweden, 1967 to date. *Proceedings of the International Occupational Hand–Arm Vibration Conference*, National Institute of Occupational Safety and Health, Cincinnati, Ohio, pp. 218–24.

140 SUGGS, C. (1970). Vibration of machine handles and controls and propagation through the hands and arms. *Proceedings of the 4th International Ergonomics Congress*, Strasbourg, France.

141 YAMAWAKI, S. (1975). Reduction of vibration in power saws in Japan. *Proceedings of the International Occupational Hand–Arm Vibration Conference*, National Institute of Occupational Safety and Health, Cincinnati, Ohio, pp. 209–17.

142 POLITSCHUK, A. P., and OBLIVIN, V. N. (1975). Methods of reducing the effects of noise and vibration on power saw operators. *Proceedings of the International Occupational Hand–Arm Vibration Conference*, National

Institute of Occupational Safety and Health, Cincinnati, Ohio, pp. 230–2.

143 KEIGHLEY, G. D. (1975). FAO/ECE/ILO Resolution on Hand–Arm Vibration for Modern Anti-vibration Chain Saws: Recommendations for Medical Monitoring. *Proceedings of the International Occupational Hand–Arm Vibration Conference*, National Institute of Occupational Safety and Health, Cincinnati, Ohio, pp. 233–5.

144 SUGGS, C. (1974). Modelling of the dynamic characteristics of the hand–arm system. In *The vibration syndrome* (ed. W. Taylor), pp. 169–86. Academic Press, London.

145 REYNOLDS, D., and KEITH, R. (1977). Hand–arm vibration, Pt. I: analytical model of the vibration response characteristics of the hand. *J. Sound Vibrat.* **51**, 237–53.

146 SUGGS, C. (1971). Mechanical impedance technique for evaluating the dynamic characteristics of biological material. *J. Agric. Engng Res.* **16**, 307–15.

147 ABRAMS, C., and SUGGS, C. (1971). Modeling the vibrational characteristics of the human hand by the driving point mechanical impedance method. *Proceedings of the Purdue Noise Control Conference*, Lafayette, In, pp. 234–8.

148 SUGGS, C. W., and MISHOE, J. W. (1975). Hand–arm vibration: implications drawn from lumped parameter models. In *Proceedings of the International Occupational Hand–Arm Vibration Conference*, National Institute of Occupational Safety and Health, Cincinnati, Ohio, pp. 136–41.

149 MELTZER, G. (1979). A vibration model for the human hand–arm-system. Paper presented at the *International Symposium on Man Under Vibration: Suffering and Protection*, Udine, Italy. Centre International des Science Mecanique, Udine, Italy.

150 REYNOLDS, D., and SOEDEL, W. (1974). Dynamic response of the hand–arm system to sinusoidal input. In *The vibration syndrome* (ed. W. Taylor) pp. 149–68. Academic Press, London.

151 ABRAMS, C., and SUGGS, C. (1975). Mechanical impedance modeling of the vibrational characteristics of human hand–arm system. *Proceedings of the 5th Annual Southeastern Symposium on System Theory* pp. 184–7. North Carolina State University, Raleigh, NC and Duke University, Durham, NC.

152 WASSERMAN, D. E., and BADGER, D. W. (1974). Industrial vibration—an overview. *Am. Soc. safety Engng J.* **19**, pp. 38–43.

153 GAGE, H. (1975). Correlation of segmental vibration with occupational disease. *Proceedings of the International Occupational Hand–Arm Vibration Conference*, National Institute of Occupational Safety and Health, Cincinnati, Ohio, pp. 239–43.

10. Biomechanical aspects of low-back pain

RICHARD OUDENHOVEN, DHANJOO N. GHISTA, AND
GAUTAM RAY

10.1. The prevalence of back pain

Back pain, although not truly a disease, is the greatest epidemic society has to face. Disability resulting from back pain continues to increase annually at an alarming rate. Statistically, low-back pain occurs in all occupations with equal frequency. Disability is not related to occupation, race, or sex and is equally as common in white- as in blue-collar workers. However, a strong correlation exists between an increase in the incidence of low-back injury and an increased level of manually handled load.

Health statistics of low-back pain are overwhelming. In 1977, treatment and compensation of low-back pain sufferers in the United States was in excess of fourteen billion dollars. This exceeded the cost of all other industrial injuries. As early as 1965, the National Center for Health Statistics reported that 6.5 million individuals were in bed on a given day because of chronic back pain. It is further projected that two million cases of low-back pain would be added each year; assuming these facts to be true, currently 34.5 million people are disabled. In June 1965 Marshall Smith (*Life* magazine) estimated that 20 million Americans suffered from 'bad backs'; he further estimated that in 1970, 30 million would suffer from similar pain. In the United States, for every thousand employed individuals, 1400 working days are lost annually as a result of back pain.

Low-back pain is not peculiar to the United States. It is characteristic of all industrialized societies. In Britain, during the course of one year, 1.3 million new patients see their family physician because of back pain. Back pain ranks second to stroke as a threat to health in middle-aged people. For every one thousand people working, it is estimated 530 days are lost annually through back pain.

In 1973 fiscal reports from Sweden revealed that 80 per cent of its population at one time or another had low-back pain. A minimum of 4 per cent of the working population between the ages of 20 and 50 are absent from work for a minimum of one month to an average of three months because of back pain. Statistics from Japan, too, report of incidence of 80 per cent.

The impact of chronic back pain is even more dramatic in Holland. In 1955, 530 people were permanently disabled because of back pain and 4500 in 1965. As a result of social change and new disability requirements, it is projected that in 1980, 100 000 (one out of every eighty Dutch citizens) would be similarly disabled.

These statistics are not presented to magnify the incidence of low-back pain, nor to imply that such pain is invariably disabling. They are presented to emphasize that back pain is far from unusual. Since the general health of patients with chronic back pain is excellent, national mortality statistics do not include low-back pain. Nevertheless, overall disability due to chronic back pain is greater than that due to cancer, heart disease, and hypertension.

In the past decade, tuberculosis and poliomyelitis have been eliminated as a threat to national health; new vaccines become available daily; total hip and knee replacements are an everyday occurrence, and open heart surgery is no longer unusual. Chronic back pain is now the greatest threat to our national health.

10.2. Functional anatomy—how the parts work

Integrated vertebral movement is the result of co-ordinated action of the soft tissues lying between the vertebrae: 'motion segment'. The primary components of a motion segment are the intervertebral disc, the apophyseal (facet) joints, and the major ligaments (Fig. 10.1). Without co-ordinated interplay of these soft tissues, spine motion would not be possible.

The intervertebral disc is the most dynamic of the three structures comprising the motion segment. The intervertebral disc is subjected to stress

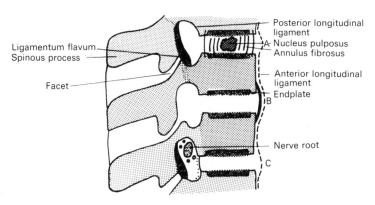

FIG. 10.1.   Schematic diagram.
A. Intervertebral disc.
B. Motion segment.
C. Intervertebral foramen.

whether the individual is upright, walking, or lying down. Its nucleus is the most specialized portion of the disc and is 'incompressible'. During rest, the nucleus has a round shape; stress or pressure deforms the nucleus, and the annulus then exercises its restraining influence, equalizing the forces of bending, twisting, or turning. Classic examples of the interplay between the components of the intervertebral disc are simple flexion and extension of the spine. In extension (bending backwards), the intervertebral disc, as a result of body weight and muscle contraction, opens up in front and shuts down in back, while the nucleus, ordinarily placed posteriorly, moves towards the middle. In flexion (bending forward), the intervertebral disc closes up in front and opens up at the back, while the nucleus moves toward the back. As a result of this interplay in movement of the nucleus within the annulus, compression forces are distributed evenly throughout the nucleus and the annulus, avoiding injury to the supporting bony structures of the vertebrae.

The apophyseal (facet) joints limit mobility between two vertebrae to a predetermined degree and direction. Their own movement is restrained by the intervertebral disc and the ligamentum flavum. Apophyseal joints and intervertebral discs are interdependent; together they limit back movement and distribute forces as equally as possible between themselves. Movement of one affects movement of the other. Put in another way, the intervertebral disc protects the body of the vertebrae from wear and tear and, at the same time, allows considerable freedom of movement. Apophyseal joints keep one vertebra from slipping or twisting on the other. As particular back muscles contract to bring about forward bending or extension of the spine, the apophyseal joints do not actually slide, but have a 'ball and socket' motion somewhat similar to the relationship of the hip to the pelvis. The upper lumbar vertebrae are capable of only flexion and extension because their articular surfaces have a sagittal or up and down direction (Fig. 10.2). Lateral flexion (bending from side to side) is permitted at the lumbosacral joint (junction of the fifth lumbar vertebra to the sacrum) because of the coronal (oblique) plane of the facets.

10.3. Back pain—what goes wrong?

Many different diseases have symptoms of low-back and extremity pain: tumors of the vertebrae (benign, malignant, and metastatic); errors of body metabolism with secondary changes in the vertebrae (gout, postmenopausal osteoporosis); inflammation of the spine (tuberculosis, viruses, bacteria); circulatory diseases; emotional disturbances; disease of distant, unrelated structures (kidney and pelvis); rheumatoid arthritis and mechanical spinal instability with disc degeneration and herniation. This chapter considers the most common causes; mechanical instability, disc degeneration, and herniation.

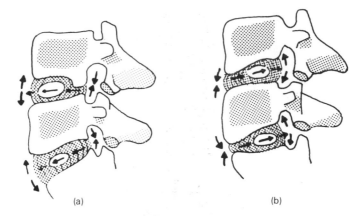

(a)   (b)

FIG. 10.2.   (a) Lumbar extension. Intervertebral spaces open anteriorly and shut down posteriorly. Nucleus and axis of movement shift forward. (b) Lumbar flexion. Intervertebral spaces open up posteriorly, shut down anteriorly. Nucleus and axis of movement shift backward.

Structural weaknesses of the vertebrae and supporting structures may be the result of congenital abnormality, developmental abnormality, or the application of continual and excessive mechanical stress to normal structures. Congential abnormalities of the lumbosacral junction are the most common abnormalities of the spine. Among these are: 'spondylolysis'—defects in the neural arch; 'spondylolisthesis'—actual separation of the body from the neural arch; 'transitional vertebrae'—incompletely developed vertebrae; and an 'abnormal number of vertebrae'—either four or six.

Developmental abnormalities are those abnormalities that occur during the course of growth and aging and are most frequent at the lumbosacral level. Spine stability demands that the lower lumbar vertebrae and sacrum be seated deeply within the pelvis. Deep seating of the sacrum and the lower lumbar vertebrae decreases stress. Should the fourth and fifth lumbar vertebrae not be protected by the brim of the pelvis, these vertebrae would be more receptive to injury. Deep-seated vertebrae are protected by this strong bony structure, its muscles and ligaments. The plane or angle of the apophyseal joints also makes them more prone to stress. The upper lumbar apophyseal joints are in a side-to-side transverse position. At the lumbosacral level their direction is more oblique (Fig. 10.3).

Finally, the junction of the fifth lumbar vertebra with the sacrum is a transitional zone, (any bony junction consisting of a mobile segment and an immobile segment). The mobile segment in this case is the fifth lumbar vertebra and the immobile segment is the sacrum and pelvis. Transitional zones are particularly vulnerable to excesses of abnormal spine loading, including poor posture, gait abnormalities, and excessive lifting or prolonged

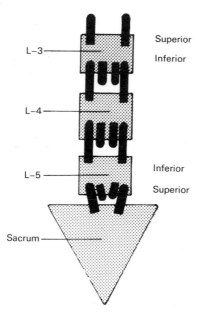

F<small>IG</small>. 10.3.   Planes of articular facets (note change in direction of facets at L-5-sacrum
as compared with L-3-4 and L-4-5).

sitting. Overloading applied to the right or the left side of the spine results in
gradual changes of the planes of these apophyseal joints. Thus, asymmetry or
variations in the direction of either the right or the left apophyseal joint, as
compared with the opposite side, occurs with unusual frequency. This results
in imbalance. Proper spine motion is dependent upon the wellbeing and
symmetry of the major structures of the vertebrae: the body and neural arches.
Any defect of the vertebral body or its supporting ring immediately
contributes to inadequate spine mobility.

Stability and strength of the spine are not the exclusive responsibility of the
vertebrae, but are also dependent upon strong supporting structures; in
particular, the motion segments (intervertebral disc and apophyseal joints),
muscles and ligaments. Improper forces applied to weak muscles, ligaments,
and joint surfaces will precipitate and aggravate spine instability. As a result
of an improper loading of a spine, which is poorly supported by weak muscles,
one vertebra no longer enjoys rhythmic, smooth motion with the adjoining
vertebrae in bending, turning, stretching, or twisting. Even an architecturally
sound back without bony abnormalities will ultimately be weakened by forces
applied, as a result of poor muscle tone, lack of exercise, and misuse of body
mechanics. Improper spine loading is not only the result of excessive lifting, it
can occur with any improperly performed body activity (such as prolonged,
unsupported sitting). Weakness or defect in any structure of the low back, be it

bone, muscle or ligament, or disc, results in changes in associated structures (a domino effect). In other words, a defect in the disc ultimately affects the apophyseal joints and vertebrae.

Initial and repeated overloading of the spine results in fractures of the endplates (Fig. 10.4). The nucleus and annulus are deprived of needed nutrition; degeneration occurs. Degeneration of the disc, as the result of spine overloading, is enhanced by spontaneous and gradual loss of the disc's water content, which occurs in all between the ages of 20 and 50 years. Heredity also plays some factor in disc degeneration. Degeneration results in loss of elasticity and viscosity of the nucleus and annulus. The annulus and nucleus lose their shock-absorbing capacity and their close interplay with the apophyseal joints is no longer effective (Fig. 10.5).

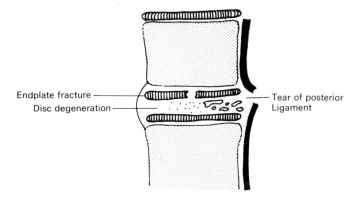

Endplate fracture ——— Tear of posterior
Disc degeneration ——— Ligament

FIG. 10.4.   Endplate fracture.

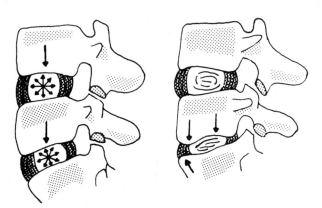

A - Normal nucleus   B - Nuclear destruction

FIG. 10.5.   Effect of nuclear destruction on motion segment:
(a) Normal transmission from nucleus to annulus and facet.
(b) Narrowed interspace produces incongruity of facets.

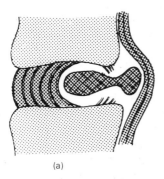 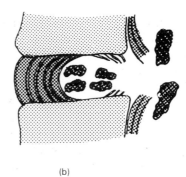

(a)   (b)

FIG. 10.6.   Comparison of nuclear protrusion and nuclear herniation.
A. Nuclear protrusion through annulus with intact posterior ligament.
B. Nuclear herniation—tear in annulus and posterior ligament.

Continual overloading of the spine does occur in sedentary or a labouring capacity; spontaneous loss of water content, compounded by the unique vulnerability of the lower lumbar spine to rotational stress and generally weak muscles and poor posture results in some inevitable degeneration of the disc. The fibres of the annulus weaken, bulge, and may tear (Fig. 10.6). Since the nucleus in the lower lumbar region has a posterior location, injury to the fibres of the annulus occur primarily in this posterior direction. Posterior changes and bulging of the annulus result in pressure on the posterior longitudinal ligament, the weaker of the two major spinal ligaments (Fig. 10.7). As a result, the sinuvertebral nerves, which supply the posterior longitudinal ligament and anterior dura, become irritated and inflamed. Back and extremity pain may follow rapidly or slowly. Should tears of the annulus and posterior longitudinal ligament develop, the gelatinous core (nucleus) will escape from the vertebral canal, resulting in the extrusion or herniation of the nucleus (the slipped disc) with pressure on the nerve root (Fig. 10.8).

There are three phases in the production of back and extremity pain: instability of the motion segment, disc degeneration, and disc herniation. Phases two and three do not invariably follow phase one. Minor injuries to the motion segment, be it the disc or apophyseal joints, will heal without continuing pain if they are given enough time and not subjected to repeated overloading.

Osteoarthritis (degenerative arthritis) is not a satisfactory explanation for chronic back pain. Degenerative arthritis is invariably a radiological and not a clinical diagnosis. It is no more diagnostic than lumbago or sciatica. Degenerative arthritis involves almost exclusively the front (anterior) or side (lateral) portions of the vertebra and rarely develops posteriorly near the nerve root or pain receptors (Fig. 10.9). These overgrowths are the result of abnormal stress applied to the spine, with degeneration of the annulus

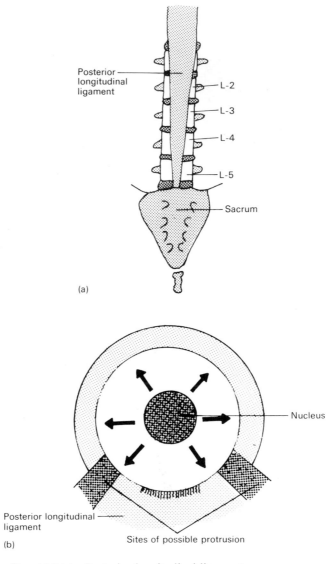

Posterior longitudinal ligament

L-2

L-3

L-4

L-5

Sacrum

(a)

Nucleus

Posterior longitudinal ligament

(b)

Sites of possible protrusion

FIG. 10.7.(a) Posterior longitudinal ligament.
(b) Posterior longitudinal ligament—cross section.

fibrosus, while the nucleus pulposus remains intact. Development is anterior at the body, because of the peculiarities of the attachment of the anterior longitudinal ligament; the anterior longitudinal ligament is fixed to the vertebral bodies and not attached at the disc, while in contrast, the posterior ligament is attached at the disc and unattached to the vertebral body. The

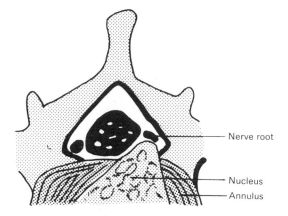

FIG. 10.8.   Nerve root compression resulting from herniated nucleus.

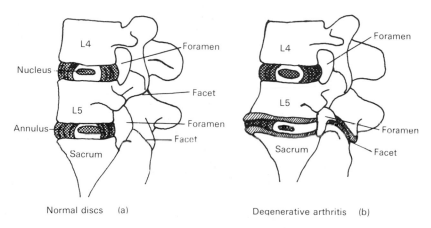

FIG. 10.9.   Degenerative arthritis.
(a) Normal motion segment.
(b) Abnormal motion segment L5-S1 with bulging of annulus. Anteriorly—anterior bone overgrowth overriding facet and normal nucleus.

anterior ligament thus offers a course for degeneration when subjected to stress. Bulging of the annulus results in bone irritation and excessive bone growth (Fig. 10.10). Ultimately, bone overgrowth results in an immobile, non-painful low back. Eighty-year-old patients with X-ray findings of extensive, abnormal bony overgrowths rarely complain of pain; but their X-ray findings are horrendous. There is a regrettable tendency to attribute chronic back pain in patients between the ages of 40 and 70 to arthritis, but generally this diagnosis is based purely on X-ray findings.

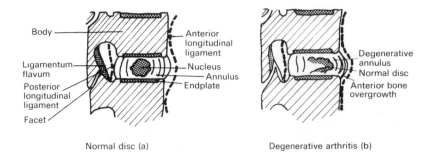

Normal disc (a)

Degenerative arthritis (b)

FIG. 10.10. Schematic drawing showing development of degenerative arthritis. (Schmorl, George, Junghanns, and Hubert (1971). *The human spine in health and disease.* Grun and Stratton, New York.

The diagnosis of 'arthritis' has become so acceptable to patients that an experienced, knowledgeable physician has great difficulty convincing his patient that arthritis is not the cause of the pain. The more advanced the degenerative arthritis is, the more stable the joint. Degenerative arthritis, when complete, must result in a stable, immobile, non-painful spine. Such a spine resists stress of twisting and turning, more so even than a normal spine.

Soft tissue injuries (muscular strain or sprain) do not play a significant role in the development of persistent pain or in the prevention of healing. The diagnoses are symptomatic and are misleading. These terms are erroneously used by some physicians when their examination reveals spasm of the back muscles and some restricted back motion. Disabling muscular strains or sprains are not present unless there is underlying mechanical instability of the spine.

10.4. Mechanics of factors aggravating and preventing back pain

Major advances have been made in the understanding of back and extremity pain. The underlying causes are poor posture, inadequate muscle tone, developmental vertebral (bony) abnormalities, sedentary existence, and increase in body weight. These underlying weaknesses accentuate the stresses in the components of the motion segment, resulting in its injury due to overloading of the spine. This overloading occurs equally as well during recreational as during occupational activity.

Early prevention

Prevention begins when a child first walks. Good posture means avoiding rounded or stooped shoulders; proper posture need not be military: stand erect, ears in line with the shoulders, back straight, feet directed forward (Fig. 10.11). In proper posture, the normal and bending stresses on the disc, due to

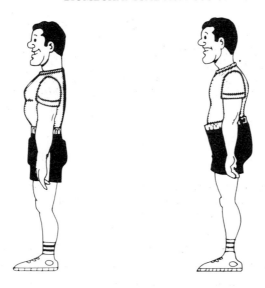

F<small>IG</small>. 10.11.   Proper posture (on the left relative to that on the right).

compressive forces and bending moments, caused by eccentrically acting body segmental weights (of the head, upper extremities, and trunk) acting eccentrically with respect to spinal neutral axis, are within allowable limits. However, malposture results in abnormal spinal curvature, undue eccentricity of body segment weight and passive extension of the back muscles, all of which together result in accentuation of bending stresses in the disc.

Likewise, the feet must be well supported. 'Duck feet', 'knock knees', and 'bow legging' should be corrected. Walking should be well balanced with toes forward, not inward or outward. Turning of the foot outward results in the hip being rotated outward and the weight bearing being carried on the inside of the foot. 'Swaying of the back' in children increases curvature and produces posterior disc bulging and increased pressure inside of the disc.

Children must sleep on firm mattresses, which support the back. Soft mattresses produce excessive spinal curvature, increased muscle tension, and bending stresses in the disc. Improperly constructed school desks also accentuate low-back muscle tension and disc stresses. The back of the seat should be inclined 120 degrees, in order to minimize the combined effect of compressive and bending stresses. Further, a low seat allowing the feet to rest on the ground results in reduction of the lumbar spinal curvature and bending stresses in the disc due to decreased end moment exerted on the sacral end of the spine. The desk surface should be inclined, rather than flat. A flat desk increases the flexion of the mid- and lower spine, exaggerating muscle excitability. In addition, with flat desks, a student's head must be kept in marked flexion with anterior displacement of the centre of gravity in relationship to the cervical spine.

Abuse of the low back is to be avoided, and proper body mechanics must be inculcated at an early age. As long as the head and neck, trunk, and upper extremities are 'stacked up straight', there is minimal stress on the back. With proper performance of back motion, the spine can be called upon, from time to time, to do excessive work without injury.

Proper conditioning and elastic muscles, smooth and co-ordinated movements, and co-ordination of the upper and lower extremities not only improve performance, but also lessen the stress on the back. Prevention of disabling back and extremity pain begins in adolescence.

Prevention in the adult

Prevention of back and extremity pain in adults is often impossible if, as children, they have either been deprived of the knowledge of proper spinal mechanics, or, if not so deprived, they have preferred to ignore these recommendations. By the age of 21, habit controls movement. Heavy lifting, or labour *per se*, is not a significant factor in producing back pain. Individuals who do heavy lifting, and do it properly, do not develop back pain. Disability or injury to the back can result from either proper lifting of weights exceeding the biomechanical limits of spinal structures (as will be of even lesser weight levels specified later) or improper handling.

Load lifting, carrying, pushing, and pulling. The starting position for proper lifting is with the back flat and the knees, hips and upper extremities bent (Fig. 10.12). We will first analyse this case and determine safe lifting capacities with a straight back. In general, a load P lifted at an eccentricity e with respect to the spine (Fig. 10.13) will develop tensile stresses in the back muscles and a compressive force along with a flexion moment on a transverse section of the vertebral-disc interface. The forces on the vertebral-disc interface will induce elevated hydrostatic pressure in the nucleus pulposus and tensile circumferential stresses in the annulus fibrosus. The limiting value of the load handled is hence deemed to be governed by the micro rupture strength of the annulus fibrosus fibres, which is estimated to be of the order of 1500 lbf in^{-2}.

Thus, for an analysis and computation of the stresses induced in annulus fibrosus, we employ (i) mechanics models for determining the compression force and bending moment acting on the vertebral-disc interface (referred to as *macro models*), in conjunction with (ii) an additional set of mechanics models for determining the tensile stresses in the annulus fibrosus (referred to *micro* or *disc models*).

One type of a macro model is the one described on p. 258. Here, (as indicated in Fig. 7.5), static equilibrium (of the load, and body segment weight above the section, abdominal pressure, muscle and disc forces) at the L5–S1 disc section yields the value of the compressive force and the moment on the surface of the disc.

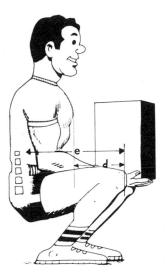

FIG. 10.12.   Proper starting position for load lifting, with the back flat and knees, hips, and arms bent.

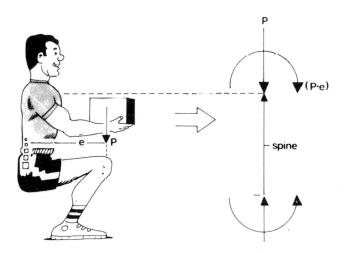

FIG. 10.13.   Equivalent compressive force P and bending moment ($P.e$) induced in the spine, due to the load P lifted at an eccentricity e.

Another *macro model*, of the spine, back muscle, and ligament structure, may be conceived of as a *curved beam in bending* (Fig. 10.14), with the back muscles and supra-spinous ligaments only capable of bearing tensile stress. With this model, the stresses on the vertebral-disc interface can be computed as a function of the spinal geometry and moment-of-inertia property of the section.

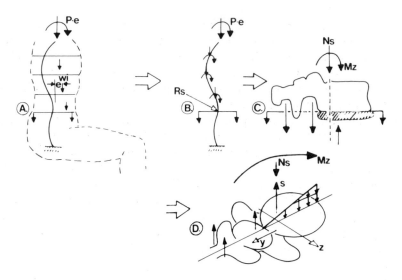

F$_{IG}$. 10.14.   'Curved beam in bending' type macro model of the spine, to determine the stresses acting on a spinal section through the vertebral body-disc interface.

Thus, the spinal-column and back muscle complex may be modelled as a curved elastic beam (with disc and muscle-ligaments have different values of the Young's modulus) in bending (about the Z axis), with varying inertia property (J_z) and radius of curvature R_s (along the geometric axis). Thereby, we can put down, for the stress σ on the disc surface (due to direct compression exerted by the load P as well as due to bending moment $P \cdot e$) as follows [2]:

$$\sigma_d = M_z \left[-\frac{1}{A_e R} + \frac{y}{J_z(1 - y/R)} \right] - \frac{N_s}{A_c} \tag{10.1}$$

where

$$N_s = P + \Sigma W_i - F_{AB}; \quad M_z = P \cdot e + \sum_i W_i \cdot e_i - F_{AB} \cdot A \tag{10.2}$$

$$J_z = \int_{A_e} \frac{y^2}{1 - y/R} \, dAe, \tag{10.3}$$

F_{AB} = the abdominal force, whose expression is given in Chapter 7, and the distance A at which F_{AB} acts is defined in Fig. 7.5 (p. 259).

A_c is the compression-bearing area of the disc; A_e is the area of the equivalent section, where (i) the lateral dimension of the back muscle cross-section is equal to its actual dimension times the ratio of its tensile modulus of elasticity to the compressive modulus of the disc; (ii) the lateral dimension of the supra-spinous ligament cross-section is equal to its actual dimension times the ratio of its modulus of elasticity to the compressive modulus of the disc; (iii) y, the distance from the centroid (or neutral axis) of the transformed section is measured positive towards the local centre of curvature of the spine and the moment M_z, is positive if it produces tension in the positive y segment, $y = h_a$ being the distance from the neutral axis to the anterior edge of the disc; and (iv) the integral is taken over the equivalent A_e.

In the sagittal plane, the spine is concave anterior (or forward) in the thoracic region and convex anterior (or forward) in the lumbar region. Thus for a spinal section normal to its axis, y is positive in (i) the anterior direction from the neutral axis, in the thoracic region, and (ii) the posterior direction from the neutral axis, in the lumbar region. For a flexion moment, as is applied in the case of a lifted load (as illustrated in Fig. 10.14), M_z is negative in the thoracic region and positive in the lumbar region. Thus if h_a denotes the anterior edge of the disc, then for the lumbar region, in eqn (10.1), M_z is taken to be positive and $y = h_a$, for the determination of the maximum compressive stress.

For determining the tensile stresses induced in the annulus fibrosus due to the above determined stresses on the disc surface, the disc is modelled [3] as made of a circular annular elastic annulus fibrosus material with a central nucleus pulposus fluid position, compressed between rigid plane plates (by means of which the macro model obtained surface stresses (σ_d) are applied); refer Fig. 10.15. Its analysis, carried out by Sonnerup [3], is for an axially symmetrical stress loading. The nucleus pulposus pressure P_n is determined experimentally to be of the order of 1.5 σ_d, whereas the maximum circumferential tensile stress (σ_{at}) on the inner edge of the annulus fibrosus is of the order of

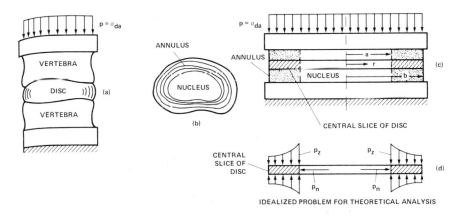

FIG. 10.15. Elasticity model of the disc, to analyse the maximum tensile stresses in the annulus fibrosus due to the maximum compressive stress (σ_{da}) acting on the vertebral body–disc interface (as determined from the curved spinal beam macro model of Fig. 10.15).

2.45 P_n. Thus the annulus fibrosus tensile stress σ_{at} is of the order of 3.6 times the applied stress σ_d.

If we wish to employ this disc model, we can adopt σ_d to be equal to the value of the compressive stress at the anterior edge (σ_{da}) by putting $y = -h_a$ in eqns (10.1–10.3). In that case, the annulus tensile stress $\sigma_{at} = 3.6\,\sigma_{da}$. If we now wish to determine the safe load that can be lifted, we need to ensure that σ_{at} ($= 3.6\,\sigma_{da}$) does not exceed the allowable annulus fibrosus rupture stress, denoted by σ_a^* (of value equal to 1500 lbf in^{-2}) we require that

$$\sigma_{at} = 3.6\,\sigma_{da} = \left[3.6\left\{M_z\left[-\frac{1}{A_e R} + \frac{y}{J_z(1 - y/R)}\right] - \frac{N_s}{A_c}\right\}\right] \le \sigma_a^* \le 1500 \text{ p.s.i.}$$

$$(10.4)$$

wherein $y = -h_a$.

In the above equation, the quantities W_i, h_a, A_e, A_c, R, and J_a are all measurable; F_{AB} is adopted to be equal to (as in Chapter 7) 35 lb and its moment arm is taken equal to 5 in. For a representative occupational worker, of weight $W = 160$ lb, we can (for representative values of spinal section as depicted in Fig. 10.14, and hence for A_e, A_c, J_z and R) plot eqn (10.4) on either the 'P vs e' coordinate plane, or preferably on the P/W vs e/H coordinate plane (where W and H are body weight and height), for various levels. Figure 10.16 shows the curve for L5 level, for a subject of height 5′10″ ft and weight 160 pounds.

It is noted, from this graph, if an average person (say of height 5 ft 10 in and weight 160 lb) lifts a load held close to his body (for which $e = 12$ in, say) while keeping the back straight (as shown in Fig. 10.12), he can lift a total of about 70 lb (or less than half his body weight); however, if we wish to invoke a factor of safety of 2, he should only attempt to lift 35 lb. On the other hand (with reference to Fig. 10.12), were he to attempt to lift the load held at, say, $d = 15$ in from his body, (for which $e = $ say, 25 in), then he can only lift 35 lb.

Now, returning momentarily to Fig. 10.14, and the manner in which the normal bending compressive and tensile stresses are developed in the size and muscle-ligaments respectively, we note from eqn (10.1) that the tensile stress in the back muscles (σ_{mt}) is given by the first term of the equation, namely

$$\sigma_{mt} = M_z\left[-\frac{1}{A_e R} + \frac{y}{J_z(1 - y/R)}\right]$$

$$(10.5)$$

wherein $y = h_m$, the distance from the neutral axis to the posteriorly located back muscle cross-section.

It is interesting to study the *relationship between the back muscle tensile stress and the maximum annulus tensile stress* at a certain spinal segment level. For instance, in the lumbar spine region we obtain, by taking (i) $y = -h_d$ and

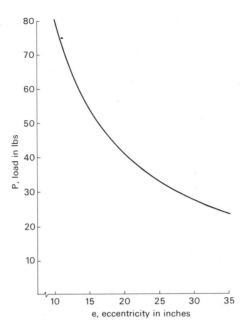

FIG. 10.16.   The graph of 'allowable' to-be-lifted load (P) vs. its eccentricity (e), as computed by means of the curved spine-beam macro model (Fig. 10.14) in conjunction with the elasticity model of the disc (Fig. 10.15).

M_z to be positive for the maximum annulus tensile stress (σ_{at}) in eqn (10.4), and (ii) $y = h_m$ and M_z to be positive, for the back muscle tensile stress (σ_{mt}), in eqn (10.5), the following relationship is obtained for the ratio of the tensile stress in the back muscle (σ_{mt}) to the tensile stress in the annulus fibrosus (σ_{at})

$$\frac{\sigma_{mt}}{\sigma_{at}} = \frac{\left\{ -\dfrac{1}{A_e R} + \dfrac{h_m}{J_z(1 - h_m/R)} \right\}}{3.6\left\{ -\dfrac{1}{A_e R} - \dfrac{h_d}{J_z(1 + h_d/R)} - \dfrac{\left[1 + \sum_i (W_i/P) - F_{AB}/P \right]}{e A_c\left[1 + \sum_i \left(\dfrac{W_i}{P}\cdot\dfrac{e_i}{e} \right) - \left(\dfrac{F_{AB}}{P}\cdot\dfrac{A}{e} \right) \right]} \right\}}$$
(10.6)

which indicates that σ_{mt} is proportional to σ_{at}. The manner in which the ratio σ_{mt}/σ_{at} varies with P and e is illustrated by Fig. 10.17, (for the case of the previously discussed individual with $W = 160$ pounds). This graph shows that an estimate of the disc stress can be obtained from an EMG-derived estimate of the muscle tension invoked in lifting the load, particularly for load P greater than 30 lb.

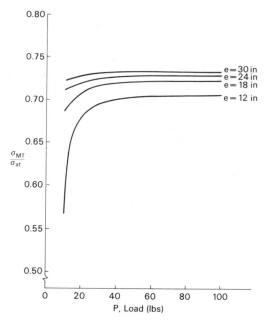

Fɪɢ. 10.17.   The graph showing the correlation of the back muscle tension (σ_{mt}) and the tensile stress in the disc annulus (σ_{at}), as depicted by the variation of the ratio σ_{mt}/σ_{at} with the load magnitude P and load eccentricity e.

Before dealing with the biomechanics of lifting the load with a bent back, let us examine one more set of *Macro–micro finite element models. The macro whole-body finite element model* is shown in Fig. 10.18, where the discretized elements of the elastic spinal structure have three degrees of freedom [4] and the back muscle elements can only take tension; this model has now been extended to 6 degrees of freedom to handle generalized 3-dimensional loadings. The nodes of the spinal structure correspond to inter-vertebral joint locations. Each element of the spinal structure has the segmental inertia and weight properties of the torso segment containing that spinal element; the values of these element properties are obtained from the experimental data of Liu and Wickstrom [5]. The special muscle elements represent (as shown in the figure) gluteus maximus, Sacro-spinalis, and the trapezius. The abdominal pressure is again as adopted in Chapter 7.

In order to exercise this model for a particular body configuration, the load weight force is transferred by conducting a finite element analysis of the upper limb to the appropriate spinal structure node as a force and a moment. The effect of the abdominal pressure is likewise transferred as extension force and moment at a node. Then the spinal and back muscle structure is analysed by applying the loading incrementally. If at a particular stage of the incremental

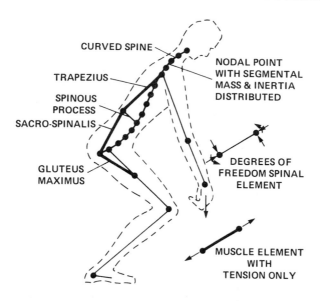

FIG. 10.18. Macro whole-body finite element model, to determine the forces exerted on the micro disc finite-element model (of Fig. 10.20).

load, the total accumulated stress in the muscle element becomes compressive, then for that particular load increment, the muscle element stiffness is taken to be zero (in the overall stiffness matrix).

The force and moment output from this model is applied to the vertebral-disc interface, as distributed axial and shear forces and a central moment on a finite element model of the disc [6], shown in Fig. 10.19. This model is made up of two components, namely, the nucleus pulposus cavity containing fluid of bulk modulus (K), and the surrounding annulus fibrosus elastic medium. The discretized annulus fibrosus is hence loaded by (i) the external forces and moment computed from the macro model at its vertebral interface, as well as (ii) the internal forces applied at the cavity nodes due to the pressure in the nucleus pulposus cavity.

This disc model is stressed by an incremental loading procedure. At each load step, the pressure increase (Δp) in the nucleus pulposus cavity is computed as

$$\Delta p = -K\left(\frac{\Delta v}{v_0}\right) \tag{10.7}$$

wherein v_0 and Δv are respectively the cavity volume at the previous loading step and the incremental volume due to Δp. The maximum principal stress (σ_{ap}) is computed at the L5 level, using the disc material property obtained from *in vitro* studies [7, 8].

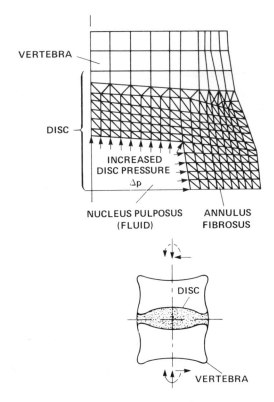

FIG. 10.19.   Micro finite-element model of the disc, showing the annulus pulposus discretization.

This macro–micro finite element model is now employed to compute the principal stress (σ_{ap}) for each stage of straight-back load lifting configuration shown in Fig. 10.20, for various values of the load (P). The anthropometric proportions of 50 percentile industrial worker are employed; these are almost analogous to those employed for the previous macro–micro model of curved spinal beam and elastic disc. At each stage or configuration (represented by the coordinates H and B), the L.5 level disc principal stress σ_{ap} is evaluated for different values of the load (P). For each set of coordinates, the limiting value of the load P is determined for which σ_{ap} equals the micro-rupture stress σ_a^* (equal to 1500 p.s.i.). Thus constant *maximal allowable P contours* are developed in the *H* vs *B* coordinate plane, shown in Fig. 10.20.

It is seen from Fig. 10.20, that the safe maximal load carrying limits obtained by this model are of the same orders of magnitude as those obtained by the earlier presented macro–micro elastic continuum model. For instance, the maximal load that can be lifted at chest level (i.e. at $H = 50$ in) close to the body

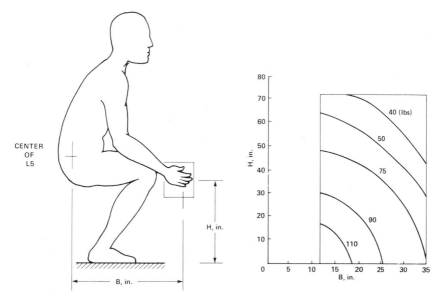

FIG. 10.20.   Results of macro–micro finite-element modeling, showing contours of recommended safe maximum weight of lift (for a 50 per cent industrial male population—back straight lift), plotted on the *H* vs. *B* coordinate plane.

is about 72 lb which is about the value obtained from Fig. 10.16; however, at 15 in away from the body (i.e. for $B = 25$ in and $H = 50$ in), the safe limiting load reduces to 50 pounds. It is particularly interesting to note, in Fig. 10.21, that the load lifting capacity levels determined by Snook [9], based on psychological tests, provide good correlation with the capacities developed by our finite element biomechanical model.

The situation changes considerably if *the weight is lifted with the back bent* (as shown in Fig. 10.22); in this case, the maximum lifting capacity is reduced considerably to about half of the previous amount, provided it is not repeated too often. Let us biomechanically analyse how this occurs, by means of our elastic curved beam macro model of the spine in interaction with the elastic annular disc micro model.

As seen in Fig. 10.22, the load *P* exerts a moment *P . e* (*e* being the moment arm of the line of action of the load about the pertinent spinal section) at a curved-beam spinal model section. The bending moment (*P . e*) induces normal compression and tensile stresses on the vertebral body–disc interface spinal cross-section, while the shear force *P* causes a shear stress (Fig. 10.23).

The normal (compressive–tensile) bending stress on the vertebral body-disc interface is given by

$$\sigma(y) = P\left[-\frac{e}{AR} + \frac{ey}{J_z(1 - y/R)} \right] \qquad (10.8)$$

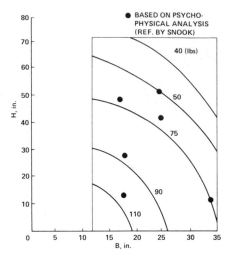

F<small>IG.</small> 10.21.   Correlation between biomechanical results (shown in Fig. 10.21) and psychophysical criterion based results, concerning the safe lifting load capacities.

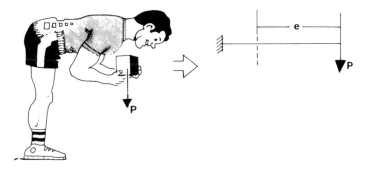

F<small>IG.</small> 10.22.   Load P lifted with the back bent, induces a bending moment $(P.e)$ as well as a shear force P in the spine.

wherein, based on the sign convention is as indicated earlier in the case of eqn (10.1), $y = -h_a$ (the distance from the neutral axis to the anterior edge of the disc) yields the maximum compressive stress.

The shear stress $\sigma_s(y)$ at the disc–vertebral body interface is given by

$$\sigma_s(y) = \frac{P}{b(1 - y/R)}\left(\frac{Q_z}{J_z} - \frac{A'}{AR}\right) \tag{10.9}$$

$$\text{where} \quad \overline{Q}_z = \int_{A'} \frac{y}{1 - y/R}\, \mathrm{d}A \tag{10.10}$$

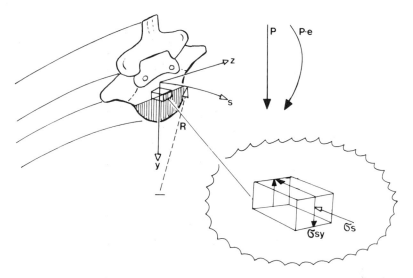

F$_{\text{IG}}$. 10.23. Transverse cross-section of the spine, showing the resulting normal bending stresses σ_s and shear stresses σ_{sy} due to the moment $(P.e)$ and shear force P (see Fig. 10.22).

A' being the shaded portion of the cross-section at the edge of which the shear stress $\sigma_s(y)$ is being determined. The maximum principal compressive stress is determined, for $y = y^*$, as follows

$$\text{Max } \sigma_{cp}(y = y^*) = \left[\left(\frac{\sigma_c - \sigma_t}{2} \right)^2 + \sigma_s^2 \right]^{1/2} \tag{10.11}$$

wherein σ_c and σ_t are given by eqn (10.8) and σ_s is given by eqn (10.9).

For the anatomically based value of e for the L5 transverse section and representative values of P, it is found that the maximum principal stress is almost equal to the maximum compressive stress value, and its orientation is almost normal to the vertebral–body disc interface. Thus the macro model will be deemed to apply normal stresses to the micro two-dimensional ring shaped annular thin disc model.

Now for determining the maximum value of the safe load carrying capacity P, we recall that the maximum annulus tensile stress σ_{at} equals 3.6 times σ_{da} (the maximum compressive stress at the anterior edge of the vertebral-disc interface). Thus, we put down the criterion that

$$\sigma_{at} = 3.6\sigma_{da} = 3.6\,P\left[-\frac{e}{AR} + \frac{ey}{J_z(1 - y/R)} \right]$$

$$\leqslant \frac{\sigma^*}{2}\,(= 500 \text{ p.s.i.}) \tag{10.12}$$

wherein $y = -h_a$.

For the previously considered L5 geometry section, the above equation yields $P = 35$ pounds, which is *half the value of the load (P) lifted close to the body with a straight back*.

For the two lifting postures of Figs. 10.13 and 10.23, women may be expected to lift maximum loads of 35 pounds with a straight back and 17.5 pounds with a bent back respectively. Lifting however, is not the primary function of the low back. The low back is busy enough supporting its superstructure, and initiating and terminating movements. Hips, knees, and upper extremities must be bent in order to keep the back straight and minimize the shear and bending stresses in the low back. Thus 'leg lifting' as opposed to 'back lifting' must become a matter of habit.

Proper lifting also involves proper carrying of the lifted object. Once lifted, the carried weight must be held close to the body rather than away from it, in order to again minimize the bending moment due to the carried weight and the resulting bending stresses. Thus, as seen from the graphs in Fig. 10.16, a 40 pound weight lifted at a distance of nine inches away from the body exerts the same pressure on the spine as a 70 pound weight carried close to the body. It is not possible for the spine to be balanced properly with a heavy object at arm's length (Fig. 10.24). Classic examples of such abnormal stress are removing groceries from a car trunk or lifting an infant from a cot (Fig. 10.25).

One must keep in mind the need for keeping the back flat when lifting. Flexion or bending forward of the spine must be avoided, since it results in an unbalanced position when lifting. Even with the lower extremities properly fixed and flexed, spinal flexion results in loss of equilibrium and imbalance. An

FIG. 10.24.   Improper and proper lifting.

FIG. 10.25.   Improper lifting (from a car boot).

unbalanced position in body motion produces increased muscle activity and overloading of the discs and supporting structures, as explained above, due to the flexion moment exerted by the eccentric loading. Extremes of flexion of the spine are limited by the supporting ligaments, and not by the supporting and active muscles of the back, since at extremes of flexion, the muscles relax. The neutral axis is thereby shifted anteriorly, resulting in increased disc compressive stress to sustain the flexion moment.

Rather than carrying an excessive weight in front, *it is better to carry it on the back*, as illustrated in Fig. 10.26. Since the biomechanics of how the load as well as the various muscles transfer the load to the spine (and reduce the flexion moment and resulting compressive bending stresses on a spinal cross-section) has not (to the knowledge of the authors) been delineated, it would be relevant to conduct such an investigation.

One final comment concerning lifting is not to lift above waist level but instead stand on a properly supported stool, ladder, or other support (Fig. 10.27). Lifting above waist level, with arms stretched to their fullest, greatly increases back muscle tension and hence disc pressure; see results of Fig. 10.17. It is also better to lift a lesser weight repeatedly than to lift a greater weight less frequently, and rapid lifting results in greater stress in the spine than lifting slowly.

Muscle weakness is compounded in middle age by obesity and the pendulous abdomen. The increased weight exerts an added compressive load and moment on the spine at the lumbar level, causing the back to sway forward, increase the lumbar curve, and disc compression. This situation is worsened by weakening of abdominal muscles.

As far as load '*pushing versus pulling*' is concerned, it is far better to push than to pull. Pulling increases back muscle tension, while pushing increases

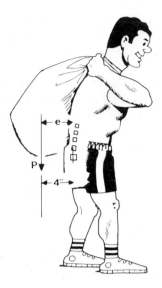

FIG. 10.26.   Proper carrying of the load (over the back).

FIG. 10.27.   Proper lifting.

abdominal muscle tension. Since the primary spine muscles are the abdominals, they should be used in preference to the back muscles. Proper body movement or positioning and decreased back tension is the direct result of leverage, and abdominal muscles provide such a leverage. Should pushing be impossible owing to the location of the object to be moved, the direction of the pull should be at waist level (and not above or below it), for which less back muscle tension is invoked to resist the applied flexion bending moment on the spine.

In order to provide a *quantitative assessment of the benefit of push vs pull* as well as other occupational load handling situations, the macro–micro finite-element model, presented earlier was employed to analyse and determine the annulus principal tensile stress for various occupational load handling situations schematized in Fig. 10.28. The values of principal tensile stress in the annulus fibrosus are computed for various levels of load handled [10], and plotted in Fig. 10.29. This graph indicates a 25 per cent increase in the level of induced disc annulus tensile stress when the same load is pulled instead of pushed.

Sleeping, sitting, standing and driving activities. Proper sleeping habits and surfaces also aleviate or diminish back pain. Firm mattresses support the back and reduce muscle tension from excessive bending of the spine (Fig. 10.30). Let us invoke a qualitative mechanics insight into the associated mechanisms.

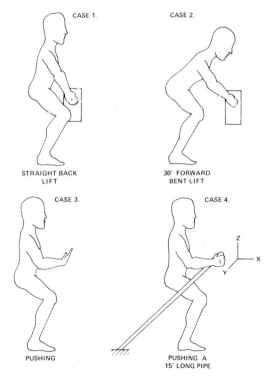

FIG. 10.28.   Various occupational load handling configurations analysed by the macro–micro finite element model.

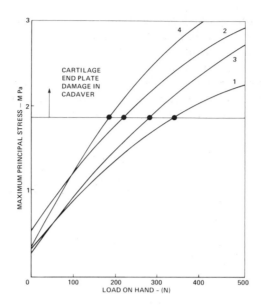

FIG. 10.29. Maximum principal stresses in the annulus fibrosis, computed for the four occupational local-handling cases shown in Fig. 10.28, by the macro–micro finite element model.

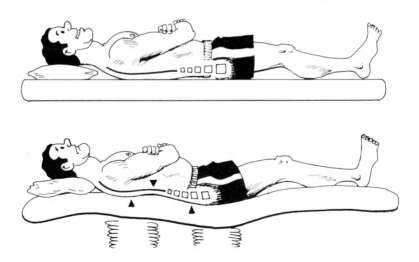

FIG. 10.30. A firm mattress supports the back and reduces back muscle tension due to excessive bending of the spine.

In order to gain a qualitative insight into the *stress induced in the spine due to lying on a mattress*, the bending of the spine on a mattress may be likened to the bending of a curved finite elastic beam (i) of continuously or discretely varying elasticity and moment of inertia, with a continuous or discretely varying distributed body weight, acting on it as well as end moments due to moments exerted by the overhanging weights of the head and lower limbs, (ii) on an elastic foundation of stiffness K (Fig. 10.31).

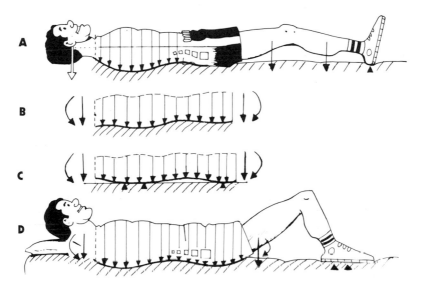

FIG. 10.31.   The mattress is simulated as an elastic foundation, and the spine as a curved beam-on-elastic foundation, loaded by distributed body weight forces and head and lower limbs' weights induced end shear forces and bending moments.

It can be seen, from Fig. 10.31 (b and c), that when such a curved beam-spinal model is supported on a rigid unyielding surface, the effect of a fairly soft mattress is to augment the spinal curvature and hence the sustained self-weight induced bending moment, with consequential elevated disc (annulus) stress and deformation into the spinal canal. This is because for a beam-on-elastic foundation bearing a distributed load, the deflection is proportional to the local value of loading. Now, since the thoracic and sacral portions of the spine have bigger proportions of body weight acting on them (due to the rib cage and pelvis weights), their existing posterior-convex curvatures are augmented; likewise, the existing posterior-concave lumbar spinal curvatures are also augmented, resulting in increased bending moments and disc stresses.

Note that the effect of the end moments, due to the overhanging weights of head and lower limbs, is to further increase the spinal curvature and the induced bending moments. However if the head and the lower limbs are supported by means of pillows or even if the legs are drawn up (as shown in Fig. 10.31d), the magnitude of the end bending moments, and consequently the disc (annulus) stress and deformation is reduced. The spinal curvature, the induced bending compressive stresses on the vertebral-disc interface, as well as the disc annulus tensile stress can all be effectively reduced by lying on a hard relatively unyielding surface (Fig. 10.31c), since the local deformation of the foundation will not then be so substantially influenced by the local distributed body-weight.

Prolonged standing in one position, with both feet placed firmly on the floor, will exaggerate the low-back pain. To decrease this postural aggravation, one should stand with either foot on a footstool or similar support (as shown in Fig. 10.32). Raising the foot in this manner reduces the psoas muscle spasm and loads on the spine, thereby diminishing the lumbar lordosis or anterior curvature of the lumbar area.

Fig. 10.32.   Proper prolonged standing.

Shoes too have an influence on the lumbar lordosis, they should be firm and well-fitted. High heels cause a forward roll of the pelvis, exaggerating the lumbar lordosis, which increases the angle between the fifth lumbar vertebra and the sacrum, resulting in a weak lumbar spine.

Let us now study the *manner in which the back supports of chairs influence the stresses in the spinal structure*. For a certain angle (θ) of the back support (as shown in Fig. 10.33), the vertical gravity forces on the spine can be resolved

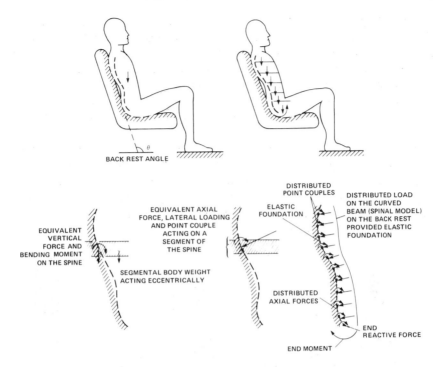

FIG. 10.33.   Mechanical behaviour of the spine during sitting likened to a beam column on elastic foundation, with distributed and end forces and moments as shown in the figure.

into (i) forces acting axially along the spine and normal to the axial length of the spine (i.e. acting laterally), as well as (ii) distributed point couples. The forces normal to the spine are resisted by the flexible back support. Now the spine with the distributed normal component of the vertical (body weight) gravity forces acts as an elastic beam-on-elastic foundation (provided by the back support), with a distributed loading and point couples acting on it (Fig. 10.33); it hence incurs bending stresses. Additionally, the axial components of the gravity forces cause increasing compression in the spine from the thoracic to the sacral level. At the sacral end, the muscle tension (caused by the extremities) induces a counter-clockwise moment on the spine (as shown in the figure).

Thus, taking into account the combined action of the axial loading and the muscle tension at the sacral end, the mechanical behaviour of the spine may be likened to a beam-column on an elastic foundation with an end moment (as shown in Fig. 10.33). Due to the combined effects of (i) the uneven distribution of the beam-type lateral loading on the spinal beam column (the loading intensity being greater in the thoracic and sacral segments), (ii) the distributed point couples, (iii) the axial or the column-type loading, and (iv) the sacral end-moment, the curvatures in the thoracic and lumbar regions differ from those in the normal state of the spine, thereby resulting in spinal bending moments and disc stresses.

For a value of angle $\theta = 90°$ (i.e. for a seat with a straight back), the gravity forces result primarily in direct compressive loading of the disc, with some bending due to the eccentricity of the body segmental weights with respect to

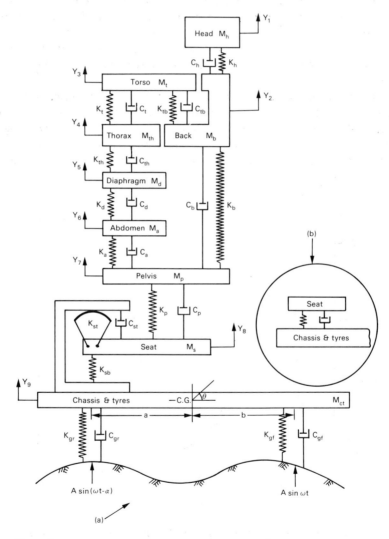

FIG. 10.34.   Occupant tractor model with a 'Patil–Palanichamy–Ghista' suspension.

the spine; however the additional bending stresses, due to the lateral loading on the spinal beam-on-elastic foundation, are absent for $\theta = 90°$. As θ increases, the axial components of the gravity force (representing compressive column loading) decrease, while the lateral components (representing distributed beam loading) increase. With the increase in the distributed beam loading, the bending stresses in the spinal beam-column increasingly come into effect. The optimal configuration or optimal is when the compressive stress on the vertebral-disc interface is minimum under the combined effects of the direct compression loading on the spine and the bending induced by the distributed beam loading and point-couple loading. This is the biomechanical rationale behind the evidence that the best chairs have a back support at 120 degrees, in which configuration the muscle activity is also found to be minimal.

The muscle activity and disc pressure are reduced with armrests which partially support the body weight when sitting, and transfer stresses from the low back to the upper extremities while getting off the seat. Also, elevation of the lower extremities while getting stool allows displacement of the hips beneath the torso, decreasing the bending stresses (and the anterior convex curvature) in the lumbar spine. In general, all people who work sitting down

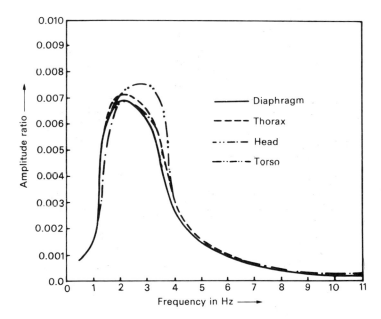

Fig. 10.35.   Variations of amplitude ratios of diaphragm, thorax, torso, and head with frequency of vibration for the PPG tractor seat suspension provided at the plane of centre of gravity of chassis.

for several hours a day must be mindful of the harm done when sitting and over-stressing the spine.

People who spend 50 per cent or more of their time driving a motor vehicle are three times as likely to develop a herniated disc as people who do not have such an occupation. If the car seat is moved closer to the steering wheel it will be possible for the driver to bend the knees and the hips, avoid leg straightening, and hence enjoy proper support of the back. Heavy-duty truck and tractor operators are further exposed to vibration at intolerable and uncomfortable intensity levels in the frequency range of 2–10 cycles per second, which makes them prone to a number of spinal disorders [11]. In

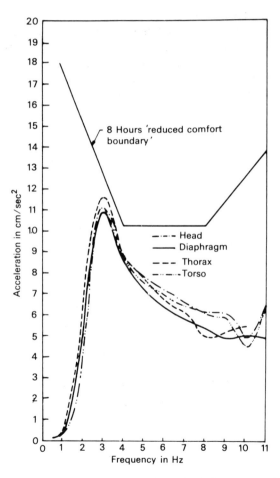

Fig. 10.36. Comparison of acceleration responses of diaphragm, thorax, torso and head with tolerance (8-hour 'reduced comfort boundary') curve when the new type of tractor seat suspension is provided at the plane of CG of chassis.

order to analyse the *efficiency of conventional tractor suspension systems*, with a view to then design a more efficient occupant vibration isolation, a hybrid lumped-parameter model of a tractor occupant and suspension system has been developed [12]. By means of this model, the ratio of the acceleration level of the occupant's anatomical components to the acceleration input at the wheels (of the order of 1.5 g to 0.5 g in the 2–10 cycles Hz range) was determined for the standard suspension systems; this acceleration amplitude ratio was found to be of the order of 2.5 to 4.5, which exceeds the specifications of the International Standards Organization.

Hence a new type of suspension system (the 'PPG' system) has been developed, as shown in Fig. 10.34(a), where the seat is held between a compression spring and a leaf spring in parallel with a dashpot acting in opposition to each other [12]. Figures 10.35 and 10.36 respectively illustrate the amplitude ratio and the acceleration levels of the body components. Note that the amplitude ratios are of the order of one-hundredth of the values attained by the standard suspension system (shown in the insert of Fig. 10.34(b)). The maximum acceleration level for the thorax is 11.8 cm s^{-2} (compared to 9 m s^{-2} in the case of the standard suspension system), and satisfies the 8-hour comfort level specification of the International Standards Organisation.

References

1 SCHMOVL, GEORG, and JUNGHANNS, HUBERT (1971). *The human spine in health and disease.* Grun and Stratton, New York.

2 ODEN, J. T. (1967). *Mechanics of elastic structures.* McGraw-Hill, New York.

3 SONNERUP, L. (1980). Stress and strain in the inter-vertebral disc in relation to spinal disorders. In *Osteoarthro mechanics* (ed. D. N. Ghista). Hemisphere, McGraw-Hill, New York.

4 LIU, Y. K., and RAY, G. (1975). Dynamic response of human torso in G$_z$ acceleration—A two dimensional analysis. *Proc. ASME Biomechanics Symp.*

5 LIU, Y. K., and WICKSTROM, J. D. (1973). Estimation of the inertia property distribution of the human torso from segmented cadaveric data. In *Perspectives in biomedical engineering* (ed. R. M. Kenedi). Macmillan, New York.

6 LIU, Y. K., and RAY, G. (1978). Systems identification scheme for the estimation of the linear viscoelastic properties of the intervertebral disc. *Aviat. space environ. Med.*, 175–7.

7 LIU, Y. K., RAY, G., and HIRSCH, C. (1976). The resistance of the lumbar spine to direct shear. *Orthop. Clins. N. Am.* **6**, 33–47.

8 Liu, H. S., Liu, Y. K., and Ray, G. (1978). Systems identification for material properties of the intervertebral joint. *J. Biomech.* **2**, 1–14.

9 Snook, S. H. (1978). Psychophysiological indices—what people will do. In *Safety in manual materials handling* (ed. C. G. Drury). US Dept of Health, Education and Welfare—National Institute for Occupational Safety and Health.

10 Ray, G., and Ghista, D. N. (1978). Disc stresses due to various occupational manual lifting tasks. *Digest of the 1st International Conference in Medicine and Biology (Aachen)*. Verlag Witzstrock, Baden Baden.

11 Radke, A. O. (1957). Vehicle vibration. *ASME paper 57-A*.

12 Patil, K. M., Ghista, D. N., Palanichamy, M. S. (1978). Human body responses to tractor vibrations and a new suspension system for maximum comfort. *Digest of the 1st International Conference in Medicine and Biology (Aachen)*. Verlag Witzstrock, Baden Baden.

Section III   Athletics and sports

11. Biomechanics of kicking

T. C. HUANG, ELIZABETH M. ROBERTS, AND YOUNGIL YOUM

11.1. Introduction

Attempts to portray the versatility, dynamic line, and aesthetic quality of human motion are as old as the graphic arts. As the centuries passed man's wonder at his own motion began to stimulate his curiosity and in time led him to investigate and explain. Yet it is only relatively recently that studies of human motion have begun to take on a quantitative character and vague explanations have given way to analytical and experimental dynamic analyses.

Bipedal locomotion has without question received the greatest share of scientific attention, probably in large part because of its medical significance. The approaches of Chao [1], Frank [2], and Townsend and Seireg [3] show promise of elucidating fundamental dynamic and control components of this all important activity. Investigation into less utilitarian but perhaps more aesthetic and versatile human motions similarly has a long, though less voluminous, history. Basic equations of kinematics and kinetics have been applied to analyses of movements in sports by Plagenhoef [4], Dillman [5], Hochmuth [6], Kane [7], and others. While these and other investigators have taken significant strides toward describing and simulating human motion, much remains to be done in the techniques of data collection and processing, system modelling, and kinematic and kinetic analyses.

Kicking is one of the basic movement skills which man utilizes in various games and sports. The kick is essentially a modification of walking and running. It differs from walking and running movements in that the primary force of the kick is generated by the swinging limb rather than by the supporting limb, and the speed of the swinging limb is greater in the kick than in a run. Any given type of kick can be thought of as consisting of two phases: (i) the development of velocity in the kicking foot and (ii) the impact phase.

In games and sports in which kicking is used to give impetus to a ball a variety of styles and speeds of kicks are executed, for example instep kicks, toe kicks, punts, and drop kicks. In some the run-up to the ball is directly in line with the intended flight as in the American style football place kick, the soccer toe kick, and the Australian drop punt; in many others the approach is at an angle to the intended flight. In the latter case the kicking motion is clearly three-dimensional with considerable rotation of the pelvic girdle and, frequently, inward rotation of the kicking femur, while in the former case the main action takes place more nearly in a plane with rotation of the hips and femur minimized.

Kicking studies have been advanced recently by Youm *et al.* [8], Roberts *et al.* [9], Zernicke [10], Zernicke and Roberts [11], and others by combining experimental programmes and dynamic analyses. In any experimental programme an important consideration is the degree of accuracy of various measuring and recording techniques. An attempt should be made to establish the confidence limit of the kinematic and kinetic quantities derived from film or tape records of various movements. These quantities are obtained partly from experiments where the movement is recorded on photographic film or magnetic tape and partly from computations performed on data derived from these records.

Human musculoskeletal systems can be modelled by joining rigid links representing limb segments to form an open linkage. Since there are no direct analytical approaches available for the open linkage, the displacement data are obtained from film or tape records. Subjects performing the motion are photographed [12] and instrumented for tape recording as shown in Fig. 11.1. The joints and centres of mass of the segments are marked according to data provided by Dempster [13].

The film or tape records are digitized. Curve-fitting techniques are used with these data to give estimates of displacements as continuous functions of time [14]. A position diagram illustrating the toe kick is shown in Fig. 11.2 [15]. Velocity and acceleration are then obtained as the first and second derivatives of displacement with respect to time.

Two types of co-ordinate systems, inertial and local, are used in kinetic analyses, and transformation matrices are constructed between the inertial co-ordinates and each local co-ordinates system. The velocities and accelerations of the centre of mass of each segment are determined in both co-ordinate systems. Free-body diagrams in which the unknown muscle forces are replaced by an equivalent force and an equivalent moment at each joint are drawn, and the equations of motion for each segment are formulated in each respective local co-ordinate system [8]. Data on segment masses are needed in these formulations [13]. Continuity conditions are prescribed at each joint based on the fact that the forces and moments at the joint between the two segments are equal in magnitude and opposite in direction.

This system of equations of motion and continuity conditions can be handled conveniently in matrix form. Solving this system of simultaneous equations gives joint forces and joint moments in terms of local co-ordinates. If desired, joint forces and joint moments in system co-ordinates can be obtained by applying transformation matrices. Some of these analytical results can be checked with experimental measurement such as force platform readings.

FIG. 11.1. Kicking experiment at the University of Wisconsin.

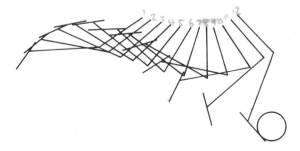

FIG. 11.2. Position diagram of a toe kick.

The knowledge gained will contribute to a basic understanding of the human dynamic system.

11.2. Kinematic analysis

Model

A conceptual model of a human body limb can be constructed by joining a series of rigid links with continuous mass distribution. In the study of kicking, the lower limb is modelled as either two segments, thigh and shank (leg and foot), or three segments, thigh, leg, and foot. In the human body the mechanical nature of the joints is quite complex. Consequently all the joints in the model are assumed to be either hinged, universal, or ball-and-socket joints, depending upon restrictions consistent with body construction. A model of the human body's musculo-skeletal segments could also consist of connected rigid and elastic bodies, as in the case of the spine. For the lower limb, the dimensions of each link can be measured on the subject, and the centre of mass and mass moment of inertia can be estimated from cadaver data [13].

There are two types of model in linkage systems: closed and open linkage systems. Human skeletal systems are basically thought of as open linkage. However, in the study of a simulated kick [8], a variable four-bar linkage model was constructed, adding a fictitious massless link. This is shown as BC in Fig. 11.3. From these variable four bar linkage models, kinematic analysis of position can be made analytically with less input data. However, this model cannot be used for velocity and acceleration analyses.

Kinematic equations

Basic kinematic equations for the motion of a point (joint, centre of mass, etc.), relative motion of points, motion of a rigid body, and relative motion of rigid bodies are listed as follows [16]:

(*a*) *motion of a point.* For the motion of a point in space, the displacement, velocity, and acceleration can be expressed in the rectangular cartesian xyz coordinates as

$$\mathbf{r} = x\mathbf{i} + y\mathbf{j} + z\mathbf{k}$$
$$\mathbf{v} = \dot{\mathbf{r}} = \dot{x}\mathbf{i} + \dot{y}\mathbf{j} + \dot{z}\mathbf{k} \tag{11.1}$$
$$\mathbf{a} = \dot{\mathbf{v}} = \ddot{\mathbf{r}} = \ddot{x}\mathbf{i} + \ddot{y}\mathbf{j} + \ddot{z}\mathbf{k}$$

where $\mathbf{r}$, $\mathbf{v}$, and $\mathbf{a}$ are the position, velocity, and acceleration vectors respectively. For curvilinear motion in a plane eqn (11.1) can be simplified as

$$\mathbf{r} = x\mathbf{i} + y\mathbf{j}$$
$$\mathbf{v} = \dot{\mathbf{r}} = \dot{x}\mathbf{i} + \dot{y}\mathbf{j} \tag{11.2}$$
$$\mathbf{a} = \dot{\mathbf{v}} = \ddot{\mathbf{r}} = \ddot{x}\mathbf{i} + \ddot{y}\mathbf{j}.$$

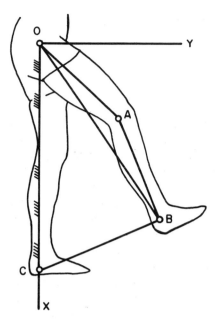

Fig. 11.3.   A four-bar linkage model of kicking.

For rectilinear motion, eqn (11.2) can be further simplified as

$$r = x$$
$$v = \dot{r} = \dot{x} \qquad\qquad (11.3)$$
$$a = \dot{v} = \ddot{r} = \ddot{x}.$$

From the experiment the components of the position vectors can be obtained at specific intervals. From these discrete data each displacement component can be derived as a continuous function of time by using curve-fitting techniques. One such technique is least squares curve fitting by a polynomial of appropriate degrees. Spline functions [17] can also be used and may prove more satisfactory for kicking analyses [10, 18].

Once the displacement as a function of time has been determined, velocity and acceleration can be derived by successive differentiation with respect to time.

b. *Relative motion of points.* Let the displacements, velocities and accelerations of point A and B be denoted by r_A, v_A, a_A, and r_B, v_B, a_B, respectively. Then the displacement, velocity, and acceleration of point B relative to point A, denoted as $r_{B/A}$, $v_{B/A}$, and $a_{B/A}$ respectively, can be expressed as follows:

$$r_{B/A} = r_B - r_A$$
$$v_{B/A} = v_B - v_A \qquad\qquad (11.4)$$
$$a_{B/A} = a_B - a_A.$$

c. Motion of a rigid body. When a rigid body rotates with angular velocity ω and angular acceleration $\dot{\omega}$, and a reference point O of the rigid body moves with velocity $\dot{\mathbf{r}}_0$ and acceleration $\ddot{\mathbf{r}}_0$, the general kinematic equations for any point of the rigid body can be expressed as

$$\mathbf{r} = \mathbf{r}_0 + \rho$$
$$\dot{\mathbf{r}} = \dot{\mathbf{r}}_0 + \omega \times \rho \qquad\qquad (11.5)$$
$$\ddot{\mathbf{r}} = \ddot{\mathbf{r}}_0 + \omega \times \dot{\rho} + \omega \times (\omega \times \rho)$$

where ρ is the position vector of the point relative to point O.

The kinematic quantities $(\omega, \dot{\omega})$ of the rigid body and $(\mathbf{r}, \dot{\mathbf{r}}, \ddot{\mathbf{r}})$ of its centre of mass are required in the equations of motion. When the kinematic quantities, namely $\mathbf{r}, \dot{\mathbf{r}}, \ddot{\mathbf{r}}$ of both ends of a link are known, eqns (11.5) will give ω and $\dot{\omega}$ of the link. The kinematic quantities of the centre of mass of the link are then obtained by applying eqns (11.5) again.

Information about the angular displacements of a rigid body can also be obtained directly from the filmed record of the two markings on a segment. These discrete data of the angular displacements are then treated by the curve-fitting techniques in the same way as linear displacements are treated. The angular velocities and angular accelerations then can be derived from the time functions of the angular displacements thus obtained.

Relative motion of rigid bodies. Let A and B be two moving (translating and rotating) rigid bodies. Kinematic equations for relative angular motions of these two rigid bodies can be expressed as

$$\omega = \Omega + \omega_r$$
$$\dot{\omega} = \dot{\Omega} + \dot{\omega}_r + \Omega \times \omega_r \qquad\qquad (11.6)$$

where ω and $\dot{\omega}$ are the angular velocity and angular acceleration respectively of rigid body B, Ω and $\dot{\Omega}$ are the angular velocity and angular acceleration respectively of rigid body A, and ω_r and $\dot{\omega}_r$ are the angular velocity and angular acceleration respectively of B relative to A. Using the results obtained from eqns (11.5) ω_r and $\dot{\omega}_r$ can be evaluated from eqns (11.6).

Experimental data are another source of relative angular displacements of two joining segments. These data can be obtained either from electro-goniometer records or from film records. The film records consist of three markings: one at the joint of these two segments and one at some location on each of the two segments. Curve-fitting techniques are applied as before in order to obtain the time functions of the relative angular displacements from which relative angular velocities and relative angular accelerations are derived.

11.3. Kinetic analysis

Equations of motion for a rigid body

Kinetics is the part of dynamics which deals with the effects of forces on the motion of body. When the motion is known the problem is then to find the force system acting on the body. When a system of connected rigid bodies is involved, the force system also consists of interacting forces and moments between the rigid bodies. In the case of human body motion these interacting forces are joint forces and joint moments. Joint forces and joint moments are unknowns in equations of motion. External forces and moments may exist also. With all the kinematic quantities known, it is possible to find the joint forces and moments from the resulting force system acting on each element. This is done by solving a system of simultaneous equations at successive time intervals.

The joint force system mentioned above is not a real joint contact force but an equivalent force including muscle force and real joint force (bony contact force). Since muscles exert an unknown force system, they are resolved into force and moment at the joint. Therefore the resolved muscle force and the real joint force are treated as a totally unknown joint force system in the analysis.

A summary of six equations of motion is listed as follows [16]. Three equations for linear motion are

$$\Sigma F_x = m\ddot{x} \qquad \Sigma F_y = m\ddot{y} \qquad \Sigma F_z = m\ddot{z}. \tag{11.7}$$

Three equations for angular motion when the co-ordinate system is attached to the rigid body are expressed either in the form of general equations or Euler equations. They are as follows.

(i) General equations:

$$
\begin{aligned}
M_x &= I_{xx}\dot{\omega}_x - (I_{yy} - I_{zz})\omega_y\omega_z - I_{xy}(\dot{\omega}_y - \omega_x\omega_z) \\
&\quad - I_{yz}(\omega_y^2 - \omega_z^2) - I_{zx}(\dot{\omega}_z + \omega_x\omega_y) \\
M_y &= I_{yy}\dot{\omega}_y - (I_{zz} - I_{xx})\omega_z\omega_x - I_{yz}(\dot{\omega}_z - \omega_y\omega_x) \\
&\quad - I_{zx}(\omega_z^2 - \omega_x^2) - I_{xy}(\dot{\omega}_x + \omega_y\omega_z) \\
M_z &= I_{zz}\dot{\omega}_z - (I_{xx} - I_{yy})\omega_x\omega_y - I_{zx}(\dot{\omega}_x - \omega_z\omega_y) \\
&\quad - I_{xy}(\omega_x^2 - \omega_y^2) - I_{yz}(\dot{\omega}_y + \omega_z\omega_x)
\end{aligned} \tag{11.8}
$$

(ii) Euler's equations:

$$
\begin{aligned}
M_x &= I_x\dot{\omega}_x - (I_y - I_z)\omega_y\omega_z \\
M_y &= I_y\dot{\omega}_y - (I_z - I_x)\omega_z\omega_x \\
M_z &= I_z\dot{\omega}_z - (I_x - I_y)\omega_x\omega_y.
\end{aligned} \tag{119}
$$

The general equations (11.8) are greatly simplified by fixing the co-ordinate axes so that they coincide with the principal axes of the rigid body. The products of inertia are then zero and the resulting equations (11.9) are called Euler's equations. Thus in Euler's equations the x, y, and z axes are the principal axes of the body with the origin at the centre of mass of the body.

Of these two sets of equations (11.8) and (11.9), the simpler set, Euler's equations, should be used. This choice is possible when local co-ordinates are used and these co-ordinates coincide with the principal axes of the simplified model of each element.

Co-ordinate systems and co-ordinate transformation

In dealing with human body motions, an inertial co-ordinate system can be set arbitrarily which is common to all the body segments. In addition, there is a member co-ordinate system attached to each segment. There are as many member co-ordinate systems as there are segments. Let $\mathbf{i}$, $\mathbf{j}$, $\mathbf{k}$ be the unit co-ordinate vectors of the inertial co-ordinate system (XYZ) and $\mathbf{i}_i$, $\mathbf{j}_i$, $\mathbf{k}_i$ be the unit co-ordinate vectors of a member co-ordinate system $(x_i y_i z_i)$ of the ith segment as shown in Fig. 11.4. Then the transformation from member to inertial co-ordinates can be performed by the transformation matrix $[T]$ as follows:

$$[T] = \begin{bmatrix} \mathbf{i}\cdot\mathbf{i}_i & \mathbf{i}\cdot\mathbf{j}_i & \mathbf{i}\cdot\mathbf{k}_i \\ \mathbf{j}\cdot\mathbf{i}_i & \mathbf{j}\cdot\mathbf{j}_i & \mathbf{j}\cdot\mathbf{k}_i \\ \mathbf{k}\cdot\mathbf{i}_i & \mathbf{k}\cdot\mathbf{j}_i & \mathbf{k}\cdot\mathbf{k}_i \end{bmatrix}. \tag{11.10}$$

It is convenient to use member co-ordinates to formulate the equations of motion for each element. However, in solving the system of equations for joint forces and joint moments it is necessary to transform these forces and moments to the common inertial co-ordinates. A consistent systematic way of defining the member co-ordinate for a segment, approximated by a body of revolution, is prescribed below.

Knowing the vector of the ith segment in the inertial co-ordinates, a member co-ordinate system as shown in Fig. 11.4 is set as follows.

The x_i axis is chosen along the link AB and the y_i axis lies on the XY plane of the inertial co-ordinates. This y_i axis will be unique in the XY plane. Then the z_i axis will be found by the right-hand rule in the rectangular cartesian co-ordinate system.

Let the vector of link AB be known as $\mathbf{l}_i$; then the unit vector $\mathbf{i}_i$ of the x_i axis in the member co-ordinate system will be

$$\mathbf{i}_i = \frac{\mathbf{l}_i}{|\mathbf{l}_i|} = a_1\mathbf{i} + a_2\mathbf{j} + a_3\mathbf{k} \tag{11.11}$$

where $\mathbf{i}$, $\mathbf{j}$, and $\mathbf{k}$ are unit vectors of the system co-ordinates.

Since the unit vector $\mathbf{j}_i$ along the y_i axis is on the XY plane of the system co-

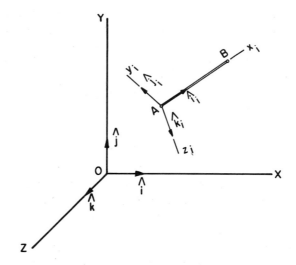

FIG. 11.4.   Inertial and member coordinate systems.

ordinate, there is no **k** term involved. Thus

$$\mathbf{j}_i = \frac{b_1 \mathbf{i} + b_2 \mathbf{j}}{(b_1^2 + b_2^2)^{1/2}} \tag{11.12}$$

where b_1 and b_2 are unknowns. For the unit vector $\mathbf{k}_i$ along the z_i axis

$$\mathbf{k}_i = \frac{c_1 \mathbf{i} + c_2 \mathbf{j} + c_3 \mathbf{k}}{(c_1^2 + c_2^2 + c_3^2)^{1/2}} \tag{11.13}$$

where c_1, c_2, and c_3 are also unknowns.

Once the unknowns b_1, b_2, c_1, c_2, and c_3 are solved, the unit vectors $\mathbf{j}_i$ and $\mathbf{k}_i$ are known. To solve these unknowns we use the orthogonal conditions of unit co-ordinate vectors. They are

$$
\begin{aligned}
\mathbf{i}_i \cdot \mathbf{j}_i &= 0 && \text{or} && a_1 b_1 + a_2 b_2 = 0 \\
\mathbf{j}_i \cdot \mathbf{k}_i &= 0 && \text{or} && b_1 c_1 + b_2 c_2 = 0 \\
\mathbf{i}_i \cdot \mathbf{k}_i &= 0 && \text{or} && a_1 c_1 + a_2 c_2 + a_3 c_3 = 0.
\end{aligned} \tag{11.14}
$$

Solving these three equations gives

$$b_2 = -\frac{a_1}{a_2} b_1$$

$$c_2 = -\frac{b_1}{b_2} c_1 = \frac{a_2}{a_1} c_1 \tag{11.15}$$

$$c_3 = -\frac{a_1^2 + a_2^2}{a_1 a_3} c_1.$$

Thus

$$\mathbf{j}_i = \frac{-a_2\mathbf{i} + a_1\mathbf{j}}{(a_1^2 + a_2^2)^{1/2}} \tag{11.16}$$

$$\mathbf{k}_i = \pm \frac{a_1 a_3 \mathbf{i} + a_2 a_3 \mathbf{j} - (a_1^2 + a_2^2)\mathbf{k}}{\{(a_1 a_3)^2 + (a_2 a_3)^2 + (a_1^2 + a_2^2)^2\}^{1/2}}.$$

Since $\mathbf{i}_i$ is a unit vector $a_1^2 + a_2^2 + a_3^2 = 1$; therefore

$$(a_1 a_3)^2 + (a_2 a_3)^2 + (a_1^2 + a_2^2)^2 = a_1^2 + a_2^2 \tag{11.17}$$

and the denominator of $\mathbf{k}_i$ can be simplified. The sign of $\mathbf{k}$ is determined from the fact that the element of a transformation matrix is equal to its own cofactor. Summarizing,

$$\mathbf{i}_i = a_1\mathbf{i} + a_2\mathbf{j} + a_3\mathbf{k}$$

$$\mathbf{j}_i = \frac{-a_2\mathbf{i} + a_1\mathbf{j}}{(a_1^2 + a_2^2)^{1/2}} \tag{11.18}$$

$$\mathbf{k}_i = \frac{-a_1 a_3 \mathbf{i} - a_2 a_3 \mathbf{j} + (a_1^2 + a_2^2)\mathbf{k}}{(a_1^2 + a_2^2)^{1/2}}.$$

Then the transformation matrix (11.10) for the ith segment is expressed as

$$[T_{ij}]_i \atop M \to S = \begin{bmatrix} a_1 & -a_2/A & -a_1 a_3/A \\ a_2 & a_1/A & -a_2 a_3/A \\ a_3 & 0 & A \end{bmatrix} \tag{11.19}$$

where $A = (a_1^2 + a_2^2)^{1/2}$ and M $\to$ S stands for co-ordinate transformation from member (local) to system (inertial) co-ordinates. The system to member co-ordinate transformation matrix will be $[T_{ij}]_i^T \atop M \to S$ because

$$[T_{ij}]_i^T \atop M \to S = [T_{ij}]_i^{-1} \atop M \to S \tag{11.20}$$

Plane motion of a three-segment kicking model

If all the kinematics at the centre of mass of the segments are known, joint force and joint moment analysis of human motion can be performed. First free-body diagrams of the segments involved are drawn; then the equations of motion are derived. Usually human body segments form an open linkage system during motion and the forces at the end of the distal segment (e.g. foot, hand) are zero. Therefore by drawing free-body diagrams starting from the distal segment one can formulate equations of motion and solve all the unknowns successively without having to solve simultaneous equations. For a three-segment kicking model in plane motion free-body diagrams and equations of the motion are derived as follows. A two-segment model is a special case.

Free-body diagrams for the foot, leg, and thigh during kicking are shown in Fig. 11.5. The equations of motion are formulated from the free-body diagrams as follows:

(i) Foot

$$F_{xB}^{(1)} - W_1 \sin \theta_1 = m_1 a_{x1}$$
$$F_{yB}^{(1)} - W_1 \cos \theta_1 = m_1 a_{y1} \tag{11.21}$$
$$M_1 + F_{yB}^{(1)} l_1' = I_1 \alpha_1$$

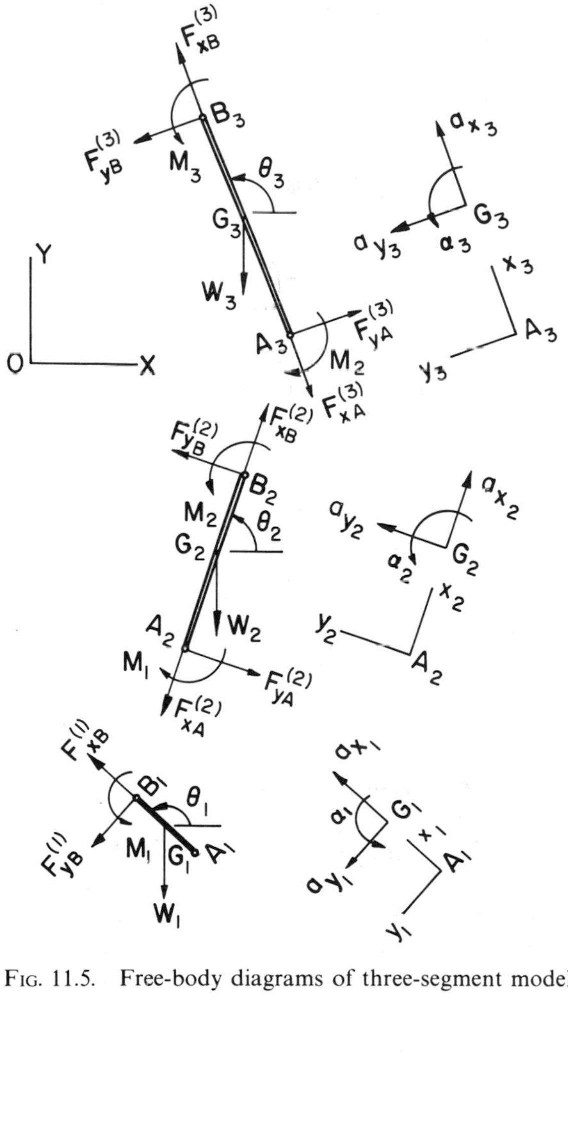

FIG. 11.5.   Free-body diagrams of three-segment model.

where

$$l_1' = G_1 B_1.$$

(ii) Leg

$$F_{xB}^{(2)} - F_{xA}^{(2)} - W_2 \sin \theta_2 = m_2 a_{x2}$$
$$F_{yB}^{(2)} - F_{yA}^{(2)} - W_2 \cos \theta_2 = m_2 a_{y2} \qquad (11.22)$$
$$M_2 - M_1 + F_{yA}^{(2)}(l_2 - l_2') + F_{yB}^{(2)} l_2' = I_2 \alpha_2$$

where

$$l_2' = G_2 B_2.$$

(iii) Thigh

$$F_{xB}^{(3)} - F_{xA}^{(3)} - W_3 \sin \theta_3 = m_3 a_{x3}$$
$$F_{yB}^{(3)} - F_{yA}^{(3)} - W_3 \cos \theta_3 = m_3 a_{y3} \qquad (11.23)$$
$$M_3 - M_2 + F_{yA}^{(3)}(l_3 - l_3') + F_{yB}^{(3)} l_3' = I_3 \alpha_3$$

where

$$l_3' = G_3 B_3.$$

The two sets of continuity conditions are as follows.

(i) At the ankle joint

$$F_{xA}^{(2)} = F_{xB}^{(1)} \cos \phi_1 + F_{yB}^{(1)} \sin \phi_1 \qquad (11.24)$$
$$F_{yA}^{(2)} = -F_{xB}^{(1)} \sin \phi_1 + F_{yB}^{(1)} \cos \phi_1$$

where

$$\phi_1 = \theta_2 - \theta_1.$$

(ii) At the knee joint

$$F_{xA}^{(3)} = F_{xB}^{(2)} \cos \phi_2 + F_{yB}^{(2)} \sin \phi_2 \qquad (11.25)$$
$$F_{yA}^{(3)} = -F_{xB}^{(2)} \sin \phi_2 + F_{yB}^{(2)} \cos \phi_2$$

where

$$\phi_2 = \theta_3 - \theta_2.$$

Plane motion of a continuous n-segment model

The equations of motion and the continuity conditions for an *n*-segment model shown in Fig. 11.6 are derived as follows.

(i) The equations of motion for the distal segment:

$$F_{xB}^{(1)} - W_1 \sin \theta_1 = m_1 a_{x1}$$
$$F_{yB}^{(1)} - W_1 \cos \theta_1 = m_1 a_{y1} \qquad (11.26)$$
$$M_1 + F_{yB}^{(1)} l_1' = I_1 \alpha_1.$$

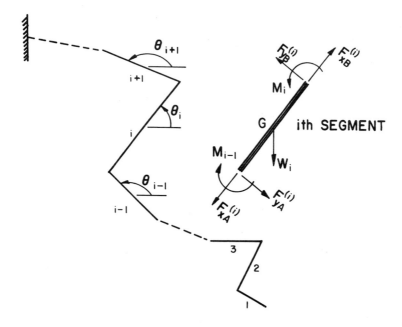

FIG. 11.6.   A continuous n-segment model.

(ii) The continuity conditions between the ith and $(i+1)$th segments:

$$F_{xA}^{(i+1)} = F_{xB}^{(i)} \cos \phi_i + F_{yB}^{(i)} \sin \phi_i$$
$$F_{yA}^{(i+1)} = -F_{xB}^{(i)} \sin \phi_i + F_{yB}^{(i)} \cos \phi_i \qquad (11.27)$$

where

$$\phi_i = \theta_{i+1} - \theta_i.$$

(iii) The equations of motion for the ith segment:

$$F_{xB}^{(i)} - F_{xA}^{(i)} - W_i \sin \theta_i = m_i a_{xi}$$
$$F_{yB}^{(i)} - F_{yA}^{(i)} - W_i \cos \theta_i = m_i a_{yi} \qquad (11.28)$$
$$M_i - M_{i-1} + F_{yA}^{(i)} (l_i - l_i') + F_{yB}^{(i)} l_i' = I_i \alpha_i$$

where

$$l_i' = G_i B_i$$
$$i = 2, \ldots, n.$$

Ground reactions for plane kicking motion

In order to check the analytical results from the previous analysis we can calculate the ground reaction forces and compare them with the force platform read-outs obtained in the experiments.

The free-body diagram for a simulated kick is shown in Fig. 11.7 alongside a sketch illustrating the force platform experiment. The equations of equilibrium are

$$R_x - F_{xB}^{(3)} \cos \theta_3 + F_{yB}^{(3)} \sin \theta_3 = 0$$
$$R_y - W' - F_{xB}^{(3)} \sin \theta_3 - F_{yB}^{(3)} \cos \theta_3 = 0 \qquad (11.29)$$
$$M_R - M_3 + R_x l = 0$$

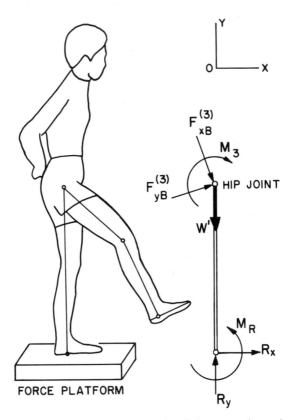

FIG. 11.7. Free-body diagram involving ground reactions.

where

$$W' = W_{\text{total}} - W_{\text{thigh + leg + foot}}$$
$$l = l_{\text{thigh}} + l_{\text{leg}} + l_{\text{ankle axis to ground}}.$$

Therefore the reaction forces and moments are solved as

$$R_x = F_{xB}^{(3)} \cos \theta_3 - F_{yB}^{(3)} \sin \theta_3$$
$$R_y = W' + F_{xB}^{(3)} \sin \theta_3 + F_{yB}^{(3)} \cos \theta_3 \qquad (11.30)$$
$$M_R = M_3 - R_x l$$

R_x, R_y, and M_R can also be obtained from the force platform as direct read-outs. These read-outs can be compared with the calculated values of R_x, R_y, and M_R.

Plane motion analysis of kicking models with inertial co-ordinates only

In the previous sections of plane motion analysis a combination of local and inertial co-ordinate systems was used. This section of plane motion analysis will use only an inertial co-ordinate system.

For a three-segment model the free-body diagrams for the foot, leg, and thigh are shown in Fig. 11.8. The equations of motion are derived as follows.

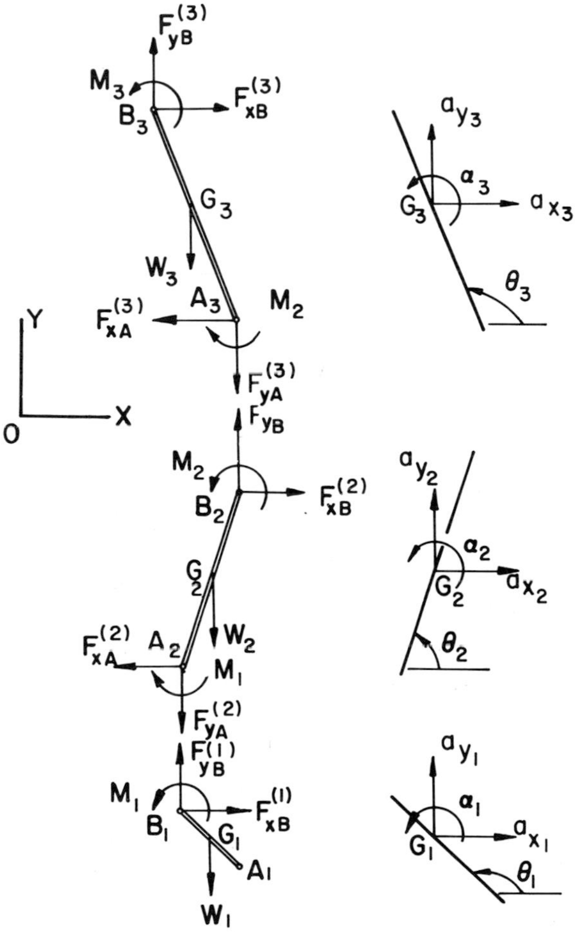

FIG. 11.8. Free-body diagrams of a three-segment model in inertial co-ordinates.

(i) Foot:

$$F_{xB}^{(1)} = m_1 a_{x1}$$
$$F_{yB}^{(1)} - W_1 = m_1 a_{y1} \tag{11.31}$$
$$M_1 - F_{xB}^{(1)} l_1' \sin \theta_1 + F_{yB}^{(1)} l_1' \cos \theta_1 = I_1 \alpha_1.$$

(ii) Leg:

$$F_{xB}^{(2)} - F_{xA}^{(2)} = m_2 a_{x2}$$
$$F_{yB}^{(2)} - F_{yA}^{(2)} - W_2 = m_2 a_{y2} \tag{11.32}$$
$$M_2 - M_1 - F_{xB}^{(2)} l_2' \sin \theta_2 + F_{yB}^{(2)} l_2' \cos \theta_2$$
$$\qquad - F_{xA}^{(2)} (l_2 - l_2') \sin \theta_2 + F_{yA}^{(2)} (l_2 - l_2') \cos \theta_2 = I_2 \alpha_2.$$

(iii) Thigh:

$$F_{xB}^{(3)} - F_{xA}^{(3)} = m_3 a_{x3}$$
$$F_{yB}^{(3)} - F_{yA}^{(3)} - W_3 = m_3 a_{y3} \tag{11.33}$$
$$M_3 - M_2 - F_{xB}^{(3)} l_3' \sin \theta_3 + F_{yB}^{(3)} l_3' \cos \theta_3$$
$$\qquad - F_{xA}^{(3)} (l_3 - l_3') \sin \theta_3 + F_{yA}^{(3)} (l_3 - l_3') \cos \theta_3 = I_3 \alpha_3.$$

From these equations and the conditions that

$$F_{xA}^{(2)} = F_{xB}^{(1)} \qquad F_{yA}^{(2)} = F_{yB}^{(1)}$$
$$F_{xA}^{(3)} = F_{xB}^{(2)} \qquad F_{yA}^{(3)} = F_{yB}^{(2)} \tag{11.34}$$

$F_{xB}^{(1)}, F_{yB}^{(1)}, M_1, F_{xB}^{(2)}, F_{yB}^{(2)}, M_2, F_{xB}^{(3)}, F_{yB}^{(3)}$ and M_3 can be solved.

It should be noted that plane motion analysis with inertial co-ordinates only is relatively uncomplicated compared with the method with both co-ordinates because continuity conditions are so simple. Nevertheless, either method may be chosen for plane motion analysis. However, in spatial motion analysis, which will be discussed later, a combination of both co-ordinate systems must be used in order to apply Euler's equations.

As a generalization, equations of motion for the ith segment of a continuous n-segment model are derived as follows:

$$F_{xB}^{(i)} - F_{xA}^{(i)} = m_i a_{xi}$$
$$F_{yB}^{(i)} - F_{yA}^{(i)} - W_i = m_i a_{yi} \tag{11.35}$$
$$M_i - M_{i-1} - F_{xB}^{(i)} l_i' \sin \theta_i + F_{yB}^{(i)} l_i' \cos \theta_i$$
$$\qquad - F_{xA}^{(i)} (l_i - l_i') \sin \theta_i + F_{yA}^{(i)} (l_i - l_i') \cos \theta_i = I_i \alpha_i.$$

From these equations all the unknowns involved can be solved.

The ground reaction force can be found in a manner similar to the one treated previously.

Spatial motion of a three-segment kicking model

In the derivations of equations for spatial motion the member co-ordinate system is adopted. This is because the simplicity of applying Euler's equations, instead of general equations, can be achieved only by using the member co-ordinate system.

Let W_{ix}, W_{iy}, and W_{iz} be the weight components of segment i, $i = 1, 2, 3$, in local co-ordinates. They are related to the corresponding components in system co-ordinates as follows:

$$\begin{Bmatrix} W_{ix} \\ W_{iy} \\ W_{iz} \end{Bmatrix}_{\text{local}} = [T_{ij}]_i \underset{S \to M}{} \begin{Bmatrix} 0 \\ -W_i \\ 0 \end{Bmatrix}_{\text{inertial}} \tag{11.36}$$

where $[T_{ij}]_i \underset{S \to M}{}$ is the transformation matrix from system (inertial) to member (local) co-ordinates. If the unit position vector of the segment is known as

$$\mathbf{l}_i = l_{ix}\mathbf{i} + l_{iy}\mathbf{j} + l_{iz}\mathbf{k}. \tag{11.37}$$

Then the transformation matrix is

$$[T_{ij}]_i \underset{S \to M}{} = \begin{bmatrix} l_{ix} & l_{iy} & l_{iz} \\ -l_{iy}/S_i & l_{ix}/S_i & 0 \\ -l_{ix}l_{iz}/S_i & -l_{iy}l_{iz}/S_i & S_i \end{bmatrix} \tag{11.38}$$

where

$$S_i = (l_{ix}^2 + l_{iy}^2)^{1/2} \tag{11.39}$$

The free-body diagram of a three-segment kicking model in spatial motion is shown in Fig. 11.9. Knowing all the kinematics at the centre of mass the equations of motion for each segment can be formulated as follows.

(i) Foot. The force equations are

$$\begin{aligned} F_{xB}^{(1)} + W_{1x} &= m_1 a_{x1} \\ F_{yB}^{(1)} + W_{1y} &= m_1 a_{y1} \\ F_{zB}^{(1)} + W_{1z} &= m_1 a_{z1}. \end{aligned} \tag{11.40}$$

The moment equations are

$$\begin{aligned} M_{xB}^{(1)} &= I_{x1}\alpha_{x1} - (I_{y1} - I_{z1})\omega_{y1}\omega_{z1} \\ M_{yB}^{(1)} - F_{zB}^{(1)} l_1' &= I_{y1}\alpha_{y1} - (I_{z1} - I_{x1})\omega_{z1}\omega_{x1} \\ M_{zB}^{(1)} + F_{yB}^{(1)} l_1' &= I_{z1}\alpha_{z1} - (I_{x1} - I_{y1})\omega_{x1}\omega_{y1} \end{aligned} \tag{11.41}$$

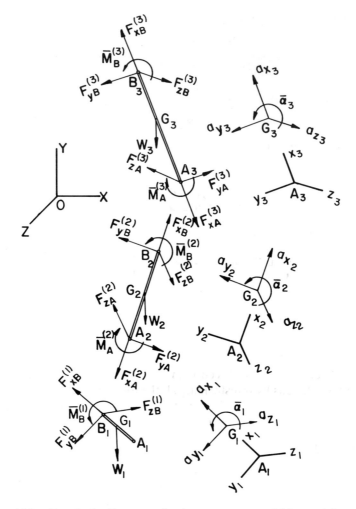

FIG. 11.9.   Free-body diagrams of a three-segment model in spatial motion.

in which ω_1 and α_1 are known from kinematic analyses and the moments of inertia are obtained from anthropometric data [13].

(ii) Leg. The force equations are

$$F_{xB}^{(2)} - F_{xA}^{(2)} + W_{2x} = m_2 a_{x2}$$
$$F_{yB}^{(2)} - F_{yA}^{(2)} + W_{2y} = m_2 a_{y2}$$
$$F_{zB}^{(2)} - F_{zA}^{(2)} + W_{2z} = m_2 a_{z2}.$$

$$(11.42)$$

The moment equations are

$$M_{xB}^{(2)} - M_{xA}^{(2)} = I_{x2}\alpha_{x2} - (I_{y2} - I_{z2})\omega_{y2}\omega_{z2}$$
$$M_{yB}^{(2)} - M_{yA}^{(2)} - F_{zB}^{(2)}l_2' - F_{zA}^{(2)}(l_2 - l_2')$$
$$= I_{y2}\alpha_{y2} - (I_{z2} - I_{x2})\omega_{z2}\omega_{x2} \qquad (11.43)$$
$$M_{zB}^{(2)} - M_{zA}^{(2)} + F_{yB}^{(2)}l_2' + F_{yA}^{(2)}(l_2 - l_2')$$
$$= I_{z2}\alpha_{z2} - (I_{x2} - I_{y2})\omega_{x2}\omega_{y2}.$$

(iii) Thigh. The force equations are

$$F_{xB}^{(3)} - F_{xA}^{(3)} + W_{3x} = m_3 a_{x3}$$
$$F_{yB}^{(3)} - F_{yA}^{(3)} + W_{3y} = m_3 a_{y3} \qquad (11.44)$$
$$F_{zB}^{(3)} - F_{zA}^{(3)} + W_{3z} = m_3 a_{z3}.$$

The moment equations are

$$M_{xB}^{(3)} - M_{xA}^{(3)} = I_{x3}\alpha_{x3} - (I_{y3} - I_{z3})\omega_{y3}\omega_{z3}$$
$$M_{yB}^{(3)} - M_{yA}^{(3)} - F_{zB}^{(3)}l_3' - F_{zA}^{(3)}(l_3 - l_3')$$
$$= I_{y3}\alpha_{y3} - (I_{z3} - I_{x3})\omega_{z3}\omega_{x3} \qquad (11.45)$$
$$M_{zB}^{(3)} - M_{zA}^{(3)} + F_{yB}^{(3)}l_3' + F_{yA}^{(3)}(l_3 - l_3')$$
$$= I_{z3}\alpha_{z3} - (I_{x3} - I_{y3})\omega_{x3}\omega_{y3}.$$

At the joint the forces and moments between two segments are equal in magnitude and opposite in direction in the inertial co-ordinate system. One can find the relations by transforming all the forces and moments from local co-ordinate systems to a single inertial co-ordinate system. The continuity conditions at the ankle and knee joints are described as follows.

(i) Ankle joint:

$$[T_{ij}]_1 \atop{M \to S} \begin{Bmatrix} F_{xB}^{(1)} \\ F_{yB}^{(1)} \\ F_{zB}^{(1)} \end{Bmatrix} = -[T_{ij}]_2 \atop{M \to S} \begin{Bmatrix} F_{xA}^{(2)} \\ F_{yA}^{(2)} \\ F_{zA}^{(2)} \end{Bmatrix}$$

or

$$\begin{Bmatrix} F_{xA}^{(2)} \\ F_{yA}^{(2)} \\ F_{zA}^{(2)} \end{Bmatrix} = -[T_{ij}]_2^{\mathrm{T}} [T_{ij}]_1 \atop{M \to S \quad M \to S} \begin{Bmatrix} F_{xB}^{(1)} \\ F_{yB}^{(1)} \\ F_{zB}^{(1)} \end{Bmatrix}. \qquad (11.46)$$

Similarly

$$\begin{Bmatrix} M_{xA}^{(2)} \\ M_{yA}^{(2)} \\ M_{zA}^{(2)} \end{Bmatrix} = -[T_{ij}]_2^{\mathrm{T}} [T_{ij}]_1 \atop{M \to S \quad M \to S} \begin{Bmatrix} M_{xB}^{(1)} \\ M_{yB}^{(1)} \\ M_{zB}^{(1)} \end{Bmatrix}. \qquad (11.47)$$

(ii) Knee joint:

$$\begin{Bmatrix} F_{xA}^{(3)} \\ F_{yA}^{(3)} \\ F_{zA}^{(3)} \end{Bmatrix} = -\underset{M \to S}{[T_{ij}]_3^{\mathrm{T}}} \; \underset{M \to S}{[T_{ij}]_2} \begin{Bmatrix} F_{xB}^{(2)} \\ F_{yB}^{(2)} \\ F_{zB}^{(2)} \end{Bmatrix}. \tag{11.48}$$

and

$$\begin{Bmatrix} M_{xA}^{(3)} \\ M_{yA}^{(3)} \\ M_{zA}^{(3)} \end{Bmatrix} = -\underset{M \to S}{[T_{ij}]_3^{\mathrm{T}}} \; \underset{M \to S}{[T_{ij}]_2} \begin{Bmatrix} M_{xB}^{(2)} \\ M_{yB}^{(2)} \\ M_{zB}^{(2)} \end{Bmatrix}. \tag{11.49}$$

All the unknown joint forces and joint moments can be solved from the above equations of motion and continuity conditions.

Spatial motion of a continuous n-segment model

The equations of motion and continuity conditions for a continuous n-segment model can be generalized as follows.

(i) *i*th segment. The force equations are

$$\begin{aligned} F_{xB}^{(i)} - F_{xA}^{(i)} + W_{ix} &= m_i a_{xi} \\ F_{yB}^{(i)} - F_{yA}^{(i)} + W_{iy} &= m_i a_{yi} \\ F_{zB}^{(i)} - F_{zA}^{(i)} + W_{iz} &= m_i a_{zi}. \end{aligned} \tag{11.50}$$

The moment equations are

$$\begin{aligned} M_{xB}^{(i)} - M_{xA}^{(i)} &= I_{xi}\alpha_{xi} - (I_{yi} - I_{zi})\omega_{yi}\omega_{zi} \\ M_{yB}^{(i)} - M_{yA}^{(i)} - F_{zB}^{(i)} l_i' &- F_{zA}^{(i)} (l_i - l_i') \\ &= I_{yi}\alpha_{yi} - (I_{zi} - I_{xi})\omega_{zi}\omega_{xi} \\ M_{zB}^{(i)} - M_{zA}^{(i)} + F_{yB}^{(i)} l_i' &+ F_{yA}^{(i)} (l_i - l_i') \\ &= I_{zi}\alpha_{zi} - (I_{xi} - I_{yi})\omega_{xi}\omega_{yi}. \end{aligned} \tag{11.51}$$

(ii) The continuity conditions between the *i*th and $(i+1)$th segments are

$$\begin{Bmatrix} F_{xA}^{(i+1)} \\ F_{yA}^{(i+1)} \\ F_{zA}^{(i+1)} \end{Bmatrix} = -\underset{M \to S}{[T_{ij}]_{i+1}^{\mathrm{T}}} \; \underset{M \to S}{[T_{ij}]_i} \begin{Bmatrix} F_{xB}^{(i)} \\ F_{yB}^{(i)} \\ F_{zB}^{(i)} \end{Bmatrix} \tag{11.52}$$

and

$$\begin{Bmatrix} M_{xA}^{(i+1)} \\ M_{yA}^{(i+1)} \\ M_{zA}^{(i+1)} \end{Bmatrix} = -\underset{M \to S}{[T_{ij}]_{i+1}^{\mathrm{T}}} \; \underset{M \to S}{[T_{ij}]_i} \begin{Bmatrix} M_{xB}^{(i)} \\ M_{yB}^{(i)} \\ M_{zB}^{(i)} \end{Bmatrix}. \tag{11.53}$$

Ground reactions for spatial kicking motion

The free-body diagram involving ground reactions in spatial kicking motion is shown in Fig. 11.10. The equations of motion are as follows.

The force equations are

$$\left\{ \begin{array}{c} R_x \\ R_y - W' \\ R_z \end{array} \right\}_{\text{inertial}} = -[T_{ij}]_i \underset{M \to S}{} \left\{ \begin{array}{c} F_{xB}^{(i)} \\ F_{yB}^{(i)} \\ F_{zB}^{(i)} \end{array} \right\}_{\text{local}} . \qquad (11.54)$$

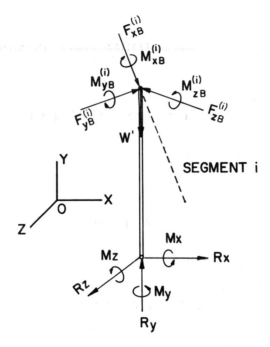

FIG. 11.10.   Free-body diagram involving ground reactions in spatial motion.

The moment equations are

$$\left\{ \begin{array}{c} M_x \\ M_y \\ M_z \end{array} \right\}_{\text{inertial}} = -[T_{ij}]_i \underset{M \to S}{} \left\{ \begin{array}{c} M_{xB}^{(i)} \\ M_{yB}^{(i)} \\ M_{zB}^{(i)} \end{array} \right\}_{\text{local}} . \qquad (11.55)$$

11.4. Experiments for validation

Attempts can be made to validate certain values calculated from the analysis.

(a) *Displacement.* Angular displacement between the thigh and the leg at the knee joint can be recorded directly by electrogoniometer. Present records indicate good agreement with film data [19].

(b) *Velocity*. The horizontal velocity of a kicked ball can be measured directly by dark field velocimetry [20]. Present data indicate good agreement with film data [10, 21].

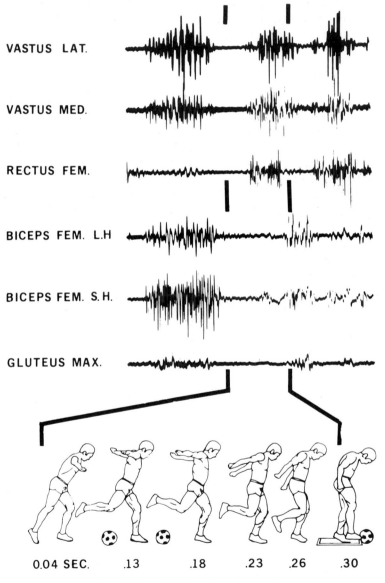

FIG. 11.11.   EMG studies of a toe kick.

(c) *Acceleration.* Linear acceleration in two planes might be measured directly by accelerometer.

(d) *Force.* Total forces and moments acting through the feet can be measured directly by force platform and compared to calculated values [10, 18].

(e) *Muscle activity.* Some information on the timing of the activity of key muscles involved in kicking can be obtained by direct surface electromyographic (EMG) recordings as shown in Fig. 11.11 [9, 10].

11.5. Numerical examples

The methods described in the previous sections have been initially applied to simulated kicks and soccer toe kicks. The simple two-segment planar model, for which the leg and foot are considered as one segment, was applied to the simulated kick. The anthropometric data of the subject is shown in Table 11.1 and its complete kinematic and kinetic results are shown in Figs. 11.12–11.14. The values of joint forces, obtained from kinetic analysis, are presented in Figs. 11.15 and 11.16. Figure 11.17 shows the validation of ground reaction force calculations for a simulated kick in comparison with force platform readings. The three-segment planar model has been applied to the toe kick [9, 10]. A sample of the results for the subject with the anthropometric data given in Table 11.2 is shown in Fig. 11.18.

It should be noted that displacement was measured with the backward horizontal axis at $-90°$ in the simulated kick and $0°$ in the toe kick. The simulated kick analysis included only the forward swing of the limb and continued into the follow-through motion during 0.27–0.4 s. While the toe kick included the backswing after the kicking foot left the ground as well as the

TABLE 11.1

Anthropometric data of the subject in simulated kicks

	FPS unit	SI unit
Weight or mass [13]		
Total body	119.6 lb	54.3 kg
Leg plus foot	7.1 lb	3.2 kg
Thigh	16.3 lb	7.4 kg
Length		
Height	5.53 ft	1.69 m
Length of leg plus foot	1.32 ft	0.40 m
Length of thigh	1.18 ft	0.35 m
Mass moment of inertia [13]		
Leg plus foot at centre of mass	0.069 slug ft^2	0.094 kg m^2
Thigh at centre of mass	0.073 slug ft^2	0.099 kg m^2

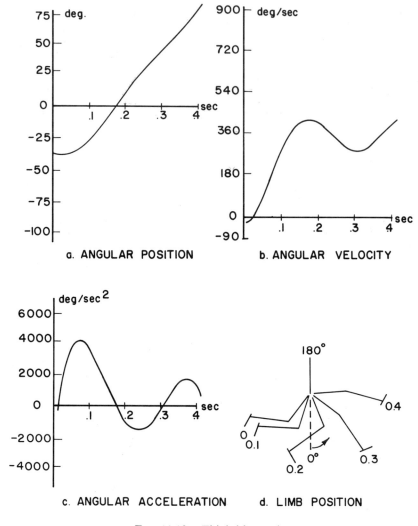

FIG. 11.12. Thigh kinematics.

forward swing up to ball impact, it did not include the follow-through. The toe kick takes less time than the simulated kick (0.3 s versus 0.4 s) and is considerably faster.

Figure 11.18(a) shows the angular displacement curves for the thigh and leg segments of a two-step toe kick from 'foot off' to one film frame (about 0.005 s) prior to ball impact. Figure 11.18(b) shows the angular accelerations derived from a sixth-order polynomial fit of displacement. An examination of the

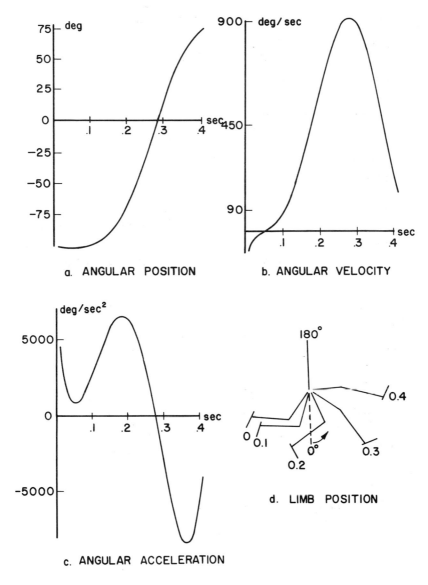

a. ANGULAR POSITION

b. ANGULAR VELOCITY

c. ANGULAR ACCELERATION

d. LIMB POSITION

Fɪɢ. 11.13.   Leg kinematics.

curves shows that the thigh starts rotating forward (positive displacement slope) while the leg is still rotating backward at about 0.16 s. The leg then begins rotating forward, at about 0.2 s, and shortly thereafter, at about 0.25 s, the thigh begins to slow down (the acceleration becomes negative). It is almost

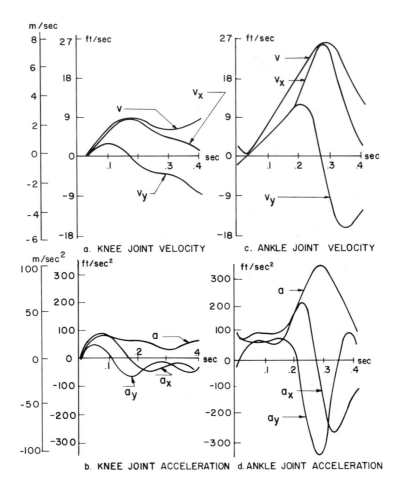

FIG. 11.14. Kinematics of knee and ankle joints.

stationary at impact while the leg and foot have reached peak velocity and zero acceleration. The phenomenon of the thigh slowing down or stopping before ball impact prior to speeding up in the follow through may not be generally known. The time during which the phenomenon occurs is so short that it is difficult to observe visually.

The angular kinematics of the thigh and leg in the simulated kick show a similar pattern (Figs. 11.12 and 11.13) [8]. This can best be seen by the angular acceleration curves of Figs. 11.12(c) and 11.13(c) between 0.025 and 0.27 s. The thigh accelerates first during 0.025–0.17 s and slows as the leg approaches peak acceleration near 0.18 s. This exchange of angular velocities between the

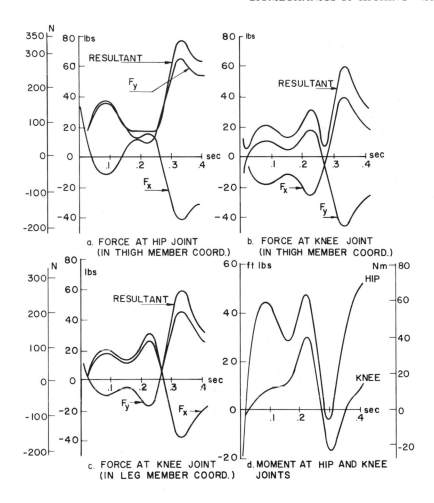

FIG. 11.15. Kinetics at hip and knee joints.

proximal and distal segments would suggest possible research into transfer of angular momentum from the more massive thigh to the less massive leg.

Knee angular acceleration reaches a maximum about the time of reversal from flexion to extension. If it is assumed, as seems fairly reasonable, that the quadricep muscles play a considerable role in achieving the knee extension angular acceleration, then the data indicate that they must be involved in slowing knee flexion prior to accelerating knee extension. In other words, backward rotation of the leg during knee flexion provides an inertial load acting against the knee extensors and some of the positive angular impulse

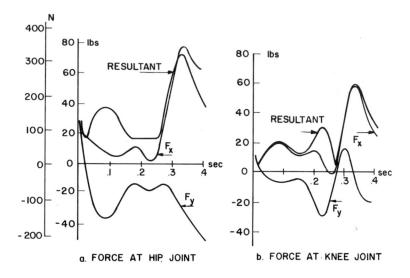

FIG. 11.16.   Kinetics at hip and knee joints in inertial co-ordinates.

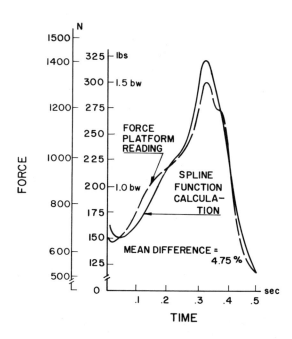

FIG. 11.17.   Vertical force comparison (bw, body weight).

TABLE 11.2

Anthropometric data of the subject in toe kicks

	FPS unit	SI unit
Weight or mass [13]		
Total body	192.5 lb	87.3 kg
Segment 1, foot	2.5 lb	1.1 kg
Segment 2, leg	8.5 lb	3.8 kg
Segment 3, thigh	24.8 lb	11.3 kg
Length		
Height	5.84 ft	1.78 m
Length of foot (ankle joint to first meta-		
tarsophalangeal joint)	0.52 ft	0.16 m
Length of leg	1.35 ft	0.41 m
Length of thigh	1.31 ft	0.40 m
Mass moments of inertia [13]		
Segment 1, foot	0.0047 slug ft^2	0.0064 kg m^2
Segment 2, leg	0.042 slug ft^2	0.057 kg m^2
Segment 3, thigh	0.14 slug ft^2	0.19 kg m^2

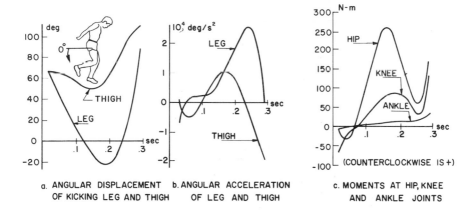

a. ANGULAR DISPLACEMENT   b. ANGULAR ACCELERATION   c. MOMENTS AT HIP, KNEE
   OF KICKING LEG AND THIGH   OF LEG AND THIGH   AND ANKLE JOINTS

FIG. 11.18.   Kinematics and kinetics of a toe kick.

developed is involved in slowing this backward motion to zero. The remainder
of the positive impulse will accelerate the leg forward prior to dropping off
very quickly as the leg approaches maximum velocity.

Preliminary recording of myoelectric potentials (Fig. 11.11) is in keeping with
the above interpretation. It must be kept in mind, however, that under dynamic
conditions such as these the latency between the appearance or disappearance

of muscle membrane potentials and the development or dissipation of tension sufficient to have an effect on the segment is difficult to assess. Corser [22] reported contraction latencies of as much as 20 to 75 ms with a mean of 44 ms for upper extremity 'rapid' elbow flexion and extension. Relaxation latencies ranged from 25 to 117 ms with a mean of 60 ms (see also ref. 23).

The observation that preliminary 'backward' motion of a segment prior to 'forward' motion occurs in rapid human motion has been reported previously [9, 15, 24–27]. Since the data show that peak acceleration is reached close to the point of direction reversal, it might be speculated that the initial backward motion provides optimal loading conditions for the development of muscle force such that the peak force is available to start the segment moving in the desired direction. Since the range of motion in the desired direction is frequently very limited, as is the case for knee extension in a kick, the peak force must be immediately applied to accelerate the limb to its maximum speed within the limited time and range available. On the other hand, Cavagna *et al.* [27] found that even though the ground reaction force developed in vertical jumping was greater when a preliminary crouch immediately preceded a maximal effort jump the velocity developed was not greater than that attained when the jump was started from a static crouch. Further investigation of the response of the lower limb masculature under dynamic loading conditions such as those observed in the kick may help to clarify the situation, (see also ref. 28). Seireg and Arvakar [29] and Miller [30] are extending their method of calculating individual muscle force patterns in quasi-static walking to dynamic conditions in bicycling. The technique has potential to advance our knowledge of muscle activity in kicking.

The calculated moments at the hip, knee, and ankle joints as shown in Fig. 11.18(c) seem reasonable. They are negative as the kicking foot leaves the ground and the limb segments are rotating backward. They then become positive, with the hip joint moment reaching a maximum value as the thigh's angular acceleration reaches a maximum value at about 0.17 s (Fig. 11.18(b)) and just after the thigh begins rotating forward (Fig. 11.18(a)). The knee moments reach a maximum at about 0.185 s as the leg reverses direction (Fig. 11.18(a)) and the leg angular acceleration is high but not yet maximum (Fig. 11.18(b)).

In the simulated kick the hip joint moment (Fig. 11.15(d)) shows two peaks: one near 0.1 s when the thigh's angular acceleration is near maximum (Fig. 11.12(c)) and one near 0.20 s when the knee joint moments and leg angular acceleration are high (Fig. 11.13(c)).

The fact that the hip moment has two peaks in the simulated kick and only one in the toe kick seems somewhat surprising in view of the similarities in kinematics noted earlier. It is possible that the peak knee joint moment in the toe kick occurs so close in time to the peak hip moment that its influence is easily smoothed by the polynomial curve-fitting procedure used in this case.

Recent research [10, 18, 21] using cubic spline function curve-fitting techniques suggests that there is, in fact, a second peak in the hip joint moment of the soccer toe kick.

The resultant forces at the hip and knee joints in the simulated kick have peak magnitudes of 0.66 times body weight (bw) and 0.5 bw respectively. These are much less than the 3.9 bw mean peak hip joint force and 3.4 bw mean peak knee joint force reported by Paul [31] for the stance limb in natural speed walking. The highest values recorded by him were 6.4 bw for the hip joint and 4.46 bw for the knee joint. Morrison's [32] values for the knee joint of the stance limb in walking are similar to Paul's, ranging from 2 to 4 bw with a mean of 3.03 bw. The hip and knee joint values in the simulated kick of 0.66 bw and 0.5 bw respectively are, however, similar to those graphically displayed by Paul for the hip (about 0.66 bw) and knee (about 0.4 bw) joints, during comparable phases, for the swing limb in normal walking. The values in fast soccer toe kicks with foot speeds averaging 20.59 m s^{-1} and ball speeds of 27.4 m s^{-1} are still considerably less than those for the stance limb in walking. Mean peak resultant values for the hip, knee, and ankle were 1.34, 0.68 and 0.46 bw respectively [10, 21]. The EMG magnitudes in Fig. 11.11 tend to be in keeping with these comparisons since they are greater when the subject is standing on, rather than kicking with, his right lower extremity.

Since the impact phase of kicking has not been studied extensively, the full implications of the timing and magnitudes of joint forces and moments to ball speed, direction, and spin are not clear. Ball speed relative to foot speed has been reported by several investigators. Roberts and Metcalfe [15] found that the foot speed 15 ms before contact was 18–24 m s^{-1} and the ball speed immediately following the impact phase was 5–7 m s^{-1} faster than the foot. The fastest foot and ball speeds were found in the American football punt. One of the faster foot speeds resulted in a relatively low ball speed because of a bad ball contact. Plagenhoef [4] reported ball speeds of 78.2 ft s^{-1} (28.8 m s^{-1}) to 95.5 ft s^{-1} (29.1 m s^{-1}) for a single individual skilled in both soccer and American football with foot speeds ranging from 53.6 ft s^{-1} (16.34 m s^{-1}) to 79.2 ft s^{-1} (24.14 m s^{-1}). He did not find good correspondence between foot speed and ball speed and concluded that the 'placement of the foot on the ball is a greater variable than is the attainment of maximum foot velocity'. On the other hand, for the soccer toe kick performed at fast (above 27.4 m s^{-1}), medium (21.34 ± 1.52 m s^{-1}), and slow (15.24 ± 1.52 m s^{-1}) speeds by skilled performers Zernicke [10, 21] found a close correlation ($r = 0.96$) between foot speed and ball speed.

Macmillan [33] analysed three of the six different types of kicks allowed for in Australian Rules Football, the drop punt, drop kick, and stab kick. All three kicks were primarily in a plane and were performed for two different distances each. The maximum velocity of the foot immediately prior to contact was found to be a function primarily of the angular velocity of knee extension at that time, and the launch angle of the ball was primarily a function of the path of the foot during its contact with the ball. While the velocity of the ball was largely related to the velocity of the foot and thus to the angular velocity of knee extension, the differences in weights of factors in the multiple regression equations (foot velocity, knee angular acceleration, knee angular jerk, and ankle angle extension) suggested that differences existed between kickers in the way in which the contributions were generated. For one subject the predictive accuracy was greater when foot velocity, the agent imparting momentum to the ball, was excluded from the regression equation. Macmillan agreed with Plagenhoef that the nature of the contact between foot and ball as momentum is transferred from one to the other needs more direct investigation.

11.6. Conclusion

The present chapter is derived from an interdisciplinary research program between the Departments of Engineering Mechanics and Physical Education and Dance, University of Wisconsin, Madison. This research deals with kinematic and kinetic analyses of the natural high speed movement exhibited by human beings in normal everyday settings as well as in special performances such as national and international athletic competitions. Continuing research in three-dimensional kinematic and kinetic analyses and application of momentum and energy methods are progressing. The equivalent force and moment systems provide insight into the time course of individual body segment kinetics and lay the ground work for further investigation into specific muscular forces and neural events related to the observed kinetics. In this respect they contribute to efforts to understand how the human mechanism generates high velocity motion in distal segments and thus provide guidance to the observer and teacher of motor skills in classrooms, athletic fields, gymnasia, hospitals, clinics, and rehabilitation centres.

Summary (physics/engineering science)

Kinematics and kinetics of a particle and a rigid body are used in this study. Linear and angular displacements obtained from experiment as discrete data are curve fitted to produce continuous time functions from which velocities and accelerations are derived. The force systems are then solved from the equations of motion of rigid bodies and continuity conditions between the rigid bodies.

Summary (medical science)

The time course and magnitudes of mechanical characteristics of human movement in terms of joint forces and moments can be derived by application of the dynamic model described in this study to kicking and certain other skills.

Such information is vital for delineation of the mechanical bounds within which the neuromuscular system controls movement not only in the skilled adult but also in the developing child and the handicapped individual.

Acknowledgements

This chapter is based on research supported in part by the WARF (Wisconsin Alumni Research Foundation) and administered by the Research Committee of the Graduate School, University of Wisconsin, Madison, Wisconsin, USA.

Glossary/Terminology

Euler's equations. Equations of motion of a rigid body for rotation with a fixed point or centre of mass as base point when the co-ordinates attached to the rigid body are the principal axes

Joint forces and joint moments. The joint force system which consists of joint forces and joint moments is an equivalent force system replacing muscle force and real joint force

Kinematics. Kinematics is the study of the geometry of motion without reference to the cause of motion. It deals with position, displacement, velocity, acceleration, and time. These quantities are known as kinematic quantities.

Kinetics. Kinetics is the study of the relationship between forces and the resulting motion of bodies on which they act.

Leg. The portion of the lower extremity between the knee and ankle.

Mass moment of inertia. The mass moment of inertia of a solid body with respect to a given axis is the limit of the sum of the products of the elementary masses into which the body may be conceived to be divided and the square of their distances from the given axis.

Rigid body. A rigid body is a mathematical model of a material body in which the distance between any two points remains constant.

Thigh. The portion of the lower extremity above the knee.

References

1 CHAO, E. Y. (1971). Determination of applied forces in linkage systems with known displacements, with special application to biomechanics. Ph.D. Thesis. University of Iowa, Iowa City, Iowa.

2 FRANK, A. S. (1970). An approach to the dynamic analysis and synthesis of biped locomotion machines. *Med. biol. Eng.* **8**, 465–76.

3 TOWNSEND, M. A., and SEIREG, A. (1972). The syntheses of bipedal locomotion. *J. Biomech.* **5**, 71–83.

4 PLAGENHOEF, S. (1971). *Patterns of human motion: a cinematographic analysis.* Prentice-Hall, Englewood Cliffs, NJ.

5 DILLMAN, C. J. (1971). A kinetic analysis of the recovery leg in sprint running. In *Selected topics on biomechanics*, pp. 137–65. The Athletic Institute, Chicago, Ill.

6 HOCHMUTH, G. (1967). *Biomechanik sportlicher Bewegungen*, Sportverlag, Berlin.

7 KANE, T. R. (1971). Self-rotation of animate beings. In *Biomechanics II*, pp. 212–18. Karger, Basel.

8 YOUM, Y., HUANG, T. C., ZERNICK, R. F., and ROBERTS, E. M. (1973). Mechanics of simulated kicking. In *Mechanics and sport*, pp. 181–95. ASME, New York.

9 ROBERTS, E. M., ZERNICKE, R. F., YOUM, Y., and HUANG, T. C. (1974). Kinetic parameters of kicking. In *Biomechanics IV*, pp. 157–62. University Park Press, Baltimore, Md.

10 ZERNICKE, R. F. (1974). Human lower extremity kinetic parameter relationships during systematic variations in resultant limb velocity. Ph.D. Thesis. University of Wisconsin, Madison, Wis.

11 ZERNICKE, R. F., and ROBERTS, E. M. (1976). Human lower extremity kinetic relationships during systematic variations in resultant limb velocity. In *Biomechanics V*, pp. 20–5. University Park Press, Baltimore, Md.

12 ROBERTS, E. M. (1971). Cinematography in biomechanical investigation. In *Selected topics in biomechanics*, pp. 41–50. The Athletic Institute, Chicago, Ill.

13 DEMPSTER, W. T. (1955). Space requirements of the seated operator, kinematic and mechanical aspects of the body with special reference to the limbs. *Wright Air Development Center Tech. Rep. 55–159.*

14 YOUM, Y., ROBERTS, E. M., and ATWATER, A. E. (1972). Three dimensional cinematographic data analysis by computer. *AAHPER Abstr.* p. 69.

15 ROBERTS, E. M., and METCALFE, A. (1968). Mechanical analysis of kicking. *Biomechanics I*, pp. 228–37 Karger, Basel.

16 HUANG, T. C. (1967). *Engineering mechanics, Vol. 2—Dynamics*, Addison-Wesley, Reading, Mass.

17 GREVILLE, T. N. E. (1970). Spline functions and applications, *MRC Orientation Lecture Series No. 8*. Mathematics Research Center, The University of Wisconsin, Madison, Wis.

18 ZERNICKE, R. F., CALDWELL, G., and ROBERTS, E. M. (1976). Fitting biomechanical data with cubic spline functions. *Res. Q. am. Ass. Health phys. Educ. Recreat.* **47**, 9–19.

19 ANDERSON, C. C. (1971). A method of data collection and processing for cinematographic analysis of human movement in three dimensions. M.S. Thesis. University of Wisconsin, Madison, Wis.

20 ROBERTS, T. W. (1972). Incident light velocimetry. *Percept. mot. Skills* **34**, 263–8.

21 ZERNICKE, R. F., and ROBERTS, E. M. (1978). Lower extremity forces and torques during systematic variation of non-weight bearing motion. *Med. Sci. Sports* **10**, 21–6.
22 CORSER, T. (1974). Temporal discrepancies in the electromyographic study of rapid movement. *Ergonomics* **17**, 389–400.
23 MIYASHITA, M., MUIRA, M., MATSUI, H., and MINANITATE, K. (1972). Measurement of the reaction time of muscular relaxation. *Ergonomics* **15**, 555–62.
24 FENN, W. O. (1930). Frictional and kinetic factors in the work of sprint running. *Am. J. Physiol.* **92**, 583–611.
25 ELFTMAN, H. (1966). Biomechanics of muscle. *J. bone jt. Surg. Am. Vol.* **48**, 363–77.
26 WARTENWEILER, J., and WETTSTEIN, A. (1971). Basic kinetic rules for simple human movements. *Biomechanics II*, pp. 134–45 Karger, Basel.
27 CAVAGNA, G. A., KOMAREK, L., CITTERIO, G., and MARGARIA, R. (1971). Power output of the previously stretched muscle. *Biomechanics II*, pp. 159–67. Karger, Basel.
28 BOUISSET, S. (1973). EMG and muscle force in normal motor activities. In *New developments in electromyography and clinical neurophysiology, Vol. 1* (ed. J. E. Desmedt), pp. 596–606. Karger, Basel.
29 SEIREG, A., and ARVIKAR, R. J. (1973). A mathematical model for evaluation of forces in lower extremities of the musculo-skeletal system. *J. Biomech.* **6**, 313–26.
30 MILLER, N. (1976). An investigation of the static and dynamic behaviour of the musculo-skeletal system during dynamic activity. Ph.D. Thesis. University of Wisconsin, Madison, Wis.
31 PAUL, J. P. (1967). Forces transmitted by joints in the human body. *Proc. Inst. Mech. Eng.* (Lond.) **181**, (3J), 8–15.
32 MORRISON, J. B. (1970). The mechanics of the knee joint in relation to normal walking. *J. Biomech.* **3**, 51–61.
33 MACMILLAN, M. B. (1975). Determinants of the flight of the kicked football. *Res. Q. am. Ass. Health phys. Educ. Recreat.* **46**, 48–57.

12. Biomechanical parametric analysis of pole vaulting and optimization of performance

H. S. WALKER and P.G. KIRMSER

12.1. Introduction

The most important single element of the dynamic sport of pole vaulting is the athlete himself. The second most important element is the pole. Other items such as the pole pit, cross bar, landing mat, etc. play a much less important role and are indeed the same for all vaulters at a meet. Not to be overlooked, however, is the coach and his contribution.

This analysis of pole vaulting is an attempt to use the kinetics of the vaulter and the pole to suggest guidelines for improving performance. Although more accurate models are being studied at present, the simple models used here illustrate the usefulness of mathematics and mechanics in the analysis of vaulting by suggesting actions which should be taken to improve an athlete's performance. Ideally mathematical analyses should be coupled with coaching to form a feedback mechanism which would help a vaulter optimize his actions. The work presented here is a step in that direction.

Nearly all of the numerous articles on pole vaulting which have appeared in journals well known in athletic circles are concerned with qualitative discussions of techniques which have been found to be good practice through experience in observing and coaching vaulters and with physical conditioning of the athlete. The flavour of these articles, the past and present history of pole vaulting (which is an ancient sport), good practices in vaulting, much factual information, and many references are probably best summarized in a single source by Ganslen [1].

Only a few more technical analyses of the mechanics of vaulting have appeared. Fletcher, Lewis, and Wilkie [2], and Hay [3] determined energy distributions during vaults from measurements made from frames of movie film, and Ganslen [4], Welsh [5], and Ballreich [6] determined the energy distributions from analyses modelling a vaulter–rigid-pole system. These analyses are elementary by today's standards of theoretical and experimental techniques.

More recently Hay [7] and Steben [8] investigated correlations between various factors affecting vaulting by running regression analyses on data obtained from motion picture films. These correlations include those between combinations of pairs of variables such as vertical velocity at take off, horizontal velocity at take off, handspread, angle at take off, height of hand hold, pole bend, degree of elbow extension, and duration of push off. There are also correlations between some of these variables and the height attained. Although knowledge of this type is of some value in coaching it provides little detailed information and can be no substitute for theoretical analysis and optimization. The correlations obtained are for existing techniques used by good vaulters. They do not give any indication as to whether or not the techniques measured were the best possible or even beneficial for specific vaulters.

The physical model for a real vaulter using a weightless flexible pole is described by a system of non-linear coupled second-order ordinary differential equations with variable coefficients which is treated as an initial value problem. Although analytical solutions are not feasible for this difficult problem, it is now possible to obtain numerical solutions using high speed digital computers.

12.2. The physical model

The physical model for the pole-vaulting system chosen for this study is shown in Fig. 12.1. The flexible pole is taken to be an elastica hinged at the pole pit and hand holds. The vaulter is represented by a mass of variable moment of inertia $I(t)$ and variable distance $r(t)$ from his centre of gravity to his hand hold.

Although this is a simple model, it is a general one. The functions $r(t)$ and $I(t)$ are determined by the body configurations of the vaulter during the jump. The significance of these functions is described for the benefit of coaches and

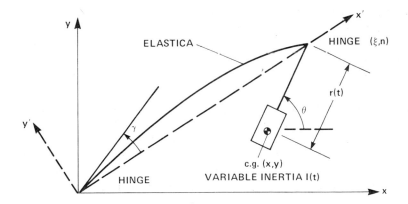

FIG. 12.1. The physical model.

others who might be unfamiliar with the field of mechanics in Appendix 12.1.

The functions $r(t)$ and $I(t)$ should be chosen to optimize the performance. Any sequence of body motions defines the functions $r(t)$ and $I(t)$ uniquely. Unfortunately the functions $r(t)$ and $I(t)$ must be chosen carefully, for, given these functions, body configurations are not defined uniquely and may not even exist.

This model is a double-pendulum model but differs from that of Ganslen [4] in that the pole is flexible and the moment of inertia and radius from the hand hold to the centre of gravity are variable.

Coaches and others who have little interest in the mathematical development are urged to move on to § 12.5 where the results of computer runs for three different vaults are presented and compared.

12.3.   The mathematical model

The mathematical model has two parts. One describes the forces and configurations of the pole and the other the motions of a mass of variable inertia which represents the vaulter.

Studies of the large deflections of rods were made by Euler in the eighteenth century. These resulted in the theory of the elastica, which is found in Love's famous book on elasticity [9].

The elastica

A slender rod pivoted at the ends and loaded by forces P is shown in a buckled position in Fig. 12.2. The co-ordinate axes $x'–y'$ are those shown in Fig. 12.1. These rotate with the pole during the vault. The forces P are those exerted on the ends of the pole by the vaulter and the pole pit during the vault.

The load P is related to the angle of the tangent to the rod at its hinged end, its stiffness, and its length by

$$P = \frac{4\beta K^2(k)}{L^2} = P(\beta, k, L) \tag{12.1}$$

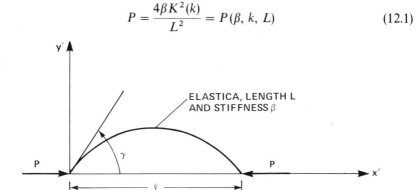

Fɪɢ. 12.2.   The elastica.

where

$$k = \sin(\gamma/2)$$

and

$$K(k) = \int_0^{\pi/2} (1 - k^2 \sin^2 \zeta)^{-1/2}\, d\zeta$$

which is a complete elliptic integral of the first kind.

The chord length of the buckled rod is related to the length of the rod by

$$l = L\left\{2\frac{E(k)}{K(k)} - 1\right\} \tag{12.2}$$

where

$$E(k) = \int_0^{\pi/2} (1 - k^2 \sin^2 \zeta)^{+1/2}\, d\zeta$$

is a complete elliptic integral of the second kind and the other variables are as given previously.

The above elliptic integrals are given in series form by

$$K(k) = \frac{\pi}{2}\left\{1 + \left(\frac{1}{2}\right)^2 k^2 + \left(\frac{1\times3}{2\times4}\right)^2 k^4 + \left(\frac{1\times3\times5}{2\times4\times6}\right)^2 k^6 + \ldots\right\} \quad |k| < 1 \tag{12.3}$$

and

$$E(k) = \frac{\pi}{2}\left\{1 - \left(\frac{1}{2}\right)^2 k^2 - \left(\frac{1\times3}{2\times4}\right)^2 \frac{k^4}{3} - \left(\frac{1\times3\times5}{2\times4\times6}\right)^2 \frac{k^6}{5} + \ldots\right\} \quad |k| < 1 \tag{12.4}$$

which have been tabulated in the *Handbook of tables for mathematics* [10].

The equations of motion

A two-dimensional free-body diagram of the vaulter is shown in Fig. 12.3. The irregular body, which is really that of the athlete, is the variable pendulum shown in Fig. 12.1. The variables (ξ, η), θ, and (x, y) are position variables. X, Y, and W are forces exerted by the pole and gravity on the vaulter.

The equations of motion with respect to the moving centre of gravity are

$$\frac{W}{g}\ddot{y} = Y - W \tag{12.5}$$

$$\frac{W}{g}\ddot{x} = X \tag{12.6}$$

and

$$I\ddot{\theta} + \dot{I}\dot{\theta} = Yr\cos\theta - Xr\sin\theta \tag{12.7}$$

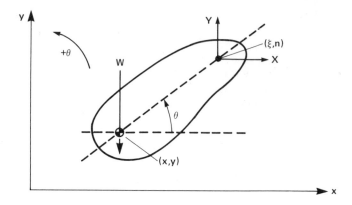

FIG. 12.3.   A free-body diagram of the vaulter.

where X and Y are the forces applied by the pole on the athlete and $I = I(t)$ and $r = r(t)$ are functions defined by his actions.

These equations, with the forces exerted by the pole on the vaulter determined from the equations of the elastica using the co-ordinates of the hand hold of the pole and the origin to define its configuration, must be solved for given initial values of x, y, θ, $\dot{x}$, $\dot{y}$, and $\dot{\theta}$, and given parameters for the stiffness β of the pole, the length L of the pole to the vaulter's hand hold, and the vaulter's weight W.

12.4. The mathematical model used for simulation

In order to use standard scientific subroutines in computation where possible the mathematical model was reduced to a system of six non-linear ordinary differential equations given by

$$\dot{x} = x_1 \tag{12.8}$$

$$\dot{y} = y_1 \tag{12.9}$$

$$\dot{\theta} = \theta_1 \tag{12.10}$$

$$\dot{x}_1 = \frac{g}{w} X \tag{12.11}$$

$$\dot{y}_1 = \frac{g}{w} Y - g \tag{12.12}$$

and

$$\dot{\theta}_1 = \frac{Y}{I} r \cos\theta - \frac{X}{I} r \sin\theta - \frac{\dot{I}}{I} \sin\theta_1 \tag{12.13}$$

and the functions

$$l = (x^2 + y^2 + r^2 + 2x r \cos \theta + 2y r \sin \theta)^{1/2} \qquad (12.14)$$

$$G(k) = \frac{E(k)}{K(k)} = \frac{1}{2}\left(1 + \frac{l}{L}\right) \qquad (12.15)$$

$$k = G^{-1}\frac{1}{2}\left(1 + \frac{l}{L}\right) \qquad (12.16)$$

$$P(\beta, k, L) = \frac{4\beta K^2(k)}{L^2} = P(x, y, \theta, r) \qquad (12.17)$$

$$X = \frac{P(x, y, \theta, r)}{l}(x + r \cos \theta) \qquad (12.18)$$

and

$$Y = \frac{P(x, y, \theta, r)}{l}(y + r \sin \theta). \qquad (12.19)$$

Integration of the system of non-linear differential equations was done numerically using a fourth-order Runge–Kutta–Gill procedure [11] and time increments of 0.05 s.

Schematically, the integration proceeds as follows. For numerical computation functions $r(t)$ and $I(t)$ must be given along with the parameters β, L, and W. Preliminary values of these functions and parameters can be obtained from physical measurements and cinematography of a successful vault. Using these and starting from given initial conditions x, y, θ, x_1, y_1, and θ_1, the initial right-hand sides of equations (12.8)–(12.19) are calculated (i.e. the values at time $t = 0$). Equations (12.8)–(12.13) are integrated one step forward by the Runge–Kutta–Gill procedure to obtain the values of x, y, θ, x_1, y_1, and θ_1 at time $t = 0.05$ s. The values of x, y, and θ thus obtained are used in eqn (12.14) to obtain the new chord length which in turn is used in the solution of eqn (12.16) to find the parameter k. This parameter is then used in eqn (12.17) to find the force P exerted by the pole when one end is at the origin (the pole pit) and the other at the co-ordinates (ξ, η), where these co-ordinates are related to those of the centre of gravity by

$$\xi = x + r \cos \theta \qquad (12.20)$$

and

$$\eta = y + r \sin \theta. \qquad (12.21)$$

The force P, chord length l, and the co-ordinates x, y, and θ providing the location of the vaulter's centre of gravity and the orientation of his hand hold with respect to it are then used in eqns (12.18) and (12.19) to calculate new values of X and Y which are the forces now exerted by the pole on the vaulter in his incremented position.

All of this provides new values for the right-hand sides of eqns (12.8)–(12.19) which are then integrated by the Runge–Kutta–Gill procedure to step forward once again. In this way the equations are integrated as the vaulter moves in the simulation from the initial conditions established at the start of the jump to those which exist when the pole is released near the end of the vault. The integration, as carried out, is a mathematical simulation of a vault by an athlete whose actions taken during the jump are known *a priori* and perfectly executed.

The performance indicated by the calculations is evaluated by looking at the height attained, the attitude at release and at clearing the bar, and the forward velocity. Perturbations in the initial conditions and the functions $r(t)$ and $I(t)$ are made, and the calculations are repeated so as to improve performance. These suggest to coaches and athletes options which, if they could be carried out after suitable training, would improve performance.

12.5. Simulations using poles of uniform stiffness

Simulations of this type will be of great value to vaulters and their coaches for the effects of changes in parameters and actions taken during the jump on the maximum height attained and the form used in passing over the bar can be determined as if they were carried out or changed exactly as planned.

In order to illustrate the possibilities, the outputs of three trial simulations are shown in Figs. 12.4–12.6. These figures were drawn from the computer output of the solution by a Calcomp plotter. This output consists of the path of the hand hold and the path of the centre of gravity of the vaulter. The points indicated on the curves are at 0.05 s time intervals. Straight lines drawn between corresponding pairs of points give rough indications of the body positions of the vaulter during the jump. The last pair of points show the position of the vaulter just before release. Several of these positions are shown in Fig. 12.4. The last two positions indicated in this figure show bad form at the release of the pole because the hand hold has already gone ahead of the centre of gravity and the vaulter is rotating clockwise at a fairly rapid rate. In addition to this the pole is falling forward into the bar and landing mat. Figure 12.5 shows good form. At the release of the pole, the pole is beginning to fall back away from the bar. Also, the vaulter's feet are slightly ahead of his hand hold and he is rotating counterclockwise just fast enough to be in good form as he passes over the bar. Figure 12.6 again shows bad form. Although the pole falls back away from the bar and the vaulter's feet are ahead of his hand hold, he is rotating counterclockwise much too fast to have good form when crossing the bar.

The figures show that part of the vault up to release of the pole. The free-fall portion, which adds height in each case, is not shown. The centre of gravity of

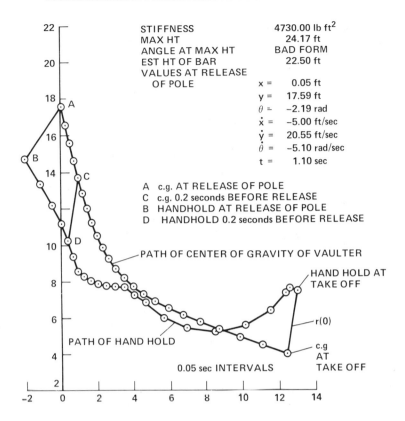

FIG. 12.4.   A vault using a softer than optimum pole for the given initial conditions: stiffness, 4730.00 lb ft²; maximum height, 24.17 ft; angle at maximum height, bad form; estimated height of the bar, 22.50 ft; values at release of pole, $x = 0.05$ ft, $y = 17.59$ ft, $\theta = -2.19$ rad, $\dot{x} = -5.00$ ft s^{-1}, $\dot{y} = 20.55$ ft s^{-1}, $\dot{\theta} = -5.10$ rad s^{-1}, $t = 1.10$ s. A, centre of gravity at release of the pole; B, hand hold at release of the pole; C, centre of gravity 0.2 s before release; D, hand hold 0.2 s before release.

the vaulter describes a parabola which is easily calculated from the data given at release. The input data for these simulations are given in Table 12.1.

Suitable bounds for values of $r(t)$ and $I(t)$ were taken from the data of Hanavan [12]. The values of $r(t)$ and $I(t)$ were found by trial simulations repeated with different values until results considered to be reasonable were obtained. This is essentially an iterative procedure in which the functions $r(t)$ and $I(t)$, which are given numerically, are perturbed from their first values (which may be taken from analyses of moving pictures of vaults) in order to optimize performance.

The simulation proved to be surprisingly realistic for it was found necessary to 'coach' the mathematical program via selection of poles, hand-hold heights,

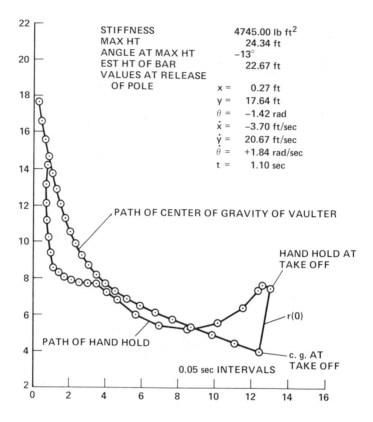

Fɪɢ. 12.5.   A vault using the optimum pole for the given initial conditions: stiffness, 4745.00 lb ft^2; maximum height, 24.34 ft; angle at maximum height, $-13°$; estimated height of the bar, 22.67 ft; values at release of pole, $x = 0.27$ ft, $y = 17.64$ ft, $\theta = -1.42$ rad, $\dot{x} = -3.70$ ft s^{-1}, $\dot{y} = 20.67$ ft s^{-1}, $\dot{\theta} = 1.84$ rad s^{-1}, $t = 1.10$ s.

and body configuration functions to achieve anything better than mediocre performance. Some of the poor simulations resembled movies of vaulters being thrown back towards their take-off points. Although detailed studies have yet to be made, it is already evident that pole vaulting is surprisingly sensitive to small changes in all parameters and configuration functions.

The simulations shown in Figs. 12.4–12.6 differ only by minor amounts in the stiffness of the pole used. All other conditions are the same. These figures show that when the best stiffness is used (that for Fig. 12.5) the vaulter ends his jump by balancing in a handstand on the end of the pole rising almost vertically but falling back towards the take-off point once the vaulter releases his hold. If the pole stiffness varies by 15 ft lb^2 one way or the other the vaulter will be given sufficient angular velocity to necessitate premature release of the

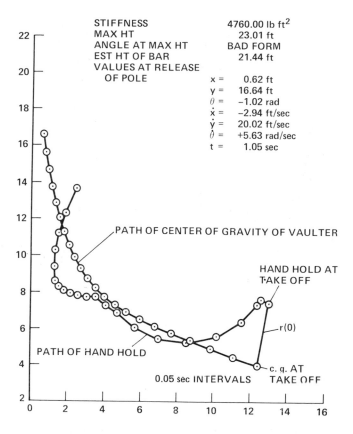

STIFFNESS   4760.00 lb ft^2
MAX HT   23.01 ft
ANGLE AT MAX HT   BAD FORM
EST HT OF BAR   21.44 ft
VALUES AT RELEASE
   OF POLE   x =   0.62 ft
                     y =   16.64 ft
                     θ =   −1.02 rad
                     $\dot{x}$ =   −2.94 ft/sec
                     $\dot{y}$ =   20.02 ft/sec
                     $\dot{\theta}$ = +5.63 rad/sec
                     t =   1.05 sec

PATH OF CENTER OF GRAVITY OF VAULTER

HAND HOLD AT
T-AKE OFF

r(0)

PATH OF HAND HOLD

0.05 sec INTERVALS

c. g. AT
TAKE OFF

FIG. 12.6.   A vault using a stiffer than optimum bar for the given initial conditions: stiffness, 4760.00 lb ft^2; maximum height, 23.01 ft; angle at maximum height, bad form; estimated height of the bar, 21.44 ft; values at release of the pole, x = 0.62 ft, y = 16.64 ft, θ = −1.02 rad, $\dot{x}$ = −2.94 ft s^{-1}, $\dot{y}$ = 20.02 ft s^{-1}, $\dot{\theta}$ = 5.63 rad s^{-1}, t = 1.05 s.

pole or to require other corrective action in order to cause himself to cross the bar in a nearly horizontal position.

12.6. Properties of poles of variable stiffness

Real poles are designed to have non-uniform stiffness in order to increase the capacity for energy storage.

The differential equation and the boundary conditions which describe the non-linear pin-ended elastica of arbitrary stiffness distribution are [13]

$$\beta(s)\frac{d\theta}{ds} = -Py(s) \tag{12.22}$$

TABLE 12.1

Input data for the simulations shown in figs.
12.4–12.6

t (s)	$r(t)$ (ft)	$I(t)$ (slugs)
0.00	3.50	12.00
0.05	3.50	12.00
0.10	3.00	1.00
0.15	2.50	10.00
0.20	2.00	9.00
0.25	1.50	8.10
0.30	1.00	7.40
0.35	0.50	8.10
0.40	0.50	9.00
0.45	0.50	10.00
0.50	1.00	11.00
0.55	1.50	12.00
0.60	2.00	12.00
0.65	2.50	12.00
0.70	3.00	12.00
0.75	3.50	12.00
0.80	3.50	12.00
>0.80	3.50	12.00

$L = 15$ ft, $W = 160$ lb, $\dot{x} = -27.3$ ft s^{-1}, $\dot{y} = +10.0$ ft s^{-1}
Note: The only change in the three simulations is pole stiffness.
Fig. 12.4, $\beta = 4730.0$; Fig. 12.5, $\beta = 4745.0$; Fig. 12.6,
$\beta = 4760.0$.

and

$$y(0) = y(L) = \frac{d\theta}{ds}(0) = \frac{d\theta}{ds}(L) = 0. \qquad (12.23)$$

In these equations, $\theta(s)$ is the angle of the tangent to the deflection curve measured counterclockwise from the horizontal, $y(s)$ is the deflection of the elastica, $\beta(s)$ is the given stiffness distribution function which is always positive, s is the arc length measured from the origin, L is the length of the elastica, and P is the applied load. This is shown in Fig. 12.7. This elastica is the same as that shown as the pole in Fig. 12.1. Although solutions in the form of elliptic integrals are known for elastica of uniform stiffness [9], none are known for variable stiffness distributions.

Equation (12.22) can be converted to the standard form for non-linear differential equations of the second order

$$\frac{d^2\theta}{ds^2} + \frac{1}{\beta(s)}\frac{d\beta}{ds}\frac{d\theta}{ds} + \frac{P}{\beta(s)}\sin\theta = 0 \qquad (12.24)$$

by differentiation. This equation could be solved by standard phase plane analysis or shooting methods [14], but neither is as convenient as a direct

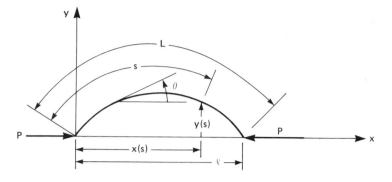

F<small>IG</small>. 12.7.   The elastica.

application of the shooting method to eqn (12.22) in the manner which follows.

Analysis

Equation (8.22) is equivalent to

$$\frac{d\theta}{ds} = -\frac{Py(s)}{\beta(s)}$$

(12.25)

which is to be solved for the boundary conditions of eqn (12.23).

This boundary value problem could be solved as an initial value problem for a given load P if the initial tangent $\theta(0)$ were known. In the shooting method, it is assumed to be known and the integration is carried out over the length of the elastica. Unless the value assumed for $\theta(0)$ was correct, the deflection curve found by the integration will be as shown in Fig. 12.8.

$y(L)$ and $x(L)$ are determined by the integration and, if $y(L) \neq 0$, $\theta(0)$ is corrected and the equation is integrated again. Successive integrations quickly lead to a value of $\theta(0)$ for which $y(L) = 0$. Equation (12.22) itself ensures that if $y(L) = 0$, $d\theta/ds = 0$, so that this one condition is sufficient to ensure that all of the given boundary conditions are satisfied.

The system of first-order equations used to carry out this procedure for given P and $\beta(s)$ was

$$\frac{d\theta}{ds} = -\frac{Py(s)}{\beta(s)}$$

(12.26)

$$\frac{dy}{ds} = \sin\theta$$

(12.27)

$$\frac{dx}{ds} = \cos\theta$$

(12.28)

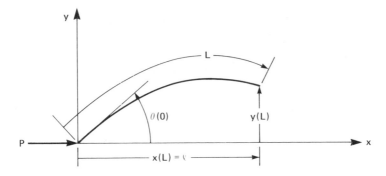

FIG. 12.8.   The shooting method.

with initial value conditions

$$x(0) = y(0) = 0 \quad \theta(0) = \theta_n. \tag{12.29}$$

These equations were integrated using a fourth-order Runge–Kutta–Gill procedure [11] to find

$$y_n(L) = \int_0^L \sin\theta \, ds \tag{12.30}$$

and

$$x_n(L) = \int_0^L \cos\theta \, ds. \tag{12.31}$$

Then the initial value $\theta(0)$ was set to θ_{n+1}, and $y_{n+1}(L)$ and $x_{n+1}(L)$ were computed. From this point on the new estimates for $\theta(0)$ were found using the recursion formula

$$\theta_{n+2} = \frac{\theta_n y_{n+1}(L) - \theta_{n+1} y_n(L)}{y_{n+1}(L) - y_n(L)}. \tag{12.32}$$

This is Newton's method [15] applied to the functional mapping of the $\theta(0)$ space to the $y(L)$ space made by the Runge–Kutta–Gill integration.

In the application which follows, convergence was rapid for the initial estimates $\theta_0 = 1.0$, $\theta_1 = 1.1$. Convergence was also rapid for small angles, but to the trivial solution which is possible for all values of P but which does not exist for those values which exceed the Euler buckling load.

An example

The program developed was used to calculate the buckling behaviour of a commercial fibreglass pole used for pole vaulting.

The stiffness function $\beta(s)$ was obtained from standard beam deflection tests and is shown in Fig. 12.9. The load–chord length curves were calculated as described above. They are shown in Fig. 12.10.

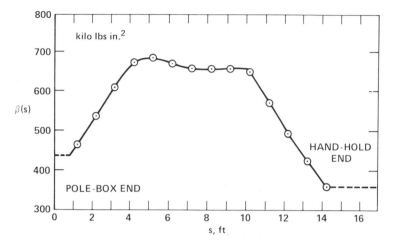

Fɪɢ. 12.9.   The stiffness function $\beta(s)$ for a pole-vault pole.

The shooting method is convenient, cheap, and easy to use for the problem considered here. It is a great improvement over conventional boundary value methods for the elastica or arbitrary stiffness distribution. The example presented enables a new display of the properties of the non-linear elastica to be made. This display is of value to vaulters and their coaches, and of importance in the computer modelling of vaulting by providing load–deflection relationships for real poles of arbitrary construction. These relationships would replace equations (12.15)–(12.17) of §12.4.

12.7. Concluding remarks

The ultimate heights reached in these simulations are so far beyond present world records that one wonders if they are indeed possible.

Simple energy considerations show that it should be possible to attain at least 20 ft. The kinetic energy of a man running 27.3 ft s^{-1} is enough to raise his centre of gravity 11.57 ft. His centre of gravity is likely to be about 3.31 ft above the ground to begin with. He should be able to add to this something approximately twice the distance from his centre of gravity to the height of his hand hold above the ground at take-off (i.e. he should be able to pull himself up and do a handstand as if his hand hold on the pole were fixed in space). This amounts to nearly 7.00 ft, making the equivalent of 21.88 ft in all. With allowances made for the thickness of his body and an efficiency less than 100 per cent, a jump of 20 ft is clearly possible. The margin between existing world records and the ultimate height attainable is great enough that new records should be set relatively easily in the near future.

Vaulting performance is extremely sensitive to small changes in nearly all of the parameters. The coaching of vaulters would be assisted by the use of instrumented take-off platforms and pole pits, recording of force–time histories on storage oscilloscopes, and slow-motion instant playback television recordings of practice jumps. These would be used to compare a vaulter's actions for conformance to optimums established by simulation and for consistency of action in repeating practice jumps.

The selection of vaulting poles should be made from simulations taking at least a vaulter's running speed, height, and weight into account. These would be kept constant in simulations while varying pole stiffness and hand-hold length until optimal values are found. Once a pole has been selected the vaulter will tend to train himself to optimize his performance for the pole selected, which is now the pole which will allow him to reach his best achievable performance.

Because the best performance is achieved at the edge of falling the wrong way from a handstand on the end of a pole more than 15 ft long, it is understandable that vaulters train conservatively—it is necessary for them to retain horizontal velocities which are too great to reach optimum heights in order to ensure landing on the mat. A vaulter must be able to make errors on the dangerous side during practice in order to develop a feeling for where the edge of danger lies. To permit this, new safety devices such as manually or automatically controlled catching nets used to eliminate the danger of falling off the mat during practice must be developed.

The poles used in the simulations shown in Figs. 12.4–12.6 are 'softer' and their greatest deflections larger than those safely withstood by the fibreglass poles currently used. New poles should be developed to make the incredible vaults which were simulated in this study possible.

The buckling load curve in Fig. 12.10 is for a pin-ended elastica. Straight columns of this type are essentially rigid until the Euler buckling load is reached. Thereafter the apparent stiffness decreases to that shown by the $l < L$ curves branching to the left from the Euler buckling load curve shown.

In actual practice the vaulter can apply a moment to the upper end of the pole. The effect of this moment is to increase or decrease the apparent stiffness of the pole. If the vaulter tends, by applying this moment, to bend the pole more, the apparent stiffness goes down. Conversely, if the vaulter tends to straighten the pole by applying the moment, the apparent stiffness increases.

Poles that are initially curved decrease the peak initial forces experienced by the vaulter in the pole plant. This is because the initially curved poles do not exhibit the essentially rigid character which the straight poles have before the Euler buckling load is reached. The use of initially curved poles should provide smaller shocks and thus allow better control for the vaulter on take-off.

Computer simulations will not replace coaches. Rather, they will serve as aids in coaching for they will enable the coach to optimize the impedance

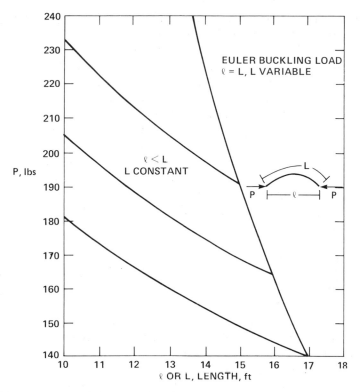

F<small>IG</small>. 12.10.   Buckling loads for pin-ended elastica of variable stiffness.

match between the vaulter and what he is trying to do and show both the coach and the vaulter what effects contemplated changes in vaulting style will produce.

This work is being continued to study the effects of parameter changes on vaulting and to make more sophisticated simulations by including more detail in modelling and possibly by extensions to three dimensions.

Nomenclature

$E(k)$ $\int_0^{\pi/2} (1 - k^2 \sin^2 \zeta)^{+1/2} \, d\zeta$, the complete elliptic integral of the second kind

g the acceleration of gravity

$G(k)$ the ratio of $E(k)$ to $K(k)$

$I(t)$ the moment of inertia of the vaulter with respect to an axis perpendicular to the x, y plane and passing through his centre of gravity, a function of time t

k $\sin(\gamma/2)$

$K(k)$ $\displaystyle\int_0^{\pi/2} (1 - k^2 \sin^2 \zeta)^{-1/2} d\zeta$, the complete elliptic integral of the first kind

L the length of the undeformed pole, pole-pit end to hand hold

l the chord length from the pole pit to the vaulter's hand hold

P the force exerted on the pole to hold it in a deformed position

$P(\beta, k, L)$ $P(x, y, \theta, r(t))$, the force P expressed in terms of the variables which can be used to define it

$r(t)$ distance from the centre of gravity of the vaulter to his hand hold, a function of time t

W the weight of the vaulter

X the component of P acting in the x direction on the vaulter at his hand hold

x, y co-ordinates of the centre of gravity of the vaulter

Y the component of P acting in the y direction on the vaulter at his hand hold

β the flexural stiffness of the pole

γ the angle between the line tangent to the pole at the pole pit and the chord from the pole pit to the vaulter's hand hold

ξ, η co-ordinates of the vaulter's hand hold on the pole

θ angle between the line from the centre of gravity of the vaulter to his hand hold and the horizontal measured counterclockwise

$\dot{\theta}$ a dot above a variable denotes differentiation with respect to time

Acknowledgements

Some of the material included here was previously published in *Mechanics and Sport*, AMD-Vol. 4, ASME, November 1973. Permission for use of this material is gratefully acknowledged.

Appendix 12.1

The effect of the functions $r(t)$ and $I(t)$ are often difficult to understand by someone unfamiliar with mechanics. They are not that difficult to describe geometrically, however.

The function $r(t)$ is simply the distance from the pivot point of the pendulum to the centre of mass of the body. If the pendulum is swinging an increase in r will tend to slow up the swing while a decrease in r will tend to speed up the swing. It should be mentioned that the location of the centre of mass of the body is dependent upon the configuration of the body at that particular time.

The function $I(t)$ is somewhat more complicated to determine but can be described as the manner in which the total mass is distributed about the centre of gravity. The further the mass is distributed from the centre of mass, the larger I becomes. Conversely, the closer each part of the mass is to its centre, the smaller I becomes.

One of the better known events in which the effect of I can be observed is in figure-skating. When the skater goes into a spin with arms outstretched (I is relatively large since the mass of the arms is located as far as possible from the centre of mass), the speed of rotation is fairly slow. As the arms are brought in (I is decreased) the speed of rotation increases. The energy during this manoeuvre remains essentially constant. If the skater now extends the arms (I goes back up) the speed of rotation will decrease again.

References

1 GANSLEN, R. V. (1963). *Mechanics of the polevault* (5th edn). John Swift, St. Louis, Mo.

2 FLETCHER, J. G., LEWIS, H. E., and WILKIE, D. R. (1960). Human power output: the mechanics of pole vaulting. *Ergonomics* **3** (1), 30–4.

3 HAY, J. G. (1968). Mechanical energy in pole vaulting. *Track Tech.* **33**, 1047–51.

4 GANSLEN, R. V. (1941). A mechanical analysis of the pole vault. *Athlet. J.* **21**, 20, 5.

5 WELSH, W. (1955). A 17 foot vault? *Athlet. J.* **35** (8), 18, 20, 67.

6 BALLREICH, R. (1962). Eine Mechanische Betrachtung des Stabhoch-sprungs. *Lehre Leichtathlet.* **41**, 995–8.

7 HAY, J. G. (1967). Pole vaulting: a mechanical analysis of factors influencing pole-bend, *Res. Quart. am. Ass. Health phys. Educ. Recreat.* **38**, 34–40.

8 STEBEN, R. E. (1970). A cinematographic study of selective factors in the pole vault, *Res. Quart. am. Ass. Health phys. Educ. and Recreat.* **41** (1), 95–104.

9 LOVE, A. E. H. (1944). *A treatise on the mathematical theory of elasticity* (4th edn) pp. 401–5. Dover Publications, New York.

10 WEAST, R. C., and SELBY, S. M. (1957). *Handbook of Tables for Mathematics* (3rd edn), pp. 757–69. Chemical Rubber Co., Cleveland, Ohio.

11 RALSTON, A., and WILF, H. S. (1960). *Mathematical methods for digital computers*, pp. 110–20. Wiley, New York.

12 HANAVAN, E. P. JR. (1964). *A mathematical model of the human body. Rep. AMRL–TR–64–102.*Behavioral Sciences Laboratory, Aerospace Medical Research Laboratories, Air Force Systems Command, Wright–Patterson Air Base, Ohio.

13 MCLACHLAN, N. W. (1956). *Ordinary non-linear differential equations* (2nd edn), p. 34. Clarendon Press, Oxford.

14 BAILEY, P. B., SHAMPINE, L. F., and WALTMAN, P. E. (1968). *Nonlinear two point boundary value problems.* Academic Press, New York.

15 HAMMING, R. W. (1962). *Numerical methods for scientists and engineers*, p. 81. McGraw-Hill, New York.

13. Biomechanical (analyses and applications) of shot put and discus and javelin throws

Tsai–Chen Soong

13.1. Introduction

There seems to be no limit to man's physical capabilities to run faster, jump higher, and throw further. Records are broken as fast as they were made. For example, the world record discus throw in 1950 was 186 ft $10\frac{7}{8}$ in (56.969 m) made by Gordien of the USA. In 1970 over 60 men surpassed that distance, the best being 220 ft 3 in (67.132 m) by Bruch of Sweden (ref. 1, p. 146). On the other hand, the margin needed to win in world games has become unbelievably slim. For example, Wolferman of the Federal Republic of Germany won the gold medal in the javelin throw at the 1972 Olympic Games with a throw of 296 ft 10 in (90.47 m) and Lusis of the USSR came in second with a throw that was just 1 in shorter. Based on a theoretical analysis, Lusis could have been champion if he had made a slight change in his throw angle or in the javelin's inclination angle.

The present study concerns the theoretical external trajectory of the discus throw, the javelin throw, and the shot put. Body movements, momentum gathering, staging, and thrust, which are matters of professional coaching and vary from coach to coach, will not be discussed.

There are three kinds of parameter in the mechanics of field events that affect the throw range. The first kind is associated with such conventional expertise as the throw force, throw angle, and other initial geometrical conditions. The optimization of these initial geometrical parameters to achieve the maximum range is a major objective of this analysis. The second kind is environmental, e.g. wind speed and direction. Their effects on trajectories will also be studied. The third kind of parameter is not controlled by the thrower and includes the physical and aerodynamic characteristics of the equipment. They are particularly important in the javelin throw but less influential in the discus and shot put.

In the analyses that follow the equations for the discus throw are developed first as they contain all the six degrees of freedom in space. The five equations for the javelin throw and the three for the shot put are special cases of the discus throw equations and their solutions are simpler.

The notations to be used in these analyses are given in the Nomenclature.

13.2. Analysis of the discus throw

The most important parameters which affect the range of a discus throw are as follows: (1) the throw angle made by the trajectory of the centre of the discus with the ground at the instant of throw; (2) the initial discus inclination angle made by the plane of the discus with respect to the ground; (3) the spinning speed of the discus; (4) the wind velocity.

There is no consensus among coaches and athletes on the optimum values of these parameters. Taylor (ref. 1, p. 149) in 1932, using wind-tunnel tests, concluded that a head wind of 7–8 mile h^{-1} (3.1–3.6 m s^{-1}) will have the most beneficial effect on the range. At 14.5 mile h^{-1} (6.5 m s^{-1}) a head wind becomes detrimental. He suggested that 35° should be used for both the throw angle and the discus inclination angle. However, Ganslan (ref. 1, p. 149) suggested in 1959 that the optimum is 35° for the throw angle and 10°–15° less for the discus inclination. As to the effect of spinning the discus, all agreed that spin keeps the discus stable in flight and is essential to a long range but no quantitative information existed.

The present analysis contains the derivation and a solution method for six non-linear differential equations of motion which describe the position of the centre of gravity and the orientations of the principal axes of a spinning discus moving in the air. The aerodynamic pressure on the discus is obtained from existing experimental data on inclined plates and disc-shaped bodies; the effect on the moment due to the spinning motion is derived from the classical hydrodynamics of a translating and rotating ellipsoid in an ideal flow field. The numerical results are presented in the context of the 1972 Olympics discus throw record. The effects of spin speed, moment of inertia of the discus, optimum throw angle and discus inclination angle, and wind are also studied.

Derivation of equations

Equations of rotational motion of a spinning discus. Figure 13.1 shows a co-ordinate system x, y, z moving with respect to the fixed co-ordinates X, Y, Z. The origin of the x, y, z system is the centre of the discus whose normal principal axis η makes an angle Ω with the relative air velocity V_R. The orientation of the discus is defined by its principal axis system ξ, η, ζ, which, after sequential rotations described by three Euler angles ξ, θ, and ψ with respect to the moving co-ordinates, will define any spatial orientation the discus may assume at any time. Figure 13.2 illustrates these three rotations. Initially the principal axes ξ, η, and ζ of the rigid body, abcO, coincide with the axes x, y, and z respectively. The point O is the pivot of rotation of the rigid body abcO. A rotation about the z axis by an angle ϕ rotates the body to the new position a$_1$b$_1$cO, which is followed by a rotation through an angle θ about axis Ob$_1$ to reach the new position a$_2$b$_1$c$_2$O. The third rotation is about axis Oc$_2$ with an angle ψ which results in the final position a$_3$b$_3$c$_2$O.

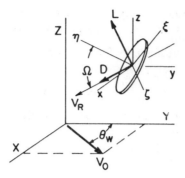

FIG. 13.1.   Notation: discus orientation with respect to the moving axes and the fixed axes.

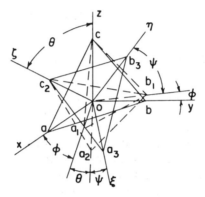

FIG. 13.2.   Sequential rotations of the Euler angles ϕ, θ, and ψ of the body Oabc to reach the position $Oa_3b_3c_2$.

The direction cosines of the principal axes and the moving co-ordinates are given in Table 13.1 (from ref. 2, p. 10).

The angular velocities of the body about its principal axes in terms of Euler angles and their time derivatives are (ref. 2, p. 16)

$$\omega_\xi = \dot{\theta} \sin\psi - \dot{\phi} \sin\theta \cos\psi$$
$$\omega_\eta = \dot{\theta} \cos\psi - \dot{\phi} \sin\theta \sin\psi \qquad (13.1)$$
$$\omega_\zeta = \dot{\psi} + \dot{\phi} \cos\theta.$$

TABLE 13.1

Directional cosines

	x	y	z
ξ	$\cos\phi \cos\theta \cos\psi - \sin\phi \sin\psi$	$\sin\phi \cos\theta \cos\psi + \cos\phi \sin\psi$	$-\sin\theta \cos\psi$
η	$-\cos\phi \cos\theta \sin\psi - \sin\phi \cos\psi$	$-\sin\phi \cos\theta \sin\psi + \cos\phi \cos\psi$	$\sin\theta \sin\psi$
ζ	$\cos\phi \sin\theta$	$\sin\phi \sin\theta$	$\cos\theta$

The time derivatives are indicated by the superscript dot. The three rotational equilibrium equations with respect to the principal axes are [3]

$$I_\xi \dot{\omega}_\xi + (I_\zeta - I_\eta)\omega_\eta \omega_\zeta = M_\xi \tag{13.2}$$

$$I_\eta \dot{\omega}_\eta + (I_\xi - I_\zeta)\omega_\xi \omega_\zeta = M_\eta \tag{13.3}$$

$$I_\zeta \dot{\omega}_\zeta + (I_\eta - I_\xi)\omega_\xi \omega_\eta = M_\zeta \tag{13.4}$$

where I and M are the moment of inertia and the applied moment respectively and the subscripts refer to the axis.

Differentiation of eqn (13.1) with respect to time and substitution into the equations (13.2)–(13.4) yields

$$\ddot{\theta} \sin \psi - \ddot{\phi} \sin \theta \cos \psi + \dot{\theta}\dot{\psi} \cos \psi - \dot{\phi}\dot{\theta} \cos \theta \cos \psi + \dot{\phi}\dot{\psi} \sin \theta \sin \psi$$

$$+ (1 - I_\eta/I_0)\omega_\eta \omega_\zeta - M_\xi/I_0 = 0 \tag{13.5}$$

$$\ddot{\theta} \cos \psi - \ddot{\phi} \sin \theta \sin \psi - \dot{\theta}\dot{\psi} \sin \psi - \dot{\phi}\dot{\theta} \cos \theta \sin \psi - \dot{\phi}\dot{\psi} \sin \theta \cos \psi$$

$$- M_\eta/I_\eta = 0 \tag{13.6}$$

$$\ddot{\psi} + \ddot{\phi} \cos \theta - \dot{\phi}\dot{\theta} \sin \theta - (1 - I_\eta/I_0)\omega_\xi \omega_\eta - M_\zeta/I_0 = 0 \tag{13.7}$$

where $I_0 = I_\xi = I_\zeta$ since the discus is rotationally symmetric about the η axis.

Applied moment due to translation. A search in earlier literature failed to find any information on the aerodynamic pressure distribution on a discus-shaped body inclined arbitrarily in the air stream with a rotatory motion. For non-rotating bodies Atsumi [4] obtained experimentally the pressure distribution on a wing (Clark Y airfoil section) of circular plan form which has a diameter of 20 cm. However, the angle of attack is only up to the stall angle which is 36°. Fage and Johnson [5] published test data on the air flow on an inclined flat plate of infinite span whose angle of attack varies from zero to 90°. The Reynolds number is 1.54×10^5, which is relatively close to the estimated Reynolds number for the discus throw of 3.9×10^5. Recently Stilley [6] published experimental data on Frisbee-like configurations. The moment and normal force coefficients of a right circular cylinder with a diameter to thickness ratio of $8:1$ and the tangential (in-plane) force coefficient of a Frisbee will be selected from Stilley's data for comparison with other similar curves.

The moment coefficients C_m from these three sources are shown in Fig. 13.3 where the points on curves A and C indicate that they are calculated values based on the pressure distribution curves given in refs. 4–6. We selected curve A from the inclined plate test as a reasonable compromise for C_m values for angles of attack ranging from 15° to 90°. The compromise is chosen because the test model that produced curve B is not exactly symmetric about its mid-plane. Also, other curves in ref. 6 not included in Fig. 13.3, such as that for a thin circular disc, exhibit C_m values higher than B for the same angle of attack. Curve C from the circular wing plan form test is chosen for angles of attack ranging from 0° to 10°. A parabolic curve fitting which joins the two pieces

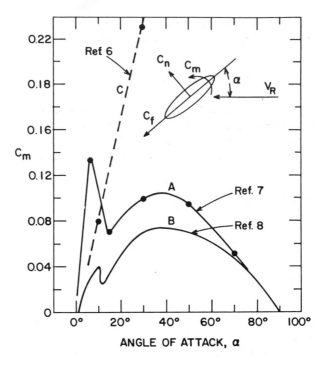

FIG. 13.3.   Three curves for the moment coefficient versus the angle of attack.

together is used from $10°$ to $15°$. This represents the moment coefficient C_m to be used in this study.

The remaining concern on the applied moment is the effect of spinning of the discus on the moment and the question of the possible effect of coupling between the translatory and the rotatory motions.

Velocity potentials. As a preliminary effort to investigate the effect of spin, a study on moment will be made which is based on the classical hydrodynamics of a rotating ellipsoid, which is equivalent to the discus in geometry, moving in an infinite stationary medium.

Laplace's equation $\nabla^2 \phi = 0$ in ellipsoidal co-ordinates can be written in the form [7]

$$(\mu - v)\left(k_\lambda \frac{\partial}{\partial \lambda}\right)^2 \phi + (v - \lambda)\left(k_\mu \frac{\partial}{\partial \mu}\right)^2 \phi + (\lambda - \mu)\left(k_v \frac{\partial}{\partial v}\right)^2 \phi = 0 \quad (13.8)$$

where λ, μ, and v are the ellipsoidal co-ordinates. They are in fact the three roots of the cubic equation $f(\theta) = 0$ expanded from the ellipse

$$\frac{x^2}{a^2 + \theta} + \frac{y^2}{b^2 + \theta} + \frac{z^2}{c^2 + \theta} = 1 \quad (13.9)$$

where a, b, and c are the semi-axes of the basic ellipsoid corresponding to $\theta = 0$. Other constant values of θ represent a central quadric of a confocal system and

$$k_\lambda = \{(a^2 + \lambda)(b^2 + \lambda)(c^2 + \lambda)\}^{1/2}. \tag{13.10}$$

k_μ and k_ν can be obtained by replacing λ with μ and ν respectively.

For an ellipsoid moving in an ideal fluid with semi-axes a, b, and c and velocity components U, V, and W, both measured along axes x, y, and z, the velocity potential for the translatory motion can be written as

$$\phi_t = abc \int_\lambda^\infty \left(\frac{Ux}{(2-\alpha_0)(a^2+\lambda)} + \frac{Vy}{(2-\beta_0)(b^2+\lambda)} + \frac{Wz}{(2-\gamma_0)(c^2+\lambda)} \right) \frac{d\lambda}{k_\lambda} \tag{13.11}$$

where $\lambda = 0$ defines the surface of the basic ellipsoid and

$$\{\alpha_0, \beta_0, \gamma_0\} = abc \int_0^\infty \left\{ \frac{1}{a^2+\lambda}, \frac{1}{b^2+\lambda}, \frac{1}{c^2+\lambda} \right\} \frac{d\lambda}{k_\lambda} \tag{13.12}$$

The velocity potential expressed by equation (13.11) satisfies the infinite field condition that when $\lambda \to \infty$, $\phi_t \to 0$.

Similarly the rotational velocity potential for the ellipsoid which rotates with velocity components $\omega_x, \omega_y, \omega_z$ and satisfies Laplace's equation can be written as

$$\phi_r = \frac{(b^2-c^2)^2 abc\,\omega_x yz}{2(b^2-c^2)+(b^2+c^2)(\beta_0-\gamma_0)} \int_\lambda^\infty \frac{d\lambda}{(b^2+\lambda)(c^2+\lambda)k_\lambda}$$
$$+ \frac{(c^2-a^2)^2 abc\,\omega_y zx}{2(c^2-a^2)+(c^2+a^2)(\gamma_0-\alpha_0)} \int_\lambda^\infty \frac{d\lambda}{(c^2+\lambda)(a^2+\lambda)k_\lambda}$$
$$+ \frac{(a^2-b^2)^2 abc\,\omega_z xy}{2(a^2-b^2)+(a^2+b^2)(\alpha_0-\beta_0)} \int_\lambda^\infty \frac{d\lambda}{(a^2+\lambda)(b^2+\lambda)k_\lambda}. \tag{13.13}$$

This velocity potential also satisfies the field condition at infinity.

Kinetic energy and moments. From equations (13.11) and (13.13), $\partial\phi_t/\partial n$ and $\partial\phi_r/\partial n$ can be calculated, where n is the normal to the surface of the ellipsoid [7]. The kinetic energy due to the motion of the ellipsoid of the medium with density ρ is given by

$$T = -\frac{\rho}{2} \int \int \left(\phi_t \Big|_{\lambda=0} + \phi_r \Big|_{\lambda=0} \right) \left(\frac{\partial\phi_t}{\partial n}\Big|_{\lambda=0} + \frac{\partial\phi_r}{\partial n}\Big|_{\lambda=0} \right) dS \tag{13.14}$$

where the elementary surface dS is taken over the entire surface of the ellipsoid.

Integration of eqn (13.14) yields

$$T = \frac{2}{3}\rho\pi abc\left[\left(\frac{\alpha_0}{2-\alpha_0}U^2 + \frac{\beta_0}{2-\beta_0}V^2 + \frac{\gamma_0}{2-\gamma_0}W^2\right)\right.$$
$$\left. - \frac{1}{5}\{G_x\omega_x^2(c^2-b^2) + G_y\omega_y^2(a^2-c^2) + G_z\omega_z^2(b^2-a^2)\}\right] \qquad (13.15)$$

where

$$G_x = \frac{(b^2-c^2)^2 abc}{2(b^2-c^2)+(b^2+c^2)(\beta_0-\gamma_0)}\int_0^\infty \frac{d\lambda}{(b^2+\lambda)(c^2+\lambda)k_\lambda} \qquad (13.16)$$

and G_y and G_z can be obtained by cyclically rotating a, b, c and α_0, β_0, γ_0.

Since the cross product of the momentum vector and the velocity vector is the moment vector on the body, we have as the moment vector

$$\mathbf{M} = \frac{\partial T}{\partial \mathbf{u}}\times\mathbf{u} + \frac{\partial T}{\partial \omega}\times\omega \qquad (13.17)$$

where $\mathbf{u}$ and ω are the transition and rotation velocity vectors respectively.

Differentiation of eqn (13.14) and substitution in eqn (13.17) yields the three moment components acting on the ellipsoid due to its translatory and rotatory motions through the medium:

$$\mathbf{M} = \mathbf{i}M_\xi + \mathbf{j}M_\eta + \mathbf{k}M_\zeta$$
$$= \frac{4\pi}{3}\rho abc\left(\mathbf{i}\left[2\frac{\beta_0-\gamma_0}{(2-\beta_0)(2-\gamma_0)}VW - \frac{1}{5}\{G_y(a^2-c^2)-G_z(b^2-a^2)\}\omega_\eta\omega_\zeta\right]\right.$$
$$-\mathbf{j}\left[2\frac{\gamma_0-\alpha_0}{(2-\gamma_0)(2-\alpha_0)}WU - \frac{1}{5}\{G_z(b^2-a^2)-G_x(c^2-b^2)\}\omega_\zeta\omega_\xi\right]$$
$$\left. +\mathbf{k}\left[2\frac{\alpha_0-\beta_0}{(2-\alpha_0)(2-\beta_0)}UV - \frac{1}{5}\{G_x(c^2-b^2)-G_y(a^2-c^2)\}\omega_\xi\omega_\eta\right]\right) \qquad (13.18)$$

where the unit vectors $\mathbf{i}, \mathbf{j}, \mathbf{k}$ and the velocity components U, V, W are along the positive directions of the principal axes ξ, η, and ζ of the ellipsoid.

Even though eqns (13.15) and (13.18) are not difficult to derive, they nevertheless have not previously been given in the literature. The result indicated that according to the inviscid incompressible theory the translatory and rotatory motions do not interfere with each other and the moment induced by the rotation possesses a gyroscopic characteristic. This conclusion is corroborated by a wind-tunnel test result on spinning Frisbee-shaped bodies [6] which showed that the pitching moment and lift would not themselves be affected by the spin but that the predominant effect would be the provision of gyroscopic stiffness.

Since $\mathbf{u}$ is the velocity vector opposite to the resultant air velocity V_R

$$\{U, V, W\} = -V_R\{\cos R_\xi, \cos R_\eta, \cos R_\zeta\} \tag{13.19}$$

where the direction cosines between the resultant wind and the principal axes are

$$\begin{aligned}
\{\cos R_\xi, \cos R_\eta, \cos R_\zeta\} &= \cos R_x\{\cos \xi_x, \cos \eta_x, \cos \zeta_x\} \\
&\quad + \cos R_y\{\cos \xi_y, \cos \eta_y, \cos \zeta_y\} \\
&\quad + \cos R_z\{\cos \xi_z, \cos \eta_z, \cos \zeta_z\}
\end{aligned} \tag{13.20}$$

and

$$\cos \Omega = \cos R_\eta. \tag{13.21}$$

The angle Ω and the resultant relative air velocity V_R are shown in Fig. 13.1. The latter is given by

$$V_R = \{(\dot{x} - V_0 \sin \theta_w)^2 + (\dot{y} - V_0 \cos \theta_w)^2 + \dot{z}^2\}^{1/2} \tag{13.22}$$

and the direction cosines of V_R with respect to the moving axes are

$$\{\cos R_x, \cos R_y, \cos R_z\} = -\frac{\{(\dot{x} - V_0 \sin \theta_w), (\dot{y} - V_0 \cos \theta_w), \dot{z})\}}{V_R}. \tag{13.23}$$

The nine direction cosines between the principal axes and the moving coordinate axes, as required in eqn (13.20), are given in Table 13.1.

Since the discus is rotationally symmetric we have $c = a$, $\gamma_0 = \alpha_0$, $G_y = 0$, and $G_z = -G_x$; eqn (13.18) is finally simplified to

$$\begin{aligned}
\frac{3}{4\pi\rho abc}\{M_\xi, M_\eta, M_\zeta\} &= \frac{2(\beta_0 - \alpha_0)V_R^2}{(2 - \beta_0)(2 - \alpha_0)}\cos \Omega \{\cos R_\zeta, 0, -\cos R_\xi\} \\
&\quad - \frac{1}{5}G_x(b^2 - a^2)\omega_\eta\{\omega_\zeta, 0, -\omega_\xi\}.
\end{aligned} \tag{13.24}$$

In eqn (13.24) we neglect the virtual mass moment terms due to translation which are included in the experimental moment coefficient C_m in Fig. 13.3. However, we add to the experimental C_m the virtual mass moment terms due to rotation to account for the effect of spinning of the discus.

Thus the final moment expression to be used in the analysis is

$$\begin{aligned}
\{M_\xi, M_\eta, M_\zeta\} &= C_m\frac{\rho}{2}V_R^2(\pi a^2)2a\{\cos N_\xi, 0, \cos N_\zeta\} \\
&\quad - \frac{4\pi}{15}\rho a^2 b(b^2 - a^2)G_x\omega_\eta\{\omega_\zeta, 0, -\omega_\xi\} \tag{13.25}
\end{aligned}$$

where the angles N_ξ and N_ζ are those between the two principal axes and the normal N to the plane that contains the η axis and the V_R vector. They are

$$\{\cos N_\xi, \cos N_\zeta\} = \cos N_x \{\cos \xi_x, \cos \eta_x\} + \cos N_y \{\cos \xi_y, \cos \eta_y\}$$
$$+ \cos N_z \{\cos \xi_z, \cos \eta_z\} \qquad (13.26)$$

where the direction of the N axis is defined by the right-hand thumb rule of the vector product of the η axis and V_R:

$$\mathbf{N} = (\mathbf{i} \cos N_x + \mathbf{j} \cos N_y + \mathbf{k} \cos N_z) = \eta X \, (V_R/|V_R|). \qquad (13.27)$$

which yields

$$\cos N_x = (\cos \eta_y \cos R_z - \cos \eta_z \cos R_y)/\sin \Omega$$
$$\cos N_y = (\cos \eta_z \cos R_x - \cos \eta_x \cos R_z)/\sin \Omega \qquad (13.28)$$
$$\cos N_z = (\cos \eta_x \cos R_y - \cos \eta_y \cos R_x)/\sin \Omega$$

When $\Omega = 0$ or π, $\cos N_\xi$ and $\cos N_\zeta$ in eqn (13.25) vanish.

This concludes the derivation of the moments on the discus in flight. It should be noted that the direction of spinning of the discus has no effect on the trajectory because when ω_η is changed to $-\omega_\eta$, at the same time changing the sign of ϕ, θ, and ψ in eqns (13.25), (13.5), and (13.7), the three equations of rotation remain the same.

Translatory equations of motion. Figure 13.4 shows the normal force coefficient C_n curves from a circular wing plan form [4], an inclined flat plate [5], and a circular plate of 8 : 1 ratio [6]. Since at an angle of attack of 90° C_n of the discus should be reasonably close to the circular plate value, which is 1.12, curve B is superior to curve A in the discus problem. For the force coefficient parallel to the plane of the discus only ref. 6 (a Frisbee-shaped body) supplied a test result. The infinite plate and the circular wing are thin; therefore the C_f is very small.

The lift and drag forces on the discus are given by

$$\{D, L\} = D_f \{\sin \Omega, -\cos \Omega\} + \text{sign}(\cos \Omega)D_p \{\cos \Omega, \sin \Omega\} \quad (13.29)$$

where sign$(\cos \Omega)$ denotes the sign of $\cos \Omega$. In eqn (13.20) D_f and D_p are the aerodynamic force components on the discus centre of gravity along and perpendicular to the plane of the discus respectively.

The force components D_f and D_p can be written in the conventional form as

$$\{D_f, D_p\} = \frac{\rho}{2} V_R^2 (\pi a^2) \{C_f, C_n\}. \qquad (13.30)$$

Based on the lift and drag forces and the direction cosines, the translatory equations of motion for the trajectory of the centre of gravity of the discus are

$$\frac{w}{g}\{\ddot{x}, \ddot{y}, \ddot{z}+g\} = D\{\cos R_x, \cos R_y, \cos R_z\} + L\{\cos L_x, \cos L_y, \cos L_z\}$$
$$(13.31)$$

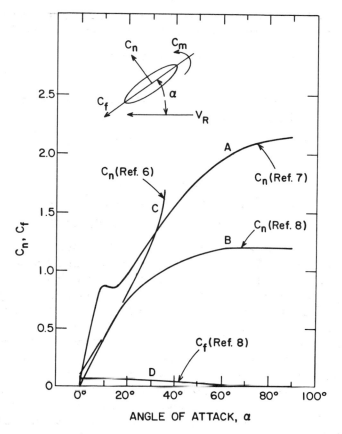

FIG. 13.4. Three normal force coefficient (C_n) curves and one tangential force coefficient (C_f) curve.

where the directional angles between the lift force L and the moving co-ordinates are given by

$$\mathbf{L}/|L| = \mathbf{i} \cos L_x + \mathbf{j} \cos L_y + \mathbf{k} \cos L_z = (\mathbf{V_R}/|V_R|) \times \mathbf{N} \qquad (13.32)$$

which yields

$$\cos L_x = \cos R_y \cos N_z - \cos R_z \cos N_y$$
$$\cos L_y = \cos R_z \cos N_x - \cos R_x \cos N_z \qquad (13.32a)$$
$$\cos L_z = \cos R_x \cos N_y - \cos R_y \cos N_x.$$

when $\Omega = 0$ or π, L vanishes.

This completes the derivation of the equations. Equations (13.5), (13.6), and (13.7) with the moment terms given by eqn (13.25) are the dynamic rotational equations of equilibrium; eqn (13.31) with the loading terms given by eqns (13.29) and (13.30) is the set of the translatory equations of equilibrium. These six non-linear differential equations will yield the spatial position of the centre of gravity of the discus in terms of x, y, z, and also the orientations of its three principal axes in terms of the Euler angles ϕ, θ, and ψ; all are functions of time.

Solution method

Solution of the six non-linear differential equations of motion will be obtained by a numerical method. The principle behind the method will be briefly described.

Assume that at a point in time t_1 the displacements and velocities of the discus are known. We want to find the corresponding quantities at the next time point t_2 based on information available at t_1. By taking three terms in a power series expansion of the solution for $t_1 \leqslant t \leqslant t_2$, where t_2 is sufficiently close to t_1, we have

$$\{x, y, z, \phi, \theta, \psi\} = \{x_1, y_1, z_1, \phi_1, \theta_1, \psi_1\} + \{v_x, v_y, v_z, v_\phi, v_\theta, v_\psi\}(t-t_1)$$

$$+ \{a_x, a_y, a_z, a_\phi, a_\theta, a_\psi\}\tfrac{1}{2}(t-t_1)^2 \quad t_1 \leqslant t \leqslant t_2 \quad (13.33)$$

where x, y, z are the spatial position of the centre of gravity of the discus and ϕ, θ, ψ are the Euler angles describing the orientations of the discus principal axes. Equation (13.33) automatically satisfies the 'initial' conditions at $t = t_1$:

$$\{x, y, z, \phi, \theta, \psi\}_{<t=t_1>} = \{x_1, y_1, z_1, \phi_1, \theta_1, \psi_1\} \quad (13.34)$$

$$\{\dot{x}, \dot{y}, \dot{z}, \dot{\phi}, \dot{\theta}, \dot{\psi}\}_{<t=t_1>} = \{v_x, v_y, v_z, v_\phi, v_\theta, v_\psi\} \quad (13.35)$$

where the right-hand side quantities of eqns (13.34) and (13.35) are known at $t = t_1$. The six unknowns a_x to a_ψ, called acceleration vectors, in eqn (13.33) are used to describe the increments necessary to make the transition from t_1 to t_2 a smooth function.

By substituting eqn (13.33) into all relevant expressions used in the equilibrium equations and assuming that they are satisfied exactly at the next point in time t_2, with $T = t_2 - t_1$ sufficiently small, the six equations (13.5), (13.6), (13.7), and (13.31) can be written in a general form as

$$F_i(a_x, a_y, a_z, a_\phi, a_\theta, a_\psi) = 0 \quad i = 1, 2, 3, 4, 5, 6 \quad (13.36)$$

with T, x_1, y_1, ..., ψ_1, v_x, v_y, ..., v_ψ, and other physical and geometrical quantities as prescribed values and a_x, a_y, ..., a_ψ as unknowns.

Numerical methods must be used to solve the six non-linear algebraic equations (eqn (13.36)) for the six unknowns. Take the Newton–Raphson method as an example. Let $(F_i)^j$ indicate the value of the ith equation in which the jth set of trial values of a_x, a_y, a_z, a_ϕ, a_θ, a_ψ has been used. If $(F_i)^j$ is identically zero, the equation is exactly satisfied. Assume that not all the six equations are exactly satisfied at the jth iteration and the $(j+1)$th iteration is needed. The increments $\Delta a_x, \Delta a_y, \Delta a_z, \Delta a_\phi, \Delta a_\theta, \Delta a_\psi$ to be added to the jth set are given by the solution of the following simultaneous equations:

$$\left(\frac{\partial F_i}{\partial a_x}\right)^j \Delta a_x + \left(\frac{\partial F_i}{\partial a_y}\right)^j \Delta a_y + \left(\frac{\partial F_i}{\partial a_z}\right)^j \Delta a_z + \left(\frac{\partial F_i}{\partial a_\phi}\right)^j \Delta a_\phi$$

$$+ \left(\frac{\partial F_i}{\partial a_\theta}\right)^j \Delta a_\theta + \left(\frac{\partial F_i}{\partial a_\psi}\right)^j \Delta a_\psi = -(F_i)^j \quad i = 1, 2, 3, 4, 5, 6$$

$$(13.37)$$

where the partial derivatives can be obtained by the finite-difference method since no analytical method is possible. The iteration can be terminated when the absolute values of the residuals of F_i are smaller than a fixed small quantity.

After the converged values of $a_x, a_y, \ldots, a_\psi$ are obtained for the time point t_2 the 'initial' condition to start the next time increment can be found from equation (13.33) and its first time derivative.

Initial conditions. As an example, if the discus is released at a height $z_0 = 1.52$ m, thrown with velocity $\overline{V}_0$, throw angle θ_f, initial discus inclination angle θ_j, and spinning speed ω_{n0}, the initial conditions are

$$\{x, y, z, \phi, \theta, \psi\}_{\langle t=0 \rangle} = \left\{0, 0, 1.52\,m, 0, \frac{\pi}{2}, \frac{\pi}{2} + \theta_j\right\} \quad (13.38)$$

$$\{v_x, v_y, v_z, v_\phi, v_\theta, v_\psi\}_{\langle t=0 \rangle} = \{0, \overline{V}_0 \cos \theta_f, \overline{V}_0 \sin \theta_f, v_{\phi0}, v_{\theta0}, v_{\psi0}\} \quad (13.39)$$

where the three initial Euler angular velocities $v_{\phi0}, v_{\theta0}$, and $v_{\psi0}$ have to be solved from eqn (13.1) by substituting the values

$$\{\phi, \theta, \psi, \omega_\xi, \omega_\eta, \omega_\zeta\} = \left\{0, \frac{\pi}{2}, \frac{\pi}{2} + \theta_j, 0, \omega_{n0}, 0\right\} \quad (13.40)$$

in eqn (13.1) and then solving for $v_{\phi0}, v_{\theta0}$, and $v_{\psi0}$ from the following equations:

$$v_{\theta0} \cos \theta_j + v_{\phi0} \sin \theta_j = 0$$
$$-v_{\theta0} \sin \theta_j - v_{\phi0} \cos \theta_j = \omega_{n0} \quad (13.41)$$
$$v_{\psi0} - v_{\phi0} \sin \theta_j = 0.$$

Convergence problem when ω_{n0} is not small. Convergence will become difficult when the spinning speed ω_{n0} is not small because then the Euler angles will change rapidly and the time step T has to be extremely small in order to maintain a reasonable continuity between successive time points. The difficulty may be resolved by substituting $\omega_n + \omega_{n0}$ for ω_n in eqns. (13.5), (13.7), and (13.25) and setting $v_{\phi0} = v_{\theta0} = v_{\psi0} = 0$. This will allow the three principal axes of the discus to rotate in space with a speed of comparable order of magnitude with respect to each other, while the effect of the initial spinning ω_{n0} is preserved. When the flight time is short and the spinning speed does not change rapidly with time this technique produces satisfactory results.

Numerical results

Geometry of the discus. Figure 13.5(a) shows a conventional discus which conforms to the National Collegiate Athletic Association (NCAA) specifications (ref. 8, p. TF-27). It weighs 2000 g and the frontal area along the ξ axis is 68.86 cm². The equivalent minor axis for a corresponding ellipsoid of the same major axis and cross-sectional area is 4.0 cm. Compared with the original thickness of the discus, which is 4.5 cm, the equivalent ellipsoid seems to be a reasonably close mathematical model for the purpose of analysis.

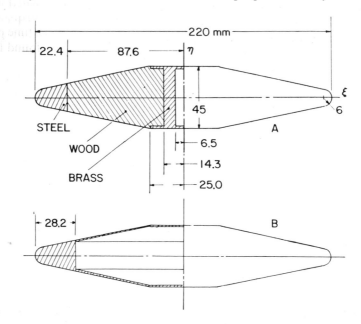

FIG. 13.5. Geometry of the discus. (a) The conventional discus: weight $W = 2000$ g; moments of inertia $I_\eta = 157.61$ g cm s^{-2}, I_ξ 78.81 g cm s^{-2}; (b) An ideal discus with the mass concentrated in the rim: $W = 2000$ g; $I_\eta = 182.5$ g cm s^{-2}; $I_\xi = 91.3$ g cm s^{-2}.

Figure 13.5(b) is an ideal discus which has the same outer geometry and weight of the conventional discus shown in Fig. 13.5(a) but its mass is concentrated at the rim. Consequently, the discus in Fig. 13.5(b) has the maximum moment of inertia about its principal axes. A comparison of their performances will be made.

Effect of the initial spinning speed $\omega_{\eta 0}$. The numerical result is put in the context of the 1972 Olympics discus throw record where Danek (Czechoslovakia) (see Table 13.2) threw a distance of 64.389 m. We assume here that Danek made the record by using the discus shown in Fig. 13.5(a). θ_f and θ_j are both 35°, the optimum suggested by Taylor (ref. 1, p. 149).

TABLE 13.2

1972 Olympics discus throw record

1	Danek (Czechoslovakia)	211 ft 3 in	(64.389 m)
2	Sylvester (U.S.A.)	208 ft 4 in	(63.500 m)
3	Bruch (Sweden)	208 ft	(63.398 m)

We shall take the spinning speed $\omega_{n0} = 36.9$ rev/sec as a standard for comparison. This is the angular speed reached if the discus is rolling along its rim on a surface without sliding while its center is travelling at the initial throw speed of $\overline{V}_0 = 25.5$ m/s. At that spinning speed, the Olympic throw range of 64.389 m can be obtained with 35°/35° throw angles. The maximum height of the trajectory is 13.93 m and the flight time is 3.794 s. At the end of the flight, the plane of the discus will be tilted slightly. In the beginning, its normal axis η is at 35° with the z-axis and 90° to the x-axis. At the time of landing, it becomes 41.09° to the z-axis and 75.45° to the x-axis. The lateral drift due to the gyroscopic action of the spinning is 2.30 m. Increasing the spinning speed from 36.9 rev/s to larger values does not change the range appreciably, however the pitching and rolling motions could be reduced.

The effect of the spinning speed to the range R can be seen from Fig. 13.6 and Table 13.3. The range will be approximately the same if the spinning speed is not less than 18 rev/s. Assuming that an ordinary thrower who spins a discus about 4 rev/s, 1.2 m in range might be gained by increasing the spin to 8 rev/s. If the discus is not spun at all, the range is reduced by 9.4 m compared with the 4 rev/sec case. Figure 13.7 shows in scale the trajectories of two cases: curve A is for $\omega_{n0} = 36.9$ and B is for $\omega_{n0} = 0$, the case of no-spin.

Effect of the moment of inertia of the discus. Table 13.3 shows the ranges of the two discuses of Figs. 13.5(a) and 13.5(b) at different spinning speeds. The result indicated that the ratio of the moment of inertia to the weight of the conventional discus (Fig. 13.5(a)) is already sufficiently high; further effort in redistributing the mass would not produce a significant improvement to the throw range.

Effect of varying throw angle and discus inclination angle. The results of Fig. 13.8 showed that with reasonable initial spinning the two delivery angles of the discus throw greatly affected the range. At zero wind the optimum combination is 35°/26° for throw angle versus the discus inclination angle. If $\theta_f = \theta_j$ and they are varied together the optimum is 33°. The former result correlates exactly with Ganslan's observation which agrees perfectly with Taylor's wind-tunnel test result (ref. 1, p. 149).

Effect of wind speed V_0. The effect on the range of wind speed can be seen in Fig. 13.9. The results show that the discus should be thrown against the wind to

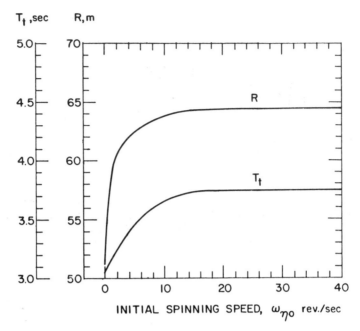

FIG. 13.6.   Effect of the initial spinning speed $\omega_{\eta 0}$ on the range R and the time T_t $(\theta_f = \theta_j = 35°)$.

TABLE 13.3

Throw range[a] as affected by the moment of inertia and the spinning speed $\omega_{\eta 0}$ of the discus

Spinning speed $\omega_{\eta 0}$ (rev s⁻¹)	Flight time T_t (s)		Range R (m)	
	Fig. 13.5(a) discus	Fig. 13.5(b) discus	Fig. 13.5(a) discus	Fig. 13.5(b) discus
0	3.026	3.037	50.630	50.761
4	3.291	3.337	60.019	60.031
8	3.515	3.566	61.221	61.668
18	3.741	3.759	63.674	63.908
36.9	3.794	3.799	64.389	64.452
74	3.808	3.810	64.576	64.592
740	3.813	3.813	64.638	64.639

[a] $\overline{V}_0 = 25.5 \, \text{m s}^{-1}$, $V_0 = 0$, $\theta_j = \theta_f = 35°$, $T = 0.005 \, \text{s}$.

take advantage of the additional lift force. When the throw angle remains at 35° the discus plane angle should be kept low if the wind is strong. A tail wind is detrimental to the throw. The head wind advantage is lost when the discus inclination angle is too high as the case of 35°/35° shows.

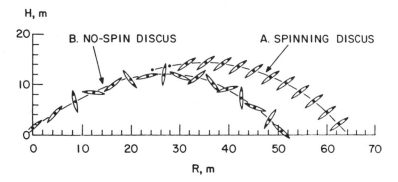

FIG. 13.7 Trajectory and spatial orientation of the discus in flight. The time between positions is 0.2 s and $\theta_f = \theta_j = 35°$. Curve A: $\omega_{\eta 0} = 36.9$ rev s^{-1}; maximum height $= 13.93$ m; $R = 64.389$ m; $T_t = 3.794$ s; lateral drift $= 2.30$ m. Curve B: $\omega_{\eta 0} = 0$ rev s^{-1}; maximum height $= 11.88$ m; $R = 50.630$ m; $T_t = 3.026$ s; lateral drift $= 0$ m.

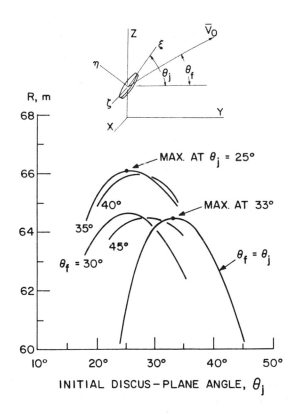

FIG. 13.8. Effect of θ_j and θ_f on the throw range R. Four curves were obtained with fixed and θ_j varying and one was obtained with $\theta_f = \theta_f$ varying together.

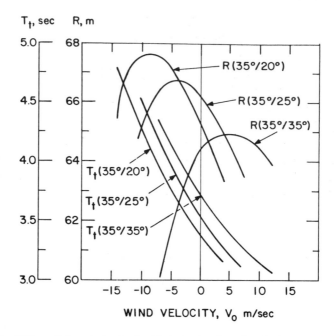

Fig. 13.9   Effect of the wind velocity V_0 (negative for a head wind, positive for a tail wind) on range R and flight time T_t for different θ_f/θ_j angles ($\omega_{\eta 0} = 36.9\,\mathrm{rev\,s^{-1}}$, $\overline{V}_0 = 25.5\,\mathrm{m\,s^{-1}}$, $T = 0.005\,\mathrm{s}$).

Conclusions. The analysis confirmed quantitatively the opinion of earlier experts that the 35°/35° throw and discus angles are among the best. Our analysis showed that 33°/33° was slightly better. We also confirmed that 35°/26° is the best combination, yielding 1.55 m more distance than the 35°/35° throw. The advantage of throwing against a head wind is also verified by the analysis.

13.3. Analysis of the javelin throw

The analysis of the javelin throw is less involved than that of the discus because the javelin is not spun about its axis of symmetry. On the other hand it is complicated by the NCAA rule that a throw is foul if the tip of the metal head does not strike the ground before any other part of the javelin. This requirement must be included in the analysis of optimum ways of throwing a javelin.

Table 13.4 shows the 1972 Olympic Games javelin throw records which will be used as a case study of the effect of various parameters on the throw range.

TABLE 13.4

1972 Olympics javelin throw records

1	Wolferman (F.R.G.)	296 ft 10 in (90.475 m)
2	Lusis (USSR)	296 ft 9 in (90.449 m)
3	Schmidt (USA)	276 ft 11 in (84.404 m)

Derivation of equations

Figure 13.10 shows the co-ordinate axes of an arbitrarily inclined javelin in space where η is the javelin's axis, O the centre of gravity, and O' the aerodynamic centre, otherwise known as the centre of pressure, through which passes the resultant of the lift and drag forces.

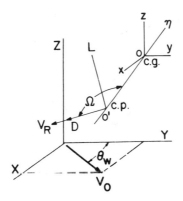

FIG. 13.10. Co-ordinates and notation for the javelin.

Equation of motion of the centre of gravity of the javelin. The resultant relative wind velocity V_R as shown in Fig. 13.10 contributes in two ways to resistance to the javelin flight. Axial-flow skin drag is produced by the velocity component of V_R along the negative direction of the η axis. A second resistance is the pressure drag perpendicular to the javelin axis. Both drags contribute to the aerodynamic resistance components L and D.

Let D_f and D_p be the friction and pressure drags respectively. Then the classical expressions are

$$D_f = C_{df} \tfrac{1}{2} \rho S_f V_R^2 \cos^2 (\pi - \Omega) \tag{13.42}$$

$$D_p = C_{dp} \tfrac{1}{2} \rho S_p V_R^2 \sin^2 (\pi - \Omega) \tag{13.43}$$

where C_{df} and C_{dp} are drag coefficients, S_f is the 'wetted' area of the javelin and S_p is the lateral projected area.

Resolving D_f and D_p into the directions of D and L as shown in Fig. 13.10 gives the aerodynamic resistance components:

$$D = \frac{\rho}{2} V_R^2 (C_{dp} S_p \sin^3 \Omega - C_{df} S_f \cos^3 \Omega) \tag{13.44}$$

$$L = -\frac{\rho}{2} V_R^2 (C_{dp} S_p \sin \Omega + C_{df} S_f \cos \Omega) \sin \Omega \cos \Omega. \tag{13.45}$$

The aerodynamic force components acting at the centre of pressure of the javelin along the positive direction of the moving co-ordinates are then

$$\{P_x, P_y, P_z\} = D\{\cos R_x, \cos R_y, \cos R_z\} + L\{\cos L_x, \cos L_y, \cos L_z\}. \tag{13.46}$$

The three translatory equations of motion of the centre of gravity of the javelin are the same as equation (13.31) where D and L for the javelin throws are given in equations (13.44) and (13.45).

Derivation of rotational equations of motion

As in the discus case the Euler angles ϕ, θ and ψ are used to define the javelin orientation in space with respect to the moving axes. The general rotational angles ϕ, θ and ψ are the same as shown in Fig. 13.2. The direction cosines are the same as in Table 13.1. However, since the javelin is rotationally symmetric about the η axis and there is no spinning speed to speak of with respect to the η axis, the spatial orientation of the javelin can be specified by any two of the Euler angles, say θ and ψ. The third angle ϕ will be determined by an arbitrary condition, say that the ξ axis be coplanar with the x and z axes. This condition yields a relationship

$$\zeta_z = \frac{\pi}{2} - \zeta_x \tag{13.47}$$

where ζ_z and ζ_x are angles formed between ζ and the z and x axes respectively. Therefore

$$\cos \zeta_z = \sin \zeta_x = (1 - \cos^2 \zeta_x)^{1/2}. \tag{13.48}$$

Substituting the direction cosine values from Table 13.1 into eqn (13.48) yields

$$\cos \theta = (1 - \cos^2 \phi \sin^2 \theta)^{1/2} \tag{13.49}$$

which yields

$$\phi = 0. \tag{13.50}$$

Consequently Table 13.1 is reduced to the special case shown in Table 13.5.

For the special case of $\phi = \dot\phi = 0$ the angular velocities of eqn (13.1) are changed to

$$\{\omega_\xi, \omega_\eta, \omega_\zeta\} = \{\dot\theta \sin \psi, \dot\theta \cos \psi, \dot\psi\} \tag{13.51}$$

TABLE 13.5

Direction cosines for the special case

	x	y	z
ξ	$\cos\theta\cos\psi$	$\sin\psi$	$-\sin\theta\cos\psi$
η	$-\cos\theta\sin\psi$	$\cos\psi$	$\sin\theta\sin\psi$
ζ	$\sin\theta$	0	$\cos\theta$

The moments about the moving axes x, y, z due to the aerodynamic force components at the centre of pressure are

$$\{M_x, M_y, M_z\} = \{P_y\cos\eta_z - P_z\cos\eta_y, P_z\cos\eta_x$$
$$- P_x\cos\eta_z, P_x\cos\eta_y - P_y\cos\eta_x\}d \qquad (13.52)$$

where the right-hand rule governs the sign of the moments, d is the distance between the centre of pressure O' and the centre of gravity O; it is positive when the centre of pressure is behind the centre of gravity as shown in Fig. 13.10.

Resolving into components about the principal axes we obtain

$$M_\xi = M_x\cos\xi_x + M_y\cos\xi_y + M_z\cos\xi_z$$
$$= -d\{P_x(-\cos\eta_y\cos\xi_z + \cos\eta_z\cos\xi_y) + P_y(-\cos\eta_z\cos\xi_x$$
$$+ \cos\eta_x\cos\xi_z) + P_z(-\cos\eta_x\cos\xi_y + \cos\eta_y\cos\xi_x)\} \qquad (13.53)$$

$$M_\xi = M_x\cos\zeta_x + M_y\cos\zeta_y + M_z\cos\zeta_z$$
$$= -d\{P_x(-\cos\eta_y\cos\zeta_z + \cos\eta_z\cos\zeta_y) + P_y(-\cos\eta_z\cos\zeta_x$$
$$+ \cos\eta_x\cos\zeta_z) + P_z(-\cos\eta_x\cos\zeta_y + \cos\eta_y\cos\zeta_x)\} \qquad (13.54)$$

$$M_\eta = M_x\cos\eta_x + M_y\cos\eta_y + M_z\cos\eta_z = 0. \qquad (13.55)$$

The rotational differential equation about η axis is trivial. The other two equilibrium equations about the principal axes ξ and ζ are eqns (13.5) and (13.7).

For the special case eqns (13.5) and (13.7) are finally reduced to

$$\ddot{\theta}\sin\psi + 2\dot{\theta}\dot{\psi}\cos\psi = \frac{-d}{I_0}(P_x\sin\theta + P_z\cos\theta) \qquad (13.56)$$

$$\ddot{\psi} - \dot{\theta}^2\sin\psi\cos\psi = \frac{d}{I_0}(P_x\cos\psi\cos\theta + P_y\sin\psi - P_z\cos\psi\sin\theta). \qquad (13.57)$$

Geometry at landing. Figure 13.11 shows that at the instant of release the javelin's centre of gravity is at point O and its axis lies in the YZ plane making an angle θ_j with the Y axis. Its direction of flight is at an angle θ_f with the horizontal. The throwing force F_0 is along the θ_f direction and no moment is applied to the javelin. With a positive tail wind V_0 blowing at an angle θ_w

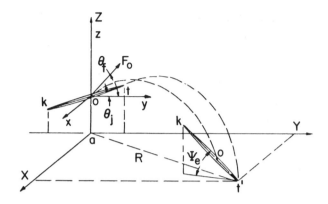

FIG. 13.11. Notation for the flight path of the javelin.

(Fig. 13.10) the tip t of the javelin finally lands at point t' with a negative landing angle ψ_e. The total range is R. According to NCAA rules the throw is valid as long as the tip of the javelin lands first, even if the landing angle ψ_e is very small.

To find the position of the tip of the javelin, consider a point (x_1^*, y_1^*, z_1^*) of a rigid body taken with respect to the principal axes ξ, η, and ζ. After the body makes an arbitrary rotation defined by the three Euler angles θ, ϕ, and ψ, the new position of the point denoted by (X_2^*, Y_2^*, Z_2^*) with respect to the fixed axes X, Y, Z can be obtained from its position with respect to the moving axes x, y, z plus the new position of the origin of the moving axes with respect to the fixed axes denoted by (X_2, Y_2, Z_2). When (x_1^*, y_1^*, z_1^*) is a general point in a rigid body performing rotations we have (ref. 2, p. 8)

$$X_2^* = (b_1^2 - b_2^2 - b_3^2 + b_4^2)x_1^* + 2(b_1 b_2 - b_3 b_4)y_1^* + 2(b_1 b_3 + b_2 b_4)z_1^* + X_2$$
$$Y_2^* = 2(b_1 b_2 + b_3 b_4)x_1^* + (-b_1^2 + b_2^2 - b_3^2 + b_4^2)y_1^* + 2(b_2 b_3 - b_1 b_4)z_1^* + Y_2$$

$$(13.58)$$

$$Z_2^* = 2(b_1 b_3 - b_2 b_4)x_1^* + 2(b_2 b_3 + b_1 b_4)y_1^* + (-b_1^2 - b_2^2 + b_3^2 + b_4^2)z_1^* + Z_2$$

where

$$b_1 = \sin\left(\tfrac{1}{2}\theta\right)\sin\left\{\tfrac{1}{2}(\psi - \phi)\right\}$$
$$b_2 = \sin\left(\tfrac{1}{2}\theta\right)\cos\left\{\tfrac{1}{2}(\psi - \phi)\right\} \qquad (13.59)$$
$$b_3 = \cos\left(\tfrac{1}{2}\theta\right)\sin\left\{\tfrac{1}{2}(\psi + \phi)\right\}$$
$$b_4 = \cos\left(\tfrac{1}{2}\theta\right)\cos\left\{\tfrac{1}{2}(\psi + \phi)\right\}.$$

For the present problem, since the tip is located at a distance d^* from the centre of gravity on the η axis, substitution of $\phi = 0$ (equation (13.50)),

$x_1^* = z_1^* = 0$, $y_1^* = d^*$, and eqn (13.59) into eqn (13.58) yields

$$X_2^* = -d^* \cos \theta \sin \psi + X_2$$
$$Y_2^* = d^* \cos \psi + Y_2 \qquad\qquad (13.60)$$
$$Z_2^* = d^* \sin \theta \sin \psi + Z_2.$$

This agrees with the direction cosine values in Table 13.5.

The range of the javelin throw is

$$R = (Y_2^{*2} + X_2^{*2})^{1/2} \qquad \text{when } Z_2^* = 0 \qquad\qquad (13.61)$$

subject to the condition $Z_2 > Z_2^* \geqslant 0$.

A case study for an NCAA official javelin

Figure 13.12 shows the geometry of a javelin fitting midway in the limit values of the NCAA javelin rules (ref. 8, p. TF-27).

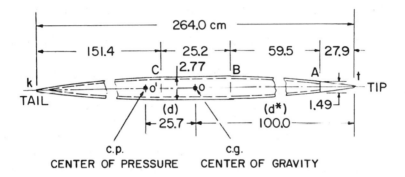

Fɪɢ. 13.12.   The geometry of a javelin that conforms to NCAA rules.

The sample javelin for the case study has a diameter at the grip of 27.7 mm (half way between the limits of 25 mm and 30 mm) and is 264 cm long (between 260 cm and 270 cm). The distance from the tip to the centre of gravity is 100 cm (between 90 cm and 110 cm) and the length of the metal head is 27.9 cm (between 25 cm and 33 cm). Its weight is 808 gm (8 gm more than the minimum requirement of 800 gm). No rules govern the centre of pressure or other aerodynamic characteristics.

The contour of the javelin's profile is such that the grip, from B to C, should be straight. The grip has a radius of 1.38 cm. The rest of the body can be either straight or smoothly curved. Assume that the profile from B to A is described by $y = 1.38 \{1 - (x^2/87.4^2)\}$ with the y axis passing through B and perpendicular to the javelin axis. Similarly, the profile from C to K is given by $y = 1.38 \{1 - (x^2/151.4^2)\}$ with the y axis passing through C.

Such a javelin has a 'wetted' surface area (S_f of eqn (13.42)) of 1601.6 cm² and a lateral projected area (S_p of eqn (13.43)) of 473.8 cm^2. The centre of the pressure drag, which in approximate aerodynamic theory for such a slender circular cylindrical body coincides with the centre of its lateral projected area, is calculated as 25.7 cm behind the centre of gravity. The moment of inertia about the centre of gravity (I_0 of eqns (13.56) and (13.57)) is calculated as 4425 gm cm s^2.

The skin friction drag coefficient (C_{df} of eqn (13.42)) is assumed to be 0.003 which is the value for a slender streamlined rotationally symmetric cylinder in turbulent flow at a Reynolds number of 5×10^6 (ref. 9, pp. 6–16). The lateral pressure drag coefficient (C_{dp} of eqn (13.43)) perpendicular to the axis is taken as 1.2 (ref. 9, pp. 3–9).

Basic data for the case study. Authorities differ in their opinions on the best throw angle of the javelin. Dyson (English) thought 35°–40° to be optimum; Kusnyetsov (Russian) proposed 28°–30° with $\theta_j = \theta_f$ (ref. 1, p. 183). We assume here that the 1972 Olympic record was reached by using the NCAA javelin described in Fig. 13.12 with both the throw angle θ_f and the initial javelin angle θ_j set at the conventional angles of 35°.

Then by the present method, assuming no initial moment, i.e. $\dot\theta_j = 0$, and with a unit time interval of T = 0.01 s, we find that an initial javelin speed of 30.45 m s^{-1} is required to reach the Olympic distance of 90.47 m. The flying time T$_t$ is 3.74 s.

That the time interval unit $T = 0.01$ s is sufficiently small for our practical purpose can be seen from Table 13.6. Consequently, an initial javelin speed of 30.45 m s^{-1} and a time interval unit of $T = 0.01$ s will be used for all calculations in the case study. With this initial speed and with both angles at 35° the NCAA javelin will reach a maximum height of 18.12 m (tip position), land at an angle of $-42.29°$, and cover the distance of 90.47 m in 3.74 s.

TABLE 13.6

Effect of time interval T on flying time T$_t$ and distance R

T (s)	T_t (s)	R (m)	Ratio $R(T)/R$ $(T = 0.003)$
0.01	3.73858	90.648	1.002099
0.05	3.73822	90.559	1.001109
0.01	3.73750	90.474	1.000172
0.005	3.73715	90.463	1.000047
0.003	3.73694	90.459	

Some important numerical results

Effect of the throw angle and javelin initial inclination angle. The throw angle (force direction) is more influential on the distance than the initial javelin angle. By maintaining the initial javelin angle at 35° and varying θ_f the optimum throw angle is 42° which produces a throw of 93.78 m with flying time 4.31 s. This is a gain of 3.31 m over the 1972 distance.

However, it is not imperative that the throw angle be different from the initial javelin angle to have a good throw. Figure 13.13 shows a study where both angles are kept the same and varied together. The distances, landing angles, and flying times are shown in curves $R(1)$, $\psi_e(1)$ and $T_t(1)$ respectively. The maximum distance occurs at $\theta_f = \theta_j = 43°$. The distance obtained is only 20 cm less than that obtained with the optimum pair of $\theta_j = 35°$ and $\theta_f = 42°$. This optimum of $\theta_j = \theta_f = 43°$ is reasonably close to that suggested by Dyson, i.e. 35° to 40°. The other curve in Fig. 13.13 will be discussed later.

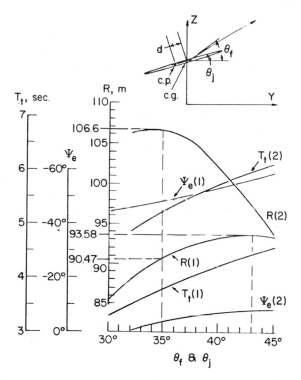

Fig. 13.13. Range R, landing angle ψ_e, and flying time T_t versus throw angle θ_f and initial javelin axis angle θ_j with $\theta_f = \theta_j$ for two cases: (1) $d = 25.7$ cm (2) $d = 0.8$ cm. c.p., centre of pressure, c.g., centre of gravity.

Effect of the distance from centre of gravity to centre of pressure (the moment arm). An important finding is the tremendous effect caused by moving the centre of pressure closer to but behind the centre of gravity position. The centre of gravity in Fig. 13.14 is fixed and the centre of pressure is brought closer to the centre of gravity. Figure 13.14 shows that the flying time will invariably be increased, the landing angle decreased, and the range increased.

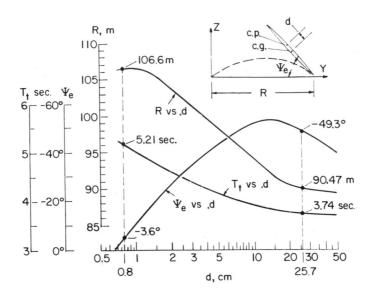

FIG. 13.14. Range R, landing angle ψ_e, and flying time T_t versus the distance d from the centre of pressure (c.p.) to the centre of gravity (c.g.) with the throw angle and the javelin axis initial angle both 35°.

Other things being equal, changing d (the moment arm) from 25.7 cm to 0.8 cm would have increased the distance of the record Olympic throw from 90.47 m to 106.60 m. The landing angle would have been very flat, i.e. 3.64°.

In Fig. 13.15 the actual trajectory of the tip of the javelin and its orientation in space are plotted against the distance travelled for the two different cases. The dots indicate the tip position and the tail behind the dot shows the inclination of the javelin axis in flight. The spacing between consecutive positions is 0.2 s. Figure 13.15 shows that the javelin with larger d flies faster but lands sooner than the one with smaller d. A shorter moment arm seems to enable the javelin to maintain a larger angle of attack during most of the flight by making the javelin harder to be pitched forward. Consequently, the javelin with the smaller d is lifted higher (19.74 m versus 18.12 m), is airborne longer (5.21 s versus 3.74 s), and has a greater range (106.60 m versus 90.47 m). However, it lands dangerously flat to the ground ($-3.64°$ versus $-49.29°$).

That one could throw a javelin with a shorter moment arm d farther than the same javelin with a longer moment arm regardless of the throw angle is clearly shown in Fig. 13.13. There, R (2) is the range curve for $d = 0.8$ cm and R (1) the corresponding curve for $d = 25.7$ cm. Since range increases as the landing angle decreases, it challenges the thrower who is using the javelin with small moment arm to obtain the maximum range without the tail touching the ground.

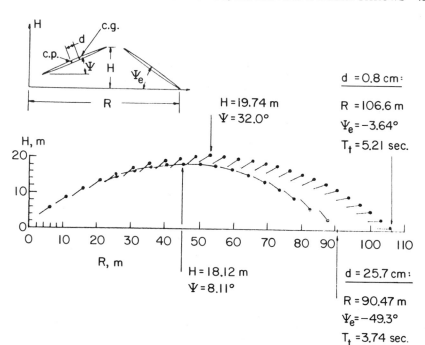

$d = 0.8$ cm:

$R = 106.6$ m

$\Psi_e = -3.64°$

$T_f = 5.21$ sec.

$H = 19.74$ m
$\Psi = 32.0°$

$H = 18.12$ m
$\Psi = 8.11°$

$d = 25.7$ cm:

$R = 90.47$ m

$\Psi_e = -49.3°$

$T_f = 3.74$ sec.

F$_{IG}$. 13.15.   Scale drawing of the trajectory of the javelin tip (dots) and axial orientation (tail) during flight for two cases: moment arm $d = 0.8$ cm and 25.7 cm.

Effect of wind. The wind effect on the range is insignificant for javelins with a large distance between the centre of pressure and the centre of gravity such as the base case of $d = 25.7$ cm. A tail wind will increase the range slightly; a head wind will have the opposite effect. It is in the order of several inches for wind speeds up to $2.7\,\mathrm{m\,s}^{-1}$.

However, the effect of wind on a javelin with $d = 0.8$ cm is opposite that with $d = 25.7$ cm. It will land tail first for a tail wind greater than $0.7\,\mathrm{m\,s}^{-1}$. At $0.7\,\mathrm{m\,s}^{-1}$ the landing angle is almost flat, i.e. $\psi_e = -0.05°$, and the distance is 0.88 m less than the no-wind distance of 106.60 m with $\theta_f = \theta_j = 35°$.

The maximum increase in range due to wind effect occurs at $V_0 = 3.1\,\mathrm{m\,s}^{-1}$. The increase is 2.03 m. The increase of maximum height and airborne time (22.69 m versus 19.74 m and 5.60 s versus 5.21 s) compared with the no-wind case are reasons for the increase of distance for a head wind of $3.1\,\mathrm{m\,s}^{-1}$. For head winds greater than $7\,\mathrm{m\,s}^{-1}$ the distance will slowly decrease.

Relationship between the aerodynamic centre and the angle of attack. In the previous analysis the position of the aerodynamic centre, i.e. the centre of pressure, has been assumed to be independent of the angle of attack α. In fact it moves as the angle of attack changes. However, we shall show that for such a slender body as the javelin the movement is small. Consequently, for practical purposes the centre of pressure may be taken as the centre of the area of the profile projected on the plane, as indicated by elementary aerodynamic theory.

Analytical procedures based on potential theory for computing the aerodynamic characteristics of bodies and missile-type configurations are limited in usefulness to very low angles of attack. Allen [10] proposed a method for predicting the forces and moments for bodies of revolution inclined to angles of attack considerably higher than those for which theories based on potential-flow concepts are known to apply. In this method a cross-flow lift force attributed to flow separation is added to the lift predicted by slender-body potential theory. Jorgensen [11] used the method to cover attack angles ranging from $0°$ to $180°$ and substantiated the analytic results with some experimental data.

Following Jorgensen's procedure, which applies to converging or diverging slender cylindrical bodies, one finally arrives at the following expressions for the pitching moment about the centre of gravity and the force normal to the axis of a javelin-type body inclined to the flow at an angle of attack α:

$$M_{cg} = q_\infty \left[\{ V_1 + V_2 + A_b (l_1 - l_2) \} \sin 2\alpha \cos \tfrac{1}{2}\alpha \right.$$
$$\left. + \{ \eta_1 A_{p1} (l_1 - x_{c1}) + \eta_2 A_{p2} (x_{c2} - l_2) \} \, sin^2\alpha \right] \qquad (13.62)$$

$$F_n = q_\infty \{ 2A_b \sin 2\alpha \cos \tfrac{1}{2}\alpha + C_{dn}(\eta_1 A_{p1} + \eta_2 A_{p2}) \, sin^2\alpha \} \qquad (13.63)$$

where a positive moment will increase α. Subscript 1 denotes the fore portion of the javelin (from the tip to the centre of gravity) and subscript 2 denotes the aft portion of the javelin (from the centre of gravity to the end of the tail). The following notations pertain to eqns (13.62) and (13.63):

$$q_\infty = \tfrac{1}{2}\rho V_R^2,$$

V_1 and V_2 are body volumes for fore and aft bodies, A_b is the body base cross-sectional area at the centre of gravity, l_1 and l_2 are the lengths of fore and aft bodies, η_1 and η_2 are the cross-flow drag proportionality factors for fore and aft bodies where $\eta_1 = 0.8$ and $\eta_2 = 0.9$, A_{p1} and A_{p2} are the plan form areas including half of the grip for the fore and aft bodies, x_{c1} and x_{c2} are the distances from the tip end to the centroid of the plan form of the fore body and from the tail end to the centroid of the aft body respectively, and $C_{dn} = 1.2$ is the cross-flow drag coefficient for the cylindrical section only. Based on eqns (13.62) and (13.63) the centre of pressure is behind the centre of gravity by an amount

$$d = M_{cg}/F_n. \qquad (13.64)$$

Table 13.7 shows the moment coefficient C_m and d as a function of α for the javelin given in Fig. 13.12. Numerical results showed that the moment is negligible for attack angles less than $10°$; between $10°$ and $90°$ the shift of the centre of pressure is never more than 13 per cent. Consequently, the value of

TABLE 13.7

Effect of the angle of attack on d and C_m for the javelin shown in Fig. 13.12

α (deg)	d (cm)	$C_m = M_{cg}/q_\infty$
0	9.7	0
2	16.4	4.0
3	17.9	7.9
5	20.0	19.7
10	22.2	71.5
15	23.2	153.7
20	23.7	263.5
30	24.4	554.7
40	24.7	905
50	25.0	1280
60	25.2	1636
70	25.3	1908
80	25.4	2088
90	25.5	2148

the centre of pressure at an angle of attack of 90 ° may be used as an invariant in a first-order analysis for a parametric study of the javelin throw.

Conclusions

The effects of other minor parameters, such as the drag coefficients (C_{df}, C_{dp}), the height of the centre of gravity of the javelin at the instant of release, the distance from the tip to the centre of gravity, and the wind angle, are found to be insignificant.

Based on the results we conclude that a javelin thrower should practice with, and use in competition, a javelin whose centre of pressure drag is closely behind its centre of gravity. Such a javelin will have flying characteristics entirely different from the same javelin with a larger moment arm. The landing angle will be much smaller for the small-moment-arm javelins; consequently, tail-end landing is a real possibility when their superiority in range is explored to the limit. Expertise in handling such a javelin with a low landing angle is clearly required.

The throw angle of a javelin depends on the magnitude of its moment arm. For small-moment-arm javelins, say d less than 3 cm, the best throw angle may be approximately 35°. For javelins with larger moment arms, including the NCAA official javelin, the optimum throw angle may be 42° or 43°. The initial javelin axis angle is not as important as the throw angle. An angle of 35°, or the same angle as the throw angle, is acceptable.

Finally, one concludes that it may be technically possible to add up to 16 m to the current Olympic record of 90.47 m by merely shifting the location of the centre of the pressure drag of the javelin. This can be accomplished legally by a slight modification of the geometry of the javelin within the current NCAA rules.

13.4. Analysis of the shot put

A literature search uncovered no previous study on the dynamics of the shot put even though the subsonic aerodynamics of a sphere have been extensively researched. The negligence may be due partly to a common belief that, since the shot is a heavy metal ball which weighs 16 lb (7.257 kg) for collegiate sport and 12 lb (5.44 kg) for high school sport and relatively slow put velocity is used, the optimum angle would be $45°$ according to elementary dynamics. This is indeed true as a first-order estimate. However, in national and international competition the margin by which a shot put medal is won or lost is often within a fraction of an inch. Consequently, a more accurate knowledge of the optimum angle is desirable. For example, at the 1972 Olympics in Munich, Komar of Poland won first place in the shot put with a record 69 ft 6 in (21.184 m). Woods of the USA was second with 69 ft $5\frac{1}{2}$ in (21.171 m), and Briesenick of the German Democratic Republic came in third with 69 ft $4\frac{1}{4}$ in (21.139 m). These distances are within $1\frac{3}{4}$ in (4.5 cm) of each other. For a range centred on 21 m the difference is only 0.2 per cent of the total distance. We shall demonstrate that a few degrees difference in the shot angle, in such a close competition, could have produced a difference in range greater than that amount.

Equations of motion

For a spherical body similar to the ball used in the shot put, there is no lift L and the drag D is

$$D = \tfrac{1}{2}\rho C_D S_p V_R^2. \tag{13.65}$$

By substituting eqns (13.22) and (13.65) into eqn (13.31) the dynamic equations of motion can be written as

$$\ddot{x} + \frac{\rho g}{2W} C_D S_p (\dot{x} - V_0 \sin\theta_w)\{(\dot{x} - V_0 \sin\theta_w)^2 + (\dot{y} - V_0 \cos\theta_w)^2 + \dot{z}^2\}^{1/2} = 0 \tag{13.66}$$

$$\ddot{y} + \frac{\rho g}{2W} C_D S_p (\dot{y} - V_0 \cos\theta_w)\{(\dot{x} - V_0 \sin\theta_w)^2 + (\dot{y} - V_0 \cos\theta_w)^2 + \dot{z}^2\}^{1/2} = 0 \tag{13.67}$$

$$\ddot{z} + \frac{\rho g}{2W} C_D S_p \dot{z}\{(\dot{x} - V_0 \sin\theta_w)^2 + (\dot{y} - V_0 \cos\theta_w)^2 + \dot{z}^2\}^{1/2} + g = 0. \tag{13.68}$$

With the prescribed initial conditions eqns (13.66), (13.67), and (13.68) can be solved without difficulty by the numerical method described earlier.

Numerical results

Geometry of the shot. NCAA rules (ref. 8, p.TF-26) specify a minimum weight of 16 lb (7.257 kg). This corresponds to an iron shot of 124 mm diameter which is within NCAA dimensional limits (minimum, 110 mm; maximum, 130 mm).

Aerodynamic constant C_D. Characteristics of spheres in a subsonic flow field have been extensively studied [12]. The sphere has an almost constant drag coefficient C_D of 0.47 obtained experimentally with Reynolds numbers between 10^4 and 5×10^5 (ref. 9, pp. 3–8). At a velocity of $45.547 \, \mathrm{ft \, s^{-1}}$ ($13.883 \, \mathrm{m \, s^{-1}}$), which will be shown later as the necessary put speed at 45° to reach the 1972 Olympics shot put record of 69.50 ft (21.184 m), the Reynolds number is 1.2×10^5. Consequently, the aerodynamic resistance of the shot can be fairly accurately determined from the experimental drag coefficient C_D of 0.47 which is independent of the speed within the range in question.

1972 Olympics shot-put record. A study was made by assuming that the 1972 Olympics record of 69 ft 6 in (21.184 m) was made with a putting angle of 45° and that the shot was released at a height of 1.829 m which is about the height of a tall man's palm whose hand is putting the shot along the 45° angle. With no wind the analysis showed that it requires an initial velocity of $13.8827 \, \mathrm{m \, s^{-1}}$ for a time step T of 0.005 s.

Table 13.8 shows the results obtained when T is varied. The convergence of the solution is extremely rapid owing to the fact that the trajectory is a smooth curve. At the T chosen for the study, i.e. 0.005 s, the results should be dependable.

TABLE 13.8
Effect of time step T on flying time T_t and distance R^a

T(s)	T_t(s)	R(m)	$R(T)/R$ ($t = 0.001$)
0.02	2.1682	21.1845	1.00005665
0.01	2.1681	21.1840	1.00003305
0.005	2.1680	21.1836	1.00001416
0.002	2.1680	21.1834	1.00000472
0.001	2.1675	21.1833	

^{a} The range is measured when any part of the shot touches the ground.

Effect of the put angle θ_f. The effect of changing the put angle is shown in Fig. 13.16. An angle of 45° is not the optimum to achieve the maximum range when the initial velocity is the same. The best angle is $42\frac{1}{2}°$. The range increase over the 45° put is 2.68 in (6.8 cm). This increment is more than would have been required (4.5 cm) to move the third place winner to the first place in the 1972 Olympics.

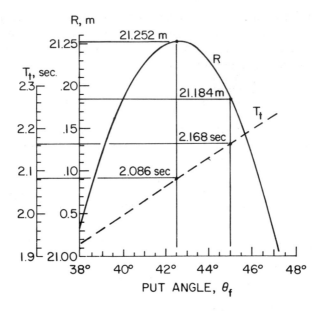

FIG. 13.16.   Effect of the put angle on the range R and the total flight time T_t: $\overline{V}_0 = 13.8827 \, \mathrm{m \, s^{-1}}$; wind speed $= V_0 = 0$.

In fact Fig. 13.16 indicates that it is better to aim at a few degrees below the $45°$ angle if a perfect aim at $42\frac{1}{2}°$ is difficult because any angle between $40°$ and $45°$ is better or at least equal to the exact $45°$ angle.

Trajectory. The trajectories of the $45°$ put and the $42\frac{1}{2}°$ put are shown in Fig. 13.17 which shows that the $45°$ put climbs higher and is airborne longer (see Fig. 13.16) but lands behind the $42\frac{1}{2}°$ put.

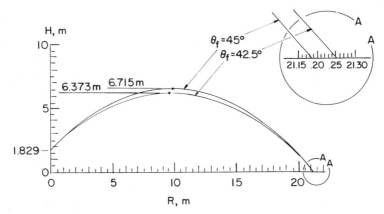

FIG. 13.17.   Trajectory of the shot for two put angles; the traditional $45°$ and the optimum $42.5°$; $R = 21.252$ m for $\theta_f = 42.5°$; $R = 21.184$ m for $\theta_f = 45°$.

Effect of the releasing height of the shot at time zero. In order to find how the aerodynamic resistance affects the optimum shot angle and the range one can assume that the air resistance is zero. Then elementary rigid-body dynamics theory yields the exact solution for the range

$$R = \frac{\bar{V}_0^2 \sin^2 \theta_f}{2g} \left\{ 1 + \left(1 + \frac{2gz_0}{\bar{V}_0^2 \sin^2 \theta_f} \right)^{1/2} \right\} \tag{13.69}$$

where z_0 is the height of the centre of gravity of the shot at release.

The optimum put angle for the maximum range is determined by the following equation:

$$2 \cot 2\theta_f \left\{ 1 + \left(\frac{2gz_0}{\bar{V}_0^2 \sin^2 \theta_f} \right)^{1/2} \right\} \left(1 + \frac{2gz_0}{\bar{V}_0^2 \sin^2 \theta_f} \right)^{1/2} - \frac{2gz_0}{\bar{V}_0^2 \sin^2 \theta_f} \cot \theta_f = 0 \tag{13.70}$$

which can be simplified to

$$(\theta_f)_{\text{optimum}} = \cot^{-1} \left(1 + \frac{z_0}{\bar{V}_0^2/2g} \right)^{1/2}. \tag{13.71}$$

Substitution of eqn (13.71) into eqn (13.69) yields

$$R_{\text{optimum}} = \frac{\bar{V}_0^2}{g} \left(1 + \frac{z_0}{\bar{V}_0^2/2g} \right)^{1/2}. \tag{13.72}$$

Based on eqns (13.71) and (13.72) the optimum angle in a vacuum is 42.56° and the corresponding range is 21.398 m for $z_0 = 1.819$ m and $\bar{V}_0 = 13.8827$ m s^{-1}. Compared with the results which include the acrodynamic resistance, i.e. 42.5° and 21.252 m, it is clear that the effect of air resistance is negligible and the optimum put angle is determined by the height of the release point above the ground. If the elevation is zero the optimum angle is 45°.

Wind effect. Figure 13.18 shows the wind effect. As expected, a tail wind (positive in Fig. 13.18) increases the range and a head wind decreases the range. However, since a tail wind also suppresses the shot during its climbing part of the trajectory, there is an optimum value of wind speed which is a tail wind at 6 m s^{-1}.

Conclusions

The effect of the aerodynamic resistance to the shot put is negligible; the optimum put angle is determined by the height at which the shot is released. A tall putter should aim at a put angle of 42½°. Any angle between 40° and 45° is superior to an angle of 45°. Within a reasonable wind range a tail wind is helpful and a head wind is disadvantageous. The best wind is a tail wind at 6 m s^{-1}. A cross wind has no discernible effect on the put distance.

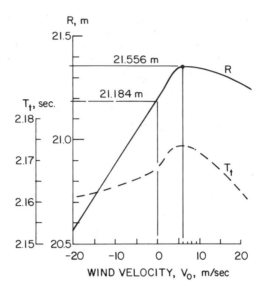

FIG. 13.18.   Effect of the wind velocity on the range and the total flight time T_t. The tail wind is positive and the head wind is negative. $\theta_w = 0$; $\theta_f = 45°$.

13.5. Remarks on throw techniques

The theoretical analyses concerning the external trajectory of the shot put, the javelin, and the discus throw have been derived and various geometrical and environmental parameters which affect the throw distance have been identified and discussed. Conclusions are based on fixed initial speed. The natural questions that follow are how to achieve the maximum initial speed and how to analyse in a scientific manner.

There is no authoritative answer to the first question. Opinions vary from coach to coach and from time to time. The governing athletic rules are a crucial constraint and dictate the relevant throw technique for maximum velocity. One current rule, for example, is that the discus thrower has to confine all his activities within a circle of diameter 2.5 m. The painstaking build-up of momentum within that small circle was described by Fox (Stanford, 1939 world discus-throw record holder) in 1941 before the National College Track Coaches Association:

> From this initial position, you lead with your back and gain momentum with an off-balance drive. . . The more centrifugal force you build up, the more you travel in an arc, rather than the old straight-line drive across the diameter of the circle. Instead of eight feet, you have about 10 feet to travel, measuring the line that the feet follow across the circle.
> Instead of moving from the right foot straight across, you move almost backwards, letting your back lead the action. In this manner you shift the drive off your right leg

onto the left, and around to the right so that you land in the throwing position in the proper place, with the left foot about one foot behind the diameter line, and the right toe on the straight diameter line. To do this you have to cut down somewhat on the drive. That is the important thing, landing in a good throwing position. . .

Fox's remarks indicated the complexity of the model for an analysis. However, with reasonable simplifications and assumptions it is not impossible to optimize the throw speed through constitutive relationships governed by applied mechanics and consistent with the constraints from biomechanical considerations.

Acknowledgments

The author acknowledges the Journal of Applied Mechanics of the American Society of Mechanical Engineers for permission to use materials published in the Journal, mainly from the following papers:
Soong, T. C. (1975). The dynamics of javelin throw. *J. appl. Mech.* **42**, 257–62.
Soong, T. C. (1976). The dynamics of discus throw. *J. appl. Mech.* **98**, 531–6.
He also acknowledges the Xerox Corporation for the use of the company's computation facilities. This chapter was written in the author's spare time and it is not related to the business of the Xerox Corporation.

Nomenclature

a, b, c	semi-axes of the ellipsoid
$a_x, a_y, a_z, a_\phi, a_\theta, a_\psi$	accelerations
d	distance from the centre of pressure to the centre of gravity of the javelin
$d*$	distance between the tip of the javelin and its centre of gravity
C_f	force coefficient parallel to the discus plane
C_m	moment coefficient
C_n	force coefficient parallel to the normal axis
D	drag force
g	gravitational acceleration
I_ξ, I_η, I_ζ	moment of inertia about the principal axes
I_0	$I_0 = I_\xi = I_\zeta$, moment of inertia of the principal axes which are perpendicular to the rotationally symmetric axis
L_x, L_y, L_z	angles between the lift force L and the moving co-ordinates
M_ξ, M_η, M_ζ	moments about the three principal axes ($\xi, \eta,$ and ζ)
n	normal vector used in the ellipsoidal co-ordinates
P_x, P_y, P_z	force components along the positive directions of the moving co-ordinates

R_x, R_y, R_z	angles between the resultant air velocity and the moving co-ordinates
T_t	total time interval from the release to the landing
T	unit time step between consecutive points in time
t	time
U, V, W	velocity components of the discus
$v_x, v_y, v_z, v_\phi, v_\theta, v_\psi$	known velocity quantities at a previous time point
V_0	wind speed
$\overline{V}_0$	initial throw speed
$v_{\phi 0}, v_{\theta 0}, v_{\psi 0}$	initial values taken for v_ϕ, v_θ, and v_ψ
x, y, z	centre of gravity position, or moving co-ordinates
z_0	height of release at time zero
$x_1, y_1, z_1, \phi_1, \theta_1, \psi_1$	known displacement quantities at a previous time point t_1
θ_f, θ_j	throw angle and initial inclination angle respectively
θ_w	direction angle of wind
α	angle of attack, angle between V_R and the η axis
ρ	air mass density
$\omega_x, \omega_y, \omega_z$	rotational velocity components about the moving axes
ϕ, θ, ψ	Euler angles
ξ, η, ζ	principal axes
$\omega_{\eta 0}$	initial spinning speed of the discus about its η-axis
ξ_x, ξ_y, ξ_z	angles between the ξ axis and the moving axes
η_x, η_y, η_z	angles between the η axis and the moving axes
$\zeta_x, \zeta_y, \zeta_z$	angles between the ζ axis and the moving axes
R_ξ, R_η, R_ζ	angles between the resultant air velocity and the principal axes
$\omega_\xi, \omega_\eta, \omega_\zeta$	rotational velocity components about the principal axes
λ, μ, ν	ellipsoidal coordinates with $\lambda = 0$ as the surface of the basic ellipsoid

A dot above a variable indicates the differential with respect to time.

References

1 DOHERTY, K. (1971). *Track and field omnibook.* TAFMOP Publishers, Swarthmore, Pa.

2 WHITTAKER, E. T. (1936). *A treatise of the analytical dynamics of rigid bodies.* Clarendon Press, Oxford.

3 TIMISHENKO, S., and YOUNG D. H. (1948). *Advanced dynamics,* p. 333. McGraw-Hill, New York.

4 ATSUMI, S. (1937). Pressure distribution on a wing with circular plan form. *J. aeronaut. Sci.,* 499–501.

5 FAGE, A., and JOHANSON, F. C. (1927). On the flow of air behind an inclined flat plate of infinite span, *Proc. R. Soc. A.* **116**, 170–97.
6 STILLEY, G. D. (1972). Aerodynamic analysis of the self sustained flare. *Rep. AD*-740 117, Naval Ammunition Depot, Crane, Indiana.
7 MILNE-THOMSON, L. M. (1960). *Theoretical hydrodynamics*, p. 511. Macmillan, New York.
8 *Official NCAA track and field rules* (1972). National Collegiate Athletic Association Publishing Service, Shawnee Mission, Kansas.
9 HOERNER, S. F. (1965). *Fluid dynamics drag*, S. F. Hoerner, c/o Membership Roster, American Institute of Aeronautics (AIAA), New York.
10 ALLEN, H. J. (1949). Estimation of the forces and moments acting on inclined bodies of revolution of high fineness ratio. *NACA RM A*9126. US Government Printing Office, Washington, DC.
11 JORGENSEN, L. H. (1973). Prediction of static aerodynamic characteristics for space-shuttle-like and other bodies at angles of attack from 0° to 180°. *NASA Tech. Note TN D*-6996. US Government Printing Office, Washington, DC.
12 *NACA TR* 185, 253, 342, 558, 581. US Government Printing Office, Washington, DC.

14. Mechanics application to sports equipment: protective helmets, hockey sticks, and jogging shoes

RENE G. THERRIEN and PAUL A. BOURASSA

Human body dynamics are often influenced by the equipment used or worn by the athlete as well as by the physical environment. This chapter deals with sports applications in which safety or performance are largely dependent upon the physical characteristics of the equipment or of the environment where the sport is being practised. The findings reported in this chapter are mainly from research programmes in ice hockey equipment and jogging footwear. However, they can be applied, as well, to other sports activities such as those (a) presenting a high risk of head impact (as in lacrosse, boxing, football, skateboarding, etc.), (b) necessitating the use of an impacting implement (as in field hockey, baseball, tennis, badminton, paddle ball, racketball, squash, etc.), and (c) including jogging-like periodic loading of the lower limbs (as in walking and running).

14.1. Protective headgear design

Protective helmets should receive more attention than any other piece of equipment in sports where there is a high risk of head impact. The task of designing adequate and effective headgear for a specific sport includes (a) a definition of the impact environment in quantitative physical terms, (b) a determination of human brain tolerance with respect to the impact environment, (c) an analysis of the physical processes taking place at impact, (d) the development of a device or a system to be placed between the player's head and the impact environment in order to bring it to tolerable level, and (e) an evaluation of the effectiveness of that device or system [1, 2]. One such intensive research programme has been undertaken at the University of Sherbrooke and applied to ice hockey helmets; the first part of the present chapter is partly based on the major results obtained to date [2, 3].

Definition of the head-board impact dynamic model

The first step in the study of the physical sport environment should include the development of a dynamic model of one or many of the most threatening impact situations likely to occur in the specific sport. In ice hockey, most severe head injuries are the result of head impact with the boards, the ice surface, or the goal posts [4, 5]. Such head impacts with the boards, for instance, are likely to occur when a player skating close to the boards is projected head first on this unyielding surface, as the result of a bodycheck from an opponent, after being tripped, or after a loss of balance.

It is highly improbable that the resulting impact will be directed exactly at the centre of mass of the head, and both translational and rotational motions of the head are likely to be evidenced at impact. Owing to the dynamic properties of the mass moment of inertia of the head with respect to the point of impact, angular acceleration of the head around the impact centre is going to take place at the same time as the centre of mass of the head is experiencing a translational deceleration. The angular acceleration produces a rotation of the skull relative to the brain, while the translational acceleration creates pressure gradients.

The former is responsible for the development of considerable tensile forces between the accelerated skull and the inert brain. Such physical processes will cause, in turn, shear strains at the brain tissue level, as clinically evidenced by the presence of tears and avulsions of veins and capillaries in superficial cortical layers and severe primary traumatic haemorrhages in the cortex as well as in the white substance [6, 7, 8]. The latter are produced by the compression of the brain particles at the impact site and the expansion of the liquid particles at the site opposite to the impact pole (the counterpole). These physical phenomena are clinically associated with intracerebral haemorrhages and traumatic necroses at the impact pole as well as with cavitation trauma and cortical contusions at the counterpole [6, 7, 8]. Both of these mechanical behaviours (rotation and translation) have been associated with closed brain injuries either jointly [7, 9, 10, 11], or separately [12, 13]. It becomes imperative, then, that the modelled situation includes both translational and rotational parameters.

Many approaches can be used in developing a dynamic model of head-board impact. Even though the ultimate realistic approach would involve the actual projection of an instrumented anthropomorphic manikin against ice rink boards, a much simpler model can be developed since the head is totally mass-controlled in impacts of such short duration. In head impacts lasting less than 20 ms, the neck structures have very little influence on the magnitude of both the linear and the angular accelerations of the head upon impact [14, 15, 16].

Such a simplified model can be developed through the use of a headform pendulum hitting a hockey board in an oblique trajectory, as illustrated in Fig. 14.1. The headform mechanical reactions upon impact should have been validated against human head behaviour. Such procedures were used for the headform illustrated in Fig. 14.2 [17, 18]. An additional way of simplifying the

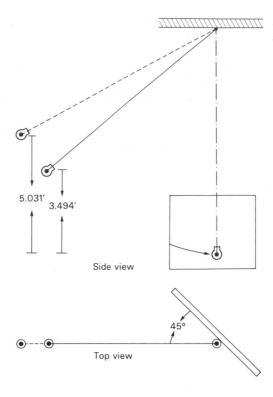

5.031'

3.494'

Side view

45°

Top view

FIG. 14.1. Heights of release and trajectory of headform pendulum.

FIG. 14.2. Side view of anthropomorphic headform and helmets.

model without altering its validity would be to statically and dynamically balance the headform and instrument it in such a way that the centre of mass of the instrumented headform, the location of the contact point with the board, and the sensitive axes of the dynamic instrumentation (like accelerometers), all lie in a same transverse plane perpendicular to the pendulum axis (Figs. 14.2, 14.3, and 14.4). Such positioning would allow both two-dimensional high-speed filming of the impact phenomena (Fig. 14.4) and the use of only four accelerometers in order to obtain both translational and rotational parameters of the impact through the use of principles from two-dimensional solid body mechanics.

The development of the equations of motion is then performed as shown in Fig. 14.5.

In such a situation, the general equation is:

$$\bar{a} = \bar{a}_{cg} + \bar{\alpha} \times \bar{\rho} + \bar{\omega}(\bar{\omega} \times \bar{\rho}) \qquad (14.1)$$

where

a: total linear acceleration

a_{cg}: linear acceleration of the centre of gravity

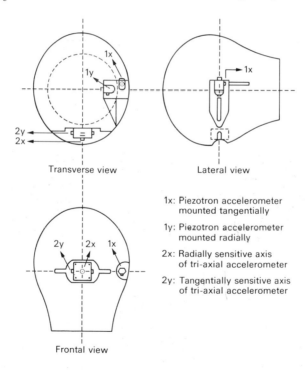

Transverse view

Lateral view

1x: Piezotron accelerometer mounted tangentially

1y: Piezotron accelerometer mounted radially

2x: Radially sensitive axis of tri-axial accelerometer

2y: Tangentially sensitive axis of tri-axial accelerometer

Frontal view

FIG.14.3.   Transverse, lateral, and frontal views of headform with accelerometers in position.

FIG. 14.4. Top view of helmet–headform assembly in board contact position.

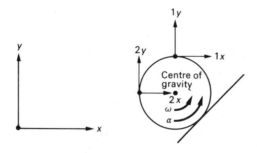

FIG. 14.5.

α: angular acceleration

ω: angular velocity

ρ: distance between the centre of gravity and a specific accelerometer

subscript 1 refers to position 1 of accelerometers

subscript 2 refers to position 2 of accelerometers

subscript x refers to x direction of measurement

subscript y refers to y direction of measurement

from the original eqn (14.1) two vector equations can be formed for each of the positions 1 and 2:

$$a_{1x} = a_{cgx} - \alpha \rho_{1x} \tag{14.2}$$
$$a_{1y} = a_{cgy} - \omega^2 \rho_{1y} \tag{14.3}$$
$$a_{2x} = a_{cgx} + \omega^2 \rho_{2x} \tag{14.4}$$
$$a_{2y} = a_{cgy} - \alpha \rho_{2y} \tag{14.5}$$

(14.2) and (14.4) yield:

$$a_{1x} - a_{2x} = -\alpha \rho_{1x} - \omega^2 \rho_{2x} \tag{14.6}$$

(14.3) and (14.5) yield:

$$a_{1y} - a_{2y} = -\omega^2 \rho_{1y} + \alpha \rho_{2y} \tag{14.7}$$

solving (14.6) and (14.7) for ω^2 and α:

$$\omega^2 = \frac{-\alpha \rho_{1x} - (a_{1x} - a_{2x})}{\rho_{2x}} \tag{14.8}$$

$$\alpha = \frac{-\omega^2 \rho_{2x} - (a_{1x} - a_{2x})}{\rho_{2x}} \tag{14.9}$$

from eqns (14.8) and (14.9) in (14.7):

$$\alpha = \frac{\rho_{2x}(a_{1y} - a_{2y}) - \rho_{1y}(a_{1x} - a_{2x})}{\rho_{1y}\rho_{1x} + \rho_{2y}\rho_{2x}} \tag{14.10}$$

$$\omega = \sqrt{\left[\frac{-\rho_{2y}(a_{1x} - a_{2x}) - \rho_{1x}(a_{1y} - a_{2y})}{\rho_{1y}\rho_{1x} + \rho_{2y}\rho_{2x}} \right]} \tag{14.11}$$

while a_{cg} was obtained through

$$a_{cg} = \sqrt{(a_{cgx}^2 + a_{cgy}^2)} \tag{14.12}$$

Equations (14.10), (14.11), and (14.12) are used to calculate the angular acceleration and the angular velocity of the headform, as well as the linear acceleration of the centre of gravity of the headform from the linear accelerations measured with linear accelerometers properly positioned on the headform.

Once the dynamic model of the impact situation is developed and the equations of motion are derived, then some form of data acquisition system has to be adapted to the specific experimental set-up, in order that the impact parameters will be continuously measured during the impact period. Figure 14.6 shows a general view of an experimental set-up developed by the writer. In this type of set-up, the release of the helmet-covered headform pendulum triggers, after a delay, the high-speed motion-picture camera located above the helmet–board contact position. Just before impact, the pendulum intercepts a laser beam activating a photo-electric cell, which triggers all pieces of the

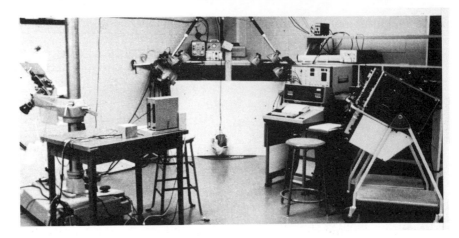

FIG. 14.6.   Overall view of experimental set-up.

analog-digital data-acquisition system and produces a synchronizing light pulse on the film record. This particular on-line system is branched to the main computer, where the data are stored and analysed through the use of appropriate computer programs.

The last step in the development of the model requires the selection of input parameters comparable to those of the actual sport situation. Non-professional adult hockey players, for instance, are capable of maximum skating speed in the range of 28 to $40\,\mathrm{ft\,s^{-1}}$ (8.5 to $12.2\,\mathrm{m\,s^{-1}}$) [19, 20]. However, outside of these short bursts of high activity, most of the skating is performed at a speed within the range of 15 to $18\,\mathrm{ft\,s^{-1}}$ (4.6 to $5.5\,\mathrm{m\,s^{-1}}$). Therefore, head–board impact velocities should cover these ranges. Such velocities can be produced in adjusting the height of release of the headform pendulum according to the desired velocity through the energy conservation equation: $mgh = \frac{1}{2}mv^2$. Figure 14.1 illustrates the heights of release producing velocities of 15 and $18\,\mathrm{ft\,s^{-1}}$ (4.57 and $5.49\,\mathrm{m\,s^{-1}}$) at impact.

Selection of human tolerance criteria

At the same time as the dynamic model of the impact situation is being developed, a selection has to be made of the criteria governing human tolerance of head impact, to which the experimental results will be compared. Even after two decades of very abundant and extensive brain injury research in the USA, the discussion is still going on among the researchers on the precise causes and mechanisms of brain injury. However, in the specific case of the oblique impact modelled, brain trauma is likely to result from two very different physical processes: skull fracture and/or pressure gradients are likely to be associated with the sudden translational deceleration taking place at impact, while rotation of the skull relative to the brain is likely to be associated with the angular acceleration of the head at impact; moreover, cerebral concussion may result from either one or both processes [7, 9, 21]. Even though severe criticisms have been formulated on the validity and reliability of the different criteria of brain damage due to translational acceleration [22, 23, 24], the Wayne State University Curve of human tolerance to translational acceleration still remains the cornerstone of the present biomechanical knowledge of human tolerance to this type of brain injury [23]. The section of the WSU Curve, which lies in the range of the experimental impact results obtained with hockey helmets, is reproduced in Fig. 14.7. Criteria of brain injury threshold in man have been derived from impact studies on sub-human primates of increasingly larger brain mass [21]. Even though Ommaya and Hirsch [21] predict that man may suffer a brain concussion in an exposure of over 1800 rad/s^2, Sano et al. [25] developed a slightly more conservative curve of human tolerance to angular acceleration. Such a curve, because of its accordance with experimental data [25], can be used as a criteria of brain

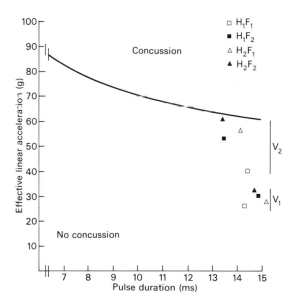

FIG. 14.7 Plot of translational accelerations on WSU tolerance curve. Identification of symbols:
    V_1: impact velocity of 15 ft s^{-1} (4.57 m s^{-1})
    V_2: impact velocity of 18 ft s^{-1} (5.49 m s^{-1})
    H_1: helmet 1, with 2-point fixed strap attachment
    H_2: helmet 2, with 6-point sliding strap attachment
    F_1: normal headform–helmet friction
    F_2: reduced headform–helmet friction

damage threshold for humans in impact situations producing an angular acceleration of the head. Figure 14.8 represents a portion of the latter curve. It should be stressed, however, that in both the translational and rotational acceleration criteria, the severity of the head impact is expressed as a function of the acceleration and the duration of impact.

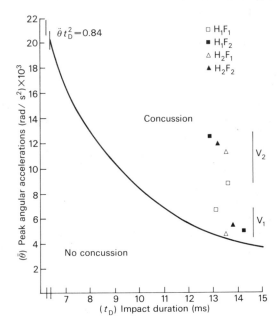

FIG. 14.8.   Plot of rotational accelerations on tolerance curve developed by Sano *et al.* [25].

Physical processes taking place at impact

Using the dynamic model developed, experimental impacts are produced and the impact phenomena analysed. Quantitative results are compared with the appropriate human tolerance criteria, while qualitative observations provide some fundamental knowledge on the sequence and nature of the processes which take place at impact. In order to derive more useful information on the effectiveness of the shock-absorption devices to be interposed between the head and the impact, the headform is covered with models of different hockey helmets for the experiment. The helmets illustrated in Fig. 14.2 have been used to obtain the results presented in this chapter. These CSA-approved helmets consist of a one-piece shell design and a shock-absorbing lining on the inside of the shell. One helmet has a strap attachment fixed at two points while the other comes with a six-point attachment of nylon straps and buckles, which provides some sliding of the straps when a certain degree of tension is developed. Other input parameters, such as impact velocity and headform–helmet friction, can be altered in order to gain more insight into the factors associated with a higher degree of protection against brain injuries.

Quantitative analysis of the average accelerations and time values from 5 impacts for each of the 8 sets of experimental conditions (2 velocities × 2 helmet models, × 2 friction levels) yields the following findings illustrated on Figs. 14.7 and 14.8:

(1) all translational accelerations measured are below the brain tolerance threshold;
(2) all rotational accelerations measured are above the brain tolerance threshold;
(3) a small increase in impact velocity (20 per cent) produces dispro- portionately high increases (close to 200 per cent overall) in both translational and rotational accelerations;
(4) helmet models have very different impact absorption characteristics which are specific to velocity and friction conditions;
(5) lower headform–helmet friction consistently produces higher accel- erations at higher velocity;
(6) experimental results were very consistent within each set of conditions;
(7) the typical shapes of the acceleration-time pulses for the lower and the higher impact velocities were as illustrated in Figs. 14.9 and 14.10 respectively;
(8) peak translational and rotational accelerations are attained 7 to 9 ms after the onset of the pulses;
(9) total duration of translational and rotational accelerations is in the 13 to 15 ms range, depending mostly upon impact velocity.

Qualitative analysis of the high speed (6000 frames s^{-1}) film records shows the following sequence of phenomena to occur during the board impact:

(1) at the onset of impact the helmet is first deformed in the area adjacent to the site of impact, producing a flattening of the outside shell;
(2) simultaneously, the helmet begins to rotate around the headform, while the latter does not yet exhibit any rotation;
(3) after some rotation (5 to 12 degrees, depending upon experimental conditions), the helmet begins to regain its original shape, while the rotation of the headform is initiated;
(4) as soon as the headform rotation takes place there is no further rotation of the helmet relative to the headform.

This particular behaviour of the helmet rotating around the headform in the first instants of impact suggests that the physical phenomena taking place at the helmet–headform interface may have a close similarity to the motion of

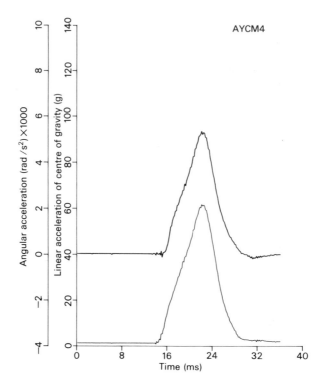

FIG. 14.9.   Typical shape of acceleration–time curves for $15\,\mathrm{ft\,s^{-1}}$ ($4.57\,\mathrm{m\,s^{-1}}$) impacts.

the skull relative to the brain, as reported by brain injury researchers to occur in eccentric head impact [7, 10].

Applications to protective headgear in sports

The studies reviewed, the principles of theoretical mechanics, as well as both the quantitative and qualitative information derived from the research programme on ice hockey helmets briefly presented in this chapter, will allow some useful applications to the design as well as the selection and use of sport protective headgear.

The results presented clearly demonstrate that present-day hockey helmets give a false illusion of protection against severe brain injuries in oblique impact situations: all angular accelerations were above the tolerance threshold to this type of acceleration, while, when using only the translational parameters of the impact, the human tolerance threshold was not attained.

Another indication from the research programme is that, at lower velocity (V_1), a lower friction between the head and the helmet seems desirable, such as the one produced by a looser attachment or a sliding strap. But at a higher

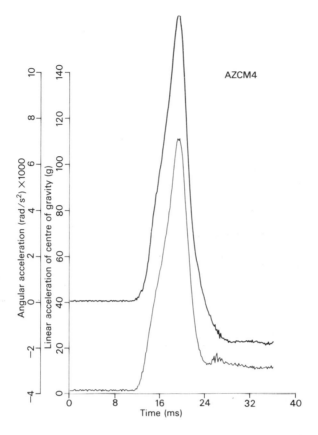

FIG. 14.10. Typical shape of acceleration–time curves for 18 ft s^{-1} (5.49 m s^{-1}) impacts.

velocity (V_2), it has an opposite effect. This could have a direct application for children whose skating speed is well below 15 ft s^{-1} (4.57 m s^{-1}). In the latter case, a looser helmet (without any exaggeration) would seem to decrease somewhat the angular components of oblique shocks.

Such safety features which seem highly desirable in the design of the ideal hockey helmet are included in Table 14.1, together with translational and rotational criteria which are suggested as minimal protective indices for headgears that would provide a serious approach to the prevention of internal head injuries in ice hockey.

Basic mechanical knowledge shows that, if the angular accelerations experienced at impact are dependent upon the moment of inertia of the head–helmet system, then more protection against brain injury is associated with helmets exhibiting lower mass and smaller outside diameter, without losing

TABLE 14.1

Desirable safety features in hockey helmets

Minimal safety criteria	Safety features associated with criteria	Special applications to children (with skating speed lower than $15\,\mathrm{ft\,s^{-1}}$)
Effective translational deceleration of the head upon impact:	More efficient padding or combination of suspension and padding between the head and the helmet shell	
Lower than 62 *g*	Additional energy-distributing mechanism between shell and padding (or suspension) More rigid (undeformable) outside shell	
Peak rotational acceleration of the head upon impact:	More spherically shaped outside shell Lower-friction outside shell	
Lower than $4000\,\mathrm{rad\,s^{-2}}$	Reduced weight for helmet Smaller outside diameter Energy-absorbing neck and/or chin attachment Head and neck stabilization mechanism (not necessarily linked to·the helmet)	Looser attachment of neck or chin straps

their ability to absorb the translational components of the impact. This would be even more important for children, who very often are seen wearing very heavy helmets with large outside diameters (due to the extra padding needed to fit the helmets on to their small heads).

However, it is highly improbable that helmet design alone, even if it included some device producing a damping of the rotational motion, could bring the rotational components of high velocity impacts well below the tolerance threshold for humans.

Bishop [26] had already adopted a similar position, and it may well be possible that his proposition of improving all safety aspects of the whole hockey environment is the most effective way of decreasing the rate of brain injuries. This means that the total environment, including boards, glass plates, goals, and especially rules, would have to be changed with one preoccupation in mind: the safety of the player.

Why not make changes in several ice rinks and in a few rules, rather than forcing millions of hockey players to look and often to feel like gladiators!

However, until these changes are performed, most of the protection against brain injury in ice hockey, as well as in other sports which involve a high risk of head impact, will come from improvements in the design and selection of better helmets.

14.2. Static and dynamic testing of hockey sticks

The hockey stick, like most sport equipment, has evolved to its present known state having gone through various stages of trial and error. Various materials, shapes, and blade curvatures, are applied examples of new design concepts that have been introduced. Manufacturing companies conduct regular tests to improve strength, quality control, and performance. Yet very little is known about the kinematic and dynamic behaviour of the hockey stick in the various situations in which it has to perform. This is precisely where experimental methods such as high-speed photography can be most helpful.

A schematic set-up used to capture the blade motion of a hockey stick under impact is shown in Fig. 14.11. Here, an air-gun, consisting of a pressurized air chamber, a solenoid valve, and a rectangular barrel can propel the puck at an exit velocity which is a function of the air pressure in the chamber. The air is supplied from nitrogen bottle and the available pressure is controlled by a pressure regulator. In order that the blade motion be fully recorded with good definition, the film must already have acquired a constant high frame rate velocity at the moment of impact. Therefore the camera is first set in motion and after a certain time, often a fraction of a second, the solenoid valve must be opened to trigger the event. In this particular case, this control is achieved with the help of a microswitch mounted inside the camera case with a lever arm resting against the film on the unloading spool. Therefore when the film

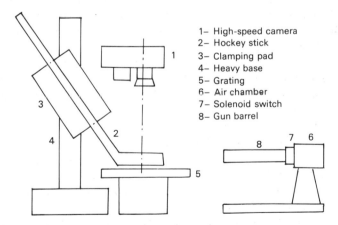

1– High-speed camera
2– Hockey stick
3– Clamping pad
4– Heavy base
5– Grating
6– Air chamber
7– Solenoid switch
8– Gun barrel

FIG. 14.11. Lab set-up for impact and high-speed photography of hockey stick performance.

has unrolled sufficiently and attained the required speed, the microswitch closes the solenoid valve circuit and the air-gun is fired.

In this experiment, the stick is firmly clamped along a central part of the handle where it is likely to be held by a hockey player. The blade motion upon impact is photographed against a grating, so that the kinematics of the motion may be quantitatively evaluated. This kind of dynamic test will of course simulate a somewhat different situation from one in which a hockey player is shooting. A hockey player, when shooting a puck, moves his own body and stick with translational and rotational velocity and acceleration. He will be varying the grip pressure in its motion as recorded on other types of equipment like golf clubs for example. These variables affect the transmitted and reflected waves motions in the stick during the contact time and ultimately control the puck motion. Meanwhile, variations in variables and parameters from one player to another and between different shots, make a laboratory type of testing more reliable, especially when an assessment of equipment performance and not of technique is required. In the set-up shown here, the puck hits an initially stationary stick and wave motion is transmitted to an inert mass through a very rigid grip. Transmitted and reflected waves proceed with an intensity ratio dependent on the inertia and elasticity of the experimental rig, but of the same nature as those encountered in the real case. Other set-ups are of course possible, all more or less elaborate. For example, a rotational impact mechanism could be employed, carrying in its motion a stick, and hitting the puck at a predetermined location, where the focus of the camera would be set.

Selected frames of a high-speed film showing puck–blade interaction are shown in Fig. 14.12 [27]. In this instance, an ice hockey puck is shot at a velocity of 70–75 mile h^{-1} (120 km h^{-1}) and the motion is photographed at the rate of 2000 frames s^{-1} at the moment of impact. A Kodak 4X reversal film 7277 under 5000 watts of light intensity was used in a Hycam high-speed camera. The number appearing below each picture in Fig. 14.12 represents frame number. Lines on the grating are spaced every inch (25.4 mm), and make possible the assessment of the incoming puck velocity in the frames preceding the impact (frames 2 to 6). The frame sequence 7 to 13 shows the first contact phase. The blade curvature initially increases, goes through a maximum, and then relaxes when coming through frame 13 where the contact between the disc and the blade is temporarily lost. A much faster filming rate would show yet more detailed action, with flexural waves propagating away from the impact site and the blade reflections superimposed. In the frame sequence 13 to 31, the blade and the hockey disc are moving without contact in a somewhat similar direction, while in frame 38, a second impact takes place through frame 48. Then occurs a motion reversal and the blade starts pushing the puck away. A second separation occurs, and subsequently at frame 60, a third contact occurs when the faster moving blade hits the disc.

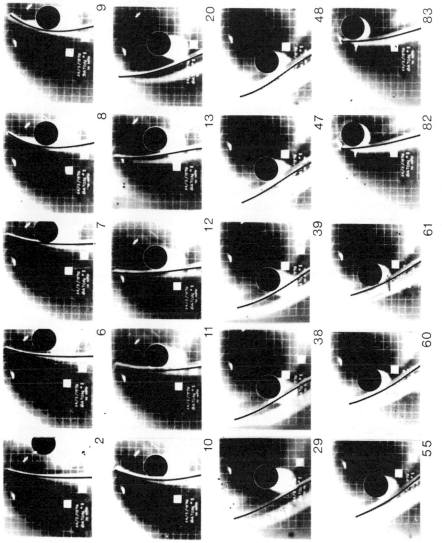

Fig. 14.12.   Selected frames of a high-speed film showing puck–blade interaction.

This last contact is maintained through frame 83. In that particular sequence, the disc executes some rolling motion as well as translation.

The continuous film projection provides much more evidence of the blade vibration in bending and torsion that occurs during and between the various impact phases. In several of the filmed impacts one can observe how a tipping motion of the hockey puck is initiated, when the blade, under torsional oscillation, impacts the disc with a large angle of incidence. Similarly, following the bending cycle of the blade, one can appreciate the effect of the incidence angle in the horizontal plane on the direction of puck rebound. Rolling of the puck is best observed at a high angle of incidence where there is a much larger component of momentum of the disc centre of mass along the blade direction. The frictional force that is developed between the puck and the blade creates the only couple that is available to accelerate the puck into rotation.

These observations suggest that the blade bending and torsional rigidity are very important parameters in the dynamic interaction of the stick blade with the puck, and consequently influence the disc motion after impact. Therefore it can be expected that these characteristics are deeply associated with puck control and the precision of the shot.

Many methods and techniques are available to the research and development specialist. Statistical testing, theoretical models and simulation, static and dynamic testing with strain-gauge and other types of transducer are more typically employed. But perhaps initially, high-speed photography remains the most profitable and economical technique. It may reveal clearly and readily the role of many factors that might otherwise have passed unnoticed. It can help one to understand the fundamentals of such complex behaviour and indicate what further testing needs to be done. High-speed photography is well indicated in the identification of the variables that need to be controlled in any statistical evaluation of hockey-stick performance.

14.3. Jogging shoes†

In sports or physical activities in which repeated cyclic loading of the lower limbs is performed for a period of time, one of the major roles of the footwear becomes the distribution and the attenuation of the concentrated ground reaction forces in such a way that musculo-skeletal tissues are not damaged either in the long run, or as a result of a single step made in adverse conditions. Another important role of the shoe is to provide adequate stability of the support leg through a complete and continuous control of foot–ground interaction. Moreover, these tasks become even more important and critical as the jogger (or runner) gets older, or heavier, or even taller.

† This study has been supported by a research grant from the 'Haut-Commissariat à la jeunesse, aux loisirs et aux sports' of the Quebec provincial government.

Need for foot–ground impact attenuation

Articular capsules, ligaments, tendons, muscles, and bones all participate in the impact-absorption process [28], but as the aging process progresses, mostly after the age of 40, the energy-absorbing capacity of these structures begins to decrease [29, 30, 31, 32]. This process is taking place at the same time as the strength of most biological tissues is progressively decreasing [33, 34] and as the fatigue resistance of compact bone to cyclic loading is also getting lower with age [35]. In older people, fatigue fractures are even observed, quite frequently, as accompanying very low levels of cyclic stresses [35].

The main shock absorption mechanisms of the human body are listed in Table 14.2, together with some of the related negative functional changes which are presently associated with the aging process.

Basic knowledge of the physical and physiological mechanisms underlying these functional changes associated with age is just beginning to develop [30], and it may still take many years before the real factors influencing these changes are found.

Meanwhile, it becomes imperative for people who are regularly performing cyclic stress loading of the lower extremities, as in jogging or running, to compensate through proper equipment the negative changes taking place in their locomotor system.

This implies the use of better shock-absorbing materials both at the heel and under the sole of the foot, as well as more caution in the selection of the footwear. Since increased local temperature has also been associated with

TABLE 14.2

Influence of the aging process on the main shock absorption mechanisms of the human body

Shock absorption mechanisms	Functional changes in related structures associated with the aging process
(a) Through controlled joint motion	Decrease in strength and elasticity of muscle tissue Decrease in strength and elasticity of tendons Decrease in strength and elasticity of ligaments Decrease in strength, elasticity, and resistance to wear of articular cartilages
(b) Through damping of vertebral column	Decrease in elasticity, tensile strength, and volume of intervertebral discs Decrease of both tensile and compressive strength as well as resistance to fatigue of vertebrae
(c) Through energy absorption properties of bone tissue	Decrease in the overall energy absorption potential of bone tissue Decrease in the resistance to fatigue of bone tissue Lowering of the stress threshold in the development of microfissures in some of the lower limb bones, especially at the level of the femoral head and neck.

lower energy absorption in the foot, well fitting shoes which produce no skin friction and which are capable of great attenuation of impact forces are desirable for most people, but are a must for older joggers.

Pertinent testing of jogging shoes

The tests reported in this section have been developed in order to provide repetitive and objective measures of the most important characteristics of jogging shoes. From the point of view of injury prevention, those characteristics are shock absorption and stability or motion control. Many millions of people are now practising the physical activity of jogging and occasionally, pains, injuries, and accidents are reported which are related to the lack of shock absorption or control. There has been heavy pressure placed on most manufacturers to produce a better shoe and show a better quality control. What is now needed is development of pertinent testing methods and research in the kinetics, kinematics, and biomechanical aspects of the body engaged in jogging.

Good shock absorption is one of the most desirable characteristics of a jogging shoe. One has only to consider, for example, that an average jogger at 8 miles per hour (13 km h^{-1}) experiences 13 000 to 15 000 impacts per hour. If one considers that shoes often transmit peak force in the order of two to three times the body weight [36, 37] it becomes obvious as to how severely the body is being stressed. For all engineering materials, for instance, it is a well known fact that the life of a sample is a function of the stress level and the number of cycles of stress level variations or stress reversals. This life is over when the material can no longer function normally, due to rupture or the extension of the crack network that has developed. This phenomenon, known as fatigue, is complex and generally irreversible. Biological material on the other hand, is different in certain aspects, one of which being its ability to heal cracks. While much research is needed on the complex nature of fatigue in bones, enough is known about other engineering materials and principles to appreciate the role of the impact strength or stress level on the rate of cumulative damage. Furthermore, the bone resistance being age dependent, it becomes more and more important to use better absorbing shoes as one gets older.

There are of course a variety of tests that can in one way or another assess the quality of shock absorption of jogging shoes; we may mention in particular, testing for shock transmitted through the shoe into the leg with sensors attached directly to the foot of a test runner [38]. While this test is very valuable from many points of view, the study of the transmission mechanism for example, it does not offer any guarantee of repeatability and variables control that are required in comparing different brands of jogging shoes. Figure 14.13 shows a set-up in which impact test on jogging shoes may be performed. In this case, a pendulum of mass M_1 and suspension cable length L, is pulled back through an angle θ and released. The shoe is mounted and held

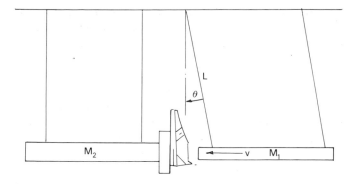

.FIG. 14.13.   Impact testing of jogging shoes.

flat on the heel portion of the sole on a force platform fixed to a second
pendulum. At impact, the following relation permits the computation of the
impacting initial velocity v_i

$$v_i^2 = 2gL(1 - \cos\theta)$$

where g is the gravity acceleration (9.8 m s^{-2}). During the time of contact, the
impacting force F_i will vary from zero load to a maximum and back to zero as
the first pendulum rebounds in the opposite direction with a velocity of scalar
value V_r. Applying the integrated form of Newton's equation over the total
impact time, the following equation is obtained

$$\int_0^t F_i \, dt = M_1(V_i + V_r).$$

The velocity V_r may be computed from the rebound angle or rebound
distance of the first pendulum. In a study [39] involving 26 pairs of different
models of jogging shoes ranging from the very best to very hard, there have
been only minor variations in the rebound velocity V_r, when holding V_i
constant. Therefore the integral on the left-hand side is very much a function of
V_i. In that study, typical force–time curves such as shown in Fig. 14.14, were
transmitted by the force platform and photographed on the screen of a CRT
oscilloscope. For V_i and V_r fixed, the force–time integral is nearly constant, and
therefore the area under the $F_i t$ curve is approximately a fixed value from one
shoe to another. The difference from one shoe to another will show up as a
different force–time distribution. A shoe with little absorption produces a high
maximum contact force and the contact time is short. A shoe with high
absorption produces a small maximum force and the contact time is long. If the
pulse can be approximated in an equivalent triangular shape of contact time or
base t_e and equivalent height or force F_e, it can then be concluded that the

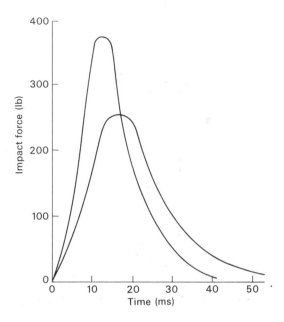

FIG. 14.14.   Typical impulses of constant value.

integral

$$\int F\,dt \cong \tfrac{1}{2}F_e t_e \cong \text{constant}$$

and therefore the maximum equivalent force is inversely proportional to the contact time. The longer the time of contact, the smaller is the maximum load F_e transmitted, and the better is the absorption. How to manufacture a shoe with good absorption is a problem for a materials engineer. He can use a wide variety of materials with various requirements and constraints. He can have recourse to various heights of sole, design shapes, etc. Whichever way is taken, only the final result is assessed here, the absorption quality of the shoe.

The importance of sufficient absorption has now been stressed and it must be answered as to which limit such needed characteristic may be extended. Research and development can be oriented in the selection, testing, or production of new materials with improved damping properties. The sole may be tapered, holes may be provided, sole thickness may be increased, various layers may be considered, etc.

Nevertheless, whatever action is taken to modify the sole absorption, another aspect of the mechanics of running, stability or motion control, must be seriously looked at. Stability study is also a complex science, and only the most simple aspect, the structural stability of a one degree of freedom model, will be illustrated in this section. A model that is suitable here is shown in Fig.

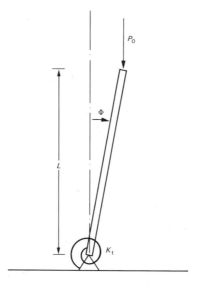

FIG. 14.15.   Stability of a one degree of freedom system.

14.15. The structure, in the form of a rigid bar of length L, is submitted to a vertical force of intensity P_o and restrained against a fixed lower base through a spring of rotational rigidity k_t. Within the limit of small displacement hypothesis, this structure may be shown to be stable [40] around the vertical position $\phi = 0$ for values of the load P_o that do not exceed the value

$$P_o \leqslant k_t/L$$

and unstable for

$$P_o > k_t/L.$$

In this model, k_t may represent the sole rotational stiffness. Such stiffness may be measured for instance as follows. The shoe is mounted on a shoe form and is rotated against a fixed base around a longitudinal central axis from toe to heel. The torque is then measured against the angular rotation and the stiffness or slope on the torque rotation curve around the origin $\phi = 0$ represents k_t. The length L may represent the distance between the body pivoting points. If structural stability around the knee is investigated, L_1 therefore represents the knee-to-sole distance. Of course, this model being a one degree of freedom system, we therefore assume that the ankle joint is well restrained and under control. For stability around the ankle, L is much shorter, while for hip stability L is very long. For a wide variety of jogging shoes, typical values of k_t have been measured to be in the range of 250 to 600 Nm rad^{-1} [39]. Taking L_1 to be 0.10 m, for instance, yields values of P_o in the range of 2500 to 6000 N at the ankle while taking L to be 0.5 m yields critical loads in the range of 500 to 1200 N at the knee. Considering an impact force of three times the body weight yields loads which exceed critical knee loading and which approach critical ankle loading.

Dynamic stability during running is a much more complex subject involving distributed inertia and body feedback reactions. With appropriate controls, many unstable systems, a bicycle for example, may be handled in a very safe way. In this sense, a very soft shoe with a low stable value of load P_o may still be worn without accident on a relatively level course during jogging. But if a shoe with a low stable load value is used in cross country jogging or in other sport activities in which fast side impact is involved such as tennis, squash, or soccer, the response time or reflex may not be fast enough to prevent unstable motion to develop and the risk of accident is potentially increased. We may conclude that there is an optimum design to be found where a shoe will have both an efficient absorption and yet provide a sufficient amount of ankle stabilization or motion control.

Conclusion

This chapter has been devoted to the presentation and discussion of some aspects of human body dynamics as they apply to selected types of sports equipment, namely protective helmets, hockey sticks, and jogging shoes.

Safety and performance in physical activities involving the use of sport equipment rely heavily on a real understanding of the dynamic interaction between the human body and the outside world through more or less adapted pieces of equipment. In the particular case of protective equipment such as safety helmets and jogging shoes, in-depth studies of the impact mechanisms and of the physical as well as physiological processes taking place at impact would allow the prevention of injuries through the design of pieces of equipment dynamically 'tuned' to the individual, and respecting the physical as well as the physiological changes associated either with human growth or the aging process. On the other hand, the precise knowledge of the relationship between weight, segment length, and stability of the shoe can ultimately permit shoe designs that will provide adequate and safe motion control of the foot–shoe entity during the interaction with the ground.

In order to pursue this task in a most efficient way, equipment designer would benefit from co-operative work with both biomechanical and biomedical people, and from a deeper knowledge of the physical environment in which the equipment is used.

References

1 Patrick, L. M. (1966). Head impact protection. *Proc. of Head Injury Conf.* pp. 41–8. Lippincott, Philadelphia.

2 THERRIEN, R. G. (1979). Sévérité des chocs obliques au hockey en fonction de la vélocité d'impact et de la friction entre la tête et le casque protecteur. Paper presented at the Association canadienne–française pour l'avancement des sciences, Annual Meeting, Montréal.

3 THERRIEN, R. G. (1977). Phénomènes physiques rotatoires accompagnant un impact excentrique entre la tête et la bande au hockey. Paper presented at the Congrès des professionnels de l'activité physique du Québec, Sherbrooke.

4 CHAPLEAU, C. (1974). Les blessures et les accidents au hockey. *Mouvement* **9**, 11–19.

5 HAYES, D. (1972). The nature, incidence, location and causes of injury in intercollegiate ice hockey. Paper presented at The Canadian Association of Sports Sciences Meeting, Vancouver, BC.

6 UNTERHARNSCHEIDT, F. J. (1971). Translational versus rotational acceleration: animal experiments with measured input. *Proc. 15th Stapp Car Crash Conf.* pp. 767–70. Society of Automotive Engineers, New York.

7 UNTERHARNSCHEIDT, F. J. (1972). Translational versus rotational acceleration: animal experiments with measured input. *Scand. J. rehabil. Med.* **4**, 24–26.

8 UNTERHARNSCHEIDT, F. J., and RIPPERGER, E. A. (1970). Mechanics and pathomorphology of impact-related closed brain injuries. In *Dynamic response of biomedical systems* (ed. N. Perrone) pp. 46–83. ASME, New York.

9 GENNARELLI, T. A., THIBAULT, L. E., and OMMAYA, A. K. (1972). Pathophysiologic responses to rotational and translational accelerations of the head. *Proc. 16th Stapp Car Crash Conf.* pp. 296–308. Society of Automotive Engineers, New York.

10 HIRSCH, Q. E., and OMMAYA, A. K. (1970). Protection from brain injury: the relative significance of translational and rotational motions of the head after impact. *Proc. 14th Stapp Car Crash Conf.* pp. 144–51. Society of Automotive Engineers, New York.

11 HODGSON, V. R., MASON, M. W., and THOMAS, L. M. (1972). Head model for impact. *Proc. 16th Stapp Car Crash Conf.* pp. 1–13. Society of Automotive Engineers, New York.

12 HODGSON, V. R., and PATRICK, L. M. (1968). Dynamic response of the human cadaver head compared to a simple mathematical model. *Proc. 12th Stapp Car Crash Conf.* pp. 280–301. Society of Automotive Engineers, New York.

13 OMMAYA, A. K., FAAS, F., and YORNELL, P. (1968). Whiplash injury and brain damage. *J. Am. med. Ass.* **204**, 285–9.

14 JOHNSON, J., SKORECKI, J., and WELLS, R. P. (1975). Peak accelerations of the head experienced in boxing. *Med. Biol. Engng* May, pp. 396–404.

15 SCHNEIDER, L. W., FOUST, D. R., BOWMAN, B. R., SNYDER, R. C.,

CHAFFIN, D. B., ABDELNOUR, T. A., and BAUM, J. K. (1975). Biomechanical properties of the human neck in lateral flexion. *Proc. 19th Stapp Car Crash Conf.* pp. 455–85. Society of Automotive Engineers, New York.

16 SNYDER, R. G., CHAFFIN, D. B., SCHNEIDER, L. W., FOUST, D. R., BOWMAN, B. M., ABDELNOUR, T. A., and BAUM, J. K. (1975). *Basic biomechanical properties of human neck related to lateral hyperflexion injury.* Final Report No. UM-HSRI-BI-75-4, Highway Safety Research Institute, University of Michigan, Ann Arbor.

17 WEBSTER, G. D. (1976). A comparison of the impact response of cadaver heads and anthropomorphic headforms. MA.Sc. Thesis, University of Ottawa.

18 WEBSTER, G. D., and NEWMAN, J. A. (1976). A comparison of the impact response of cadaver heads and anthropomorphic headforms. *Proc. 20th Conf. Am. Ass. Automotive Medicine, Detroit* pp. 221–40.

19 CHAO, E. Y., SIM, F. H., STAUFFER, R. N., and JOHANNSON, K. J. (1973). Mechanics of ice hockey injuries. *Proc. Mech. Sport*, pp. 143–54. American Society of Mechanical Engineers, Detroit.

20 LARIVIÈRE, G. (1973). Analyse technique du démarrage avant au hockey sur glace. *Mouvement*, **8**, 189–97.

21 OMMAYA, A. K., and HIRSCH, A. E. (1971). Tolerance for cerebral concussion from head impact and whiplash in primates. *J. Biomech.* **4**, 13–21.

22 FAN, W. R. S. (1971). Internal head injury assessment. *Proc. 15th Stapp Car Crash Conf.* pp. 645–65. Society of Automotive Engineers, New York.

23 NEWMAN, J. A. (1975). On the use of the head injury criterion (HIC) in protective headgear evaluation. *Proc. 19th Stapp Car Crash Conf.* pp. 615–40. Society of Automotive Engineers, New York.

24 VERSACE, J. (1971). A review of the Severity Index. *Proc. 15th Stapp Car Crash Conf.* pp. 771–96. Society of Automotive Engineers, New York.

25 SANO, K., NAKAMURA, N., HIRAKAWA, K., and HASHIZUME, K. (1972). Correlative studies of dynamics and pathology in whiplash and head injuries. *Scand. J. rehabil. Med.* **4**, 47–54.

26 BISHOP, P. J. (1976). Head protection in sport with particular application to ice hockey. *Ergonomics* **19**, 451–64.

27 BOURASSA, P., THERRIEN, R., CHEVRIER, J., DELISLE, D., and LAFLAMME, C. (1977). Hockey stick analysis with high-speed photography. Internal Report, Dept of Mech. Eng., University of Sherbrooke, Sherbrooke, Québec, Canada.

28 RADIN, E. L., and PAUL, I. L. (1970). Does cartilage compliance reduce skeletal impact loads? *Arthritis Rheum.* **13**, 139–44.

29 CURREY, J. D. (1969). The mechanical consequence of variation in the mineral content of bone. *J. Biomech.* **2**, 1–11.

30 CURREY, J. D. (1979). Changes in the impact energy absorption of bone with age. *J. Biomech.* **12**, 459–69.

31 RADIN, E. L., and PAUL, I. L. (1971). Response of joints to impact loading. 1. *In vitro* wear. *Arthritis Rheum.* **14**, 356–62.

32 SIMON, S. R., and RADIN, E. L. (1972). The response of joints to impact loading II. *In vivo* behavior of subchondral bone. *J. Biomech.* **5**, 267–72.

33 WALLS, J. C., CHATERJI, S., and JEFFERY, J. W. (1972). Human femoral cortical bone: a preliminary report on the relationship between strength and density. *Med. Biol. Engng* **10**, 673–6.

34 YAMADA, H. (1973). *Strength of biological materials* (ed. F. G. Evans). Robert E. Krieger Publishing Company, Huntington, NY.

35 CARTER, D. R., and HAYES, W. C. (1977). Compact bone fatigue damage-I residual strength and stiffness. *J. Biomech.* **10**, 325–37.

36 The 1979 *Runner's World* Shoe Survey (1978). *Runner's World*, October 1978, pp. 63–107.

37 The 1980 *Runner's World* Shoe Survey (1979). *Runner's World*, October 1979, pp. 47–75.

38 Running Shoes; A comparative test of injury prevention capabilities (1978). *Running Times*, October 1978, pp. 12–20.

39 BOURASSA, P., THERRIEN, R., and LAMBERT, C. (1979). On the performance of jogging shoes. *Proc. 7th Canadian Cong. Appl. Mech., Sherbrooke, May 1979* pp. 851–2.

40 TAUCHERT, T. R. (1974). *Energy principles in structural mechanics.* McGraw Hill, New York.

41 CAVANAGH, P. R. (1978). A technique for averaging center of pressure path from a force platform. *J. Biomech.* **11**, 487–91.

42 CAVANAGH, P. R., and LAFORTUNE, M. (1980). Ground reaction forces in distance running. *J. Biomech.* **13**, 397–406.

15. Human body mechanics in some sports and games

W. JOHNSON, S. R. REID, and A. G. MAMALIS

15.1. Introduction

The need to engage in competitive physical activity is an elemental fact of early adult behaviour which serves certain deep biological requirements. In later life, there is as well a need to engage in strenuous activity for the purpose of helping maintain fitness and of pleasurably absorbing the leisure that is a product of modern industry. These activities have been artificially developed especially in the last fifty years and are covered by the terms sports and games. Usually these activities are performed outdoors and involve an artefact, such as a ball or golf club, for the manufacture and provision of which specialist companies exist.

Below, several sports or games are reviewed and various mechanical aspects of them are briefly discussed. Readers will be able easily to add to the short accounts given, and they may feel encouraged to read more deeply relevant articles and books, the better to develop their own special interest. There are many aspects of mechanics in these sports which relate principally to the behaviour of the equipment (for example, the deformation of a ball during impact). In this chapter such topics are only dealt with briefly, reference being made to source material which provides the detail which has been omitted. In the main, emphasis is placed on those aspects of sports and games in which the response and limitations of the human body are the prime concerns.

15.2. Golf

Some basic facts

The driver head of a golf club of mass M and speed u just before it impinges against a stationary golf ball of mass m, has after impact a speed u' and the ball a speed of v'; the coefficient of restitution is e. The equation for the conservation of linear momentum and the empirical relationship between the speeds which involve e are,

$$Mu = Mu' + mv', \quad (v' - u') = eu \quad \text{and} \quad \overline{F}.\overline{t} = mv'. \tag{15.1}$$

The mean force between golf ball and club is denoted by $\overline{F}$ and the time of contact by $\bar{t}$. Thus,

$$\frac{v'}{u} = \frac{(1+e)}{1+(m/M)} \quad \text{and} \quad \frac{u'}{u} = \frac{1-e(m/M)}{1+(m/M)} . \tag{15.2}$$

Typically, $Mg = 7.2$ oz (204 g), $mg = 1.6$ oz (45 g), and $e = 0.6$. Assuming $u = 150\,\mathrm{ft\,s^{-1}}$ ($45\,\mathrm{m\,s^{-1}} = 160\,\mathrm{km/h}$) then $v' = 196\,\mathrm{ft\,s^{-1}}$ ($59\,\mathrm{m\,s^{-1}}$) and $u' = 106\,\mathrm{ft\,s^{-1}}$ ($32\,\mathrm{m\,s^{-1}}$). The face of a driver is in contact with the ball for about about a time $\bar{t} = \frac{1}{2}$ ms so that $\bar{F} \simeq 5.8$ kN (1300 lbf). It is known that the peak force during the delivery of the blow is about 8.9 kN (2000 lbf) and that the club face travels about 19 mm ($\frac{3}{4}$ in) during the impact. Note that in golf $M/m \simeq 5$, in cricket it is 7 and in tennis 8, and at these mass ratios, take-off speed does not vary greatly with M/m. One basic aim is to maximize u and to this end great emphasis has been laid on the mechanics of the swing of the club, some details of which are given below.

A comprehensive and thorough study of the mechanics of golf is to be found in the book *Search for the perfect swing* [1]. This text carefully records the results of experimental and theoretical work obtained by its authors who were once employed to investigate the topic.† Many of the following samples of facts available about golf derive from this work.

Coefficient e for some balls varies with temperature from 0.64 at $0°$C to 0.76 at $30°$C and as the take-off speed of the ball is proportional to $(1 + e)$, greater range is to be had by warming-up the ball before play. Other things being the same, a 185 yard (169 m) drive at $0°$C become a 200 yard (183 m) drive at $30°$C.

It is maintained that to 'carry' 280 yards (256 m) before hitting the ground, the ball must take-off at 175 mile h^{-1} (282 km h^{-1}) and this demands a speed of 134 mile h^{-1} (216 km h^{-1}) for a 6 oz (170 g) club head or 127 mile h^{-1} (204 km h^{-1}) from an 8 oz (227 g) one. A formula ($150 < v' < 280\,\mathrm{ft\,s^{-1}}$) for overall range R, i.e. 'carry' plus 'run', for squarely struck drives is $R = 1.25v' - 27$, where ball speed v' is in ft s^{-1} and R is in yards. Also for the 'carry', c (in yards), over 'average' ground, $c = 1.5v' - 103$.

A 7 oz (198 g) driver head with a 'loft' of $11°$ to $12°$ swung at 100 mile h^{-1} (160 km h^{-1}) sends the ball away at $9°$ or $10°$ above the horizontal with a *backspin* of 60 rev s^{-1}; a typical '7-iron' which has a loft of $39°$ sends the ball off at $26°$ elevation and about 130 rev s^{-1} backspin. The club face when driving, passes below the ball on impact and then applies a downward frictional force to the ball causing a backspin; the slightly glancing blow of the club results from its face being set back about $10°$. The accuracy of alignment of a top class golf player is said to be within $2°$.

For a heavy pivot mass on one end of a uniform flexible shaft, the other end

---

† The promoters hoped (forlornly) that a profound knowledge of the mechanics of the game would help the UK recover the Ryder Cup from the USA!

of which is clamped, the frequency of vibration is

$$f = \frac{1}{2\pi} \sqrt{\left[\frac{3EI}{(M + 0.24\mu)l^3} \right]}.$$ (15.3)

μ is the shaft mass $\simeq 0.4$ oz (11 g). The second moment of area, I, varies along the shaft, but inserting typical values for the constants ($l = 39$ in (~ 1 m) and $I = 0.006$ in^4), $f \simeq 4.3$ Hz.

It is thought that $\frac{2}{3}$ s is required for the shock to travel from the point of impact of the club head with the ball to the hands and 10 m s^{-1} for the message then to be appreciated by the brain—by which time the ball is 1 ft (30 cm) clear of the club head.

Space prevents considerable further description of ball mechanics. An excellent book for all interested in mechanics and games is the book by Daish [2] in which he examines, amongst other things, the influence of spin on the flight and bouncing of balls, a topic of interest in other sports including baseball, cricket, football, and table tennis. An example of the sort of simple calculation that can be produced concerning the flight of a ball is included in the section on football.

Analysis of the golfswing

The maximum club head speed is reached in a swing after about 0.25 s and this represents a working rate of about 4 h.p.; up-swing time is usually about 0.55 s, the up-and-down swing circles being different largely owing to (downwards) increased centrifugal effects. A golfer functions as a double pendulum, his shoulder acting as a pivot, his arms and shoulders constituting an upper lever, his wrist a hinge and the club a lower lever.

The dynamics of the golfswing have been analysed by Williams [3] and Pyne [4] on the basis of the double pendulum model using rather different assumptions for the phase of the stroke immediately prior to impact between the clubhead and the ball.

Pyne describes the downswing as being comprised of two parts or phases. During the first phase the arms are considered to be rigidly attached to the body and the whole system rotates about an axis through the upper part of the chest. The power is derived by driving the knees towards the target whilst keeping the head back; the clubshaft and left arm are at right angles, muscular effort being required by the hands to maintain this angle. The arm muscles must also exert sufficient force to keep the left elbow straight and the other arm, shoulder, and back muscles must produce sufficient force to permit the arms–club system to 'track' the tilting shoulders.

For the second phase Williams considered that, once the arms begin to rotate independently of the trunk, they continue to rotate at constant angular velocity, this requiring a force (called, by Pyne, the 'Williams' force) to be exerted by the hands to overcome the inertial reaction arising from the motion of the clubhead. Pyne replaces this assumption with one in which the total force exerted by the hands (i.e. the Williams force + the 'primary acceleration force', an inertia force arising from the motion of the arms) is zero. For cases in which club head mass is very much less than the mass of the arms, this leads to the proposal that both angular momentum and energy of the arms–club system are constant in this second phase. The analysis leading to this hypothesis and the consequent calculation of the increase in the clubhead speed during this phase and how it relates to the lengths of the club and the arm and the clubhead to arm-mass ratio can be found in Pyne's paper.

15.3. Football (soccer)

The impact and rebound of a soccer football is discussed at length in the papers by Johnson et al. [5] and Percival [6] and many of the ideas can be extended to cover the response of other balls which are essentially inextensional, flexible shells having an internal pressure greater than atmospheric. Simulated kicking has been analysed by Youen et al. [7].

The laws of the game prescribe that the ball weight $Mg = 4.45$ N (1 lbf), be of radius $R = 110$ mm (0.36 ft) and have an internal pressure p_0 of 68.7 kN m^{-2} (10 lbf m^{-2}) gauge. On normal impact with a rigid, flat surface, by treating the ball as an elastic shell undergoing only a small amount of compression, assuming no shape change and neglecting any adiabatic increase in internal pressure, its motion is simple harmonic. The time of contact T is given by,

$$T = \sqrt{\left(\frac{\pi M}{2R\,p_0}\right)};$$

(15.4)

for an Association football this is about 10 ms^{-1}. The maximum area of contact is $2.14 \times 10^{-3} u_0$ m^2 where u_0 is the impact speed in m s^{-1}. Both these results are approximately verified by experiment, though T reduces slowly with increase in u_0.

Relatively simple analysis provides a reasonable estimate of the force involved in kicking a ball and in heading. An account of the latter is given below in the context of a discussion of the possibly injurious aspects of some sports.

In flight, a football is subjected to gravitational force W, aerodynamic lift L, and drag force, D, respectively.

For a non-spinning ball, following a fairly flat trajectory, we have

$$D = -Mf = \tfrac{1}{2}C_D\rho u^2 A,$$

(15.5)

where f denotes horizontal acceleration, C_D the drag coefficient, ρ the air density and A the ball cross-sectional area.

Experimental values of C_D for a smooth sphere are readily available being given as a function of the Reynolds number, $R_e = ud/v$, where v is the kinematic viscosity of the air. There is a sudden change in C_D at a critical value R_e of about 2.15×10^5; spheres travelling at slightly super-critical Reynold's numbers experience less drag than at sub-critical ones. It is observed that,

(i) C_D is nearly constant over the range

$$10^3 < R_e < 2.15 \times 10^5, \text{ i.e. for } 0.07 < u < 15.2 \, m \, s^{-1} \qquad (15.6a)$$

$(50 \, ft \, s^{-1})$ it has a value of 0.4.

(ii) C_D is nearly constant for $R_e > 2.15 \times 10^5$,

i.e. for $u > 15.2 \, m \, s^{-1}$ $(50 \, ft \, s^{-1})$ it has a value of 0.2. $\qquad (15.6b)$

When $R_e = 10^6$, $u = 70.7 \, m \, s^{-1}$ $(232 \, ft \, s^{-1})$.

By taking account of these two regimes in [5] and starting from eqn (15.5), expressions can be derived between initial u_0 and current ball speed, u, time of flight, t, and distance travelled, s, which make it possible to produce the graph in Fig. 15.1. If it is required to determine how long it takes a football to travel 23 m (a 25 yard 'drive') starting from a speed $u_1 = 109 \, km \, h^{-1}$ $(68 \, mile \, h^{-1})$ or $30 \, m \, s^{-1}$, next on Fig. 15.1 look up the value of s at $30 \, m \, s^{-1}$, i.e. 153 m, and then subtract 23 to give 130 m; the value of u after this distance is $24.4 \, m \, s^{-1}$ or $87.8 \, km \, h^{-1}$ $(54.5 \, mile \, h^{-1})$. The difference between the times at these two

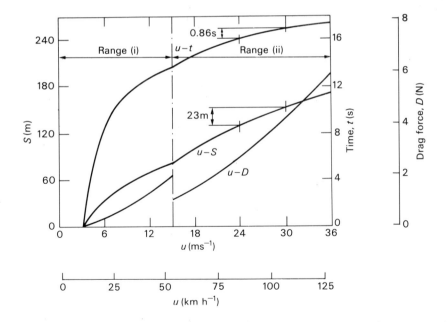

FIG. 15.1.   Speed–time–distance–drag curves for a football in horizontal flight.

speeds (actually using Table 15.3 of ref. [5]) 0.86 s, is the time taken by the ball to travel this distance. The time for a 33 m (36 yard) 'drive', the ball starting at 120.7 km h^{-1} (75 mile h^{-1}) may be checked to be 1.16 s; its speed reduces to 87.8 km h^{-1} (54.5 mile h^{-1}) at the end of this distance.

The drag on a football at 114 km h^{-1} (71 mile h^{-1}) is equal to its weight. Using Maccoll's experimental results [5, 8] for the aerodynamic lift generated by a rotating sphere, we may estimate the 'bending' or 'curving' undergone by the ball when 'taking a corner'; similar calculations would apply for a 'banana shot'. By assuming a corner kick to be delivered, about 36 m ($\sim$ 120 ft) from the goal mouth, in a horizontal plane at a constant speed of 18 m s^{-1} ($\sim$ 60 ft s^{-1}) and with a rotational speed of 15 rev s^{-1} about a vertical axis, it may be shown that the aerodynamic lift manifests itself as a horizontal force giving to the ball a transverse acceleration which can deflect it from its original direction by as much as 3 m in its flight-time of 2 s. When crossing the goal mouth, the lateral speed developed would be about 3 m s^{-1}, so that moving at 18 m s^{-1} ($\sim$ 60 ft s^{-1}) across 7.2 m (24 ft) between goal posts, the ball would move transversely by about 1.2 m (4 ft) if delivered as an 'in-swinger'. It is thus easy to appreciate how a goal may be scored from a corner, see Fig. 15.2.

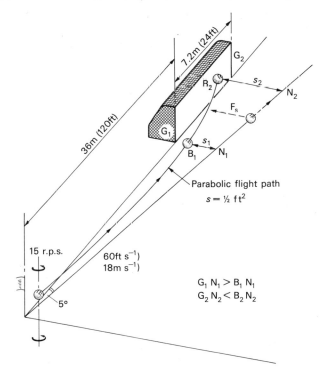

FIG. 15.2.   Scoring a goal from a corner kick. Initially the ball is shot at 5° to the dead-ball line.

15.4. Rock climbing

Stances

... 90 years ago ... Haskett Smith dismissed the rope, with other aids, as one of a group of illegitimate devices for use by bad climbers [9].

Successful rock climbing is essentially concerned with controlled progression from one equilibrium position to another, each of which can usually be easily analysed in terms of elementary statics. Much reliance is placed on friction and thus 'rubbers' on the feet are frequently used. Ref. [10] discusses many common stances adopted in climbing and we select just one for illustration. Fig. 15.3 shows the stationary equilibrium position for a climber who is 'chimneying'—also called 'back-and-footing'. The chimney is formed by opposing rock walls, the cleft being wide enough to admit the climber. By stiffening his legs the climber increases the normal contact forces at his feet, N_A, and back, N_B, so that ensuing friction forces F_A and F_B support his weight W and thus he can remain aloft. From the triangle of forces in Fig. 15.3, we have $N_A = N_B = N$ and $F_A + F_B = W$ or $N = W/(\mu_A + \mu_B)$; also

$$\frac{h+i}{k} = \frac{F_A}{N} = \mu_A \quad \text{and} \quad \frac{i}{l} = \frac{F_B}{N} = \mu_B. \qquad (15.7a)$$

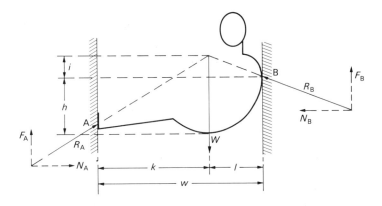

FIG. 15.3.   The triangle of forces for a back and footing climber.

Hence,

$$h = k\mu_A - l\mu_B = k(\mu_A + \mu_B) - W\mu_B, \qquad (15.7b)$$

where w is the chimney width. For given values of μ_A and μ_B, eqn (15.7b) shows that the climber's point of back rest, B, is the higher the more he straightens his back.

Rope strength

We can make an estimate of the permissible weight of a leader if his fall is to be successfully arrested by a given rope. The leader must fall a considerable distance before his 'second' can stop him and indeed he has most chance of surviving unscathed on a vertical or overhanging face.

The force–extension curve for both sisal and nylon are concave upwards, see Fig. 15.4(a). By testing, the maximum load that could be carried by a 16 mm ($\frac{5}{8}$ in) diameter sisal rope was found to be 20 kN (4500 lbf), and by a 16 mm ($\frac{5}{8}$ in) diameter nylon rope 57.8 kN (13000 lbf); each had a corresponding engineering strain of 0.17 and 0.5 respectively. The equation of the tension-engineering strain curve for these materials is

$$\frac{T}{T_M} = \left(\frac{e}{e_M}\right)^n, \tag{15.8}$$

where strain $e = x/h$ = rope extension/initial length, and n is an empirical constant, found to be about 2; suffix M refers to a greatest value.

The strain energy U which can be stored in a rope is, see Fig. 15.4(b),

$$U = \int_0^X T dx = \frac{T_M \cdot h}{e_m^n} \cdot \frac{e^{n+1}}{n+1} \tag{15.9}$$

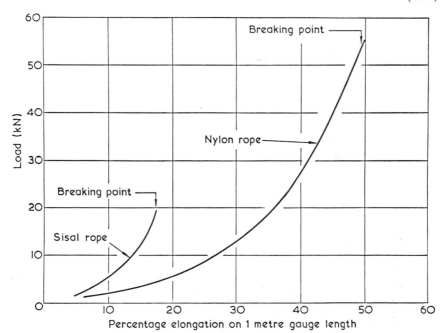

FIG. 15.4 (a)

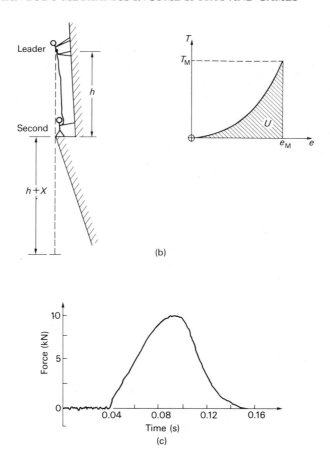

FIG. 15.4.   (a) Force–extension curves for 2 in (51 mm) circumference sisal and nylon ropes. (b) The height of fall of a leader. (c) Force–time oscillogram for a dynamically loaded rope.

or

$$U_{max} = \frac{T_M \cdot h}{n+1} \cdot e_M.$$

Now considering Fig. 15.4(b) in which a leader of weight W is at height h above his second and is brought to rest after falling a distance $(2h + X)$ where X is the rope extension under maximum load, we may assume that a successful arrest of his fall may be made if his kinetic energy (or lost potential energy) can just be totally converted into strain energy in the rope. (We assume this situation is too slow to be treated as a stress wave problem.) Hence, the limiting condition is,

$$T_M \cdot h \frac{e_M}{n+1} = W(2h + X)$$

or

$$W = T_M \cdot e_M / (n+1)(e_M + 2). \qquad (15.10)$$

Inserting into eqn (15.10) the values earlier given, the leader is thus protected no matter what height he falls from with the nylon rope if $W < 3.56$ kN (800 lbf) and with the sisal rope if $W < 0.53$ kN (118 lbf). These magnitudes include, among other things, no factor of safety and no allowance for the weight of the rope itself or for friction over a belay.

The dynamic loading of ropes of many different kinds has been a subject of investigation at the British National Engineering Laboratory in recent years and the article by Borwick on Mountaineering Ropes is very relevant to our present topic [9]. The Dodero test 'simulates the impact of a rope over an edge during the arrest of a fall' and two drops of 80 kg (176 lb) with a peak force of less than 11.9 kN (2660 lbf) must be survived to qualify for certification. Borwick describes the NEL linear-motor dynamic rope-testing device and reports that dynamic breaking loads are less than static ones. A knot (or any bending out of natural inclination) has a breaking effect due to the tensile stresses created; especially loading over an edge seriously weakens a rope. Mountaineering ropes in use have been shown to deteriorate with time (using strain energy absorption as a criterion). See also the article by Kosmath [11] and earlier ones referred to in ref. [12].

A force–time oscilligram given by the manufacturer of Edelrid kernmantle rope is shown in Fig. 15.4(c).

Finally, with reference to equipment, Griffin [13] made a very useful study of the various designs of karabiner which are available. These are snap links used to attach the main climbing rope to the climber's waistband or as a fixed or running belay.

15.5. Pole vaulting [14]

This sport is supposed to have derived from jumping over ditches, fences, and brooks with the help of a stiff iron pole. For competitive sport bamboo was introduced at an early stage, then succeeded by hollow aluminium and steel tube and now solid highly flexible fibre-glass poles are employed. Predominantly, the pole facilitates the conversion of the run-up kinetic energy of the athlete, in about 1 s, to potential energy; if 100 metres was run in 10 s, the height achieved would be $v^2/2g \simeq 5$ m. If the athlete's centre of gravity is originally 1 m from the ground, then if he adds no work to the system during his climb, with the aid of some body contortion an expert should be able to clear about 6 m. There are various inefficiencies in the act of vaulting so that 6 m is not reached; the current world record by Roberts (USA) in 5.70 m.

Different actions are required of the vaulter when he uses a 'rigid' aluminium pole and a solid fibre glass one, see Fig. 15.5, reproduced from ref. [14]. In the former, Fig. 15.5(a), the hands are brought together and the athlete 'rides' the pole; he does not perform a straight-swing but rotates his drawn-up body and first pulls and then pushes to project himself over the bar. The other technique, Fig. 15.5(b), requires the hands to be apart by about $\frac{1}{2}$ m to bend the pole when it is driven-in or 'planted' so that they can efficiently bend the pole. Not only can the athlete then supply energy by bending with his arms but also, less energy is lost in the act of planting a flexible pole than a rigid one.

The flexible pole absorbs energy equivalent to raising the athlete's weight through about $\frac{2}{3}$ m. Measured pole flexural rigidities are about 150 kgm^2 for fibre glass and about 300 kgm^2 for metal tubes. Important but difficult to assess contributions to height during the vaulting action are, the value of the vaulter's spring, the work done on the pole during the vault and the energy loss during the 'plant' of the pole.

15.6. Injuries in boxing, heading, skiing, and diving

Certain kinds of injury are frequently encountered in certain sports and it is instructive to clarify ideas about them, with quantitative information where possible. An engineer's review of this topic is given below but specific medical aspects are discussed in ref. [15].

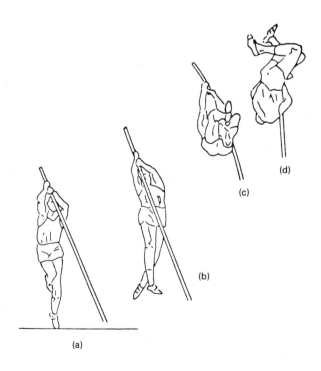

(d)

(c)

(b)

(a)

(i)

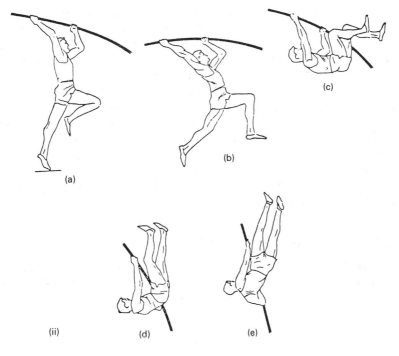

FIG.15.5.   Pole vaulting.

Boxing

There is great controversy over the long-term injuries sustained in boxing, a sport which one group sees as 'the noble art of self defence' and another as 'a brutal and degrading spectacle which should be stopped'. A recent report of the Royal College of Physicians, London, on 224 ex-professional boxers found that 37, or 17 per cent showed evidence of brain damage. The most thorough study of brain and skull injuries due to boxing and other sports is that of Unterharnscheidt [16]. Central blows (coup) to the head cause a linear acceleration of the skull which promotes pressure at the site of the blow due to the inertia of the brain and a cavitation tendency in it at its opposite side (contre-coup); a wave due to the blow passing through the material of the brain also yields the same notions of local pressure and far-side reduced pressure. Oblique blows ('hooks' and 'upper-cuts' or 'whip-lash' effects) which cause skull rotation lead to stretching and tearing of bridging and connecting veins to the brain among other things. Unterharnscheidt states that repeated translational blows to the head of low intensity (e.g. 7 m s^{-1}) produces more (i.e. chronic or long term) damage than a single intense knock-out blow (8 m s^{-1}); but 75 per cent of all acute brain damage is due to rotational acceleration.

It follows that any measurement of the tendency of blows to inflict damage would therefore be useful. Non-penetrating concussion (i.e. unconsciousness and loss of memory and therefore, by implication, brain damage) has been studied especially in connection with car accidents to find a measure of the severity of a blow. Head impact is a multi-parameter situation, but linear acceleration of the head has been shown to be one very important parameter to consider for car accidents and therefore experiments have been carried out on human beings, monkeys, human cadavers, and other animals. A curve defining the limit of tolerance to impact blows obtained at Wayne State University has been used to define a severity index—the Gadd Severity Index, I—thus,

$$I = \int_0^T a^{2.5} \, dt. \tag{15.11}$$

If f is the linear acceleration of the head, $a = f/g$ and T is the period of the impact which for eqn (15.11), to apply, must be less than 60 ms. This Index has been used mainly in studies of automobile safety, see Chapters 1 and 4, and indeed incorporated into US Automobile Safety Regulations, requiring $I < 1000$. Experiment has shown that when I exceeds 1000, brain damage is very likely to occur.†

In recent tests the head reaction of a human being subject to low intensity blows has been used to simulate boxing [17], using an accelerometer fastened to the back of the head. From a parallel test on a dummy or punch ball, information was deduced about human head reaction to heavy and fast blows. The blows delivered were straight and directed at the centre-line of the forehead to cause only translation. A $5\frac{1}{2}$ kg mass (a likely figure for a heavyweight boxer) was thrown against the human head at between 0.7 and 3.7 m s^{-1}.

The natural frequency of the head–neck system for fore and aft excitation obtained using a mechanical shaker attached to a human subject is between 2 and 3 Hz. Since the impulse duration when boxing using 6 oz (0.17 kg) gloves is found experimentally to lie between 10 and 50 m s^{-1}, it follows that in the boxing situation head response is governed almost entirely by head mass. Using film, it is observed that boxers deliver 'haymakers' at 8 m s^{-1} and 'jabs' at 7 m s^{-1}. With a typical fist speed of 8 m s^{-1}, by one approach the above method gave an extrapolated peak acceleration of 260 g for an impact time of 13 ms and another approach gave a peak acceleration of 130 g for 17 ms; the low level of the latter amateur blow is probably due to the slow speed of delivery. These figures appeared to the authors to represent the most pessimistic and the most optimistic cases. The Gadd Severity Index as calculated for these two figures were, respectively, 3700 and 980. The

† In recent literature, less simplistic interpretations from calculated values of I have been considered.

unmistakable conclusion was that when the head is unprotected in boxing the safety limit is exceeded may be by as much as a factor of 4.

The beneficial effect of headgear is due principally to its cushioning action or its ability to lengthen the time of delivery of a blow and much less so to its effective increase in the mass of the head. If $\overline{P}$ denotes mean force and t the time of its application, then for two impulses or blows of equal magnitude, for which

$$\overline{P}_1 t_1 = \overline{P}_2 t_2,$$

but in which $\overline{P}_1 > \overline{P}_2$ and therefore $t_2 > t_1$ we have,

$$\frac{(GSI)_1}{(GSI)_2} = \frac{\left(\dfrac{\overline{P}_1}{Mg}\right)^{2.5} \cdot t_1}{\left(\dfrac{\overline{P}_2}{Mg}\right)^{2.5} \cdot t_2} = \left(\frac{t_2}{t_1}\right)^{1.5}$$

M is the mass of the head. Thus if protective gear lengthens the time of a blow by $2\frac{1}{2}$, the GSI is reduced by a factor of nearly 4.

Heading a football [18]

A good footballer heads a ball by receiving it in a 'restricted area in the midline in the frontal region' of his head, with neck muscles contracted.

In a newspaper article Hughes [19] reported that it had been proposed that heading was conducive to the premature onset of conditions of arthritis in the neck. He also noted that of the 55 players reported in the *Sunday Times* who had died in football since 1951, 26 had head injuries and that 8 were attributed at coroners' inquests solely to heading the ball. In 1972, Professor W. B. Matthews, writing in the *British Medical Journal*, reported cases of 'classical migraine, including incapacitating visual field defects, repeatedly developed in five young men immediately after blows to the head while playing football'.

Some appreciation of the intensity of the blow to the head from heading a ball may be made using the same approach as described above for boxing. In ref. [18] by assuming that a blow due to the ball causes purely linear acceleration of the head, a theoretical calculation was made of the Gadd Severity Index for various impinging ball speeds. At a speed of about $129\,\mathrm{km\,h^{-1}}$ ($80\,\mathrm{mile\,h^{-1}}$) the Index was calculated to be about 460 assuming a human head of weight $4.65\,\mathrm{kg}$ ($10\frac{1}{2}\,\mathrm{lb}$); this figure was arrived at after determining the force–time pulse generated by a compressible spherical ball—or inflated shell—of given mass and speed impinging against a stationary head of given mass which possessed a flat forehead. However, as a result of experiments, the equation arrived at for the GSI was found to be $I = 0.46\,a_{max}^{2.5}T$, assuming a sinusoidal pulse is delivered; a is the maximum head acceleration and T is the pulse time-constant at about $10.3\,\mathrm{ms}$. This gave $I \simeq 230$ for an $129\,\mathrm{km\,h^{-1}}$ (80

mile h^{-1}) impact (the greatest speed with which a footballer can drive the ball) with $a_{max} \simeq 75$ g.

Evidently, when correctly heading a ball at any measurable speed at up to say about 71 km h^{-1} (44 mph or 19.7 m s^{-1}) the GSI is very low at about 26 and any damage due to frontal impact seems extremely unlikely. Of course, the effect of repeated blows is not assessed or of the glancing blows to the temporal region of a head unprepared for a blow. Arthritis in the neck may well be a long-term result of impulses due to heading being transferred to the cervicle vertebrae; this topic remains for further investigation.

The ability of competent players to stiffen the neck muscles and to transfer the blow of the ball somewhat to the body by taking it through the upper vertebrae, accounts in large measure for the absence of head injury—together with experience and judgement about the level of speed which can be successfully sustained.

Helmets in American football

In the USA, the National Operating Committee on Standards for Athletic Equipment (NOCSAE) Football Helmet Impact Standard has helped to greatly reduce serious head injuries. Fatalities have decreased by well over 50 per cent in the eight years that NOCSAE have been influencing design. Testing without facemasks has been justified mainly because it represents the worst case. Random location test impacts were conducted to expose all possible weaknesses. Elevated temperature playing conditions tend to reduce protection by softening helmet lining material and thus the Standard includes a partial test immediately after four hours in a 120 °F oven.

The Severity Index tolerance limit chosen is 1500 (for a non-concentrated impact), game performance generally being at one half this level, on average. Drop tests from a height of 10 ft to realise a SI of 1500, imply an impact speed of about 25 ft s^{-1}, which is equivalent to a 40 yard dash in 5 s. Similar considerations pertain to the design of helmets for cricketers although here design considerations are perhaps more like those relevant to motor cycle crash helmets (see Chapter 13).

Tibia fracture due to ski accident

Ski mechanics is reviewed extensively in ref. [20] but our interest here is restricted to ski injury. Most fractures caused by ski accidents happen to the tibial or fibular shafts,† see Fig. 15.6, and tibial shaft fractures are nearly always found to be spiral, thus indicating torsional overload. Under normal conditions, because of the short length of the foot, no large torque is transmitted to the tibia, but fastening the foot in a boot to a long length of ski greatly predisposes this long, relatively brittle bone to fracture. Fracture, or injury, can frequently be avoided by calibrating ski release bindings so that

†Presumably the new sport of grass-skiing will bring its own special kind of injuries!

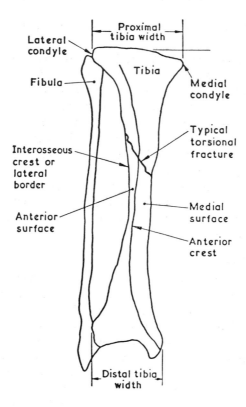

Definitions related to
lower right leg.

FIG. 15.6.   The tibia and (diagrammatically) the shape of a typical skiing fracture.

they release the foot from the ski before the load on the foot reaches a certain
level and the lower leg bones become overloaded. This release load must be set
to have a minimum level, of course, as well as a maximum in order that ski
manoeuvres can be performed.

The torsional moment that a typical tibia can withstand cannot be simply
calculated using engineering formulae because the tibia shape is far from
prismatic. However, in ref. [21] after examining X-ray photographs of broken
tibia and using the section area in the vicinity of a fracture, a correlation
between proximal tibia width (ω cm), see Fig. 15.6, and torque to fracture in
kg. cm was found as $T = 250\omega - 1250$, ($6 < \omega < 10$). (The tensile strength of
bone is about $55000\,\text{kN m}^{-2}$ ($8000\,\text{lbf in}^{-2}$).)

From instrumented ski tests it has been shown that tibia torsional strength is
several times that needed for normal skiing and thus that injury is clearly due

to non-skiing conditions, e.g. spills. Bearing in mind other factors, safety against tibia torsional fracture can be promoted by calibrating release bindings in proportion to proximal tibia width.

The article by Piziali [22] discusses dynamic torsional response of the leg in skiing and the performance of ski release bindings is treated at length in refs. [23] and [24]. Sack and Albrecht have analysed curved turns [25].

Diving

A few natives at Acapulco, Mexico regularly dive into a $4\frac{1}{2}$ m deep sea from rocks at heights of 30 m and 41 m above it [26]. The hands are held together with the thumbs interlocked so that at impact, with the body rigid, there is a nose-cone-like shape presented to the water with the blow to the head and neck being well dissipated because it is retracted into a 'solid collar' (The speed at impact with the water from a height of 41 m is about $28\,\mathrm{m\,s^{-1}}$ and if this is reduced to nought in $3\frac{1}{2}$ m, the mean retardation rate required is 11 g over about $\frac{1}{4}$ s. The effective GSI, taken only over 50 ms, is then 20—which is of negligible significance). Successful diving probably depends greatly on avoiding whiplash effects on impact with the water and, as in football heading, having the line of any blow pass along the vertebral axis.

Medical examination of the divers showed some cervical spine change but no fractures or dislocations. Performers with long necks (who dive with arms outstretched and hands apart and suffer a blow to the head) show severe bone changes. This damage is probably due to *repeated* high intensity loading. Competitive divers in the USA in recent years have jumped from heights of about 150 ft, but enter the water with a rigid body, feet first. It is reported by Sugden and Snow [27] that of 305 suicide jumps from the Golden Gate Bridge of San Francisco (impact speed with the water of about $33\,\mathrm{m\,s^{-1}}$) there were only two survivors and this is known to be about the extreme limit of the human impact with water survival range.

Karate

Feld, McNair, and Wilk [28] have given a popular exposition of various aspects of the mechanics associated with karate. Their main concern is to explain how the karate expert can break wood and concrete blocks with bare hands. It appears that the key to doing this lies in the ability to develop a peak velocity in the region of 10 to $14\,\mathrm{m\,s^{-1}}$ on impact and to be able to exert a force of more than 3000 N (675 lbf). Provided the hand and arm are held rigid, this force can be withstood. Injury is avoided, in part, by developing with practice a thickening of hard flesh on the hand at the region of contact.

The impulsive nature of the force is clearly very significant as is the accuracy with which the blow can be applied to the centre of a typical block which usually rests on two supports near its ends. The authors contend that the duration of the blow is sufficiently long for only the fundamental mode of vibration to be excited. The inertia of the block is such that it reduces the estimate of the energy required to break blocks to one-sixth of the values obtained on the basis of bending the blocks statically until fracture occurs.

It is estimated that energy of 100 J is available in a karate strike. This compares with estimates of 1.6 J and 5.3 J for the critical energies required to break concrete blocks (40 cm × 19 cm × 4 cm, weighing 6.5 kg) and wooden blocks (28 cm × 15 cm × 1.9 cm, weighing 0.28 kg) respectively. The higher energy required to break the wooden block stems from the fact that, whilst the force required is less (670 N compared with 3100 N for concrete), the deflection required to achieve fracture is 16 times greater.

The authors of reference [28] used a computer model for the impact process and concluded that the hand needed to reach a speed of 6.1 m s^{-1} to break wood and 10.6 m s^{-1} to break concrete. Such speeds are in agreement with their observation that beginners can break wooden blocks whereas the higher speed required to break the concrete blocks calls for training and practice and can only be achieved by an experienced kareteka (practitioner of karate). The prediction that concrete blocks fracture in less than five milliseconds also agrees with the observations they made using high-speed photography. As well as providing an interesting range of photographs of various numbers of blocks being broken the authors also discuss the manner in which the hand deforms during impact.

References

1 COCHRANE, A., and STOBBS, J. (1968). *Search for the perfect swing*. Heinemann, London.
2 DAISH, C. B. (1972). *The physics of ball games*. English Universities Press, London.
3 WILLIAMS, D. (1967). The dynamics of the golf swing. *Q. J. Mech. appl. Math.* **20**, 247.
4 PYNE, I. B. (1977). A clubhead speed coefficient for golf swing. *Q. J. Mech. appl. Math.* **30**, 269.
5 JOHNSON, W., REID, S. R., and TREMBACZOWSKI-RYDER, R. E. (1972–3). The impact, rebound and flight of a well inflated pellicle as exemplified in association football. *J. Manchr. Ass. Engrs*, No. 5.
6 PERCIVAL, A. L. (1976–7). The impact and rebound of a football. *J. Manchr. Ass. Engrs*, No. 5.
7 YOUM, Y., HUANG, T. C., ZERNICKE, R. F., and ROBERTS, E. M. (1973). Mechanics of simulated kicking. *Mechanics and sport* Vol. 4, pp. 183–95. ASME.
8 GOLDSTEIN, S. (ed) (1938). *Modern developments in fluid mechanics* Vols. I and II. Clarendon Press, Oxford.
9 BORWICK, G. R. (1973). Mountaineering ropes. *Alp. J.* **82**, 62.

10 HUDSON, R. R., and JOHNSON, W. (1976). Rock climbing mechanics. *Int. J. Mech. Eng. Educ.* **4**, 357–68.

11 KOSMATH, E. (1968). Die Alterung der Bergseile. *Band Flechtindustrie* **5**, 84.

12 ANON (1953). Equipment and safety. *Mountaineering* **7**, 53.

13 GRIFFIN, L. J. (1964) The strength of karabiners. National Engineering Laboratory, *Report No. 162.*

14 JOHNSON, W., AL-HASSANI, S. T. S., and LLOYD, R. B. (1975). Aspects of pole vaulting mechanics. *Proc. Inst. Mech. Engrs.* **189**, 507–18.

15 WILLIAMS, J. G. P., and SPERRYN, P. W. (1976). *Sports Med.* Arnold, London.

16 UNTERHARNSCHIEDT, F. J. (1975). Injuries due to boxing and other sports. *Handbook of clinical neurology* Vol. 23, Ch. 26. North Holland, Amsterdam.

17 JOHNSON, J., SCORECKI, J., and WELLS, R. P. (1975). Boxing and the Gadd Severity Index. *Med. Biol. Eng.* **13**, 396.

18 JOHNSON, W., SCORECKI, J., and REID, S. R. (1975). The Gadd Severity Index and measurements of acceleration when heading an association football. International Research Committee on Biokinetics of Impacts Conference, Birmingham.

19 HUGHES, R. (1974). Head damage. *The Sunday Times*, 3 November.

20 VAGNERS, J., JORGENSEN, J. E., and DEAK, A. L. (1973). On the mechanics of skis: a review of analytical and experimental studies. *Mechanics and sport* Vol. 4, pp. 9–33. ASME.

21 OUTWATER, J. O., and WOODWARD, M. S. (1967). Ski safety and tibial forces. ASME Paper 66-WA/BHF-14.

22 PIZIALI, R. L. (1973). Dynamic torsional response of the human leg relative to skiing. *Mechanics and sport*, Vol. 4, pp. 305–15. ASME, New York.

23 MOTE, C. D., JR. *et al.* (1973). Remarks on the dynamic performance of ski release bindings. *Mechanics and sport*, Vol. 4, pp. 251–66. ASME, New York.

24 BAHNIUK, E., NELSON, R. A., and ZIVIC, J. (1973). Analytical studies of the biomechanics of contemporary ski bindings. *Mechanics and sport*, Vol. 4, pp. 221–36. ASME.

25 SACK, R. L., and ALBRECHT, R. G. (1973). The mechanics of ski–snow interaction during a curved turn. *Mechanics and sport*, Vol. 4, pp. 155–73. ASME.

26 SCHNEIDER, R. C., PAPO, M., and ALVAREZ, C. S. (1962). The effects of chronic recurrent spinal trauma in high diving. *J. Bone Jt Surg.* **44**, 648.

27 SUGDEN, R. C., and SNOW, C. C. (1967). Extreme water impact. *Aerospace Med.* **38**, 379.

28 FELD, M. S., MCNAIR, R. E., and WILK, S. R. (1979). The physics of karate. *Scient. Am.* **240**, 110.

Index